Undergraduate Lecture Notes in Physics

Series Editors

Neil Ashby, University of Colorado, Boulder, CO, USA

William Brantley, Department of Physics, Furman University, Greenville, SC, USA

Michael Fowler, Department of Physics, University of Virginia, Charlottesville, VA, USA

Morten Hjorth-Jensen, Department of Physics, University of Oslo, Oslo, Norway

Michael Inglis, Department of Physical Sciences, SUNY Suffolk County Community College, Selden, NY, USA

Barry Luokkala, Department of Physics, Carnegie Mellon University, Pittsburgh, PA, USA

Undergraduate Lecture Notes in Physics (ULNP) publishes authoritative texts covering topics throughout pure and applied physics. Each title in the series is suitable as a basis for undergraduate instruction, typically containing practice problems, worked examples, chapter summaries, and suggestions for further reading.

ULNP titles must provide at least one of the following:

- An exceptionally clear and concise treatment of a standard undergraduate subject.
- A solid undergraduate-level introduction to a graduate, advanced, or non-standard subject.
- A novel perspective or an unusual approach to teaching a subject.

ULNP especially encourages new, original, and idiosyncratic approaches to physics teaching at the undergraduate level.

The purpose of ULNP is to provide intriguing, absorbing books that will continue to be the reader's preferred reference throughout their academic career.

More information about this series at https://link.springer.com/bookseries/8917

Teruo Matsushita

Electricity and Magnetism

New Formulation by Introduction of Superconductivity

Second Edition

Springer

Teruo Matsushita
Department of Computer Science
and Electronics
Kyushu Institute of Technology
Iizuka, Fukuoka, Japan

ISSN 2192-4791 ISSN 2192-4805 (electronic)
Undergraduate Lecture Notes in Physics
ISBN 978-3-030-82149-4 ISBN 978-3-030-82150-0 (eBook)
https://doi.org/10.1007/978-3-030-82150-0

This Springer imprint is published by the registered company Springer Nature Switzerland AG
The registered company address is: Gewerbestrasse 11, 6330 Cham, Switzerland

Preface to the Second Edition

The ***E*-*B*** analogy is mainly adopted in the present treatment of electromagnetism. The magnetic phenomena in superconductors in the perfect diamagnetic state were compared with the electric phenomena in conductors in the zero electric field state to strengthen the analogy between the electric and magnetic phenomena in the first edition. The concept of the equivector potential surface, which corresponds to the equipotential surface, was also introduced for the same purpose. The description of the equivector potential surface was insufficient, however, and a full comparison with the equipotential surface was not included in the first edition.

In the second edition, the direction of the vector potential itself and that of the magnetic flux density are discussed on the equivector potential surface, to compare it with an electric field that is perpendicular to an equipotential surface. The concept of the equivector potential surface is widened for the purpose, and the equivector potential surface is classified based on the condition of the magnetic flux density on the surface, which arises from the difference in the geometrical structure of the equivector potential surface.

It is possible to analyze the magnetic phenomena by using the magnetic potential, which is a scalar. Even so, the magnetic potential is not useful to describe the physical situation, while the use of the vector potential is reasonable for the purpose. The two potentials are compared in some examples and exercises. In addition, the number of examples and exercises has been increased to assist the readers to understand the content of the book through comparing similar electric and magnetic phenomena. The theoretical proof of zero resistivity for a material that shows perfect diamagnetism, i.e., the proof of superconductivity, is also added in the second edition.

The author expects that such revisions will assist the readers in gaining a deep understanding of electricity and magnetism. He would like to acknowledge Mr. Seiji Shinyama at Kyushu Institute of Technology for making new electronic figures and Dr. Tania M. Silver at Wollongong University for correction of the English in the book.

Iizuka, Fukuoka, Japan | Teruo Matsushita

The original version of the book was revised: Updated the figures and content corrections provided by Author and Friedhilde Meyer. The correction to the book is available at https://doi.org/10.1007/978-3-030-82150-0_13

Preface to the First Edition

Electromagnetism is an important subject in today's physics. The number of textbooks on electromagnetism is much larger than those on other subjects. This is because abstract concepts are frequently used and therefore it is not easy for students to come to a complete understanding of electromagnetism, although various phenomena are concisely described with mathematics. For this reason, many textbooks have been published to assist students to understand electromagnetism better. Why, then, is a new textbook on electromagnetism necessary now?

Electromagnetism is a classical subject that was almost completely formulated in the nineteenth century. However, concerning its theoretical description, there is still room for further progress. In addition, textbooks are required to describe their topics adequately within a limited space. Therefore, there is also room for improvement in textbooks from the technical point of view.

In principle, there is a beautiful formal analogy between static electric and magnetic phenomena, as will be shown in this textbook. However, the analogy is not necessarily perfect in existing textbooks because of the lack of an important concept. Electric materials are classified into conductors and dielectric materials, but only magnets are studied as magnetic materials. While it is known that electric phenomena in dielectric materials and magnetic phenomena in magnets are analogous to each other, no one has discussed magnetic materials that correspond to electric conductors. However, we have to note superconductors. In a superconductor, a current flows on its surface to shield the inside against an external magnetic field, so that the magnetic flux density $\boldsymbol{B}$ is zero in the superconductor. This is analogous to the electric phenomenon of a conductor in an external electric field. That is, an electric charge appears on its surface to shield the inside against an external electric field, so that the electric field $\boldsymbol{E}$ is zero in the conductor. This is one of the remarkable analogies in the present $\boldsymbol{E}$-$\boldsymbol{B}$ analogy.

Thus, the introduction of the superconductor into electromagnetism, which has not yet been tried systematically, seems to be quite useful for understanding electromagnetism. That is, the analogy between electricity and magnetism can be completed by the introduction of the superconductor. There can be various ways of education without such a comprehensive analogy, and this is another reason why many textbooks on electromagnetism have appeared.

From another point of view, superconductivity is a general phenomenon that appears in many single elements and most metallic compounds, if the cases of pressurization and thin films are included. The intrinsic property of superconductivity, the breaking of Ohm's law, may seem to be peculiar. However, superconductivity is a purely physical phenomenon that can be derived from minimizing free energy. In contrast, the empirical Ohm's law associated with energy dissipation cannot be derived theoretically, and electromagnetic theory is incomplete for other current-carrying materials in this sense.

Usually, students learn about static magnetic energy after they study electromagnetic induction. One of the appreciable advantages of using a superconductor is the direct derivation of magnetic energy as mechanical work done by magnetic force, similar to the electric energy resulting from the electric force. This is because the magnetic flux is conserved in a superconducting circuit disconnected from any electric sources. As a result, the electromagnetic induction can be predicted for a usual electric circuit using the relationship between the energy and magnetic force.

In electromagnetism, the magnetic moment of a magnet caused by spins and orbital motions of electrons is described using a virtual magnetizing current. However, the magnetic moment of a superconductor comes from a real current flowing in it. Hence, the introduction of the superconductor is also beneficial with regard to persuasion of the appropriateness of the virtual magnetizing current.

It should be noted that the definition of magnetization is different for magnets and superconductors. That is, magnetization comes directly from the magnetization $\boldsymbol{M}$ in magnets, while it comes from the magnetic field $\boldsymbol{H}$ in superconductors. This arises from the difference in the origin of the magnetic moments. According to the definition used for magnets, superconductors are classified as non-magnetic materials. On the other hand, the analogous electric phenomena are electrostatic shielding in conductors and electric polarization in dielectrics. These are similar electric shielding mechanisms caused by electric charges and polarization charges, but the above-mentioned different terms are used. Such comparison between electricity and magnetism is also useful for education.

The final merit of the introduction of superconductors is application of the analysis method of electromagnetic phenomena in superconductors. The continuity equation of magnetic flux used for superconductors is useful for estimating the velocity of magnetic flux lines under a magnetic field varying with time. This enables us to unify the magnetic flux law and the motional law for electromagnetic induction, which usually have been treated separately.

The purpose of this textbook is to show the remarkable analogy between static electric phenomena, described in Part I, and static magnetic phenomena, described in Part II. Hence, a comparison between the corresponding chapters in each part, such as Chap. 2 on conductors and Chap. 7 on superconductors, will assist in understanding electromagnetism. Dynamic electromagnetic phenomena are described in Part III.

I would like to express my sincere acknowledgment to Prof. Klaus Lueders at the Berlin Free University for the useful discussion we had. In addition, I would like to thank Tomoko Onoue, Etsuko Shirahasi, and Kaori Ono for assistance in making electronic files and drawing electronic figures.

Iizuka, Fukuoka, Japan Teruo Matsushita

Contents

Explanation of the Figures

The upper figures show the structure of electric flux lines when a uniform electric field is applied to a sphere of a conductor (left) and a dielectric (right), and the lower figures show the structure of magnetic flux lines when a uniform magnetic field is applied to a sphere of a superconductor (left) and a magnet (right). These will be covered in Chaps. 2, 4, 7 and 9, respectively. The reason why the electric flux lines are used instead of the more important electric field lines is to emphasize the analogy between dielectrics and magnets by showing continuous lines at the interfaces with vacuum. In the case of electric field lines, the number of lines inside the dielectric is smaller than that outside because of the shielding by polarization charges (see Chap. 4). This situation is similar when we draw the magnetic field lines instead of the magnetic flux lines for the lower right figure.

The manner of perfect shielding is different between the conductor and the superconductor. This comes from the different nature of the corresponding fields. Electric charges on the conductor surface absorb the lines directly, while currents on the superconductor surface push the lines to outside.

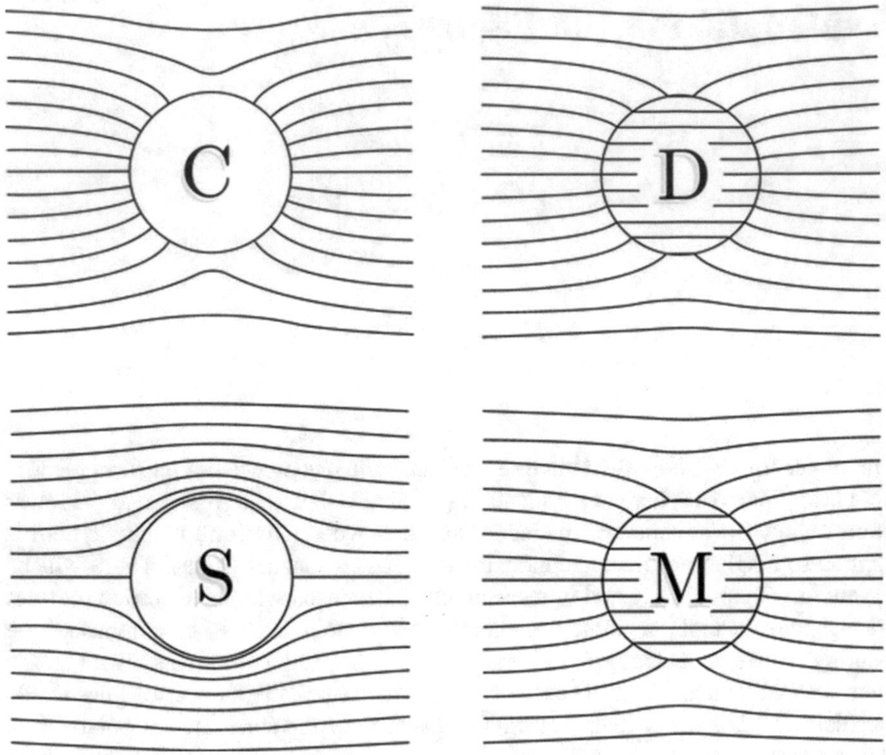
C
D
S
M

Part I
Static Electric Phenomena

Chapter 1
Electrostatic Field

Abstract This chapter covers the electrostatic phenomena caused by electric charges in a vacuum. First, we learn about the Coulomb force that works between electric charges. This force on a given charge is caused by an electrical distortion in space that is produced by other electric charges. This distortion is represented by the electric field. The local electric field produced by electric charges is described by Coulomb's law. On the other hand, Gauss' law describes the global relationship between the electric charge and the electric field. The electric field is expressed by the gradient of the electric potential and has a nature of a field with no rotation. The equipotential surface is a virtual surface on which the electric potential is the same. The electric field is perpendicular to the equipotential surface. We also learn the electrical properties of the electric dipole, i.e., a pair of positive and negative electric charges of the same magnitude, which appears under various conditions in electric phenomena.

1.1 Electric Charge in Vacuum

When we touch a metal doorknob after walking on a carpet on a dry day, we sometimes feel a shock on the fingertips as a small crackle. If it is dark, we can see a spark when inserting a key into a keyhole. This is the same phenomenon as thunder. This phenomenon is brought about by **electric charge** in substances. The usual frictional electricity we experience also comes from electric charges.

Electric charge buildup in a substance causes various kinds of electric phenomena, including the above examples. Matter is a substance that obeys universal gravity laws, and its magnitude is quantitatively described in terms of mass. In the case of electric phenomena, an amount of electric charge quantitatively describes the phenomena, and the same term, "electric charge", is also used to mean the amount of electric charge.

Unlike mass, there are two kinds of electric charge, positive and negative. The components of the electric charge are the proton with positive charge and electron with negative charge. The electric charge of a proton is called the **elementary electric charge,** and its magnitude is

T. Matsushita, *Electricity and Magnetism*, Undergraduate Lecture Notes in Physics,
https://doi.org/10.1007/978-3-030-82150-0_1

$$e = 1.602\,176\,6 \times 10^{-19}\ \mathrm{C},$$

where the unit [C] is **coulomb**. The electric charge of an electron is $-e$. The elementary electric charge is the minimum amount of electric charge, and any electric charge is its integral multiple. Since e is sufficiently small, electric charge can be regarded as a continuous quantity in many cases. This is similar to the fact that an amount of water can be regarded as a continuous quantity in usual cases.

On an atomic scale, the nucleus of an atom is composed of protons and neutrons, which are electrically neutral, and electrons stay in orbits around the nucleus. There are innumerable positive and negative electric charges in substances. Since the size of each atom is very small, electrons and protons can be regarded as being in the center of each atom on a macroscopic scale. As a result, the positive and negative charges cancel each other to yield an electrically neutral state. Ionic crystals composed of equal amounts of positive and negative ions can also be regarded as electrically neutral on the macroscopic scale, since the distance between these ions is sufficiently small. Sometimes the electric charge is not balanced. In such a case, the electric charge that remains after cancelation causes various electric phenomena.

There are two kinds of electric charge that cause electric phenomena: One is **true electric charge**, which can be transferred outside a substance, and the other is **polarization charge**, which is locally bound around a nucleus and cannot be transferred outside. The former charge appears on the surface of a conductor and will be covered in Chap. 2, and the latter appears on the surface of a dielectric and will be covered in Chap. 4. These charges that contribute to electric phenomena are called **free electric charge**.

Electric charge is generally distributed with some density in the interior or on the surface of matter. Electric charge small enough to be regarded as a point is called **point charge**. This is similar to a material particle in mechanics. Electric charge distributed along a thin line with negligible cross-sectional area is **line charge**, and electric charge distributed on a surface with negligible thickness is **surface charge**.

The **principle of conservation of charge** is a fundamental principle for electric charge, which is similar to the law of conservation of mass in mechanics. It states that the amount of electric charge is constant in a closed system. Even when positive and negative electric charges cancel each other, resulting in an electrically neutral state, the algebraic sum of electric charge is unchanged.

1.2 Coulomb's Law

Electric force works between electric charges, and this force is called the **Coulomb force**. This force is analogous to universal gravitation between two particles with masses. The Coulomb force on two point charges in vacuum is expressed as follows:

- The force between two electric charges of the same kind (i.e., both positive or both negative) is repulsive, and the force between electric charges of different kinds (i.e., one positive and one negative) is attractive.
- The magnitude of the force is proportional to the product of the two electric charges.
- The magnitude of the force is inversely proportional to the square of the distance between the two electric charges.
- The direction of the force lies on the straight line connecting the two electric charges.

The first property is different from the property of universal gravitation whereby the force between two masses is always attractive. The Coulomb force between two point charges, q and q', separated by distance d is mathematically expressed as

$$F = \frac{qq'}{4\pi\epsilon_0 r^2}, \tag{1.1}$$

where ϵ_0 is a constant called the **permittivity of vacuum**,

$$\epsilon_0 = \frac{10^7}{4\pi c_0^2} = 8.854\,2 \times 10^{-12}\ \mathrm{C^2/Nm^2} \tag{1.2}$$

with $c_0 = 2.998 \times 10^8$ m/s denoting the speed of light in vacuum. The force in Eq. (1.1) is repulsive when $F > 0$ and attractive when $F < 0$. This equation is called **Coulomb's law**.

Since force is a vector, the Coulomb force can be expressed as a vector. We denote the direction vector of point charge q measured from the position of q' as $\boldsymbol{r}$, as shown in Fig. 1.1. Then, its magnitude is $r = |\boldsymbol{r}|$, and the unit vector pointing from q' to q is $\boldsymbol{i}_r = \boldsymbol{r}/r$. Hence, the force that works on q is

$$\boldsymbol{F} = \frac{qq'\boldsymbol{i}_r}{4\pi\epsilon_0 r^2} = \frac{qq'\boldsymbol{r}}{4\pi\epsilon_0 r^3}. \tag{1.3}$$

The force on q' is given by $-\boldsymbol{F}$, and the law of action and reaction is satisfied.

When there are more than two material particles, the gravitational force on one particle is the linear sum of the gravitational forces exerted on it by all other particles, and the principle of superposition holds. The same principle holds also for the Coulomb force. Assume that n point charges, $q_1, q_2, \ldots, q_n$, are distributed in vacuum, as shown in Fig. 1.2. The total Coulomb force on another point charge, q, is given by the sum of each individual Coulomb force exerted by each point charge.

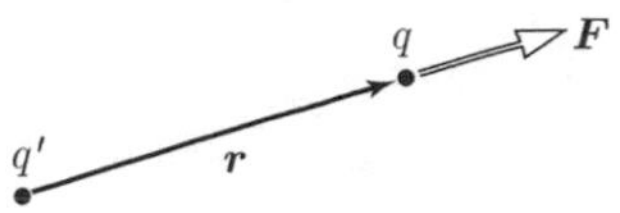

Fig. 1.1 Coulomb force exerted on point charge q by q'

Fig. 1.2 Coulomb force exerted on point charge q by more than one point charge

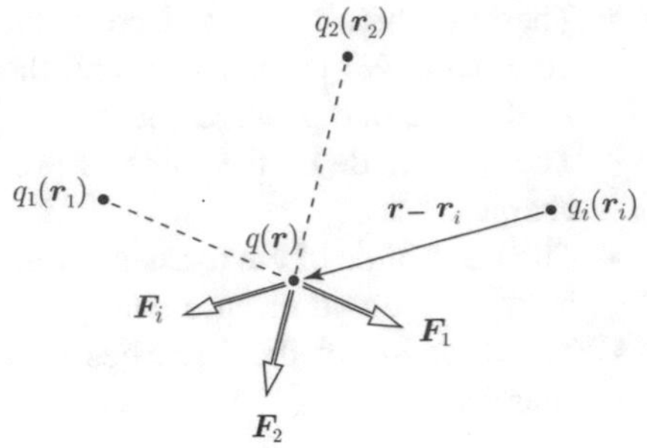

If the direction vectors of point charges q_i and q are denoted by $\boldsymbol{r}_i$ $(i = 1, 2, \ldots, n)$ and $\boldsymbol{r}$, respectively, the Coulomb force exerted by q_i on q is

$$F_i = \frac{qq_i(\boldsymbol{r}-\boldsymbol{r}_i)}{4\pi\epsilon_0|\boldsymbol{r}-\boldsymbol{r}_i|^3}. \tag{1.4}$$

Hence, the total Coulomb force on q is

$$\boldsymbol{F} = \sum_{i=1}^{n} \boldsymbol{F}_i = \frac{q}{4\pi\epsilon_0}\sum_{i=1}^{n}\frac{q_i(\boldsymbol{r}-\boldsymbol{r}_i)}{|\boldsymbol{r}-\boldsymbol{r}_i|^3}. \tag{1.5}$$

This result can be extended to the case where electric charge is continuously distributed. Suppose that electric charge is distributed with the density ρ within a region, V, in vacuum, as shown in Fig. 1.3. We treat the electric charge, $\mathrm{d}q' = \rho \mathrm{d}V'$, in an infinitesimal volume $\mathrm{d}V'$ as a point charge at the position $\boldsymbol{r}'$. The Coulomb force this charge exerts on the point charge, q, at $\boldsymbol{r}$ is given by

$$d\boldsymbol{F} = \frac{q(\rho \mathrm{d}V')(\boldsymbol{r}-\boldsymbol{r}')}{4\pi\epsilon_0|\boldsymbol{r}-\boldsymbol{r}'|^3}. \tag{1.6}$$

Hence, the Coulomb force from all electric charges is

$$\boldsymbol{F} = \frac{q}{4\pi\epsilon_0}\int_V \frac{\rho(\boldsymbol{r}')(\boldsymbol{r}-\boldsymbol{r}')}{|\boldsymbol{r}-\boldsymbol{r}'|^3}\mathrm{d}V'. \tag{1.7}$$

In the above, $\int \mathrm{d}V'$ denotes a volume integral with respect to $\boldsymbol{r}'$.

Fig. 1.3 Coulomb force exerted on point charge q by electric charge inside small volume

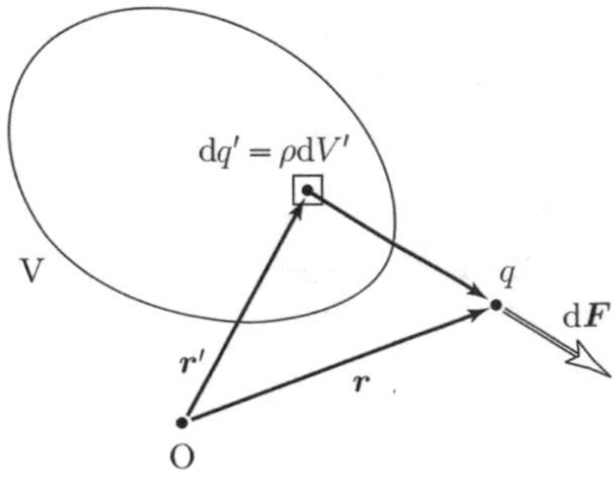

Example 1.1. Electric charge is uniformly distributed with a linear density, λ, along a semicircle of radius a. Determine the Coulomb force on a point charge, Q, placed at the center of curvature of the semicircle.

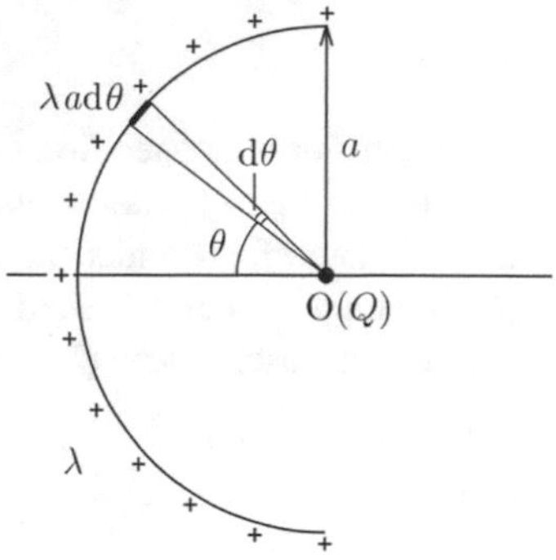

Fig. 1.4 Electric charge distributed uniformly with linear density λ on a semicircle and point charge Q placed at center O

Solution 1.1. An angle is defined as shown in Fig. 1.4. We treat an electric charge $\lambda a \mathrm{d}\theta$ between θ and $\theta + \mathrm{d}\theta$ as a point charge. The Coulomb force it exerts on charge Q at the center is

$$\mathrm{d}F = \frac{Q\lambda a \mathrm{d}\theta}{4\pi\epsilon_0 a^2} = \frac{Q\lambda \mathrm{d}\theta}{4\pi\epsilon_0 a}.$$

From symmetry, the vertical components of the Coulomb forces exerted by infinitesimal arc elements on Q cancel out, and only the horizontal component remains. This component is $\mathrm{d}F' = \mathrm{d}F\cos\theta$. Hence, the total Coulomb force is

$$F = \frac{Q\lambda}{4\pi\epsilon_0 a}\int_{-\pi/2}^{\pi/2} \cos\theta \mathrm{d}\theta = \frac{Q\lambda}{4\pi\epsilon_0 a}[\sin\theta]_{-\pi/2}^{\pi/2} = \frac{Q\lambda}{2\pi\epsilon_0 a}.$$

This force is directed to the right in the figure.

◇

1.3 Electric Field

When a point charge, Q, is placed at the origin, the Coulomb force on another point charge, q, at the position $\boldsymbol{r}$ is given by Eq. (1.3) as

$$\boldsymbol{F} = \frac{qQ\boldsymbol{r}}{4\pi\epsilon_0 r^3}. \tag{1.8}$$

This expression for force holds for arbitrary q and for arbitrary position $\boldsymbol{r}$. Hence, the space can be regarded as exerting the following force on a point charge, q:

$$\boldsymbol{F} = q\boldsymbol{E}. \tag{1.9}$$

We can consider the operation $\boldsymbol{E}$ to be caused by an electrical distortion of the space due to the charge, Q. That is, the Coulomb force on q can be understood as a force that this charge feels under the electrical distortion of the space. This operation on q, $\boldsymbol{E}$, is called the **electric field**, and its magnitude is called **electric field strength**. In the above case, where Q is at the origin, the electric field strength is

$$\boldsymbol{E} = \frac{Q\boldsymbol{r}}{4\pi\epsilon_0 r^3}. \tag{1.10}$$

The unit of electric field strength is [N/C]. This is also expressed as [V/m] using the unit [V] (**volt**) of electrostatic potential, which will be defined later. The **electrostatic field**, which is the title of this chapter, is the electric field that comes from electric charges and does not include the electric field caused by electromagnetic induction (see Chap. 10).

The electric field strength is the Coulomb force on a unit electric charge. Hence, calculating the electric field strength is equivalent to calculating the Coulomb force.

Here, we calculate the electric field strength for electric charges distributed in space. When an electric charge, q_i, is placed at position $\boldsymbol{r}_i$ ($i = 1, 2, \ldots, n$), the electric field strength at $\boldsymbol{r}$ is given by Eq. (1.5) as

$$\boldsymbol{E} = \frac{1}{4\pi\epsilon_0}\sum_{i=1}^{n}\frac{q_i(\boldsymbol{r}-\boldsymbol{r}_i)}{|\boldsymbol{r}-\boldsymbol{r}_i|^3}. \tag{1.11}$$

When an electric charge is continuously distributed in a region, V, with density $\rho(\boldsymbol{r}')$, as shown in Fig. 1.3, the electric field strength at $\boldsymbol{r}$ from Eq. (1.7) is

$$\boldsymbol{E} = \frac{1}{4\pi\epsilon_0}\int_{\mathrm{V}}\frac{\rho(\boldsymbol{r}')(\boldsymbol{r}-\boldsymbol{r}')}{|\boldsymbol{r}-\boldsymbol{r}'|^3}\mathrm{d}V'. \tag{1.12}$$

Equations (1.11) and (1.12) for the electric field strength are also called Coulomb's law.

The electric field is generally complicated depending on the distribution of electric charges. However, we can visualize the field using **electric field lines (lines of electric force)**, which help us to understand the field easily. When a point charge is put in an electric field of strength $\boldsymbol{E}$, the charge experiences the Coulomb force directed parallel to $\boldsymbol{E}$. If this charge is sufficiently small, its movement driven by the force will not appreciably change the electric field. The direction of the electric field

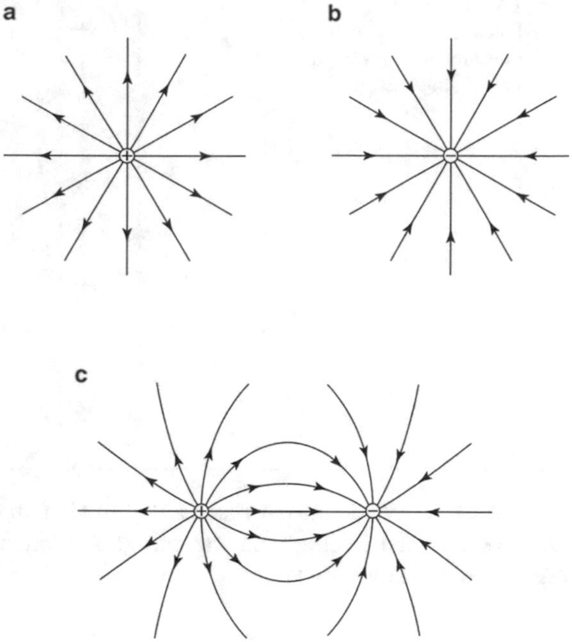

Fig. 1.5 Electric field lines of **a** single positive charge, **b** single negative charge and **c** pair of positive and negative charges

line is defined as the direction of the force on a positive charge. Therefore, we can refer to a line of electric force as an electric field line and take the tangent to an electric field line at an arbitrary point as being parallel to the direction of the electric field at this point.

It can be shown that electric field lines never cross each other. That is, if two field lines cross at a certain point, two forces must work on a point charge put on this point. However, the electric field at any point has a single strength.

Figure 1.5 shows examples of electric field lines. A field line always starts from a positive charge and ends at a negative charge. One can see that a field line never begins or ends at a point with no charge. We will cover mathematical expressions of these facts in Sect. 1.4.

In principle, we can draw an arbitrary number of field lines. Hence, we define $E = |\boldsymbol{E}|$ as the number of field lines through a unit area perpendicular to the electric field. Thus, the electric field strength can be expressed through the density of field lines. For example, if the distance from the point charge is doubled in Fig. 1.5a, the spacing between two field lines is also doubled. This means that the field strength becomes one quarter as great, which is directly derived from Eq. (1.10).

Example 1.2. Electric charge is uniformly distributed with a linear density, λ, on a straight segment of length $2a$ parallel to the y-axis, as shown in Fig. 1.6. Determine the electric field strength at a point, A, at distance b from the center of the segment to the direction of the x-axis.

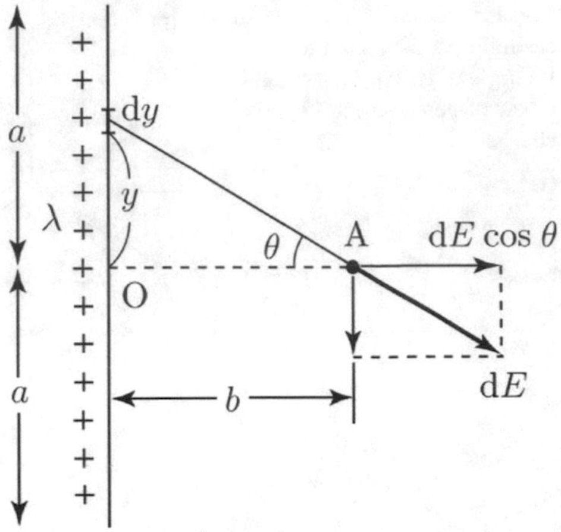

Fig. 1.6 Electric charge distributed uniformly with linear density λ on a straight segment of length $2a$

Solution 1.2. We define the y-axis along the length of the segment with the origin at its center. The electric field strength at A produced by the charge, $\lambda \mathrm{d}y$, in the region from y to $y + \mathrm{d}y$ is

$$\mathrm{d}E = \frac{\lambda \mathrm{d}y}{4\pi\epsilon_0 (y^2 + b^2)}.$$

The angle θ is defined as shown in the figure. From symmetry, the y-component of the electric field is canceled, and only the x-component, $\mathrm{d}E\cos\theta$, remains. The relationship $y = b\tan\theta$ gives $\mathrm{d}y = b\mathrm{d}\theta / \cos^2\theta$ and $y^2 + b^2 = b^2 / \cos^2\theta$. The electric field strength is given by

$$E = \frac{\lambda}{4\pi\epsilon_0 b} \int_{-\theta_a}^{\theta_a} \cos\theta \mathrm{d}\theta$$

with $\theta_a = \tan^{-1}(a/b)$. After a simple calculation, we have

$$E = \frac{\lambda}{4\pi\epsilon_0 b} [\sin\theta]_{-\theta_a}^{\theta_a} = \frac{\lambda a}{2\pi\epsilon_0 b (a^2 + b^2)^{1/2}}.$$

For an infinitely long line ($a \to \infty$), this result gives

$$E = \frac{\lambda}{2\pi\epsilon_0 b}.$$

◇

Example 1.3. Suppose that electric charge is uniformly distributed on a circle of radius a with a linear density, λ, as illustrated in Fig. 1.7. Determine the electric field strength at a point P at distance z from the center of the circle.

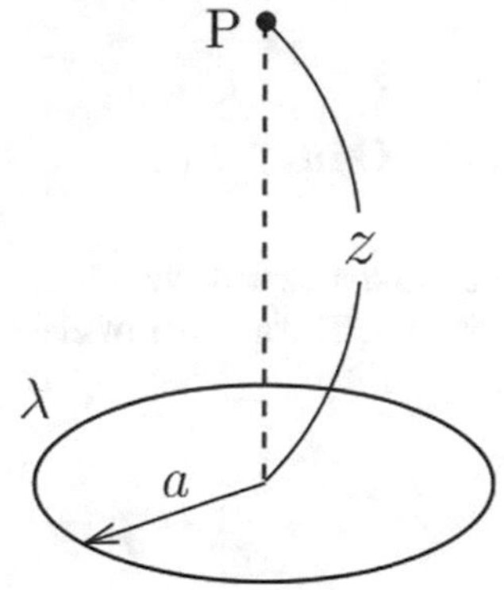

Fig. 1.7 Circle with uniformly distributed electric charge and point P above the center

Solution 1.3. The azimuthal angle, φ, is defined from the center of the circle on the plane in which the circle exists. The electric field strength at P produced by the electric charge on a small segment of length $a\mathrm{d}\varphi$ is

$$\mathrm{d}E = \frac{a\lambda\mathrm{d}\varphi}{4\pi\epsilon_0(z^2+a^2)}.$$

As shown in Fig. 1.8, only the vertical component, $[z/(z^2+a^2)^{1/2}]\mathrm{d}E$, remains from symmetry. The electric field strength is determined to be

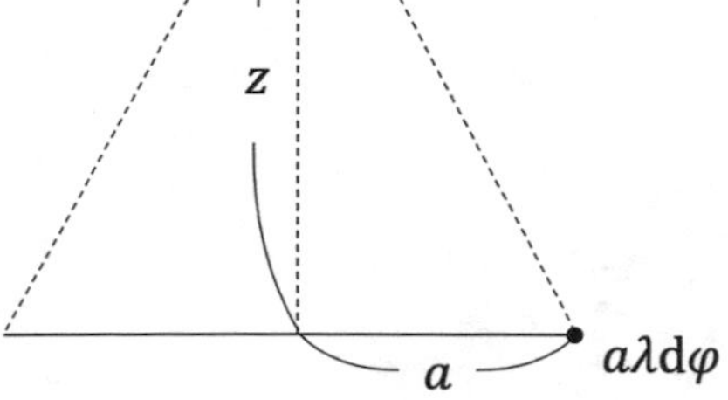

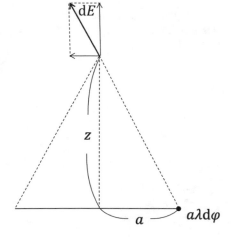

Fig. 1.8 Electric field $\mathrm{d}E$ produced by electric charge $a\lambda\mathrm{d}\varphi$ on a small segment of the circle

$$E = \int_0^{2\pi} \frac{az\lambda}{4\pi\epsilon_0 (z^2 + a^2)^{3/2}} d\varphi = \frac{az\lambda}{2\epsilon_0 (z^2 + a^2)^{3/2}}.$$

◇

1.4 Gauss' Law

Suppose a closed surface, S, which includes a point charge, q, inside, as shown in Fig. 1.9. The number of electric field lines produced by this charge that penetrate S is given by

$$N = \int_{\mathrm{S}} \boldsymbol{E} \cdot \mathrm{d}\boldsymbol{S}. \quad (1.13)$$

Since these lines do not terminate halfway, N is also the number of lines that penetrate the sphere, S_0, of radius r_0 with its center on q. Thus, we have

$$N = \int_{\mathrm{S}_0} \boldsymbol{E} \cdot \mathrm{d}\boldsymbol{S}. \quad (1.14)$$

Since $\boldsymbol{E}$ is parallel to $\mathrm{d}\boldsymbol{S}$ and $|\boldsymbol{E}| = q/\left(4\pi\epsilon_0 r_0^2\right)$ is constant on S_0, a simple calculation gives

$$N = \frac{q}{4\pi\epsilon_0 r_0^2} \int_{\mathrm{S}_0} \mathrm{d}S = \frac{q}{\epsilon_0}. \quad (1.15)$$

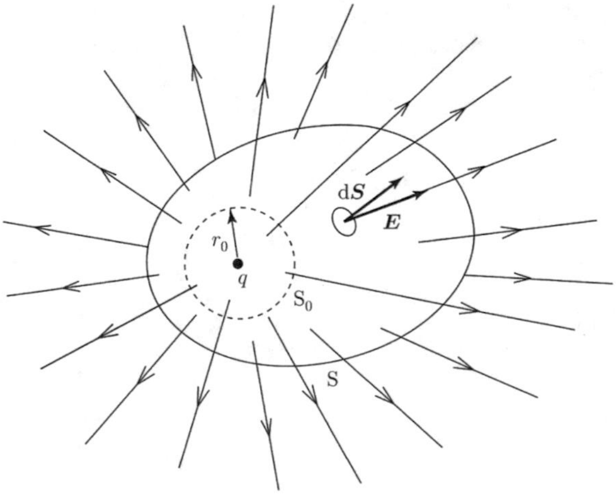

Fig. 1.9 Point charge q and closed surfaces S and S_0 containing q

In the above, $\int_{S_0} dS = 4\pi r_0^2$ is the surface area of S_0. This surface area divided by r_0^2, i.e., the surface area of a unit sphere, is equal to 4π, the full solid angle. The reason why the full solid angle appears in Coulomb's law, Eq. (1.1), is to make Gauss' law simple as in Eq. (1.15). Thus, the relationship

$$N = \int_S \boldsymbol{E} \cdot d\boldsymbol{S} = \frac{q}{\epsilon_0} \qquad (1.16)$$

holds for S in Fig. 1.9. Thus, when $q > 0$ and $N > 0$, field lines go outside S, and when $q < 0$ and $N < 0$, field lines go inside S.

Now, consider the case where the shape of sphere S is complicated in such a way that points on S and points on S_0 do not correspond one to one, as shown in Fig. 1.10. Suppose a thin cone with a top on the point charge, q. We denote small areas on S cut by the cone as dS_1, dS_2, and dS_3. If the number of field lines that go out of S through the area, dS_1, is dN, the number of field lines that go out through the areas dS_2 and dS_3 is $-dN$ and dN, respectively. Hence, the number of field lines inside the cone is dN and is equal to the number of field lines through the area of S_0 cut by the cone. Thus, we can see that Eq. (1.16) holds when the point charge is included in S, even if the shape of S is complicated.

Next, consider the case where a point charge, q, is placed outside a closed surface, S, as shown in Fig. 1.11. From the above discussion, the number of field lines that enter S through dS_1 is equal to the number of field lines that go out of S through dS_2. Hence, we have

$$N = \int_S \boldsymbol{E} \cdot d\boldsymbol{S} = 0. \qquad (1.17)$$

Based on the above discussion and the principle of superposition, we can easily obtain the number of field lines when point charges are distributed. When point charges $q_1, q_2, \cdots, q_n$ are distributed inside S and $q_{n+1}, q_{n+2}, \cdots, q_{n+m}$ are distributed outside S, the number of field lines that go out of S is

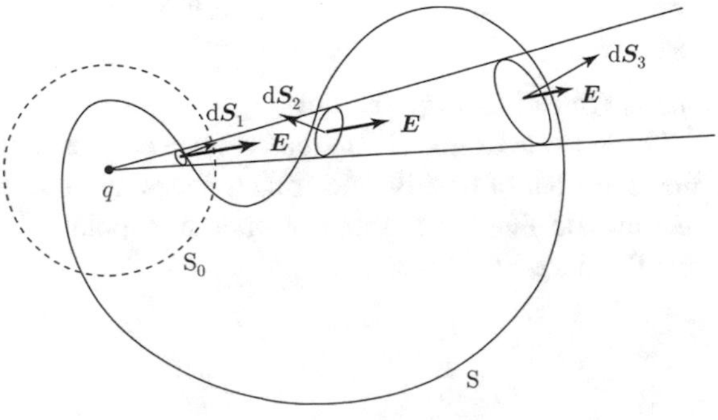

Fig. 1.10 Case where closed surface S containing point charge q has a complicated shape

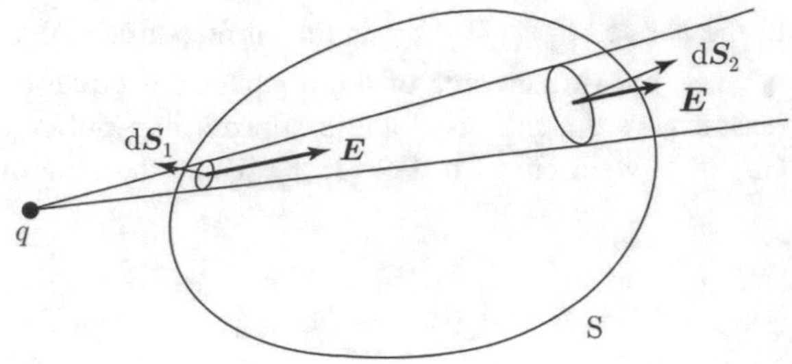

Fig. 1.11 Case where point charge q is outside closed surface S

$$N = \int_S \boldsymbol{E} \cdot \mathrm{d}\boldsymbol{S} = \frac{1}{\epsilon_0} \sum_{i=1}^{n} q_i. \tag{1.18}$$

When the electric charge is continuously distributed with the density, $\rho(\boldsymbol{r})$, the number of field lines is

$$N = \int_S \boldsymbol{E} \cdot \mathrm{d}\boldsymbol{S} = \frac{1}{\epsilon_0} \int_V \rho(\boldsymbol{r}) \mathrm{d}V, \tag{1.19}$$

where V is the region surrounded by S. Equations (1.18) and (1.19) show that the number of field lines that go out of S is equal to the total sum of electric charge inside S divided by ϵ_0. These equations are called **Gauss' law**. Gauss' law describes the global relationship between the distributed electric charge and the electric field, while Coulomb's law describes the local electric field caused by individual electric charges. These laws are equivalent to each other.

Using Gauss' theorem, we rewrite Eq. (1.19) as

$$\int_V \nabla \cdot \boldsymbol{E} \mathrm{d}V = \frac{1}{\epsilon_0} \int_V \rho(\boldsymbol{r}) \mathrm{d}V \tag{1.20}$$

for continuously distributed electric charge. Since this equation holds for arbitrary V, we obtain the relation

$$\nabla \cdot \boldsymbol{E} = \frac{\rho(\boldsymbol{r})}{\epsilon_0}. \tag{1.21}$$

This is **Gauss' divergence law**.

The left side of Eq. (1.21) represents a source of electric field lines. That is, field lines come out of positive electric charges and go into negative electric charges. An electric field line never starts or ends at a point where there is no electric charge ($\rho = 0$), since $\nabla \cdot \boldsymbol{E} = 0$.

Example 1.4. Electric charge is uniformly distributed with a density, ρ, inside an infinitely long cylinder of radius a. Determine the electric field strength inside and outside the cylinder.

Solution 1.4. It is possible to calculate the electric field strength using Coulomb's law, Eq. (1.12). However, the calculation is not easy even if we use the result in Example 1.2. Gauss' law can be used to calculate the electric field strength when the geometry is highly symmetric as in this problem.

We apply Gauss' law, Eq. (1.19), to an imaginary infinite cylindrical closed surface, S, of radius R and length l with a common axis with the infinite cylinder (see Fig. 1.12). The electric field, $\boldsymbol{E}$, is directed radially from the central axis and perpendicular to it. Hence, $\boldsymbol{E}$ is perpendicular to the elementary surface vector, d$\boldsymbol{S}$, on the top and bottom surfaces, and there is no contribution to the surface integral of the electric field strength from these surfaces. On the other hand, $\boldsymbol{E}$ is parallel to d$\boldsymbol{S}$ on the side surface, and the strength, E, is constant and depends only on the distance from the axis. Hence, the surface integral in Eq. (1.19) gives

$$\int_{\mathrm{S}} \boldsymbol{E} \cdot \mathrm{d}\boldsymbol{S} = 2\pi R l E.$$

The total charge inside S is $\pi R^2 l \rho$ for $R < a$ and $\pi a^2 l \rho$ for $R > a$. Hence, the electric field strength is

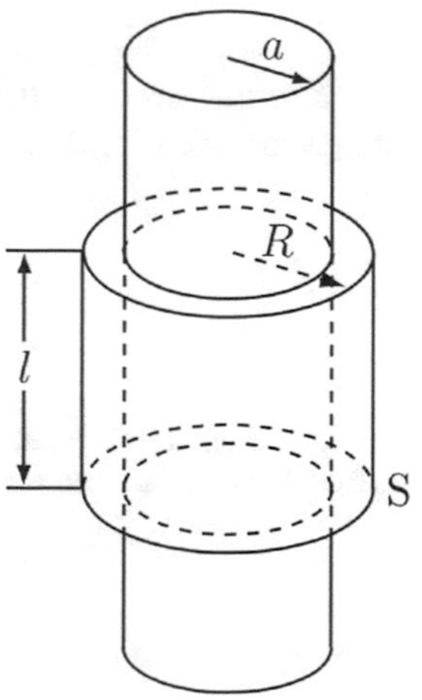

Fig. 1.12 Cylinder with distributed charge and cylindrical closed surface S (case for $R > a$)

$$E = \frac{\rho}{2\epsilon_0} R; \qquad R < a,$$
$$= \frac{\rho a^2}{2\epsilon_0 R}; \qquad R > a.$$

The result for $R > a$ is identical with the result of Example 1.2 for an infinitely long line charge with $\lambda = \pi a^2 \rho$, i.e., the case where all electric charge is concentrated on the central axis. When the length of the cylinder is finite, the electric field strength is not uniform along the length, and Gauss' law cannot be used to calculate the strength.

◇

Example 1.5. Electric charge is uniformly distributed with a density, σ, on a wide flat plane. Determine the electric field strength at a point, A, at distance h from the plane.

Solution 1.5. This problem can also be easily solved using Gauss' law. From symmetry, we can assume that the electric field, $\boldsymbol{E}$, is directed normally to the plane with its strength dependent only on the distance from the plane. Assume a closed cylindrical surface, S, of radius a and length $2h$, as shown in Fig. 1.13: Its side surface is normal to and the top and bottom surfaces are parallel to the plane, and A is on the top surface. We apply Gauss' law to this cylindrical surface. The field, $\boldsymbol{E}$, is parallel to the side surface, and there are no field lines passing through this surface. The numbers of field lines that go out of the top and bottom surfaces are the same from symmetry. Thus, we have

$$\int_{\mathrm{S}} \boldsymbol{E} \cdot \mathrm{d}\boldsymbol{S} = 2ES$$

with $S = \pi a^2$ denoting the area of the top or bottom surface. Since the total electric charge included inside S is σS, the electric field strength is

$$E = \frac{\sigma}{2\epsilon_0}. \tag{1.22}$$

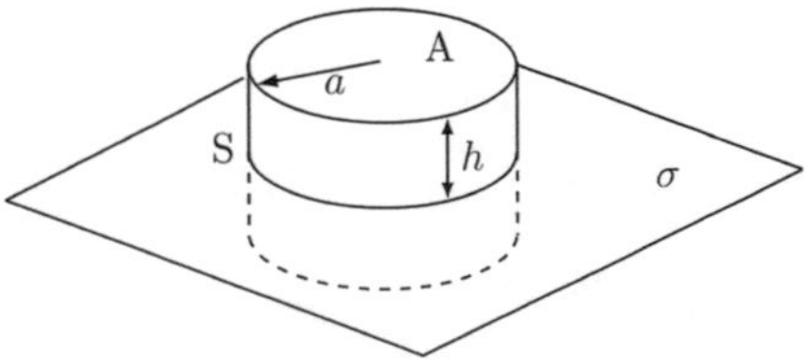

Fig. 1.13 Cylindrical closed surface S with point A on top surface

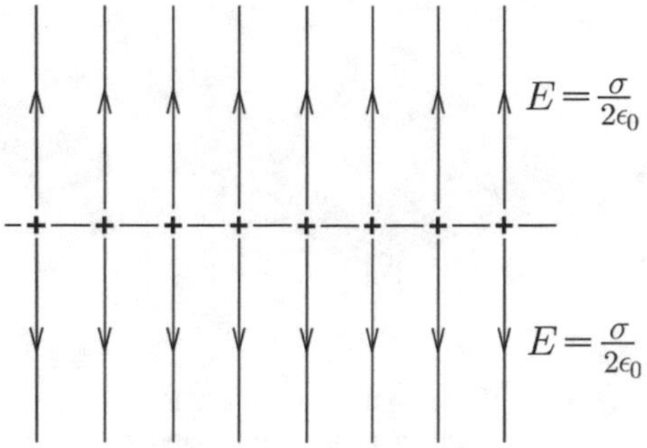

Fig. 1.14 Electric field lines from uniformly distributed electric charge on a plane

This result shows that the electric field strength does not change with the distance from the plane. This can be understood from the fact that the distance between field lines does not change with the distance from the plane, as shown in Fig. 1.14.

◇

1.5 Electric Potential

The Coulomb force works between any electric charges, and any electric field originates from electric charges. Hence, the general nature of the electric field can be deduced from one point charge. This nature can be extended to any case using the principle of superposition. The electric field strength caused by an isolated point charge, Q, placed at the origin is given by Eq. (1.10). If we note the expression,

$$\nabla\left(\frac{1}{r}\right) = -\frac{\boldsymbol{r}}{r^3}, \tag{1.23}$$

using polar coordinates, we write this electric field strength as

$$\boldsymbol{E} = -\nabla\phi \tag{1.24}$$

with

$$\phi = \frac{Q}{4\pi\epsilon_0 r}. \tag{1.25}$$

This scalar function, ϕ, is called the **electric potential** or **electrostatic potential**. The unit of electric potential is [Nm/C] and is defined as [V] (volt).

The electric potential produced by a positive electric charge is illustrated in Fig. 1.15a. This shows the electric potential on a plane that contains the electric charge. Figure 1.15b illustrates the electric potential given by a pair of positive and negative electric charges of equal amounts on a plane that contains both charges.

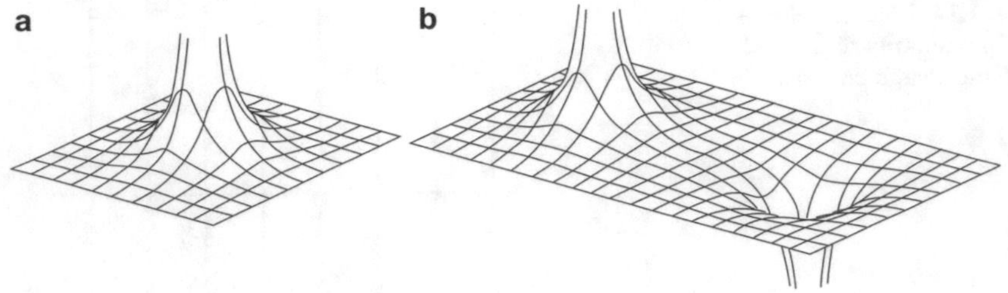

Fig. 1.15 Electric potential on a plane with **a** positive electric charge and **b** pair of positive and negative electric charges of equal amounts

When there are two or more electric charges, the principle of superposition holds for the electric field strength. A similar principle of superposition holds for the electric potential. While the superposition is with vector quantities for the electric field, it is with scalar quantities for the electric potential. For discontinuously distributed electric charges that provide the electric field strength of Eq. (1.11), the electric potential is given by

$$\phi(\boldsymbol{r}) = \frac{1}{4\pi\epsilon_0}\sum_{i=1}^{n}\frac{q_i}{|\boldsymbol{r}-\boldsymbol{r}_i|}. \tag{1.26}$$

For continuously distributed electric charge that provides the electric field strength of Eq. (1.12), the electric potential is given by

$$\phi(\boldsymbol{r}) = \frac{1}{4\pi\epsilon_0}\int_{\mathrm{V}}\frac{\rho(\boldsymbol{r}')}{|\boldsymbol{r}-\boldsymbol{r}'|}\mathrm{d}V'. \tag{1.27}$$

Since the electric field obeys Eq. (1.24), we obtain

$$\nabla \times \boldsymbol{E} = 0 \tag{1.28}$$

using Eq. (A1.45) in Sect. A1. This shows that the electrostatic field is an irrotational field, that is, a field without a vortex. Equations (1.21) and (1.28) describe the fundamental properties of the electrostatic field.

Since the inverse operation of a gradient is a curvilinear integral, the electric potential is generally given by

$$\phi(\boldsymbol{r}) = -\int_{\boldsymbol{r}_0}^{\boldsymbol{r}}\boldsymbol{E}\cdot\mathrm{d}\boldsymbol{s}, \tag{1.29}$$

where $\boldsymbol{r}_0$ is a reference point satisfying $\phi(\boldsymbol{r}_0) = 0$ and is usually taken at infinity. It should be noted that Eqs. (1.25)–(1.27) satisfy this requirement. Suppose a closed

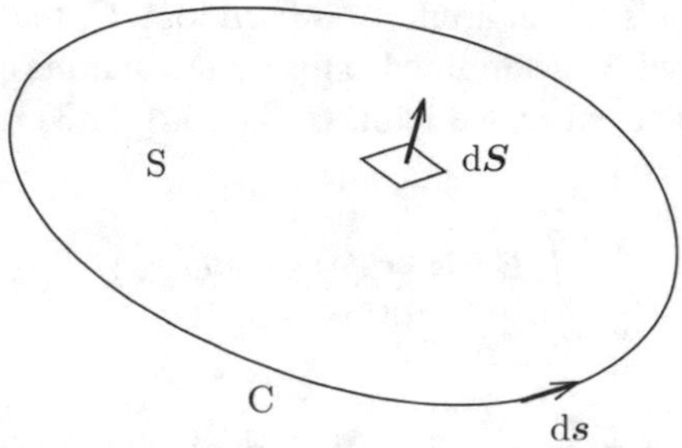

Fig. 1.16 Closed loop C with surface S in it

loop, C, with the surface in it denoted by S (see Fig. 1.16). The curvilinear integral of the electric field strength on C is

$$\oint_{\mathrm{C}} \boldsymbol{E} \cdot \mathrm{d}\boldsymbol{s} = \int_{\mathrm{S}} (\nabla \times \boldsymbol{E}) \cdot \mathrm{d}\boldsymbol{S} = 0. \tag{1.30}$$

The above uses Stokes' theorem given by Eq. (A1.73) in Sect. A1 and Eq. (1.28). This holds for any closed loops and can also be derived from Eqs. (1.24) and (A1.55).

Next, assume two points, A and B, on a closed loop, C, as shown in Fig. 1.17, and the positions of these points are denoted by $\boldsymbol{r}_{\mathrm{A}}$ and $\boldsymbol{r}_{\mathrm{B}}$, respectively. We denote the line from A to C on one side as C_1 and that on the other side as C_2. Thus, Eq. (1.30) gives

$$\int_{\mathrm{C}_1(\mathrm{A}\to\mathrm{B})} \boldsymbol{E} \cdot \mathrm{d}\boldsymbol{s} = \int_{\mathrm{C}_2(\mathrm{A}\to\mathrm{B})} \boldsymbol{E} \cdot \mathrm{d}\boldsymbol{s}. \tag{1.31}$$

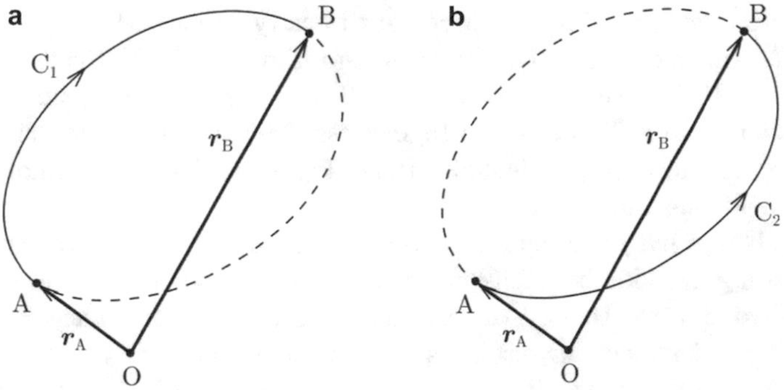

Fig. 1.17 Curvilinear integral of electric field strength from point A to point B over closed loop C through **a** C_1 and **b** C_2

Since this equation holds for an arbitrary closed loop C, the curvilinear integral of the electric field strength is determined only by the starting point, A, and terminal point, B, and is independent of the path. In fact, Eq. (1.31) is given by

$$\int_{\boldsymbol{r}_\mathrm{A}}^{\boldsymbol{r}_\mathrm{B}} \boldsymbol{E} \cdot \mathrm{d}\boldsymbol{s} = \phi(\boldsymbol{r}_\mathrm{A}) - \phi(\boldsymbol{r}_\mathrm{B}), \tag{1.32}$$

using Eq. (1.29). Such a vector is called a conservative vector. The electric field belongs to this kind of vector, and the electric field is also called conservative field.

Here, assume that an electric charge is forced to move in the electric field. The Coulomb force, $Q\boldsymbol{E}$, is exerted on a charge, Q, in the electric field of strength $\boldsymbol{E}$. Hence, it is necessary to apply an opposite force, $-Q\boldsymbol{E}$, to the electric charge to prevent it from moving and then to add an infinitesimal force to move it in a desired direction. The work necessary to carry slowly the electric charge from point A to point B is

$$W = -Q \int_{\boldsymbol{r}_\mathrm{A}}^{\boldsymbol{r}_\mathrm{B}} \boldsymbol{E} \cdot \mathrm{d}\boldsymbol{s} = Q[\phi(\boldsymbol{r}_\mathrm{B}) - \phi(\boldsymbol{r}_\mathrm{A})]. \tag{1.33}$$

Thus, the work necessary to carry the electric charge from point A to point B is proportional to the difference in electric potential between the two points but is independent of the particular path. Specifically, the work needed to carry the electric charge from a reference point such as infinity to position $\boldsymbol{r}$ is

$$W = Q\phi(\boldsymbol{r}). \tag{1.34}$$

This work can be regarded as the potential energy of the electric charge, Q.

A virtual surface composed of points with the same electric potential is called an **equipotential surface**. The work necessary to carry an electric charge, q, a small distance, $\delta\boldsymbol{r}$, on an equipotential surface is zero from Eq. (1.33). Since this work is given by $-q\boldsymbol{E} \cdot \delta\boldsymbol{r}$, we have $\boldsymbol{E} \cdot \delta\boldsymbol{r} = 0$. That is, $\boldsymbol{E}$ vector is normal to the equipotential surface. This can also be expressed by saying that the electric field lines are normal to the equipotential surface. Figure 1.18 shows examples of field lines and equipotential surfaces.

Human beings live in the earth's gravitational field. The gravity we experience is the universal gravitation between matter and the earth and is similar to the electric force discussed here. Hence, we can define a surface on which gravitational potential is constant, and this surface is spherical and concentric with the earth. The direction of gravity, i.e., the direction of free motion of matter, is toward the center

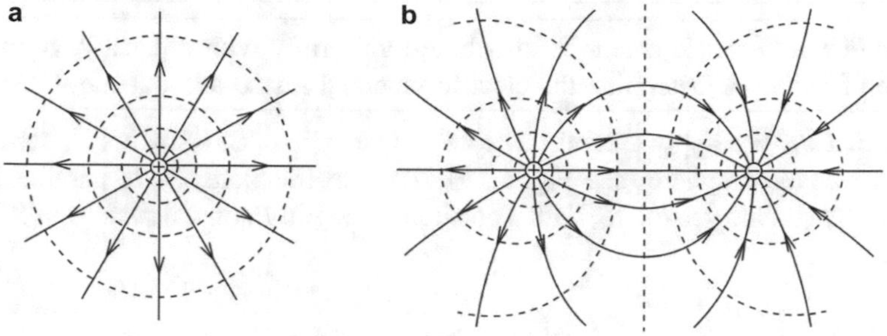

Fig. 1.18 Electric field lines (*solid lines*) and equipotential surfaces for **a** positive electric charge and **b** pair of positive and negative electric charges of equal amounts. These illustrate the conditions on a plane containing the electric charge(s), and the *dotted lines* are the crossing lines between the plane and equipotential surfaces

of the earth and is perpendicular to the sphere. Contour lines in maps are interfacial lines between spheres with different potential energies and the surface of the earth.

Elimination of the electric field strength, $\boldsymbol{E}$, using Eqs. (1.21) and (1.24) gives

$$\nabla \cdot (\nabla \phi) = -\frac{\rho}{\epsilon_0}. \tag{1.35}$$

The left side is written as $\nabla^2 \phi$. This is expressed as $\Delta\phi$, and the operator Δ is called **Laplacian**. This operator is

$$\Delta = \nabla \cdot \nabla = \frac{\partial^2}{\partial x^2} + \frac{\partial^2}{\partial y^2} + \frac{\partial^2}{\partial z^2} \tag{1.36}$$

in Cartesian coordinates. Sections A1.17 and A1.18 in the Appendix give its expressions in other coordinates. Using this operator, we rewrite Eq. (1.35) as

$$\Delta\phi = -\frac{\rho}{\epsilon_0}. \tag{1.37}$$

This equation is called **Poisson's equation**, which the electrostatic potential satisfies. When there is no electric charge ($\rho = 0$), Eq. (1.37) gives

$$\Delta\phi = 0, \tag{1.38}$$

which is called **Laplace's equation**. We can directly prove that the electric potential given by Eq. (1.27) satisfies Eq. (1.37) (see Sect. A2.1). Chapter 2 describes the method of solving Laplace's equation.

Example 1.6. Electric charge is distributed uniformly with a density, ρ, inside a sphere of radius a. Determine the electric potential inside and outside the sphere.

Solution 1.6. We apply Gauss' law to a virtual sphere, S, of radius, r, concentric with the charged sphere (see Fig. 1.19). The electric field, $\boldsymbol{E}$, is parallel to the elementary surface vector, d$\boldsymbol{S}$, with a constant strength, E, on the surface of S. Thus, we have

$$\int_{\mathrm{S}} \boldsymbol{E} \cdot \mathrm{d}\boldsymbol{S} = 4\pi r^2 E.$$

Since the total electric charge inside S is $(4\pi/3)r^3\rho$ and $(4\pi/3)a^3\rho$ for $r < a$ and $r > a$, respectively, we obtain the electric field strength as

$$\begin{aligned} E &= \frac{\rho}{3\epsilon_0} r; \qquad r<a, \\ &= \frac{\rho a^3}{3\epsilon_0 r^2}; \qquad r > a. \end{aligned}$$

Substituting these results into Eq. (1.29) gives

$$\begin{aligned} \phi &= -\int_{\infty}^{r} \frac{\rho a^3}{3\epsilon_0 r^2} \mathrm{d}r = \frac{\rho a^3}{3\epsilon_0 r}; \qquad r > a, \\ &= -\int_{a}^{r} \frac{\rho}{3\epsilon_0} r \mathrm{d}r + \frac{\rho a^2}{3\epsilon_0} = \frac{\rho}{2\epsilon_0}\left(a^2 - \frac{r^2}{3}\right); \qquad r<a. \end{aligned}$$

Fig. 1.19 Virtual sphere S with radius r smaller than a

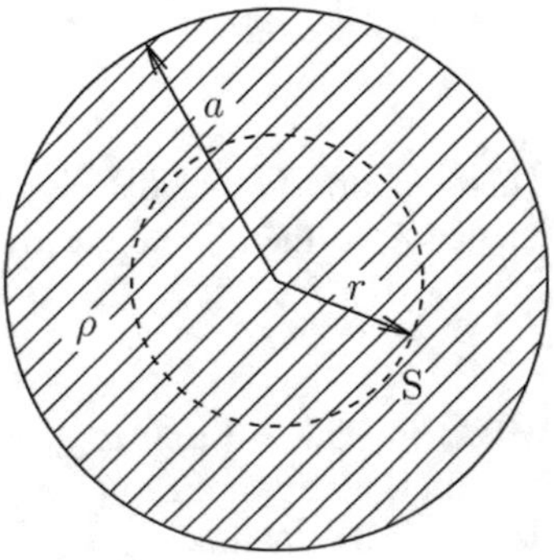

◇

Example 1.7. Electric charge is uniformly distributed with a linear density, λ, around a circle of radius a, as shown in Fig. 1.20. Determine the electric potential at the center, O.

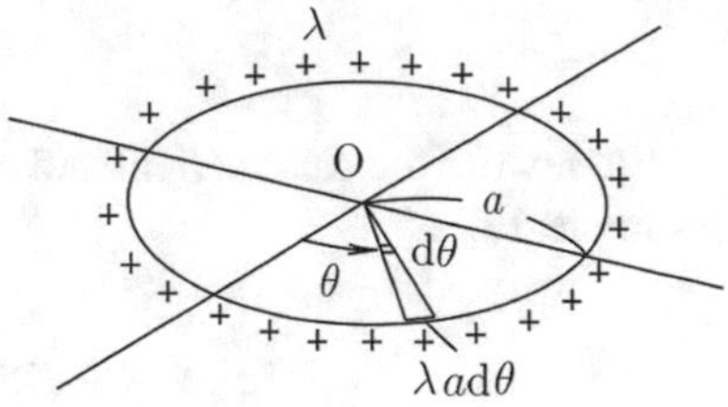

Fig. 1.20 Electric charge distributed uniformly around a circle of radius r

Solution 1.7. Here, we directly calculate the electric potential with Eq. (1.27). The electric charge in an infinitesimal region between θ and $\theta + \mathrm{d}\theta$, $\lambda a \mathrm{d}\theta$, is regarded as a point charge, $\rho \mathrm{d}V'$, with θ denoting the azimuthal angle. Thus, we have

$$\phi = \frac{1}{4\pi\epsilon_0}\int_0^{2\pi} \lambda \mathrm{d}\theta = \frac{\lambda}{2\epsilon_0}.$$

◇

Example 1.8. Determine the electric potential inside and outside the cylinder in Example 1.4 when the position $R = R_0 (> a)$ outside the cylinder is the reference point with zero electric potential.

Solution 1.8. Under the given condition, the electric potential outside the cylinder is

$$\phi(R) = -\int_{R_0}^{R} \frac{\rho a^2}{2\epsilon_0 R}\mathrm{d}R = \frac{\rho a^2}{2\epsilon_0}\log\frac{R_0}{R}; \qquad R > a \tag{1.39}$$

Then, the electric potential inside the cylinder is given by

$$\phi(R) = \phi(a) - \int_a^R \frac{\rho R}{2\epsilon_0}\mathrm{d}R = \frac{\rho a^2}{2\epsilon_0}\log\frac{R_0}{a} + \frac{\rho}{4\epsilon_0}(a^2 - R^2); \quad R < a. \tag{1.40}$$

The equipotential surface is a cylindrical surface inside and outside the cylinder.

The reason why the reference point is not infinity is that the electric potential diverges because of the infinite total electric charge. In practice, we easily find that the electric potential directly estimated from Eq. (1.27) diverges. This divergence comes from assuming an infinitely long cylinder; the theory of electromagnetism itself does not contain any defect.

◇

We rewrite the electric field and electric potential on the surface of the sphere in Example 1.6 as

$$E(r = a) = \frac{Q}{4\pi\epsilon_0 a^2}, \tag{1.41}$$

$$\phi(r = a) = \frac{Q}{4\pi\epsilon_0 a} \tag{1.42}$$

with the total electric charge $Q = (4\pi/3)a^3\rho$. Both of them diverge in the limit of a point charge, i.e., $a \to 0$, indicating an abnormal situation. However, this comes from the problematic assumption that finite electric charge exists in an infinitesimal volume, and hence, it is not an essential problem in the theory itself. In practice, the space in which electric charge exists is finite. The concept of "point charge" is an approximation of the fact that the size of the electric charge is much smaller than the treated system.

1.6 Electric Dipole

Most materials are electrically neutral with equal amounts of positive and negative electric charges, as mentioned in the beginning of this chapter. When an electric field is applied to such materials, positive and negative charges are displaced in and against the direction of the electric field, respectively, resulting in local deviations from neutrality. As a result, some electric phenomena are observed, as will be discussed in Chaps. 2 and 4. The fundamental element for such electric phenomena is a pair of positive and negative point charges that are separated by a small distance. This is called an **electric dipole**.

Suppose that electric charges q and $-q$ are displaced by $d/2$ in the positive and negative directions from the origin along the z-axis, as shown in Fig. 1.21. We determine the electric potential due to the pair of charges at a point, P, sufficiently far from the origin. In polar coordinates, the distance of P from the origin is denoted by $r(\gg d)$, and the angle of P from the z-axis is denoted by θ. The distance from the positive point charge to P is $r' = [r^2 + (d/2)^2 - rd\cos\theta]^{1/2} \simeq r - (d/2)\cos\theta$, and the electric potential at P due to this charge is

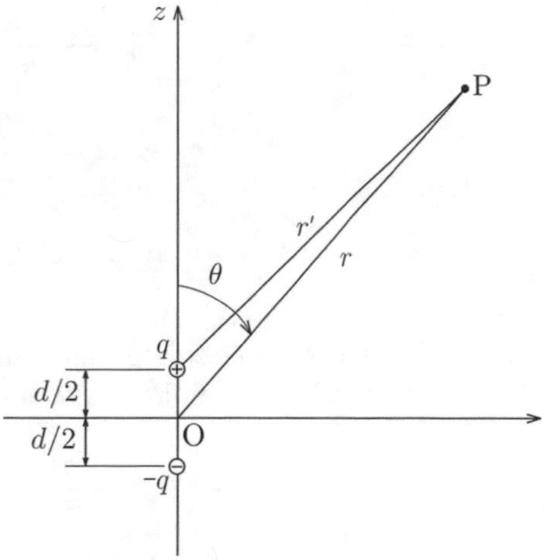

Fig. 1.21 Pair of positive and negative electric charges near the origin and point P sufficiently far from the charges

$$\phi_+(r) \simeq \frac{q}{4\pi\epsilon_0} \cdot \frac{1}{r - (d/2)\cos\theta} \simeq \frac{q}{4\pi\epsilon_0 r^2}\left(r + \frac{d}{2}\cos\theta\right). \tag{1.43}$$

Similarly, the electric potential due to the negative charge is

$$\phi_-(r) \simeq -\frac{q}{4\pi\epsilon_0} \cdot \frac{1}{r + (d/2)\cos\theta} \simeq -\frac{q}{4\pi\epsilon_0 r^2}\left(r - \frac{d}{2}\cos\theta\right). \tag{1.44}$$

Thus, the electric potential caused by the electric dipole is given by

$$\phi(r) = \phi_+(r) + \phi_-(r) \simeq \frac{qd\cos\theta}{4\pi\epsilon_0 r^2}. \tag{1.45}$$

Here, the **electric dipole moment** is defined as the product of the direction vector from the negative electric charge to the positive electric charge, $d\boldsymbol{i}_z$, and the amount of charge, q:

$$\boldsymbol{p} = qd\boldsymbol{i}_z. \tag{1.46}$$

In terms of the electric dipole moment, we rewrite the electric potential caused by the dipole as

$$\phi(r) = \frac{\boldsymbol{p}\cdot\boldsymbol{r}}{4\pi\epsilon_0 r^3} = \frac{p\cos\theta}{4\pi\epsilon_0 r^2}, \tag{1.47}$$

where $p = qd$ is the magnitude of the electric dipole moment. We calculate the resultant electric field to be

$$E_r = -\frac{\partial \phi}{\partial r} = \frac{p\cos\theta}{2\pi\epsilon_0 r^3}, \tag{1.48a}$$

$$E_\theta = -\frac{1}{r}\cdot\frac{\partial \phi}{\partial \theta} = \frac{p\sin\theta}{4\pi\epsilon_0 r^3}, \tag{1.48b}$$

$$E_\varphi = -\frac{1}{r\sin\theta}\cdot\frac{\partial \phi}{\partial \varphi} = 0. \tag{1.48c}$$

Figure 1.22 shows the electric potential and electric field produced by the dipole. The surface on which the following relationship holds is the equipotential surface:

$$\frac{\cos\theta}{r^2} = \text{const.} \tag{1.49}$$

Suppose that electric charge is distributed in a region, V, around the origin, O. We determine the electric potential caused by this charge at a point, P, with a position vector $\boldsymbol{r}$ with $r = |\boldsymbol{r}|$ sufficiently large compared with V (see Fig. 1.23). The electric potential is formally given by Eq. (1.27). We denote by θ the angle between the vector $\boldsymbol{r}'$ from the origin to a point, A, in V and $\boldsymbol{r}$. Since $r' = |\boldsymbol{r}'| \ll r$, we have

$$\frac{1}{|\boldsymbol{r}-\boldsymbol{r}'|} \simeq \frac{1}{r}\left(1 + \frac{r'}{r}\cos\theta\right) + \cdots = \frac{1}{r} + \frac{\boldsymbol{r}\cdot\boldsymbol{r}'}{r^3} + \cdots.$$

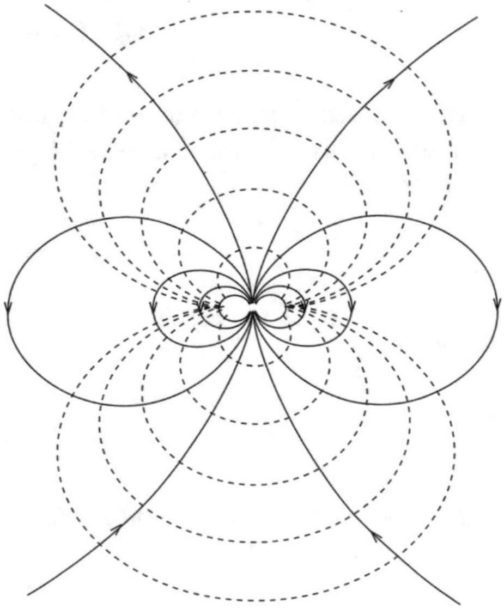

Fig. 1.22 Equipotential surfaces (*dotted lines*) and electric field lines (*solid lines*) produced by an electric dipole on a plane including positive and negative electric charges

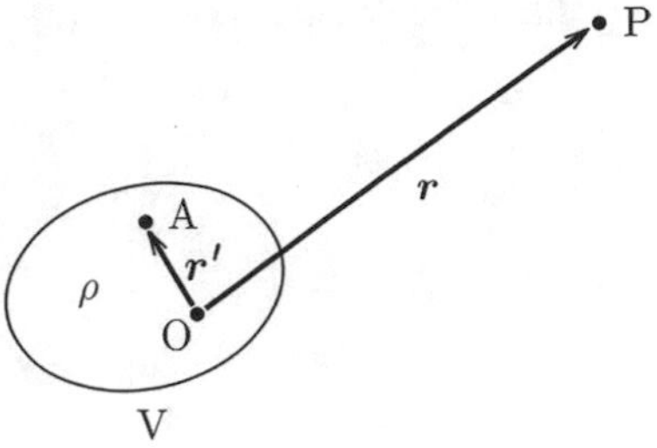

Fig. 1.23 Region V in which electric charge is distributed and point P sufficiently far from it

Hence, we rewrite the electric potential at P as

$$\phi = \frac{Q}{4\pi\epsilon_0 r} + \frac{\boldsymbol{p}\cdot\boldsymbol{r}}{4\pi\epsilon_0 r^3} + \cdots. \tag{1.50}$$

In the above,

$$Q = \int_V \rho(\boldsymbol{r}')\mathrm{d}V' \tag{1.51}$$

is the total amount of electric charge, and

$$\boldsymbol{p} = \int_V \boldsymbol{r}'\rho(\boldsymbol{r}')\mathrm{d}V' \tag{1.52}$$

is the total electric dipole moment caused by non-uniform distribution of electric charge. An expansion such as that in Eq. (1.50) is called a **multipole expansion**.

For a single electric dipole, shown in Fig. 1.21, the position vectors of charges q and $-q$ are $(d/2)\boldsymbol{i}_z$ and $-(d/2)\boldsymbol{i}_z$, respectively. Thus, we can easily show that the electric dipole moment of Eq. (1.46) is derived from Eq. (1.52).

Example 1.9. Electric charge is uniformly distributed with linear densities λ and $-\lambda$ on the lines at $x = d/2$ and $x = -d/2$ parallel to the z-axis, respectively, as shown in Fig. 1.24. Determine the electric potential at a point sufficiently far from these lines. Such a pair of electric charges is called an **electric dipole line**.

Solution 1.9. We define the cylindrical coordinates as in Fig. 1.24, where the azimuthal angle φ is measured from the x-axis. The distance between the point, P, and the line $x = d/2$ is denoted by R_+. The electric field at P produced by this line charge is

$$E = \frac{\lambda}{2\pi\epsilon_0 R_+}.$$

Fig. 1.24 Electric dipole line

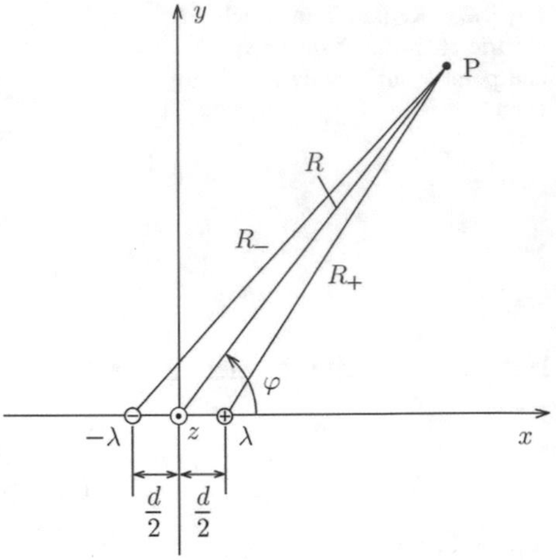

Electric field lines start radially from the line charge. The resultant electric potential is

$$\phi_+ = \frac{\lambda}{2\pi\epsilon_0}\log\frac{R_0}{R_+},$$

where R_0 is the distance to the reference point with zero electric potential. In fact, the above result satisfies the condition $\phi_+ = 0$ at $R_+ = R_0$.

Similarly, the electric potential due to the line charge at $x = -d/2$ is

$$\phi_- = -\frac{\lambda}{2\pi\epsilon_0}\log\frac{R_0}{R_-}$$

with R_- denoting the distance from the line charge to P. Thus, the total electric potential is given by

$$\phi = \phi_+ + \phi_- = \frac{\lambda}{2\pi\epsilon_0}\log\frac{R_-}{R_+}.$$

When $R \gg d$, we approximate R_+ as

$$R_+ = \left[R^2 + \left(\frac{d}{2}\right)^2 - Rd\cos\varphi\right]^{1/2} \simeq R\left(1 - \frac{d}{2R}\cos\varphi\right)$$

and have $R_- \simeq R\{1 + [d/(2R)]\cos\varphi\}$. Thus, the electric potential is reduced to

$$\phi(R,\varphi) \simeq \frac{\lambda}{2\pi\epsilon_0}\log\frac{1 + [d/(2R)]\cos\varphi}{1 - [d/(2R)]\cos\varphi} \simeq \frac{\lambda}{2\pi\epsilon_0}\cdot\frac{d}{R}\cos\varphi.$$

Using the **moment of an electric dipole line** given by

$$\hat{p} = \lambda d, \tag{1.53}$$

we write the electric potential as

$$\phi(R, \varphi) \simeq \frac{\hat{p} \cos \varphi}{2\pi\epsilon_0 R}. \tag{1.54}$$

The equality holds at the far distance ($R \gg d$). In this case, the total electric charge is 0, and the electric potential is 0 at infinity. Hence, there is no problem of divergence. The surface on which the following relationship holds is the equipotential surface:

$$\frac{\cos \varphi}{R} = \text{const.} \tag{1.55}$$

We determine the electric field strength to be

$$E_R = \frac{\hat{p} \cos \varphi}{2\pi\epsilon_0 R^2}, \tag{1.56a}$$

$$E_\varphi = \frac{\hat{p} \sin \varphi}{2\pi\epsilon_0 R^2}, \tag{1.56b}$$

$$E_z = 0. \tag{1.56c}$$

Figure 1.25 shows interfacial lines between the equipotential surfaces and the sheet and electric field lines. Both are circles.

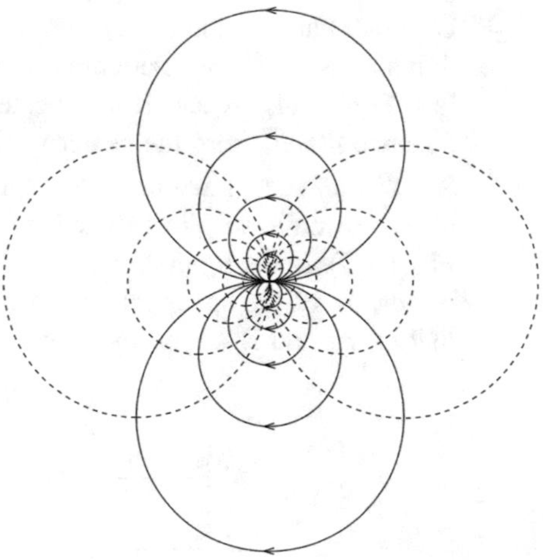

Fig. 1.25 Equipotential surfaces (*dotted lines*) and electric field lines (*solid lines*) in a plane normal to the line charges

◇

Column: Electric Charge Density in Electric Phenomena

Here, we estimate the amount of electric charges involved in a particular electric phenomenon. Equation (1.22) gives the relationship between surface electric charge and resultant electric field strength in Example 1.5. For a conductor surface, the electric field strength is doubled, as Sect. 2.1 will show.

Here, we examine the dielectric breakdown of air as an extreme case. The dielectric breakdown of air occurs when the applied electric field strength is as high as 3×10^6 V/m. Realizing such a field strength requires a surface charge density of $\sigma = \epsilon_0 E \simeq 2.7\times10^{-5}$ C/m^2. Since the electric charge of one electron is 1.6×10^{-19} C, we estimate the number density of electrons or holes on the surface to be 1.7×10^{14} m^{-2}.

The surface number density of metal atoms such as copper is about 0.6×10^{20} m^{-2}. If one of every 3.6×10^5 atoms on the surface gets or loses one electron, such a high electric field is produced. The number density of electric charge involved in usual electric phenomena is much lower.

Exercises

1.1 Electric charge, Q, is uniformly distributed on a bar, AB, of length L, as shown in Fig. E1.1. Determine the force on a point charge, q, put at point P.

1.2. Electric charge is uniformly distributed with a linear density, λ, on a bar of length a parallel to the y-axis. Determine the electric field strength at point A at distance b along the x-axis from the bottom of the bar (see Fig. E1.2).

1.3 Electric charge is uniformly distributed with a linear density, λ, around a square of width a. Determine the electric field strength at point P at distance z above the center of the square (see Fig. E1.3).

1.4 Electric charge is uniformly distributed with a surface density, σ, on a long thin slab of width $2a$. Determine the electric field strength at points A and B (see Fig. E1.4). A and B are located at distance b above the center and at distance $d(> a)$ from the center in the same plane as the slab, respectively.

1.5 A solid sphere of radius a contains a spherical void of radius b. Electric charge is uniformly distributed in the solid portion with a density, ρ (see Fig. E1.5). In the figure, $a > b + d$. Determine the electric field strength at the center, A, of the spherical void and at point B outside the sphere. Suppose that O, A, and B are on the same line.

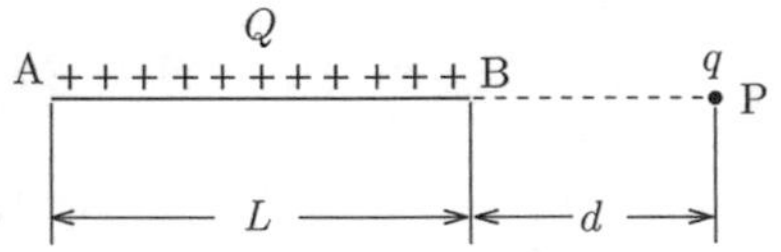

Fig. E1.1 Uniformly distributed charge Q on a bar and point charge q put at a position extrapolated from the bar

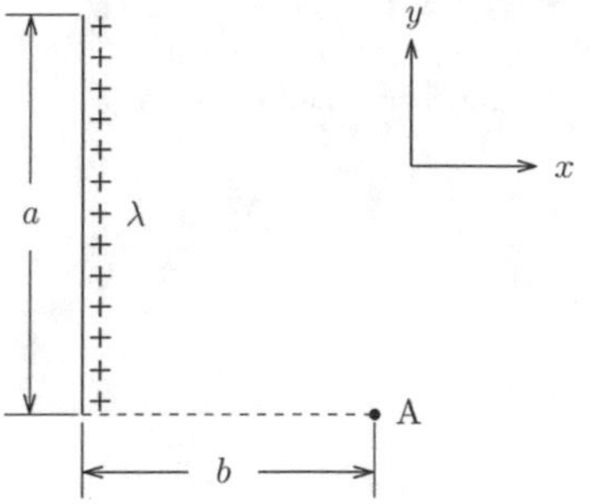

Fig. E1.2 Bar with uniformly distributed electric charge and point A

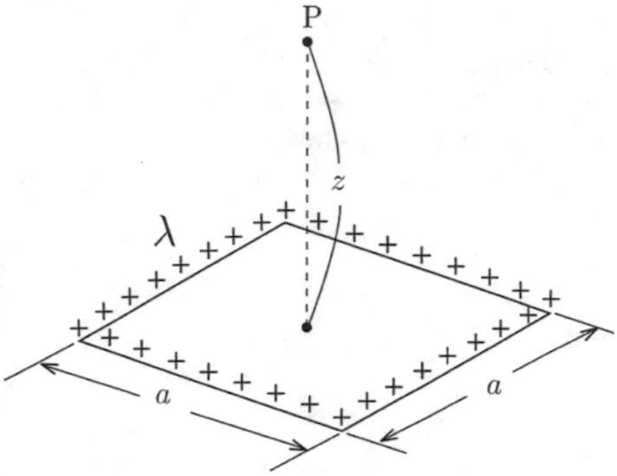

Fig. E1.3 Square with uniformly distributed electric charge and point P above the center

1.6 Determine the electric potential at point A in Example 1.2.

1.7 Electric charge is uniformly distributed with a density, σ, on the surface of a sphere of radius a. Determine the electric potential at the center of the sphere.

1.8 Determine the electric potential at point P in Example 1.3.

1.9 Electric charge is uniformly distributed on a circular plane of radius a with a surface density, σ, as illustrated in Fig. E1.6. Determine the electric field strength and electric potential at a point P at distance z from the center of the circular plane.

1.10 It is assumed that electric charge is uniformly distributed with density ρ_0 inside a wide slab of thickness $2a(-a<x<a)$ parallel to the y-z plane. Determine the electric field strength inside and outside the slab. Then, prove that the electric charge is distributed only inside the slab with density ρ_0, as assumed in the beginning, using the obtained electric field strength and Eq. (1.21).

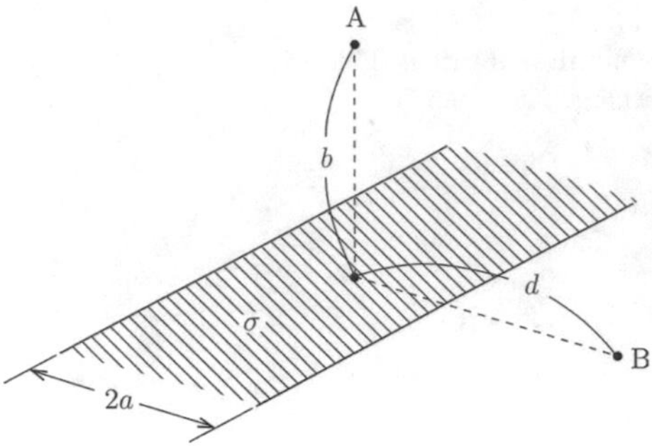

Fig. E1.4 Long thin slab with uniformly distributed electric charge and points A and B

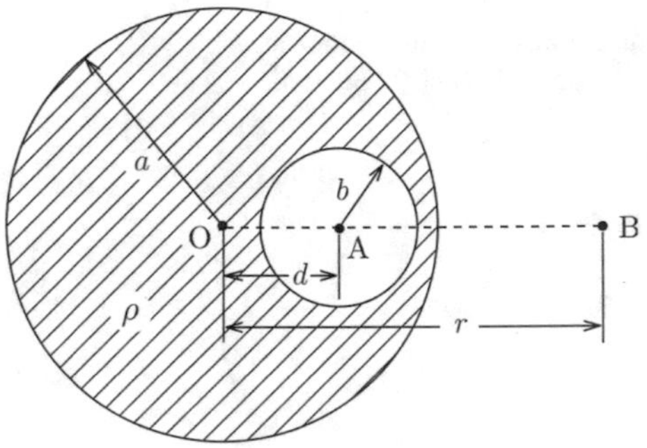

Fig. E1.5 Solid sphere containing a spherical void, uniformly distributed electric charge in the solid portion

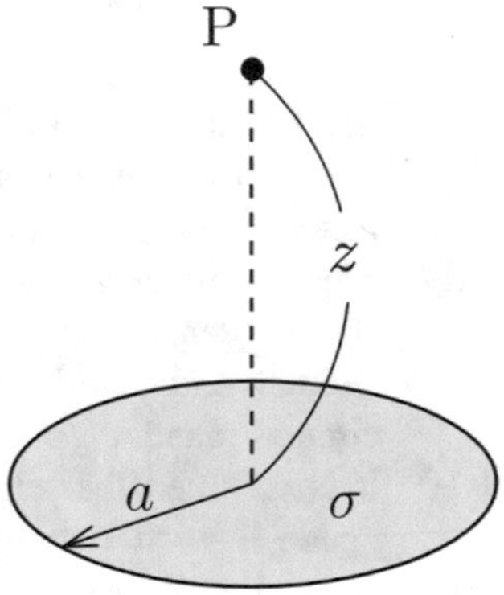

Fig. E1.6 Circular plane with uniformly distributed electric charge and point P above the center

1.11 Electric charge, Q, is uniformly distributed in a sphere of radius a. Determine the work necessary to carry a point charge, q, from point A to point B. These points are outside the sphere, and the distances of A and B from the sphere's center are r_A and r_B, respectively. Suppose that the electric charge distribution in the sphere is not influenced by the movement of the point charge.

1.12 Determine the equipotential surface given by Eq. (1.55).

Chapter 2
Conductors

Abstract This chapter covers electric phenomena inside and outside a conductor, in which the electric field is zero in the static condition. As a result, there is no electric charge inside the conductor, and the conductor is equipotential. When an electric charge is given to the conductor, it is distributed only on the conductor surface, in such a way that zero electric field strength is attained. The same thing occurs when an external electric field is applied to the conductor. This is the electrostatic shielding. The electric field on the outside is directed normal to the surface of the conductor. A special method, i.e., the method of images, is introduced to determine the electric field outside the conductor and the distribution of electric charges on the surface under such given conditions.

2.1 Electric Properties of Conductors

In terms of electric properties, materials are roughly classified into **conductors**, which can easily transport electric current, and **insulators**, which can hardly do so. The classification is based on electric conductivity, as shown in Chap. 5. Metals are conductors, and their electric property originates from free electrons that can move freely in the material. On the other hand, electrons in insulators such as mica and glass cannot move because of their bonding to atomic nuclei. Hence, the electric behavior of conductors and insulators is very different. This chapter describes the electric behavior of conductors. Chapter 4 describes that of insulators, which are also called **dielectrics** or **dielectric materials** because of their other electric properties.

The electric behavior of conductors is defined as follows; the electric field and the electric charge density inside the conductor are zero in the static condition after the conductor is put in an external electric field. That is,

$$\boldsymbol{E} = 0 \tag{2.1}$$

T. Matsushita, *Electricity and Magnetism*, Undergraduate Lecture Notes in Physics,
https://doi.org/10.1007/978-3-030-82150-0_2

and

$$\rho = 0. \tag{2.2}$$

The properties given by the above two equations are not independent of each other. Namely, Eq. (2.2) is derived from Eq. (2.1) with Eq. (1.21). From Eq. (2.1), we have

$$\phi = \text{const.} \tag{2.3}$$

Thus, we can also say that conductors are equipotential.

Here, we mention the relationship between electrical conductivity and the above definition of a conductor. If some electric field remains in the conductor, free electrons in the material will be driven by this field, which contradicts the assumption of a static condition. Thus, there is no electric field in a static conductor.

Suppose that an isolated conductor is placed in an electric field. The field forces the free electrons in the conductor to move. These electrons cannot go outside the conductor, and some of them accumulate on the surface of the conductor. The electric field produced by the electric charges on the surface exactly cancels the external electric field, resulting in a zero electric field inside the conductor. This realizes the situation assumed above for a conductor.

The appearance of electric charge on the surface of a conductor placed in an electric field is called **electrostatic induction**. The free electrons that appear on the surface are true charges in electromagnetism. It is possible to make an electric field stay inside the conductor. In this case, electric charges move inside the conductor, resulting in electric current, as will be described in Chap. 5. Hence, it is not a static situation. It should be noted that, even if the electric current does not change with time in a steady state, it is different from a static situation. This chapter describes static electric phenomena without movement of electric charges.

Here, we investigate the electric field in the vicinity of the conductor surface. Suppose that a small closed pellet-shaped surface includes the interface between the conductor and vacuum, as shown in Fig. 2.1a. We denote the height of the pellet and the area of the conductor surface inside the pellet by Δh and ΔS, respectively. Suppose that the density of electric charge on the surface of the conductor is σ. We apply Gauss' law, Eq. (1.19), to the pellet. In this case, the electric field vector, $\boldsymbol{E}$, is perpendicular to the surface of the conductor because of the orthogonality between the electric field and equipotential surface, since the surface of the conductor is equipotential. Hence, the electric field lines that go out of the side surface of the pellet are negligible if Δh is sufficiently small. Thus, all the electric field lines go out from the outer surface (see Fig. 2.1b). Since $\boldsymbol{E}$ is perpendicular to this surface, we have

$$\int_{\Delta S} \boldsymbol{E} \cdot \mathrm{d}\boldsymbol{S} = \int_{\text{outersurface}} E\mathrm{d}S = E\Delta S. \tag{2.4}$$

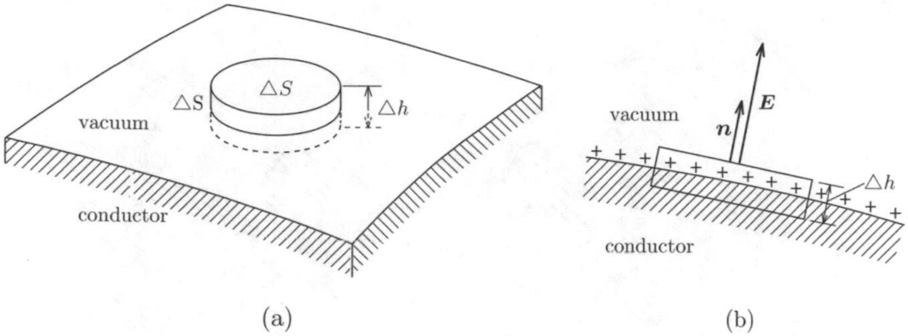

Fig. 2.1 **a** Small closed surface that includes part of the conductor surface and **b** electric field vector around the closed surface. The vector ***n*** is a unit vector normal to the surface

The total electric charge inside the pellet is $\sigma\Delta S$, and the left side of Eq. (1.19) is $\sigma\Delta S/\epsilon_0$. This gives

$$E = \frac{\sigma}{\epsilon_0}. \tag{2.5}$$

That is, the electric field strength on the surface of the conductor is equal to the surface electric charge density divided by ϵ_0.

Suppose we apply an electric charge, Q, to a spherical conductor of radius a. This determines the electric field and electric potential inside and outside the conductor. Since the electric charge stays on the surface and charges repel each other, the charge is uniformly distributed on the surface. Hence, the surface electric charge density is $\sigma = Q/(4\pi a^2)$. We apply Gauss' law to a supposed spherical surface, S, of radius r with the same center as that of the conductor. Since the electric charge distribution has spherical symmetry, we can also assume the electric field to have spherical symmetry. Hence, the electric field is directed normally to S, and its strength is uniform on S. If its strength is denoted by E, the surface integral of the electric field strength in Eq. (1.19) is $4\pi r^2 E$. For $r > a$, as shown in Fig. 2.2a, all the electric charge stays inside S, and the right side of Eq. (1.19) is Q/ϵ_0. Thus,

$$E(r) = \frac{Q}{4\pi\epsilon_0 r^2}; \quad r > a. \tag{2.6}$$

The electric field outside the conductor is the same as that when all the electric charge is concentrated on the center. For $r < a$, as shown in Fig. 2.2b, the total electric charge inside S is zero. This gives

$$E(r) = 0; \quad 0 \leq r < a. \tag{2.7}$$

Thus, Eq. (2.1) is fulfilled inside the conductor. It can also be shown that Eq. (2.6) satisfies Eq. (2.5) on the surface of the conductor ($r = a$) with the surface electric charge density determined above.

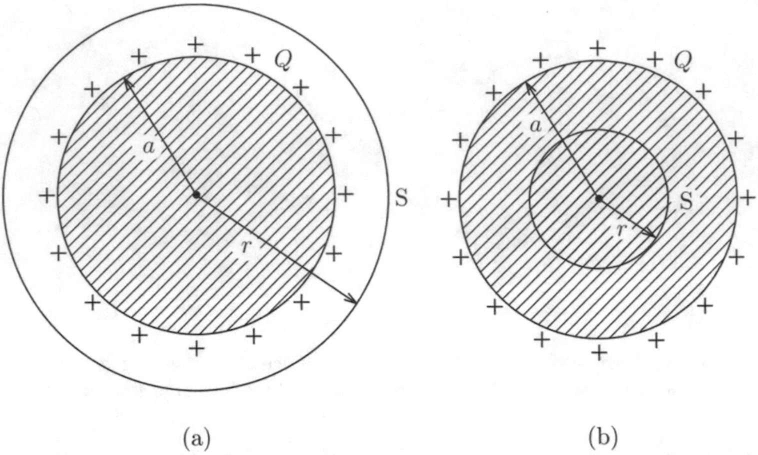

Fig. 2.2 Spherical conductor and virtual spherical surface, S: **a** case where S is outside the conductor and **b** case where S is inside the conductor

We determine the electric potential from

$$\phi(r) = -\int_{\infty}^{r} E(r)\mathrm{d}r \tag{2.8}$$

with Eqs. (2.6) and (2.7) and the condition that the electric potential is zero at infinity. This gives

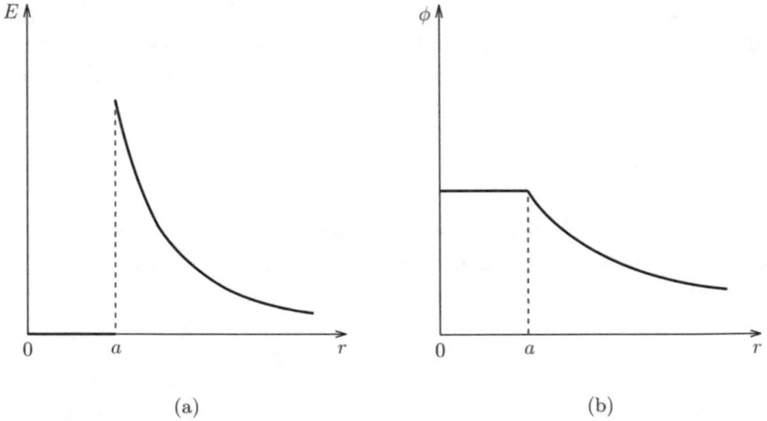

Fig. 2.3 **a** Electric field strength and **b** electric potential inside and outside the charged spherical conductor

$$\phi(r) = \frac{Q}{4\pi\epsilon_0 r}; \quad r > a, \tag{2.9a}$$

$$= \frac{Q}{4\pi\epsilon_0 a}; \quad 0 \leq r < a. \tag{2.9b}$$

Figure 2.3a, b shows the determined electric field strength and electric potential, respectively.

Example 2.1. Suppose that electric charge is distributed uniformly with surface density σ on a thin sheet separated by b from a slab conductor of thickness a, as shown in Fig. 2.4. Determine the surface density of the electric charge on each surface of the conductor and the electric field strength in each region of $x > -b$.

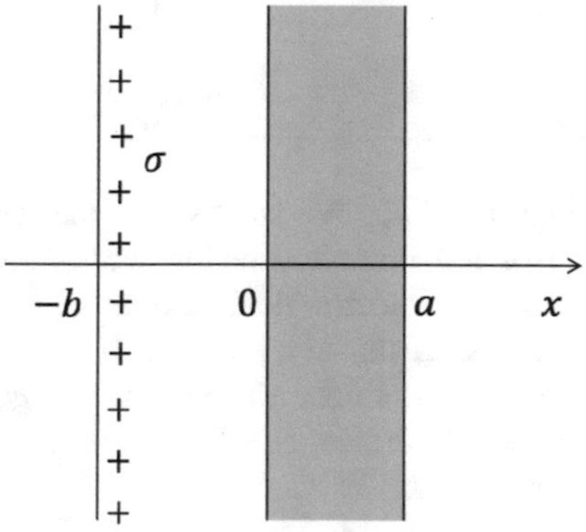

Fig. 2.4 Sheet electric charge with surface density σ and slab conductor

Solution 2.1. The electric field strength produced by the sheet charge on the left side of the conductor ($-b<x<0$) is $\sigma/2\epsilon_0$ (see Example 1.5) in the direction of the positive x-axis. The electric charge of surface density $-\sigma/2$ is induced on the left surface ($x=0$) of the conductor from Eq. (1.19) or Eq. (2.5). Then, an electric charge of surface density $\sigma/2$ is induced on the right surface ($x=a$) of the conductor from the principle of conservation of charge. Thus, the electric field strength in each region is

$$\begin{aligned} E &= \frac{\sigma}{2\epsilon_0}; & -b<x<0 \\ &= 0; & 0<x<a \\ &= \frac{\sigma}{2\epsilon_0} & x>a \end{aligned}$$

◇

Example 2.2. Suppose a pair of concentric spherical conductors, as shown in Fig. 2.5. Determine the electric field strength and electric potential in all regions when the electric charge, Q, is given on the inner conductor.

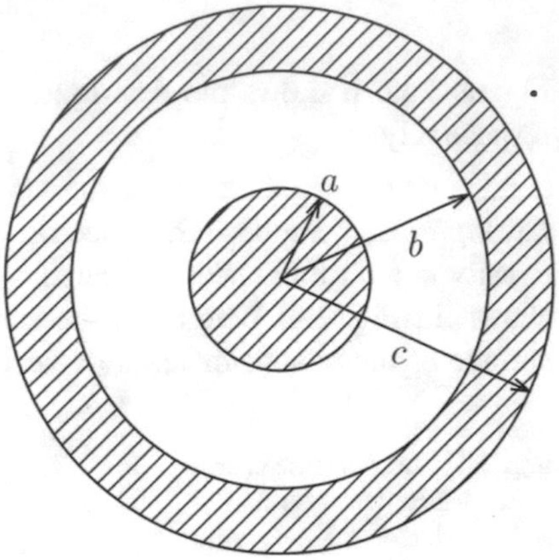

Fig. 2.5 Isolated concentric spherical conductors

Solution 2.2. We can assume that Q is uniformly distributed on the surface ($r = a$) of the inner conductor because of the spherical symmetry. This distribution makes the electric field zero inside the inner conductor ($r < a$). The electric charge appears on the inner surface ($r = b$) of the outer conductor because of the electrostatic induction. This electric charge is denoted by Q_b. We apply Gauss' law to a spherical surface, S, of radius r ($b < r < c$) with the same center as that of the conductors:

$$\int_{\mathrm{S}} \boldsymbol{E} \cdot \mathrm{d}\boldsymbol{S} = \frac{Q + Q_b}{\epsilon_0}.$$

Since $\boldsymbol{E} = 0$ on S, we obtain $Q_b = -Q$. Since no electric charge is given to the outer conductor, the electric charge that appears on the outermost surface ($r = c$) is $-Q_b = Q$.

If the total electric charge inside the virtual sphere, S, of radius r is denoted by Q_r, Gauss' law gives

$$E(r) = \frac{Q_r}{4\pi\epsilon_0 r^2}.$$

Since Q_r is equal to Q, 0, and Q for $a < r < b$, $b < r < c$, and $r > c$, respectively, we determine the electric field strength to be

$$
\begin{aligned}
E &= 0; & 0<r<a,\\
&= \frac{Q}{4\pi\epsilon_0 r^2}; & a<r<b,\\
&= 0; & b<r<c,\\
&= \frac{Q}{4\pi\epsilon_0 r^2}; & r>c.
\end{aligned}
$$

Then, we obtain the electric potential as

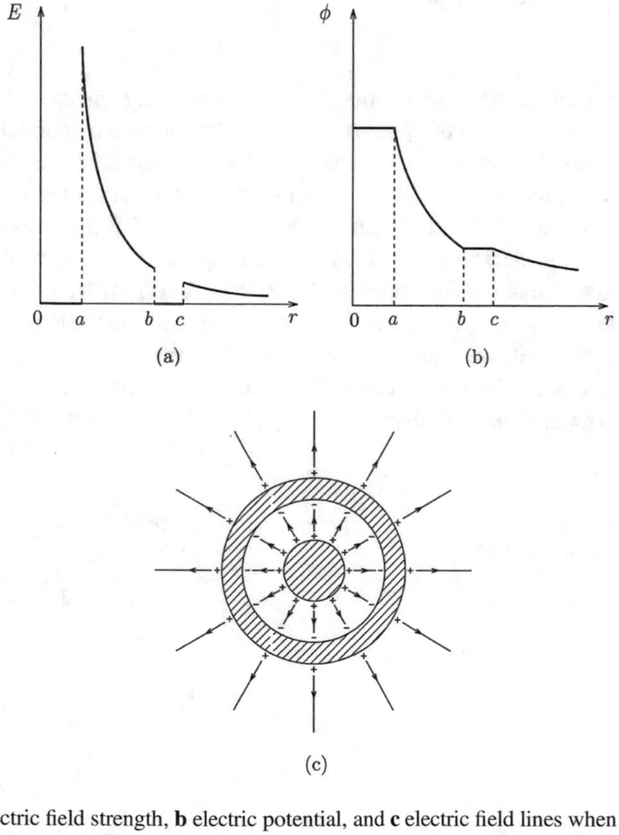

Fig. 2.6 **a** Electric field strength, **b** electric potential, and **c** electric field lines when electric charge is given to the inner conductor of a set of concentric spherical conductors

$$\phi(r) = -\int_{\infty}^{r} E(r)\mathrm{d}r = \frac{Q}{4\pi\epsilon_0 r}; \qquad r > c,$$
$$= \frac{Q}{4\pi\epsilon_0 c}; \qquad b<r<c,$$
$$= \phi(b) - \int_{b}^{r} E(r)\mathrm{d}r = \frac{Q}{4\pi\epsilon_0}\left(\frac{1}{r} - \frac{1}{b} + \frac{1}{c}\right); \qquad a<r<b,$$
$$= \frac{Q}{4\pi\epsilon_0}\left(\frac{1}{a} - \frac{1}{b} + \frac{1}{c}\right); \qquad 0\leq r<a.$$

Figure 2.6a–c shows the obtained electric field strength, electrical potential, and electric field lines, respectively.

◇

Here, we suppose that the outer conductor in Example 2.2 is grounded. **Grounding** is a method to make the electric potential of a conductor zero by connecting it to the ground. It sometimes accompanies transfer of electric charge. In the above case, the electric charge on the outer surface ($r = c$) of the outer conductor transfers to the ground through the grounding. This occurs because of the repulsive Coulomb interaction between electric charges on the outer surface. This can also be understood from the fact that the free electric charge transfers from the position of higher electric potential, $\phi = Q/(4\pi\epsilon_0 c)$, to the position of lower electric potential, $\phi = 0$. The electric charge on the inner surface ($r = b$) of the outer conductor does not transfer to the ground. This is because it is attracted by the electric charge on the surface ($r = a$) of the inner conductor (see Fig. 2.7). When some area is surrounded

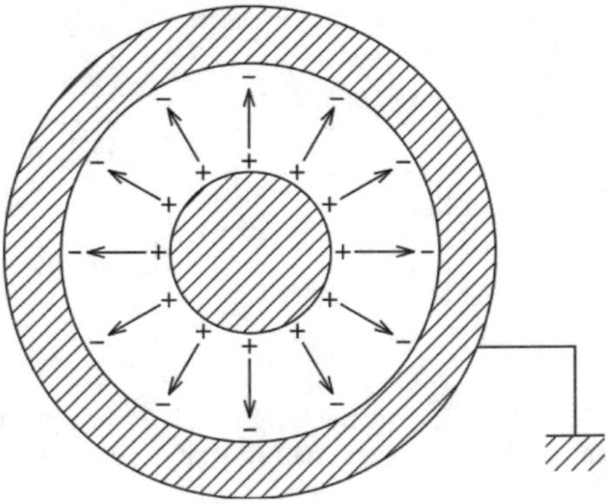

Fig. 2.7 Electric field lines when electric charge is given to the inside of a set of concentric spherical conductors and the outside is grounded

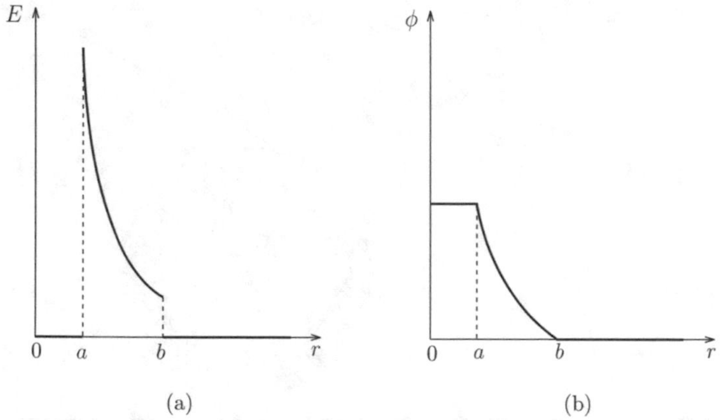

Fig. 2.8 **a** Electric field strength and **b** electric potential when electric charge is given to the inner conductor of a set of concentric spherical conductors and the outer conductor is grounded

by a grounded conductor, changes in the outside do not influence the electric field inside the grounded conductor. Such shielding from outside influence is called **electrostatic shielding**.

In this case, Q_r is Q and 0 for $a < r < b$ and $r > b$, respectively. This gives

$$
\begin{aligned}
E &= 0; & 0 \le r < a, \\
&= \frac{Q}{4\pi\epsilon_0 r^2}; & a < r < b, \\
&= 0; & r > b.
\end{aligned}
$$

Thus, we determine the electric potential to be

$$
\begin{aligned}
\phi(r) &= \frac{Q}{4\pi\epsilon_0}\left(\frac{1}{a} - \frac{1}{b}\right); & 0 \le r < a, \\
&= \frac{Q}{4\pi\epsilon_0}\left(\frac{1}{r} - \frac{1}{b}\right); & a < r < b, \\
&= 0; & r > b.
\end{aligned}
$$

Figure 2.8a, b shows the obtained electric field strength and electric potential, respectively.

Example 2.3. Suppose a pair of long coaxial conductors, as shown in Fig. 2.9. Determine the electric field strength and electric potential in all regions when an electric charge, λ, is given to the inner conductor of a unit length. The electric potential is defined to be zero at a point at distance $R_0(> c)$ from the central axis.

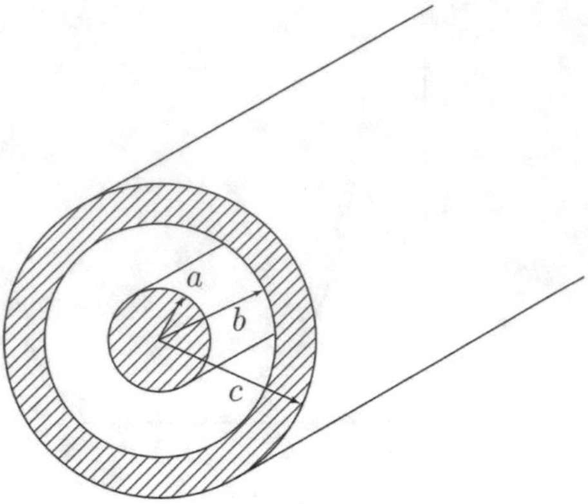

Fig. 2.9 Isolated long coaxial conductors

Solution 2.3. The electric charge is uniformly distributed on the surface ($R = a$) of the inner conductor with value λ in a unit length. The induced electric charges on the inner ($R = b$) and outer ($R = c$) surfaces of the outer conductor are $-\lambda$ and λ in a unit length, respectively. We determine the electric field strength to be

$$
\begin{aligned}
E(R) &= 0; && 0 \le R < a, \\
&= \frac{\lambda}{2\pi\epsilon_0 R}; && a < R < b, \\
&= 0; && b < R < c, \\
&= \frac{\lambda}{2\pi\epsilon_0 R}; && R > c.
\end{aligned}
$$

From the definition, the electric potential is given by

$$
\begin{aligned}
\phi(R) &= -\int_{R_0}^{R} E(R)\mathrm{d}R = \frac{\lambda}{2\pi\epsilon_0}\log\frac{R_0}{R}; && c < R < R_0, \\
&= \frac{\lambda}{2\pi\epsilon_0}\log\frac{R_0}{c}; && b < R < c, \\
&= \frac{\lambda}{2\pi\epsilon_0}\log\frac{bR_0}{cR}; && a < R < b, \\
&= \frac{\lambda}{2\pi\epsilon_0}\log\frac{bR_0}{ac}; && 0 \le R < a.
\end{aligned}
$$

The reason why infinity is not defined as the reference point of zero electric potential is that the total electric charge is infinite because of the infinite length, as mentioned in Example 1.8.

◇

Example 2.4. Two wide slab conductors are parallel to each other, as shown in Fig. 2.10, and electric charges Q and $-Q$ are given to the left and right conductors, respectively. The area of each surface is S. Determine the electric charge that appears on each conductor surface, the electric field strength, and the electric potential inside and outside the conductors.

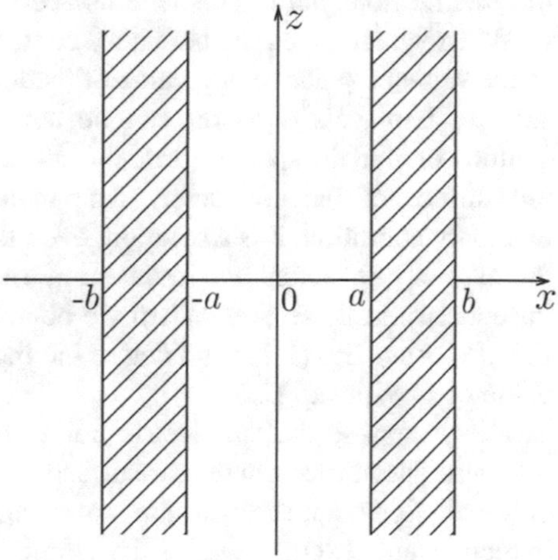

Fig. 2.10 Two parallel slab conductors

Solution 2.4. We denote the electric charge on the surface at $x = b$ by Q_b. Then, the electric charge on the surface at $x = a$ is $-Q - Q_b$. So that the electric field does not penetrate the right conductor, the electric charge at $x = b$ (Q_b) must be the same as the electric charge in the region $x \leq a$, i.e., $Q - Q - Q_b = -Q_b$. Thus, we have $Q_b = 0$, and the electric charge at $x = a$ is $-Q$. The total electric charge in the region $-b < x < b$ must be zero from Eq. (1.19) so that the electric field is zero in the two conductors. Hence, the electric charge at $x = -a$ is Q and that at $x = -b$ is zero. Thus, the electric field strength along the x-axis is

$$
\begin{aligned}
E &= 0; \qquad x < -a, \\
&= \frac{Q}{\epsilon_0 S}; \qquad -a < x < a, \\
&= 0; \qquad x > a.
\end{aligned}
$$

The electric potential is

$$
\begin{aligned}
\phi &= \frac{Qa}{\epsilon_0 S}; \qquad x < -a, \\
&= -\frac{Qx}{\epsilon_0 S}; \qquad -a < x < a, \\
&= -\frac{Qa}{\epsilon_0 S}; \qquad x > a.
\end{aligned}
$$

◇

2.2 Special Solution Method for Electrostatic Field

Suppose we need to determine the density of electric charge on the surface of a conductor or the electric field strength around the conductor when the conductor is in an external electric field. The electric potential in the conductor is constant in space, as shown in Eq. (2.3). Outside the conductor, there is no electric charge and the electric potential ϕ satisfies Laplace's Equation (1.38).

When we are given the boundary condition on the surface of a treated area, such as the value of ϕ above or a value of its derivative along the direction normal to the surface, Laplace's equation can be solved uniquely. Hence, there is only one solution of ϕ in the space outside the conductor, which becomes a constant value on the surface of the conductor. This means that, if some function satisfies the boundary condition, it is a solution, even though it may be obtained by intuition. In the case of conductors, we know some methods to solve problems. These will be introduced in this section. When we obtain a solution for ϕ, we obtain the electric field, E, using Eq. (1.24) and determine the surface electric charge density from the value of E with Eq. (2.5).

Here, suppose that an electric point charge, q, is put at a position at distance a from a flat infinite conductor surface, as shown in Fig. 2.11a. Electric charge of different signs appears on the conductor surface because of the electrostatic induction and exerts an attractive force on q. The x-y plane is defined on the conductor surface with the origin, O, at the foot of a perpendicular line from the electric charge. The electric potential is constant on the conductor surface ($z = 0$), as discussed in Sect. 2.1. Figure 1.18b shows that such an electric potential can be realized in the following way: The conductor is virtually removed, and then an electric charge, $-q$, is put at the point $(0, 0, -a)$, the point symmetric to the location of q with respect to the conductor surface, as shown in Fig. 2.11b. We now check the validity of this speculation. The electric potential that the two electric charges produce outside the conductor ($z > 0$) is

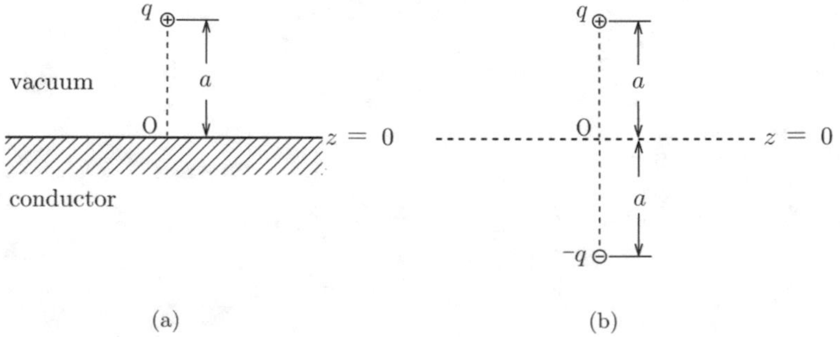

Fig. 2.11 **a** Point charge, q, at distance a from a wide flat conductor surface and **b** image charge, $-q$, put at the symmetric point of the given charge with respect to the conductor surface

$$\phi(x, y, z) = \frac{1}{4\pi\epsilon_0}\left\{\frac{q}{[x^2+y^2+(z-a)^2]^{1/2}} - \frac{q}{[x^2+y^2+(z+a)^2]^{1/2}}\right\}. \tag{2.10}$$

It is easily found that this satisfies the condition, $\phi = 0$, on the conductor surface ($z = 0$). Since this satisfies Laplace's equation outside the conductor and the boundary condition of Eq. (2.3) on the conductor surface, this is the solution. This shows that the above intuitive method is useful. In the conductor ($z < 0$), the electric potential is not given by Eq. (2.10) but by $\phi = 0$. This solution method is called the **method of images** and the virtual electric charge is called an **image charge**.

From Eq. (2.10), we obtain the electric field strength outside the conductor as

$$\begin{aligned} E_x &= -\frac{\partial\phi}{\partial x} = \frac{q}{4\pi\epsilon_0}\left\{\frac{x}{[x^2+y^2+(z-a)^2]^{3/2}} - \frac{x}{[x^2+y^2+(z+a)^2]^{3/2}}\right\}, \\ E_y &= -\frac{\partial\phi}{\partial y} = \frac{q}{4\pi\epsilon_0}\left\{\frac{y}{[x^2+y^2+(z-a)^2]^{3/2}} - \frac{y}{[x^2+y^2+(z+a)^2]^{3/2}}\right\}, \\ E_z &= -\frac{\partial\phi}{\partial z} = \frac{q}{4\pi\epsilon_0}\left\{\frac{z-a}{[x^2+y^2+(z-a)^2]^{3/2}} - \frac{z+a}{[x^2+y^2+(z+a)^2]^{3/2}}\right\}. \end{aligned} \tag{2.11}$$

On the conductor surface, this reduces to

$$E_x(x, y, 0) = E_y(x, y, 0) = 0, \quad E_z(x, y, 0) = -\frac{qa}{2\pi\epsilon_0(x^2+y^2+a^2)^{3/2}}. \tag{2.12}$$

Figure 2.12 shows the electric field lines. Then, from Eq. (2.5) we obtain the density of electric charge induced on the conductor surface as

$$\sigma = -\frac{qa}{2\pi(x^2+y^2+a^2)^{3/2}}. \tag{2.13}$$

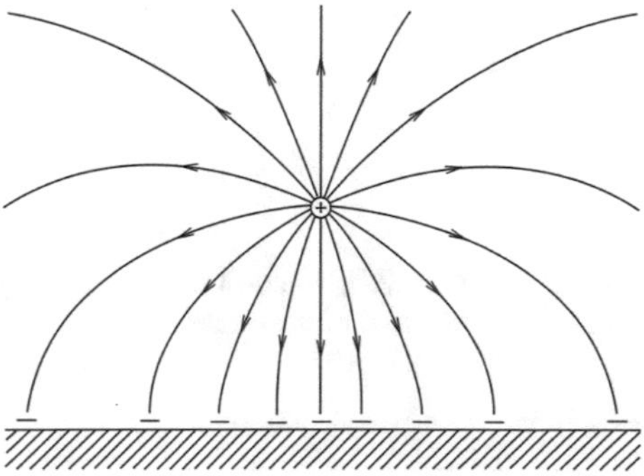

Fig. 2.12 Electric field lines between point charge and electric charges induced on the conductor surface

Now we determine the total electric charge. Using the two-dimensional polar coordinates $(x = r\cos\varphi, y = r\sin\varphi)$, we have

$$\int \mathrm{d}x \int \mathrm{d}y\sigma = -\frac{qa}{2\pi}\int_0^{2\pi}\mathrm{d}\varphi\int_0^{\infty}\frac{r}{(r^2+a^2)^{3/2}}\mathrm{d}r = -qa\left[-\frac{1}{(r^2+a^2)^{1/2}}\right]_0^{\infty} = -q. \tag{2.14}$$

That is, the total electric charge is equal to the amount of the image charge. The Coulomb force exerted on the electric charge q by the electric charge induced on the conductor surface is equal to that exerted by the image charge:

$$F = -\frac{q^2}{4\pi\epsilon_0(2a)^2} = -\frac{q^2}{16\pi\epsilon_0 a^2}. \tag{2.15}$$

This force is attractive ($F < 0$). This force is called **image force**.

The electric field strength inside the conductor ($z < 0$) produced by the electric charge on the conductor surface is equal to that produced by the electric charge $-q$ placed at the position of q, (0, 0, a). Since the latter electric field absolutely cancels out the electric field produced by q, the electric field in the conductor can be shown to be zero.

Suppose that a point charge, q, is placed at point A at distance d from the center, O, of a grounded spherical conductor of radius a ($d > a$), as illustrated in Fig. 2.13a. Now we determine the electric potential outside the spherical conductor. Assume that the conductor is removed and an image charge, Q, is placed at point B at

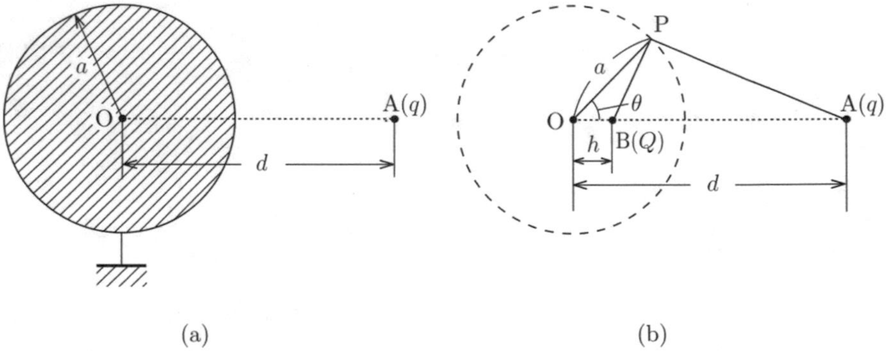

Fig. 2.13 **a** Grounded spherical conductor and point charge at point A and **b** image charge at point B after removal of the conductor

distance h from the center, as shown in Fig. 2.13b. The quantities Q and h are unknown and need to be determined. Then, the electric potential on the conductor surface is

$$\phi = \frac{1}{4\pi\epsilon_0}\left[\frac{q}{(a^2+d^2-2ad\cos\theta)^{1/2}} + \frac{Q}{(a^2+h^2-2ah\cos\theta)^{1/2}}\right]$$
$$= \frac{1}{4\pi\epsilon_0}\left\{\frac{q/\sqrt{a^2+d^2}}{[1-2ad\cos\theta/(a^2+d^2)]^{1/2}} + \frac{Q/\sqrt{a^2+h^2}}{[1-2ah\cos\theta/(a^2+h^2)]^{1/2}}\right\}, \tag{2.16}$$

where angle ∠POA is represented by θ. Hence, $\phi = 0$ is realized at any point on the conductor surface ($r = a$), and the boundary condition is satisfied if the following conditions are fulfilled:

$$\frac{q}{\sqrt{a^2+d^2}} + \frac{Q}{\sqrt{a^2+h^2}} = 0 \tag{2.17}$$

and

$$\frac{2ad}{a^2+d^2} = \frac{2ah}{a^2+h^2}, \tag{2.18}$$

which reduce to

$$h = \frac{a^2}{d}, \quad Q = -\frac{aq}{d}. \tag{2.19}$$

Thus, the electric potential at point (r, θ) outside the conductor is given by

$$\phi(r,\theta) = \frac{q}{4\pi\epsilon_0}\left\{\frac{1}{(r^2+d^2-2rd\cos\theta)^{1/2}} - \frac{a}{d[r^2+(a^2/d)^2-2(a^2r/d)\cos\theta]^{1/2}}\right\}. \tag{2.20}$$

The electric field strength can be calculated with this electric potential (see Exercise 2.8). Figure 2.14 shows the electric field lines.

We obtain the density of electric charge on the conductor surface as

$$\sigma(\theta) = \epsilon_0 E_r(r=a) = -\epsilon_0\left(\frac{\partial\phi}{\partial r}\right)_{r=a} = -\frac{q(d^2-a^2)}{4\pi a(a^2+d^2-2ad\cos\theta)^{3/2}}. \tag{2.21}$$

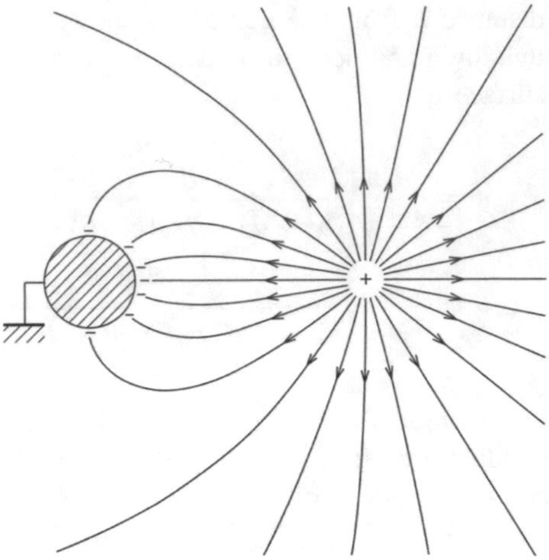

Fig. 2.14 Electric field lines between the point electric charge and grounded spherical conductor

The total electric charge is

$$\int_0^\pi \sigma(\theta) \cdot 2\pi a^2 \sin\theta \, \mathrm{d}\theta = -\frac{qa(d^2 - a^2)}{2} \int_0^\pi \frac{\sin\theta \, \mathrm{d}\theta}{(a^2 + d^2 - 2ad\cos\theta)^{3/2}}$$
$$= \frac{q(d^2 - a^2)}{2d} \left[(a^2 + d^2 - 2ad\cos\theta)^{-1/2} \right]_0^\pi = -\frac{a}{d} q, \tag{2.22}$$

which is equal to the image charge, Q. This charge is transferred from the ground to the conductor because of attraction by the point charge q. The reason it is smaller by factor a/d than in the case shown in Fig. 2.11 is that the size of the conductor is finite. For an infinitely long cylindrical conductor and line charge, the electric charge induced in a conductor of unit length is equal to the density of the given line charge (see Exercise 2.9).

Example 2.5. Suppose that the conductor is not grounded in the problem shown in Fig. 2.13a. Determine the electric potential outside the conductor.

Solution 2.5. In this case, the total electric charge on the conductor surface is zero. This problem is solved using the method of superposition. That is, this situation is obtained by a superposition of the electric charge $-aq/d$, which is distributed according to Eq. (2.21), and the charge aq/d, which is uniformly distributed on the surface. The distributed charge $-aq/d$ and point charge q give the zero electric potential of the conductor, and the distributed charge aq/d makes the conductor equipotential, $q/(4\pi\epsilon_0 d)$. Hence, this situation satisfies the conductor condition. If the electric potential given by Eq. (2.20) is denoted by $\phi_1(r, \theta)$, the electric potential outside the conductor is.

$$\phi(r, \theta) = \phi_1(r, \theta) + \frac{aq}{4\pi\epsilon_0 dr}.$$

◇

Example 2.6. Electric charge Q is placed at a point at a distance h from the center, O, of a hollow spherical conductor, as shown in Fig. 2.15. Determine the electric potential in the vacuum and the electric charge density on the inner surface of the conductor.

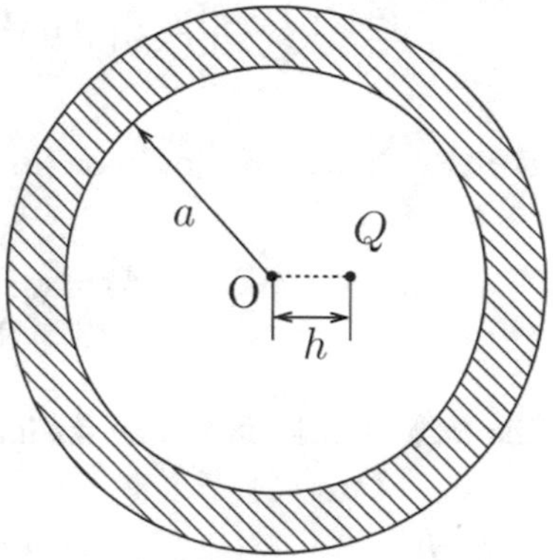

Fig. 2.15 Hollow spherical conductor and electric charge at a position inside the conductor

Solution 2.6. We remove the conductor and place an image electric charge, q, on a line extending from the center to the electric charge Q (see Fig. 2.16). We denote the distance of this point from the center by d. The electric potential at point P on the inner surface of the conductor is

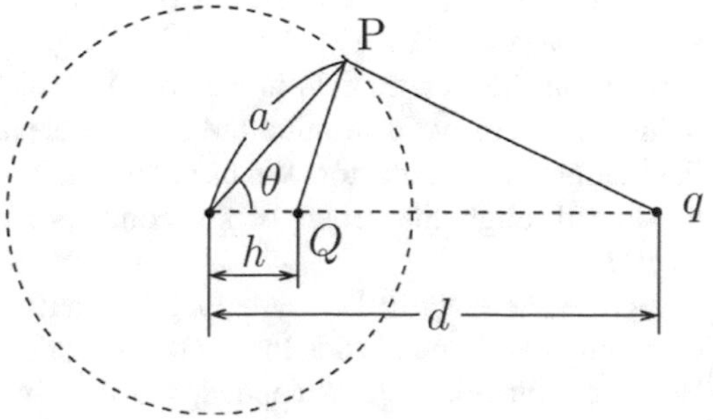

Fig. 2.16 Image electric charge q after removal of the conductor

$$\phi(a,\theta) = \frac{1}{4\pi\epsilon_0}\left[\frac{q}{(a^2+d^2-2ad\cos\theta)^{1/2}} - \frac{Q}{(a^2+h^2-2ah\cos\theta)^{1/2}}\right].$$

The conditions that satisfy $\phi(a,\theta) = 0$ are

$$d = \frac{a^2}{h}, \quad q = -\frac{dQ}{a} = -\frac{aQ}{h}.$$

Thus, the electric potential in the hollow is

$$\phi(r,\theta) = \frac{Q}{4\pi\epsilon_0}\left\{-\frac{a}{h\left[r^2+(a^2/h)^2-2(a^2r/h)\cos\theta\right]^{1/2}} - \frac{1}{(r^2+h^2-2rh\cos\theta)^{1/2}}\right\}.$$

The electric charge density on the inner surface is

$$\sigma(\theta) = -\epsilon_0 E_r(r=a) = \epsilon_0\left(\frac{\partial\phi}{\partial r}\right)_{r=a} = -\frac{Q(a^2-h^2)}{4\pi a(a^2+h^2-2ah\cos\theta)^{3/2}}.$$

The total electric charge on the inner surface is

$$\int_0^\pi \sigma(\theta)2\pi a^2\sin\theta d\theta = -\frac{Qa(a^2-h^2)}{2}\int_0^\pi \frac{\sin\theta d\theta}{(a^2+h^2-2ah\cos\theta)^{3/2}} = -Q.$$

◇

2.3 Electrostatic Induction

Suppose that a spherical conductor of radius a is put in a uniform electric field of strength $\boldsymbol{E}_0$ (see Fig. 2.17). An electric charge appears on the conductor surface and cancels out the electric field in the conductor. This phenomenon is the electrostatic induction. Here we determine the surface electric charge density and the electric field around the conductor. We use cylindrical coordinates and define the z-axis as the line through the center of the conductor along the direction of the applied electric field.

Before the electric field is applied, positive and negative electric charges are uniformly distributed inside the conductor, and the conductor is electrically neutral. When the electric field is applied, the positive and negative electric charges are

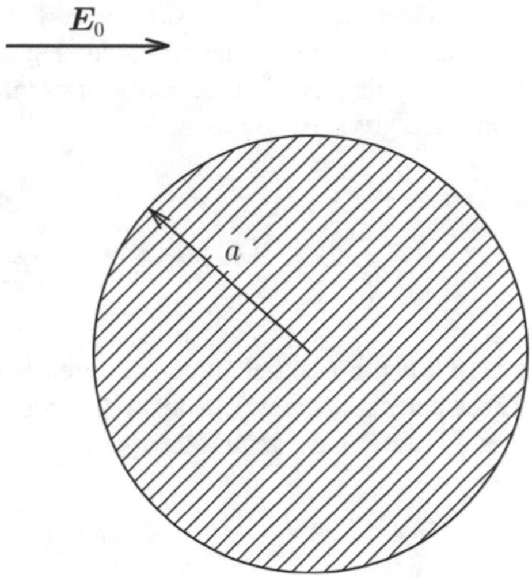

Fig. 2.17 Spherical conductor put in a uniform electric field

displaced in and against the direction of the electric field, respectively, as illustrated in Fig. 2.18a. This leads to a surface distribution of electric charge that keeps the inside electrically neutral (see Fig. 2.18b). Hence, this seems to realize the proper condition of a spherical conductor. We determine the displacement of the electric charges. In this case, we find the electric field produced by the positive charge to be the same as that produced when all the positive charge is concentrated at the center, as predicted by Gauss' law. The electric field produced by the negative charge is also the same as that produced by all the negative charge if concentrated at the center. As a result an electric dipole appears at the center of the conductor (see Fig. 2.18c). The electric dipole moment, p, that satisfies the electric potential is to be determined.

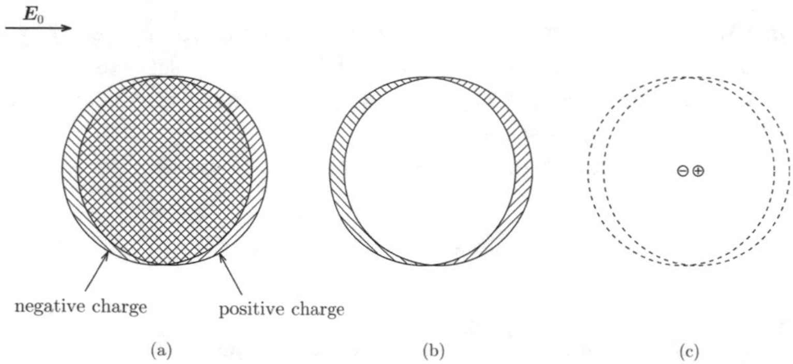

Fig. 2.18 Electrostatic induction in spherical conductor in uniform electric field: **a** displacement of positive and negative electric charges driven by the electric field, **b** electric charge that appears on the surface, and **c** electric dipole at the center

The electric potential outside the conductor is composed of the electric potential, ϕ_f, due to the applied electric field, $\boldsymbol{E}_0$, and the electric potential, ϕ_d, due to the electric dipole placed at the center after virtually removing the spherical conductor. These are given by

$$\phi_\mathrm{f} = -E_0 r \cos\theta, \tag{2.23}$$

$$\phi_\mathrm{d} = \frac{p\cos\theta}{4\pi\epsilon_0 r^2}, \tag{2.24}$$

where θ is the zenithal angle measured from the direction of applied electric field. These potentials are independent of the azimuthal angle, φ. Now prove for yourself that ϕ_f satisfies the requirements

$$-\frac{\partial\phi_\mathrm{f}}{\partial r} = E_0\cos\theta, \quad -\frac{1}{r}\cdot\frac{\partial\phi_\mathrm{f}}{\partial\theta} = -E_0\sin\theta.$$

The electric potential is given by

$$\phi = \phi_\mathrm{f} + \phi_\mathrm{d} = \left(-E_0 r + \frac{p}{4\pi\epsilon_0 r^2}\right)\cos\theta. \tag{2.25}$$

We determine the electric dipole moment, p, to be

$$p = 4\pi\epsilon_0 a^3 E_0 \tag{2.26}$$

so that the condition, $\phi(r = a) = 0$, is satisfied independently of the angle θ. Thus, we have

$$\phi = -E_0\left(r - \frac{a^3}{r^2}\right)\cos\theta. \tag{2.27}$$

Since this satisfies the boundary condition on the conductor surface ($r = a$) and satisfies Laplace's equation (note that each component of ϕ satisfies it), this is the unique solution. Thus, we can say the above speculation is valid. The electric potential inside the conductor is $\phi = 0$.

The electric field strength outside the conductor is given by

$$E_r = -\frac{\partial\phi}{\partial r} = E_0\left(1 + \frac{2a^3}{r^3}\right)\cos\theta, \tag{2.28a}$$

$$E_\theta = -\frac{1}{r}\cdot\frac{\partial\phi}{\partial\theta} = -E_0\left(1 - \frac{a^3}{r^3}\right)\sin\theta, \tag{2.28b}$$

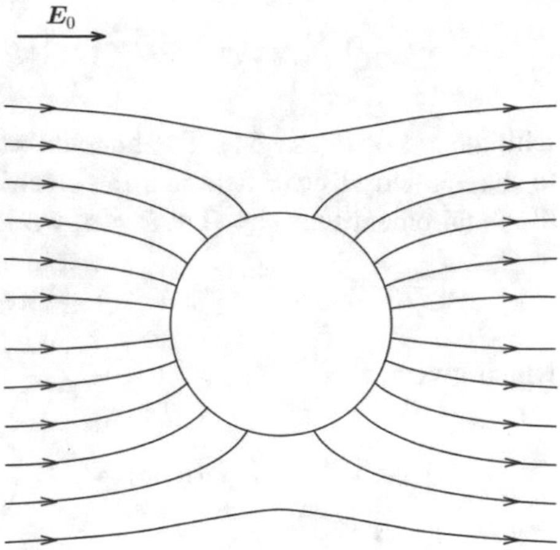

Fig. 2.19 Electric field lines outside the spherical conductor

$$E_\varphi = -\frac{1}{r\sin\theta}\cdot\frac{\partial\phi}{\partial\varphi} = 0. \tag{2.28c}$$

Figure 2.19 shows electric field lines on the plane that includes the z-axis. We can see that $E_\theta(r = a) = 0$ from Eq. (2.28b). This shows that the electric field vector is normal to the conductor surface. Equation (2.28a) shows that the electric field strength has the maximum value, $3E_0$, at both poles ($\theta = 0, \pi$). We determine the surface electric charge density to be

$$\sigma = \epsilon_0 E_r(r = a) = 3\epsilon_0 E_0 \cos\theta. \tag{2.29}$$

The electric dipole moment in a unit volume of the spherical conductor is

$$P = 3\epsilon_0 E_0, \tag{2.30}$$

which corresponds to the electric polarization in dielectric materials.

Example 2.7. A long cylindrical conductor is placed in a uniform normal electric field of strength E_0. Determine the electric potential and electric field outside the conductor and the density of electric charge on the conductor surface.

Solution 2.7. We use cylindrical coordinates and define the z-axis as the central axis of the conductor, and measure the azimuthal angle, φ, from the direction of the applied electric field. The electric potential outside the conductor can be determined by putting the electric dipole line, as shown in Example 1.9, on the central axis after removing the conductor similarly to what we did in the above analysis. We denote by $\hat{p}$ the moment of the electric dipole line in a unit length along the z-axis. Then, the electric potential outside the conductor is given by

$$\phi(R,\varphi) = \left(-E_0 R + \frac{\hat{p}}{2\pi\epsilon_0 R}\right)\cos\varphi$$

with the aid of Eq. (1.54). The first and second terms are the electric potential due to the applied electric field and the electric dipole line, respectively. Hence, from the requirement that $\phi = 0$ at $R = a$, we have

$$\hat{p} = 2\pi\epsilon_0 a^2 E_0,$$

which gives

$$\phi(R,\varphi) = -E_0\left(R - \frac{a^2}{R}\right)\cos\varphi.$$

Thus, we obtain the electric field strength as

$$\begin{aligned} E_R &= -\frac{\partial\phi}{\partial R} = E_0\left(1 + \frac{a^2}{R^2}\right)\cos\varphi, \\ E_\varphi &= -\frac{1}{R}\cdot\frac{\partial\phi}{\partial\varphi} = -E_0\left(1 - \frac{a^2}{R^2}\right)\sin\varphi, \\ E_z &= 0. \end{aligned}$$

We can see that the electric field is normal to the conductor surface from $E_\varphi(R = a) = 0$. The surface electric charge density is

$$\sigma = \epsilon_0 E_R(R = a) = 2\epsilon_0 E_0 \cos\varphi.$$

The electric dipole moment in a unit volume of the conductor is

$$P = 2\epsilon_0 E_0.$$

◇

Column: Applicability of Method of Images
The method of images is useful for solving problems when a conductor is put in an electric field, as shown in the Examples and Exercises. Now, suppose that an electric charge is given on a spherical conductor placed at some distance from a wide flat conductor surface. Can we also use the method of images in this case?

This problem can be compared with Exercise 2.10. Following the solution for that exercise, we first remove the spherical conductor and then place a point charge equal to the given charge at a point at some distance from the center. Next, we remove the wide flat conductor and put the same electric charge at the point symmetric to the location of the former charge with respect to the flat conductor surface. At first, this may seem useful in determining the electric potential outside the two conductors.

However, the electric potential cannot be determined with this method. How can we prove it? To satisfy the boundary condition on the infinitely wide conductor surface, all electric charges distributed on the surface must be connected to the electric charges on the spherical conductor surface through the electric field lines. In fact, from a superposition of point electric charges, we can show that the total amount of electric charge induced on the flat conductor surface is equal to the electric charge given to the spherical conductor. However, to satisfy the boundary condition on the spherical conductor surface after virtually concentrating all the electric charge at the image point, the absolute value of the electric charge must be smaller than the point charge assumed inside the infinite flat conductor [see Eq. (2.22)].

The method shown in Example 2.5 seems useful for making the spherical conductor surface equipotential with the same electric charge. However, the boundary condition on the infinitely wide conductor surface is not satisfied between one point charge and two separate point charges. For this reason, we cannot obtain an analytic solution. To get a solution, it is necessary to distribute the image charge in such way that the boundary condition is satisfied with the given electric charge.

For an infinitely long cylindrical conductor as in Exercise 2.10, even if the electric charge is concentrated on an infinitely thin line, the two conductors have equal total amounts of electric charge. Hence, we can obtain an analytic solution that simultaneously satisfies the two boundary conditions using the method of images. The method of images is useful also for dielectric materials, and even for magnetic phenomena in superconductors and magnetic materials. Consider the possibility of solving other problems using this method.

Exercises

2.1. Determine the electric field strength and electric potential when electric charges Q_1 and Q_2 are given to the inner and outer conductors, respectively, of the concentric spherical conductors in Fig. 2.5.

2.2. In a pair of concentric spherical conductors, an electric charge, Q, is given to the outer conductor and the inner conductor is grounded, as shown in Fig. E2.1. Determine the electric charge induced on the inner conductor surface. (Hint: Use the condition that the electric potential is also zero at infinity.)

Fig. E2.1 Concentric spherical conductors with grounded inner conductor

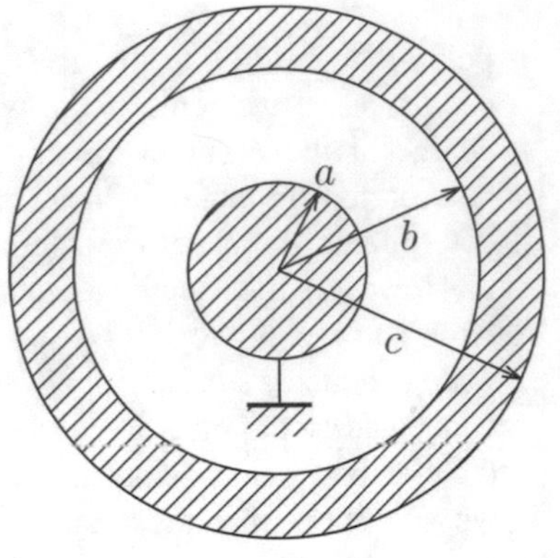

2.3. An electric charge Q is given to the left conductor in Example 2.4. Determine the electric charges that appear on each conductor surface and the electric field strength inside and outside the conductors.

2.4. When an electric charge is uniformly distributed with a surface density σ on a thin flat plane, the electric field strength near the plane is given by Eq. (1.22). However, Eq. (2.5) yields double this electric field strength near the conductor surface with the same charge density. Discuss the reason for the difference.

2.5. When an electric charge, q, is put at a point at distance a from a wide conductor surface, the electric charge induced on the conductor surface is given by Eq. (2.13). Prove that the Coulomb force exerted on q by the induced electric charge is given by Eq. (2.15).

2.6. Point charge Q is placed at a point at distances a and b from two flat conductor surfaces that are perpendicular to each other, as shown in Fig. E2.2. Determine the electric potential and electric field strength in the vacuum.

2.7. Suppose that electric charge of uniform linear density λ is put on a line at distance a from a flat infinite conductor surface, as shown in Fig. E2.3.

Fig. E2.2 Two perpendicular flat conductor surfaces and point charge

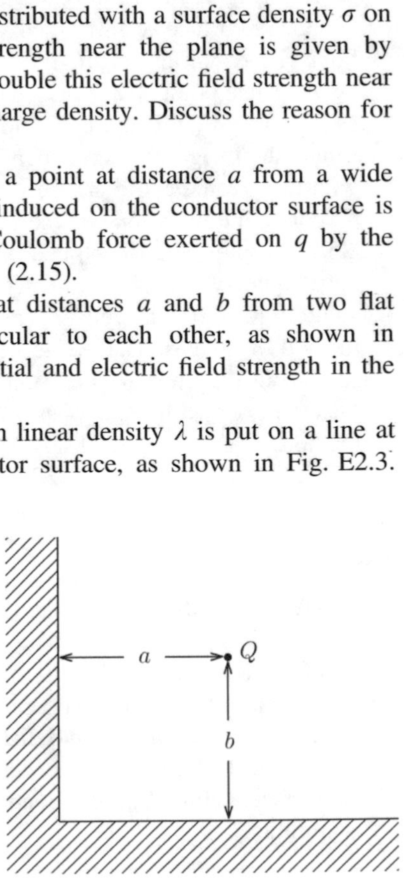

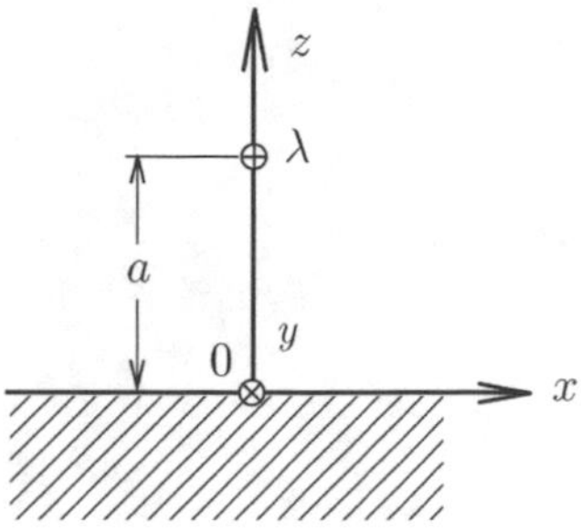

Fig. E2.3 Linear charge of density λ placed above a flat infinite conductor surface

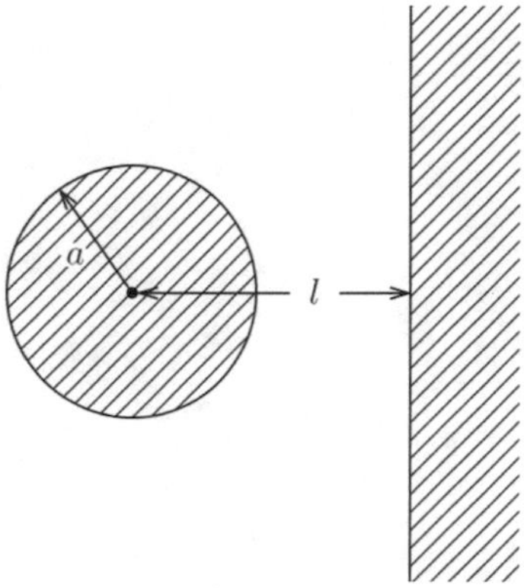

Fig. E2.4 Cylindrical conductor parallel to infinite flat conductor surface

Determine the electric potential and electric field strength in the vacuum and the density of electric charge on the conductor surface. Discuss the equipotential surface.

2.8. Determine the electric field strength in the space around a spherical conductor using the electric potential given by Eq. (2.20).

2.9. A long line of electric charge of uniform linear density λ is placed at distance d from the central axis of a grounded parallel long cylindrical conductor of radius $a(< d)$. Determine the electric charge induced on the conductor surface.

2.10. A long cylindrical conductor of radius a is placed at distance $l(> a)$ from an infinite flat conductor surface, as shown in Fig. E2.4, and an electric charge of linear density λ is given to the cylindrical conductor. Determine the density of electric charge on the surfaces of the two conductors.

2.11. Derive Eqs. (2.26) and (2.29) for a conducting sphere in a uniform electric field using Eq. (2.1) with the boundary condition that the electric field is perpendicular to the conductor surface and its value is given by Eq. (2.5).

Chapter 3
Conductor Systems in Vacuum

Abstract This chapter covers electric phenomenaņ in a conductor system composed of more than one conductor. The relationship between the electric charge and the electric potential is generally described using the coefficients of electric potential or the capacity coefficients. The simplest conductor system is a capacitor composed of two conductors. The characteristic of the capacitor is given by the capacitance, i.e., the electric charge that can be stored by a unit electric potential difference between the two conductors. The electrostatic energy of the conductor system can be determined by the mechanical work needed to carry electric charges from infinity until the final distribution of electric charge is attained and is described using the electric potential and the coefficients of electric potential.

3.1 Coefficients in Conductor System

We learned that conductors are equipotential in the static condition in the last chapter. We determined the electric potential and electric field strength around a conductor, and the distribution of electric charge was determined for simple cases. Most cases involve more than one conductor rather than a single conductor. In this chapter, we cover a **conductor system**, which consists of two or more conductors.

For a single conductor, as shown in Fig. 2.2, electric potential ϕ is generally proportional to electric charge Q according to

$$\phi = p_{\mathrm{c}} Q, \tag{3.1}$$

where p_{c} is a proportional constant called the **coefficient of electric potential** and is uniquely determined when the shape of the conductor is given. Its unit is $[\mathrm{V/C}]$.

When we rewrite Eq. (3.1) as

$$Q = C\phi \tag{3.2}$$

focusing on the electric charge, $C = 1/p_{\mathrm{c}}$ is called **capacity** or **capacitance**, which is equal to the electric charge stored by a unit electric potential, and its unit is [F] (**farad**). For the spherical conductor of radius a in Fig. 2.2, we have $C = 4\pi\epsilon_0 a$.

T. Matsushita, *Electricity and Magnetism*, Undergraduate Lecture Notes in Physics,
https://doi.org/10.1007/978-3-030-82150-0_3

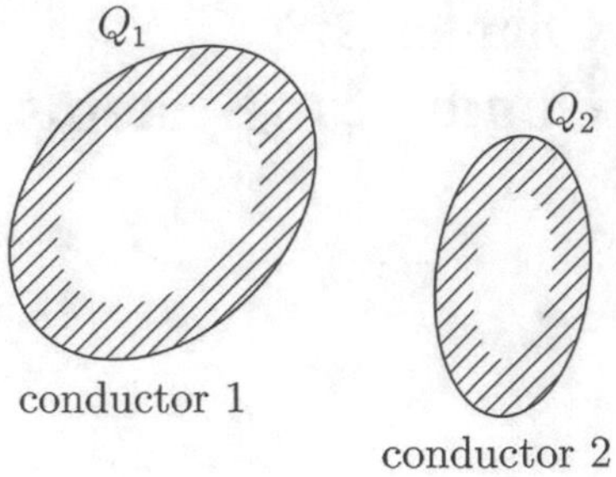

Fig. 3.1 Two charged conductors

Suppose that there are two conductors and electric charges Q_1 and Q_2 are given to conductors 1 and 2, respectively (see Fig. 3.1). When Q_1 only is given to conductor 1, the electric potential of conductor 1, ϕ_1, is given by

$$\phi_1 = p_{11}Q_1. \tag{3.3}$$

The electric potential of conductor 2, ϕ_2, is similarly given by

$$\phi_2 = p_{21}Q_1. \tag{3.4}$$

The constants p_{11} and p_{21} are also coefficients of electric potential and now elements of a matrix. While p_{11} is determined only by the shape of conductor 1, p_{21} depends both on the shape of conductor 2 and the relative position of the two conductors. When conductor 1 is charged, an electric charge appears on the surface of conductor 2 to shield its interior, resulting in an equipotential state under the condition that total electric charge is zero. This will be easily understood from Example 2.3. The distribution of electric charge on the surface of conductor 1 is also influenced by the electric charge induced on conductor 2. Thus, it should be noted that the electric charge is distributed in a way that makes the two conductors equipotential.

When Q_2 is given to conductor 2, the electric potentials in the two conductors are

$$\phi_1 = p_{11}Q_1 + p_{12}Q_2, \tag{3.5a}$$

$$\phi_2 = p_{21}Q_1 + p_{22}Q_2. \tag{3.5b}$$

Between the non-diagonal coefficients, the following relationship holds, as will be described later:

$$p_{12} = p_{21}. \tag{3.6}$$

Extending the above case, we consider a system composed of n conductors schematically shown in Fig. 3.2. Suppose that conductor i has electric charge, Q_i,

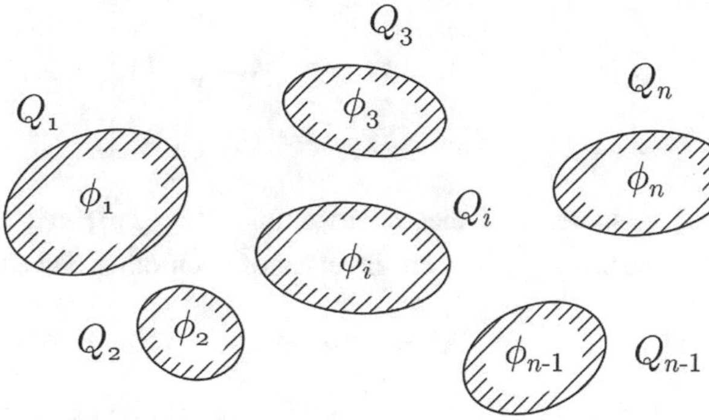

Fig. 3.2 System of *n* conductors

and electric potential, $\phi_i(i = 1, 2, \cdots, n)$. The electric potential of conductor *i* is expressed as

$$\phi_i = \sum_{j=1}^{n} p_{ij} Q_j. \tag{3.7}$$

The p_{ij}'s are the coefficients of electric potential determined by the geometrical arrangement of the conductors and fulfill the following conditions:

$$p_{ii} > 0, \quad p_{ij} = p_{ji} \geq 0 (i \neq j), \quad p_{ii} \geq p_{ij}. \tag{3.8}$$

The relationships between the electric potentials and electric charges take the matrix form

$$\begin{bmatrix} \phi_1 \\ \phi_2 \\ \cdot \\ \phi_n \end{bmatrix} = \begin{bmatrix} p_{11} & p_{12} & \cdots & p_{1n} \\ p_{21} & p_{22} & \cdots & p_{2n} \\ \cdot & \cdot & \cdots & \cdot \\ p_{n1} & p_{n2} & \cdots & p_{nn} \end{bmatrix} \begin{bmatrix} Q_1 \\ Q_2 \\ \cdot \\ Q_n \end{bmatrix}. \tag{3.9}$$

The electric charge is given by the inverse of Eq. (3.7):

$$Q_i = \sum_{j=1}^{n} C_{ij} \phi_j. \tag{3.10}$$

The C_{ij}'s in Eq. (3.10) are **capacity coefficients** or **capacitance coefficients** and have the same unit as the capacitance. These coefficients fulfill the following conditions:

$$C_{ii} > 0, \quad C_{ij} = C_{ji} \leq 0 (i \neq j). \tag{3.11}$$

Equation (3.10) can also be expressed as

$$\begin{bmatrix} Q_1 \\ Q_2 \\ \cdot \\ Q_n \end{bmatrix} = \begin{bmatrix} C_{11} & C_{12} & \cdots & C_{1n} \\ C_{21} & C_{22} & \cdots & C_{2n} \\ \cdot & \cdot & \cdots & \cdot \\ C_{n1} & C_{n2} & \cdots & C_{nn} \end{bmatrix} \begin{bmatrix} \phi_1 \\ \phi_2 \\ \cdot \\ \phi_n \end{bmatrix}. \tag{3.12}$$

The matrix of coefficients of electric potential, $\hat{P} = \{p_{ij}\}$, and the matrix of capacitance coefficients, $\hat{C} = \{C_{ij}\}$, are inverses of each other and satisfy

$$\hat{P}\hat{C} = \hat{C}\hat{P} = \hat{E}, \tag{3.13}$$

where $\hat{E}$ is a unit matrix. The above relationship can also be expressed as

$$\sum_{k=1}^{n} p_{ik}C_{kj} = \sum_{k=1}^{n} C_{ik}p_{kj} = \delta_{ij}, \tag{3.14}$$

where δ_{ij} is the Kronecker delta,

$$\begin{aligned} \delta_{ij} &= 1; \quad & i = j, \\ &= 0; \quad & i \neq j. \end{aligned} \tag{3.15}$$

The equalities $p_{ij} = p_{ji}$ and $C_{ij} = C_{ji}$ in Eqs. (3.8) and (3.11) are called the reciprocity theorem.

Here, we prove the inequalities $p_{ii} > 0, p_{ii} \geq p_{ij}$ and $p_{ij} \geq 0 (i \neq j)$, in Eq. (3.8). Suppose that a unit electric charge, $Q_i = 1$, is given only to conductor i and no electric charge is given to other conductors $\left(Q_j = 0; j \neq i\right)$. In this case, we have $\phi_i = p_{ii}$, and we can easily show that this potential is positive. The electric potential of conductor i is the highest. From the relationship $\phi_i \geq \phi_j = p_{ji}$, we can prove the inequality $p_{ii} \geq p_{ij}$. In the assumed space, there is only positive unit electric charge, and the electric potential at infinity has the lowest value, 0. Thus, we can show the condition $p_{ij} \geq 0$ holds.

The coefficients of electric potential or the capacitance coefficients are useful in practical cases. For example, we apply the coefficients of electric potential to the example in Fig. 2.13 to determine the electric charge induced on the spherical conductor by the point charge. The spherical conductor is named conductor 1, and a very small imaginary conductor placed at the position of the point charge is called conductor 2. Under the conditions $Q_1 = 1$ and $Q_2 = 0$, we have

$$\phi_1 = p_{11} = \frac{1}{4\pi\epsilon_0 a}, \qquad \phi_2 = p_{21} = \frac{1}{4\pi\epsilon_0 d}. \tag{3.16}$$

Under the conditions $Q_1 = Q$ and $Q_2 = q$, the electric potential of conductor 1 is

$$\phi_1 = p_{11}Q + p_{12}q. \tag{3.17}$$

Since conductor 1 is grounded, $\phi_1 = 0$. This and the reciprocity theorem $p_{12} = p_{21}$ give

$$Q = -\frac{p_{21}}{p_{11}}q = -\frac{a}{d}q, \tag{3.18}$$

which agrees with Eq. (2.19). Thus, the coefficients of electric potential can sometimes be used to determine the induced electric charge more easily than with the method discussed in Sect. 2.2.

Example 3.1. Prove the inequalities, $C_{ii} > 0$ and $C_{ij} \leq 0 (i \neq j)$, in Eq. (3.11).

Solution 3.1. Suppose that conductor i has a unit electric potential $(\phi_i = 1)$ by having some electric charge and the electric potentials of other conductors are forced to be zero $(\phi_j = 0)$ by grounding. In this case, the electric charge on conductor i must be positive, and we can easily prove $Q_{ii} = C_{ii} > 0$. Part of the electric field lines start from conductor i and go to infinity, but the remaining lines reach other conductors (see Fig. 3.3). Hence, the electric charges induced in these conductors are negative or zero, which proves the inequality $Q_j = C_{ji} \leq 0$.

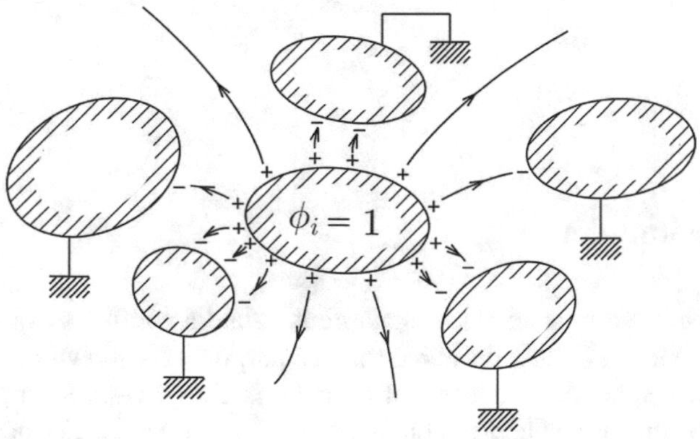

Fig. 3.3 Electric field lines when unit electric potential is given to conductor i and other conductors are grounded

◇

Example 3.2. An electric charge, Q, is given to the inner conductor of the set of concentric spherical conductors in Fig. 3.4, and then the outer conductor is grounded. Determine the electric charge induced on the outer conductor using the coefficients of electric potential. Neglect the thickness of the outer conductor.

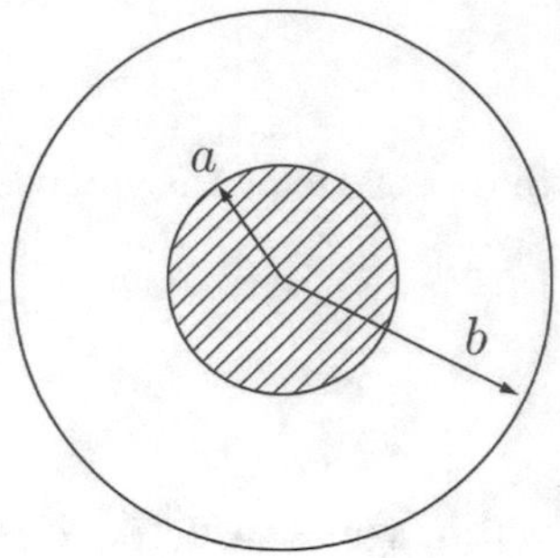

Fig. 3.4 Concentric spherical conductors

Solution 3.2. The inner and outer conductors are called conductors 1 and 2, respectively. Under the conditions $Q_1 = 0$ and $Q_2 = 1$, we have

$$\phi_2 = p_{22} = \frac{1}{4\pi\epsilon_0 b}, \quad \phi_1 = p_{12} = p_{21} = \frac{1}{4\pi\epsilon_0 b}.$$

Under the conditions $Q_1 = Q$ and $Q_2 = q$, the electric potential of conductor 2 is

$$\phi_2 = p_{21}Q + p_{22}q = \frac{Q+q}{4\pi\epsilon_0 b}.$$

When conductor 2 is grounded, the condition $\phi_2 = 0$ gives

$$q = -Q.$$

◇

3.2 Capacitor

The component used to store electric charge is called a **condenser** or a **capacitor**. Larger capacitance is usually desirable for this purpose. As shown in Sect. 3.1, the capacitance of a spherical conductor of radius a is $C = 4\pi\epsilon_0 a$. It amounts only to 5.6×10^{-12} F even for a spherical conductor 10 cm in diameter. Hence, a different shape is needed to obtain a larger capacitance.

Common capacitors consist of two parallel sheet conductors separated by a small distance, as schematically shown in Fig. 3.5. Each sheet conductor is connected to an outer circuit with a lead line for transporting electric charge. Such a capacitor is called a **parallel-plate capacitor**. The sheet conductor that stores the electric charge is called an electrode. Suppose that the surface area of each electrode is S and the distance between the two electrodes is d, which is very small compared with the electrode size: $d \ll \sqrt{S}$. When an electric charge, Q, is given to one electrode and $-Q$ is given to the other, the electric charges are distributed uniformly

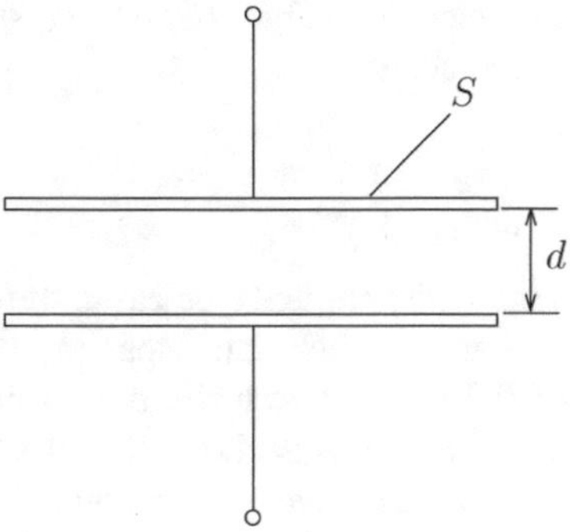

Fig. 3.5 Parallel-plate capacitor

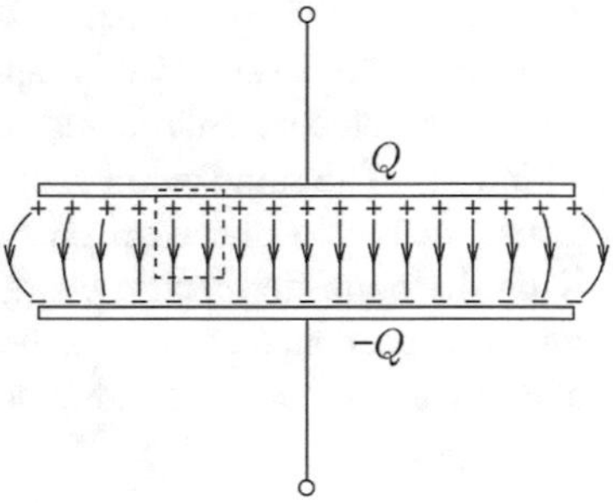

Fig. 3.6 Distribution of electric charge and electric field in parallel-plate capacitor

on the inner surfaces of the electrodes (facing each other) but do not appear on the outer surfaces (see Example 2.4). Hence, the electric field is almost concentrated in the space between the electrodes and perpendicular to them except at the edges (see Fig. 3.6).

The surface density of positive electric charge is

$$\sigma = \frac{Q}{S}. \tag{3.19}$$

Applying Gauss' law to the closed region shown by the dotted line in Fig. 3.6, we find the electric field strength in the space between the two electrodes to be

$$E = \frac{\sigma}{\epsilon_0} = \frac{Q}{\epsilon_0 S}. \tag{3.20}$$

Thus, the electric field strength is uniform in this space. The electric potential difference between the two electrodes is

$$V = Ed = \frac{Qd}{\epsilon_0 S}. \tag{3.21}$$

The electric charge that can be stored in the capacitor by a unit electric potential difference is

$$C = \frac{Q}{V} = \frac{\epsilon_0 S}{d}. \tag{3.22}$$

This is the **capacity** or **capacitance** of the capacitor.

For a parallel-plate capacitor with square electrodes 10 cm in size and a distance of 0.2 mm between the two electrodes, the capacitance is estimated to be 4.4×10^{-10} F, which is about 80 times larger than that of the spherical conductor 10 cm in diameter mentioned above. This shows that the structure of the parallel-plate capacitor is quite effective for storing electric charge. In practical capacitors, a dielectric material is used in the space between the two electrodes to enhance the capacitance further (see Chap. 4).

In the above example of a spherical conductor, we discussed the relationship between the electric charge and electric potential difference between the conductor and infinity. The same result is obtained for the parallel-plate capacitor by giving an electric charge to one electrode and by grounding the other electrode instead of giving the opposite electric charge. The grounding process gives the same definition for capacitance as for a spherical conductor. The reason for the small capacitance in the spherical conductor is that a weak electric field is spread over the whole space.

***Example* 3.3.** Determine the capacitance of the concentric spherical capacitor in Fig. 3.7.

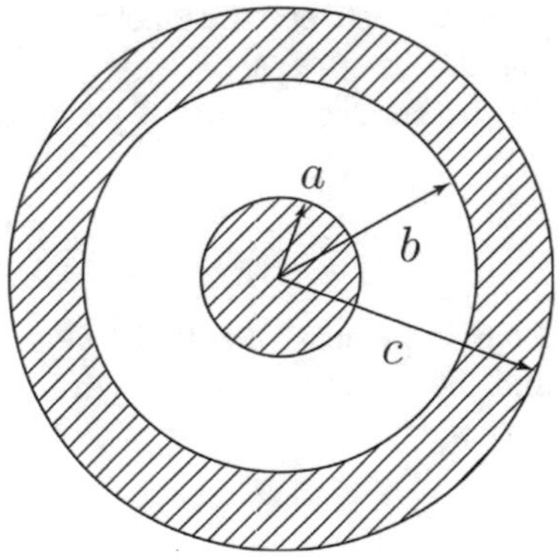

Fig. 3.7 Concentric spherical capacitor

Solution 3.3. An electric charge, Q, is given to the inner conductor and the outer conductor is grounded. Then, an electric charge, $-Q$, appears on the inner surface of the outer conductor. Applying Gauss' law, we determine the electric field strength to be

$$E = \frac{Q}{4\pi\epsilon_0 r^2}; \qquad a<r<b,$$
$$= 0; \qquad 0 \leq r<a,\ r>b,$$

where r is the distance from the center. The electric potential difference between the electrodes is

$$V = \int_a^b \frac{Q}{4\pi\epsilon_0 r^2}\mathrm{d}r = \frac{Q}{4\pi\epsilon_0}\left(\frac{1}{a}-\frac{1}{b}\right).$$

Hence, the capacitance is given by

$$C = \frac{4\pi\epsilon_0 ab}{b-a}.$$

◇

Example 3.4. A pair of conductors is regarded as a capacitor. Describe the capacitance in terms of the coefficients of electric potential or the capacitance coefficients.

Solution 3.4. Electric charges Q and $-Q$ are given to conductors 1 and 2, respectively. Since the electric potential of each conductor is written as

$$\phi_1 = p_{11}Q - p_{12}Q,$$
$$\phi_2 = p_{21}Q - p_{22}Q,$$

using the coefficients of electric potential, its difference is

$$V = \phi_1 - \phi_2 = (p_{11} + p_{22} - p_{12} - p_{21})Q.$$

Thus, the capacitance is given by

$$C = \frac{Q}{V} = \frac{1}{p_{11} + p_{22} - p_{12} - p_{21}} = \frac{1}{p_{11} + p_{22} - 2p_{12}}.$$

The electric charges in the two conductors are written as

$$Q = C_{11}\phi_1 + C_{12}\phi_2,$$
$$-Q = C_{21}\phi_1 + C_{22}\phi_2,$$

using the capacitance coefficients. Thus, we obtain the electric potential difference as

$$V = \phi_1 - \phi_2 = \frac{C_{11} + C_{22} + C_{12} + C_{21}}{C_{11}C_{22} - C_{12}C_{21}} Q.$$

This gives the capacitance,

$$C = \frac{C_{11}C_{22} - C_{12}C_{21}}{C_{11} + C_{22} + C_{12} + C_{21}} = \frac{C_{11}C_{22} - C_{12}^2}{C_{11} + C_{22} + 2C_{12}}.$$

◇

We connect capacitors in series, apply an electric charge, Q, to the top capacitor, and then ground the lowest capacitor, as shown in Fig. 3.8. An electric charge, $-Q$, is induced on the inner surface of the lower electrode of the top capacitor to shield the lower electrode from the electric field produced by Q on the upper electrode. From the principle of conservation of charge, an electric charge, Q, appears on the upper electrode of the next capacitor. This is repeated, and finally electric charges $\pm Q$ appear on both electrodes of each capacitor. When the capacitance of each capacitor is $C_i (i = 1, 2, \cdots, n)$, the electric potential difference in each capacitor is

$$V_i = \frac{Q}{C_i}. \tag{3.23}$$

Hence, the total electric potential difference through the series of capacitor is

$$V = \sum_{i=1}^{n} V_i = Q \sum_{i=1}^{n} \frac{1}{C_i}. \tag{3.24}$$

If the capacitance of capacitors connected in series is denoted by C, we have

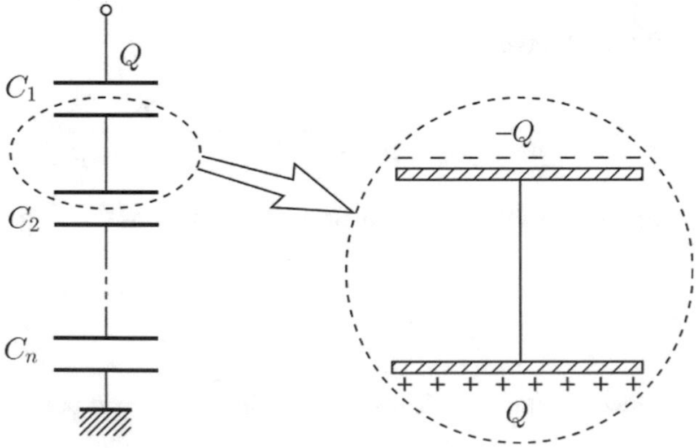

Fig. 3.8 Capacitors connected in series

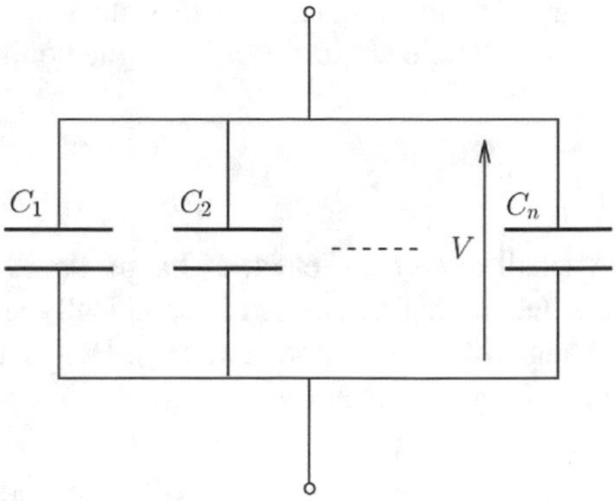

Fig. 3.9 Capacitors connected in parallel

$$\frac{1}{C} = \sum_{i=1}^{n} \frac{1}{C_i}. \tag{3.25}$$

Next, we connect capacitors in parallel, as shown in Fig. 3.9. When we apply a voltage, V, between the terminals, the electric charges that appear in the capacitor of capacitance C_i are $\pm Q_i = \pm C_i V$. Hence, the total amount of positive electric charge is

$$Q = \sum_{i=1}^{n} Q_i = V \sum_{i=1}^{n} C_i. \tag{3.26}$$

If the capacitance of capacitors connected in parallel is denoted by C, we have

$$C = \sum_{i=1}^{n} C_i. \tag{3.27}$$

3.3 Electrostatic Energy

Suppose there is a conductor system with electric charges. The electric charges produce an electric field in the space outside the conductors. Hence, we can regard the conductor system as having some kind of energy. This is called **electrostatic energy** or **electric energy**. It is reasonable to assume that there is no electrostatic energy in this system when there is no electric charge. Thus, the electrostatic energy is equivalent to the mechanical work necessary to bring electric charges from infinity until we attain the desired distribution of electric charge.

Now, we determine the electrostatic energy for an isolated conductor of capacitance C and electric charge Q. Suppose that the conductor has an electric charge, q, in an intermediate state while electric charge is being brought from infinity. The electric potential of the conductor is

$$\phi(q) = \frac{q}{C}.$$

A small amount of electric charge, $\mathrm{d}q$, is additionally carried from infinity to the conductor. If this amount is sufficiently small, the transfer of this charge does not change the electric potential $\phi(q)$. Hence, the mechanical work needed for carrying this charge is given by

$$\mathrm{d}W = \phi(q)\mathrm{d}q = \frac{q}{C}\,\mathrm{d}q.$$

Thus, the total work needed to carry all electric charge, Q, is

$$W = \int_0^Q \frac{q}{C}\,\mathrm{d}q = \frac{1}{2C}Q^2. \tag{3.28}$$

This gives the electrostatic energy, U_e. In terms of the electric potential, $\phi = Q/C$, this can also be written as

$$U_\mathrm{e} = \frac{1}{2C}Q^2 = \frac{1}{2}Q\phi = \frac{1}{2}C\phi^2. \tag{3.29}$$

Next, consider a system composed of two conductors. Suppose that conductor 1 has electric charge Q_1 and electric potential ϕ_1, and conductor 2 has electric charge Q_2 and electric potential ϕ_2. Electric charge is brought from infinity to conductor 1, and then from infinity to conductor 2. Now suppose an intermediate condition where conductor 1 has electric charge q_1 and conductor 2 has no electric charge, as shown in Fig. 3.10a. Under this condition, the electric potential of conductor 1 is $p_{11}q_1$ in terms of the coefficient of electric potential. Hence, the work needed to carry Q_1 to conductor 1 is

$$W_1 = \int_0^{Q_1} p_{11}q_1\mathrm{d}q_1 = \frac{1}{2}p_{11}Q_1^2. \tag{3.30}$$

In the next intermediate condition where conductor 1 has Q_1 and conductor 2 has q_2 (see Fig. 3.10b), the electric potential of conductor 2 is $p_{21}Q_1 + p_{22}q_2$. We similarly obtain the work needed to carry Q_2 to conductor 2 as

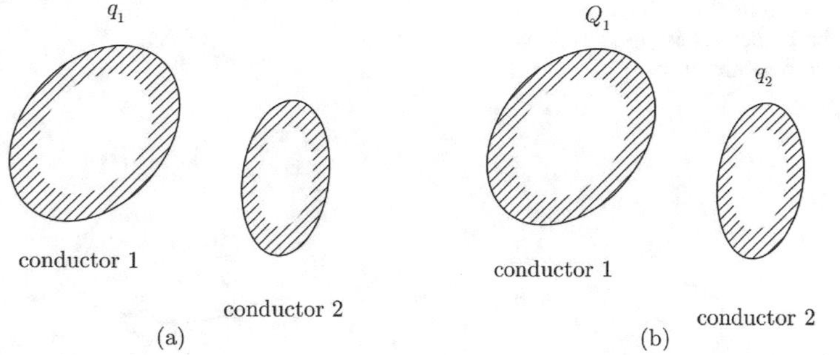

Fig. 3.10 System composed of two conductors in the intermediate condition where **a** conductor 1 has q_1 and conductor 2 has no electric charge and **b** conductor 1 has Q_1 and conductor 2 has q_2

$$W_2 = \int_0^{Q_2} (p_{21}Q_1 + p_{22}q_2)\mathrm{d}q_2 = p_{21}Q_1Q_2 + \frac{1}{2}p_{22}Q_2^2. \tag{3.31}$$

Finally, the electrostatic energy of this system is

$$U_\mathrm{e} = W_1 + W_2 = \frac{1}{2}p_{11}Q_1^2 + p_{21}Q_1Q_2 + \frac{1}{2}p_{22}Q_2^2. \tag{3.32}$$

If the order of carrying electric charge is reversed, the electrostatic energy becomes

$$U_\mathrm{e} = \frac{1}{2}p_{11}Q_1^2 + p_{12}Q_1Q_2 + \frac{1}{2}p_{22}Q_2^2 \tag{3.33}$$

by exchanging subscripts 1 and 2. Since Eqs. (3.32) and (3.33) must coincide with each other, we derive

$$p_{12} = p_{21}. \tag{3.34}$$

Thus, the reciprocity theorem holds in this simple case. Using this relationship, we rewrite the electrostatic energy as

$$\begin{aligned} U_\mathrm{e} &= \frac{1}{2}Q_1(p_{11}Q_1 + p_{12}Q_2) + \frac{1}{2}Q_2(p_{21}Q_1 + p_{22}Q_2) \\ &= \frac{1}{2}(Q_1\phi_1 + Q_2\phi_2). \end{aligned} \tag{3.35}$$

We can extend the above result to a conductor system with n conductors. Assume that conductor i has an electric charge Q_i and an electric potential $\phi_i (i = 1, 2, \cdots, n)$. The electrostatic energy of this system is given by

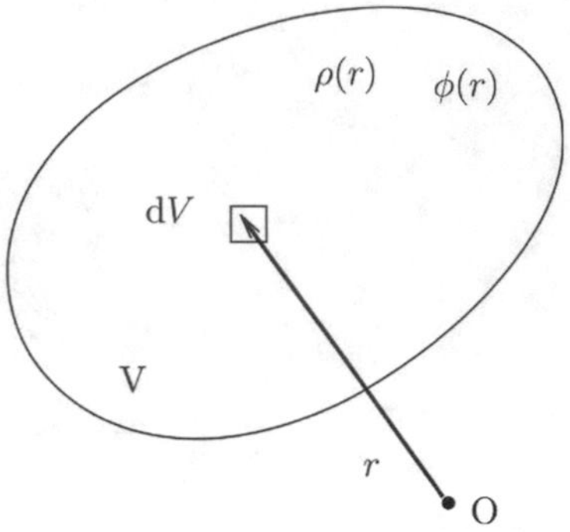

Fig. 3.11 Region V in which electric charge density is $\rho(\boldsymbol{r})$ and electric potential is $\phi(\boldsymbol{r})$

$$U_e = \frac{1}{2}\sum_{i=1}^{n} Q_i\phi_i = \frac{1}{2}\sum_{i=1}^{n}\sum_{j=1}^{n} p_{ij}Q_iQ_j. \tag{3.36}$$

We can extend this result to the case where electric charge is continuously distributed with the density $\rho(\boldsymbol{r})$ and the electric potential is given by $\phi(\boldsymbol{r})$ in region V (see Fig. 3.11). Regarding an electric charge $\rho \mathrm{d}V$ in a small region $\mathrm{d}V$ as a point charge, Eq. (3.36) gives

$$U_e = \frac{1}{2}\int_V \rho(\boldsymbol{r})\phi(\boldsymbol{r})\mathrm{d}V \tag{3.37}$$

for the electrostatic energy in V.

Example 3.5. When electric charges $\pm Q$ are given to a capacitor of capacitance C, the electric potential difference between the two electrodes is V. Prove that the electrostatic energy of this capacitor is

$$U_e = \frac{1}{2C}Q^2 = \frac{1}{2}QV = \frac{1}{2}CV^2. \tag{3.38}$$

Solution 3.5. We apply Eq. (3.35) to the electrostatic energy of this capacitor. The positive and negative electrodes are named capacitors 1 and 2, respectively. Then, we have $Q_1 = Q$, $Q_2 = -Q$ and $\phi_1 = \phi_2 + V$. Substituting these relationships into Eq. (3.35) gives

$$U_e = \frac{1}{2}QV.$$

Other expressions are obtained using Eq. (3.22).

◇

When there is an electric field in space, the space is distorted electrically as mentioned in Sect. 1.3. Hence, the electrostatic energy is understood as the energy associated with the electrical distortion in the space. Now, we discuss the electrostatic energy for the parallel-plate capacitor in Sect. 3.2. The capacitance of this capacitor with surface area S and distance d between the electrodes is given by Eq. (3.22). From this and Eq. (3.38), we have

$$U_{\mathrm{e}} = \frac{1}{2}\epsilon_0 E^2 Sd. \tag{3.39}$$

In the above, $V = Ed$ is used for the electric potential difference. Since Sd is the volume of the region in which the electric field is concentrated with a constant strength E, we can regard that electrostatic energy of density

$$u_{\mathrm{e}} = \frac{1}{2}\epsilon_0 E^2 \tag{3.40}$$

as filling this region. This is called **electrostatic energy density** or **electric energy density**.

We can show that Eq. (3.40) holds generally also for cases where the electric field strength is not uniform in the space. Substituting Eq. (1.21) into Eq. (3.37) gives

$$\begin{aligned} U_{\mathrm{e}} &= \frac{1}{2}\epsilon_0 \int_{\mathrm{V}} \phi \nabla \cdot \boldsymbol{E}\, \mathrm{d}V \\ &= \frac{1}{2}\epsilon_0 \int_{\mathrm{V}} \nabla \cdot (\phi \boldsymbol{E}) \mathrm{d}V - \frac{1}{2}\epsilon_0 \int_{\mathrm{V}} \boldsymbol{E} \cdot \nabla \phi\, \mathrm{d}V, \end{aligned} \tag{3.41}$$

where Eq. (A1.40) is used. Applying Gauss' theorem to the first integral, we have

$$\frac{1}{2}\epsilon_0 \int_{\mathrm{S}} \phi \boldsymbol{E} \cdot \mathrm{d}\boldsymbol{S}, \tag{3.42}$$

where S is the surface area of region V. If we select a sphere with a sufficiently large radius r as V, we have $\phi \propto r^{-1}, E \propto r^{-2}$, and $\int \mathrm{d}S \propto r^2$ on S. Hence, we can show that the surface integral is proportional to r^{-1}, resulting in zero in the limit $r \to \infty$. Thus, we can disregard the surface integral. Using Eq. (1.24) for the second term, Eq. (3.41) gives

$$U_{\mathrm{e}} = \frac{1}{2}\epsilon_0 \int_{\mathrm{V}} \boldsymbol{E}^2 \mathrm{d}V. \tag{3.43}$$

Thus, the electrostatic energy density is generally given by Eq. (3.40).

Example 3.6. Determine the electrostatic energy when an electric charge Q is given to a spherical conductor of radius a.

Solution 3.6. The electrostatic energy can be obtained by more than one method. First, we do this using the electric potential. The electric field strength is

$$E(r) = 0; \qquad 0 \le r < a,$$
$$= \frac{Q}{4\pi\epsilon_0 r^2}; \qquad r > a,$$

and the electric potential of the conductor is given by

$$\phi(a) = \frac{Q}{4\pi\epsilon_0 a}.$$

Hence, from Eq. (3.29), we obtain the electrostatic energy as

$$U_{\mathrm{e}} = \frac{1}{2}Q\phi(a) = \frac{Q^2}{8\pi\epsilon_0 a}.$$

We then calculate the electrostatic energy using the electrostatic energy density. From Eq. (3.40) with the above electric field strength, the electrostatic energy density is

$$u_{\mathrm{e}}(r) = 0; \qquad 0 \le r < a,$$
$$= \frac{Q^2}{32\pi^2\epsilon_0 r^4}; \qquad r > a.$$

Hence, we obtain the same result,

$$U_{\mathrm{e}} = \int_a^\infty u_{\mathrm{e}} \cdot 4\pi r^2 \,\mathrm{d}r = \frac{Q^2}{8\pi\epsilon_0 a}.$$

◇

Example 3.7. Electric charges Q_1 and Q_2 are given to the inner and outer spherical conductors shown in Fig. 2.5, respectively. Determine the electrostatic energy using Eq. (3.36).

Solution 3.7. We denote the inner and outer conductors as conductors 1 and 2, respectively. When we give a unit charge only conductor 1 ($Q_1 = 1, Q_2 = 0$), the electric potential of each conductor is

$$\phi_1 = p_{11} = \frac{1}{4\pi\epsilon_0}\left(\frac{1}{a} - \frac{1}{b} + \frac{1}{c}\right),$$

$$\phi_2 = p_{21} = p_{12} = \frac{1}{4\pi\epsilon_0 c}.$$

When we give a unit charge only conductor 2 ($Q_1 = 0, Q_2 = 1$), the electric potential of conductor 2 is

$$\phi_2 = p_{22} = \frac{1}{4\pi\epsilon_0 c}.$$

The electrostatic energy when Q_1 and Q_2 are given to the inner and outer spherical conductors is

$$\begin{aligned} U_\mathrm{e} &= \frac{1}{2}p_{11}Q_1^2 + p_{12}Q_1Q_2 + \frac{1}{2}p_{22}Q_2^2 \\ &= \frac{1}{8\pi\epsilon_0}\left[Q_1^2\left(\frac{1}{a} - \frac{1}{b}\right) + \frac{(Q_1+Q_2)^2}{c}\right]. \end{aligned}$$

Since the electric field strength is $(Q_1 + Q_2)/4\pi\epsilon_0 r^2$ for $r > c$ and $Q_1/4\pi\epsilon_0 r^2$ for $a<r<b$, the same result is obtained from Eq. (3.43).

◇

3.4 Electrostatic Force

The electrostatic force between charged conductors is the sum of the Coulomb force on individual electric charges distributed on the conductor surface. We can also determine this force using the principle of virtual displacement and the electrostatic energy.

Suppose that part of an isolated conductor system is forced to move a small distance $\Delta\boldsymbol{s}$ by an electrostatic force $\boldsymbol{F}$. The work done by the system, $\boldsymbol{F}\cdot\Delta\boldsymbol{s}$, is the energy that the system loses. Hence, if we use ΔU_e to denote the variation in electrostatic energy caused by the movement, the work is equal to $-\Delta U_\mathrm{e}$. Thus, we have

$$\boldsymbol{F}\cdot\Delta\boldsymbol{s} + \Delta U_\mathrm{e} = 0. \tag{3.44}$$

In the limit $|\Delta\boldsymbol{s}| = \Delta s \to 0$ for displacement along the direction of the electrostatic force, this gives

$$F = -\frac{\partial U_\mathrm{e}}{\partial s}. \tag{3.45}$$

We calculate the force on the electrodes of the parallel-plate capacitor in Fig. 3.5 using this method, when electric charges $\pm Q$ are given. The surface area and the distance between the electrodes are S and d, respectively. Assume that the distance is changed to x. In this case, the electric charge Q and the electric field strength $E = Q/\epsilon_0 S$ are unchanged, and it is reasonable to describe the electrostatic energy of Eq. (3.39) as

$$U_e = \frac{Q^2 x}{2\epsilon_0 S} \tag{3.46}$$

in terms of Q. It should be noted that the voltage V changes with the distance, and it is not suitable to describe U_e in terms of V. Hence, we determine the electrostatic force on the electrode as

$$F = -\left.\frac{\partial U_e}{\partial x}\right|_{x=d} = -\frac{Q^2}{2\epsilon_0 S}. \tag{3.47}$$

Since this force is negative for expansion (increasing x), it is attractive.

Example 3.8. Suppose that the electric potential difference V of the parallel-plate capacitor is kept constant by connecting an electric power source, as shown in Fig. 3.12. The surface area of the electrode is S, the distance between the electrodes is d, and the electric charges are $\pm Q$. Determine the electrostatic force between the electrodes.

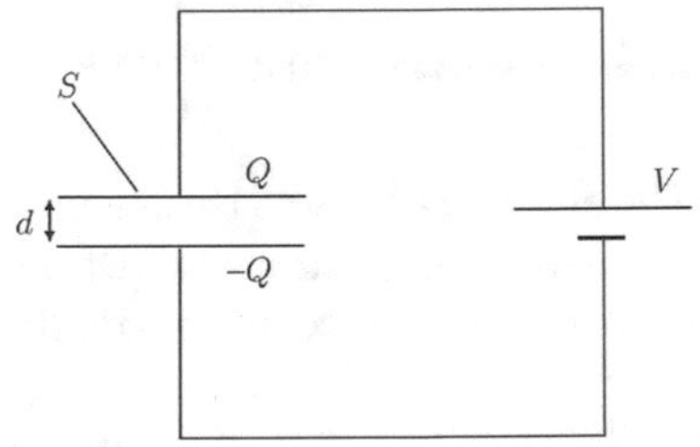

Fig. 3.12 Parallel-plate capacitor connected with electric power source of voltage V. The electric charges are $\pm Q$ in the initial condition

---

Solution 3.8. It should be noted that the electric charge and the electric field strength change when the distance between the electrodes changes. That is, when the distance between the electrodes changes from d to $d + \Delta x$, the electric field strength and electric charge change to $V/(d + \Delta x)$ and $\epsilon_0 SV/(d + \Delta x)$, respectively. This is caused by a transfer of electric energy between the capacitor and the power source. From Eq. (3.39), we obtain the variation in electrostatic energy as

$$\Delta U_e = \frac{\epsilon_0 S V^2}{2}\left(\frac{1}{d + \Delta x} - \frac{1}{d}\right).$$

The amount of electric charge that transfers to the power source is

$$\Delta Q = -\epsilon_0 SV \left(\frac{1}{d+\Delta x} - \frac{1}{d} \right).$$

Hence, the energy that flows into the power source during the displacement is

$$\Delta W = V\Delta Q = \epsilon_0 SV^2 \left(\frac{1}{d} - \frac{1}{d+\Delta x} \right).$$

If we denote the force on the electrode by F, the energy that the system loses is equal to the sum of the work done to the outside, $F\Delta x$, and ΔW. Since this is equal to $-\Delta U_{\mathrm{e}}$, we have

$$F = -\lim_{\Delta x \to 0} \frac{\Delta W + \Delta U_{\mathrm{e}}}{\Delta x} = -\frac{\epsilon_0 SV^2}{2d^2}.$$

This is also an attractive force. It should be noted that this force is the same as that in Eq. (3.47).

◇

Column: Case of Infinite Amount of Electric Charge

The surface integral of Eq. (3.42) must be zero so that Eqs. (3.37) and (3.43) coincide with each other. Is this condition also fulfilled when there is an infinite amount of electric charge in the whole space as discussed in Example 1.9? Here, we treat the case where an electric charge Q' is given to a unit length of an infinitely long cylindrical conductor of radius a. The electric field strength and electric potential are

$$\begin{aligned} E(R) &= 0; \qquad 0 \le R \le a, \\ &= \frac{Q'}{2\pi\epsilon_0 R}; \quad R > a, \end{aligned}$$

and

$$\begin{aligned} \phi(R) &= \frac{Q'}{2\pi\epsilon_0} \log\frac{R_0}{a}; \quad 0 \le R \le a, \\ &= \frac{Q'}{2\pi\epsilon_0} \log\frac{R_0}{R}; \quad R > a. \end{aligned}$$

In the above, R_0 is the distance from the central axis to the reference point at which the electric potential is zero. We obtain the electrostatic energy in a unit length from Eq. (3.29) as $U'_{\mathrm{e}} = Q'\phi(a)/2 = (Q'^2/4\pi\epsilon_0)\log(R_0/a)$.The electrostatic energy can also be calculated from Eq. (3.43). Restricting the region within $a < R < R_0$ for integration, we have

$$U'_e = \frac{1}{2}\epsilon_0 \int_a^{R_0} \left(\frac{Q'}{\pi\epsilon_0 R}\right)^2 2\pi R \mathrm{d}R = \frac{Q'^2}{4\pi\epsilon_0}\log\frac{R_0}{a},$$

which agrees with the above result. Thus, it is consistent with the assumption that ϕ is zero on the surface of $R = R_0$, resulting in a zero surface integral in Eq. (3.42). This agreement holds even in the limit of $R_0 \to \infty$. This means that there is no contradiction in the theoretical framework, although there is the problem of divergence of electrostatic energy because of the infinite amount of electric charge.

Exercises

3.1. Determine the electric potential of the spherical conductor in Example 2.5 using the coefficients of electric potential and confirm that the result agrees with the solution in this example. The spherical conductor and a small imaginary conductor placed at the position of point charge q are named conductors 1 and 2, respectively.

3.2. In Exercise 2.9, we supposed that a line charge of uniform linear density λ is placed at distance d from the central axis of a grounded parallel long cylindrical conductor of radius $a(< d)$. Determine the electric charge induced in the cylindrical conductor using the coefficients of electric potential. (Hint: The reference point of the electric potential should be sufficiently far away following the original definition.)

3.3. Determine the electric charge induced in the internal spherical conductor in Exercise 2.2 using the coefficients of electric potential.

3.4. An electric charge λ in a unit length is uniformly distributed on the inner conductor of the coaxial conductors in Fig. E3.1, and the outer conductor is grounded. Determine the electrostatic energy and capacitance in a unit length of the coaxial conductors.

3.5. Determine the electrostatic energy when electric charges $\pm Q$ are given to the concentric spherical conductors in Fig. 3.7 using the following methods: (1) the electrostatic energy density, (2) the electric potential, and (3) the coefficients of electric potential.

3.6. Two long and thin cylindrical conductors of radius a are placed parallel to each other with distance d, as shown in Fig. E3.2. Determine the capacitance in a unit length of these conductors. Assume that a is sufficiently smaller than d.

3.7. We denote the inner, middle, and outer spherical conductors in Fig. E3.3 as conductors 1—3. (a) Determine the coefficients of electric potential, and

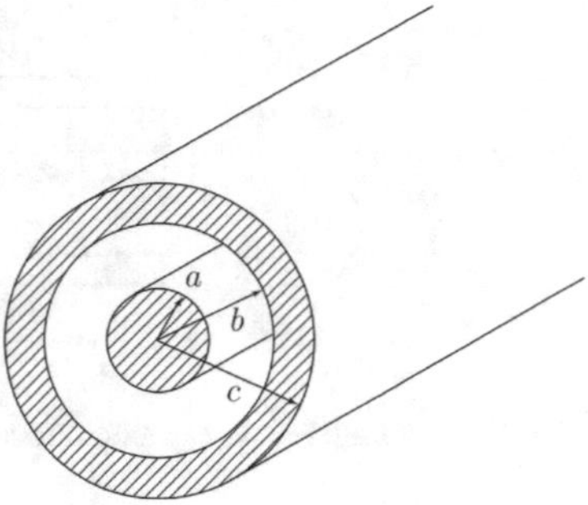

Fig. E3.1 Cylindrical coaxial conductors

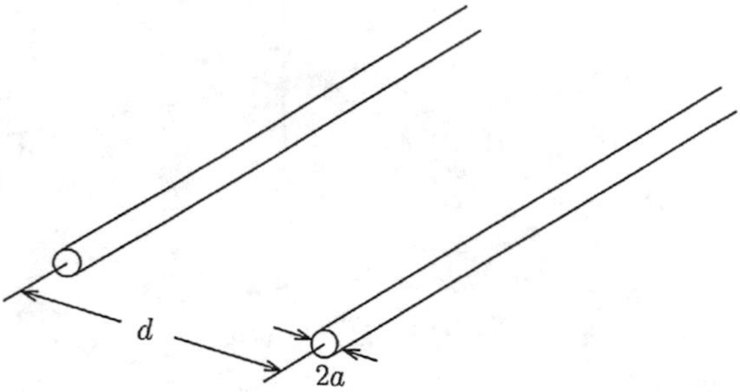

Fig. E3.2 Two long and thin cylindrical conductors parallel to each other

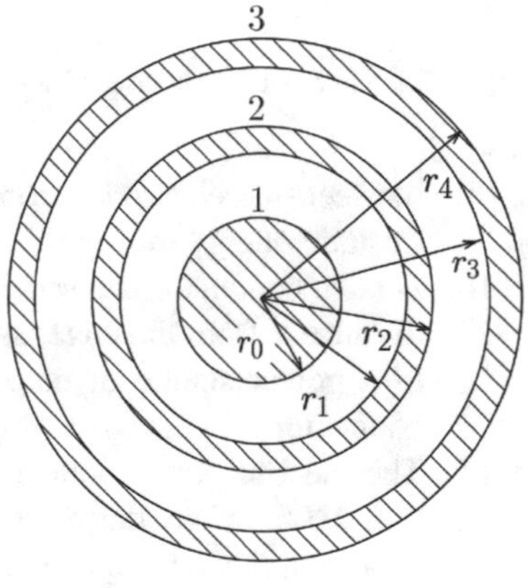

Fig. E3.3 Cross section of coaxial structure comprising spherical conductors

(b) determine the electrostatic energy using the coefficients of electric potential, when we give electric charges Q_1, Q_2, and Q_3 to conductors 1, 2, and 3, respectively.

3.8. Determine the electrostatic energy and the force on the cylindrical conductor in a unit length in Exercise 2.10.

3.9. Electric charges $\pm Q$ are given to each electrode of the capacitor shown in Fig. E3.4. Determine the mechanical work needed to change the distance

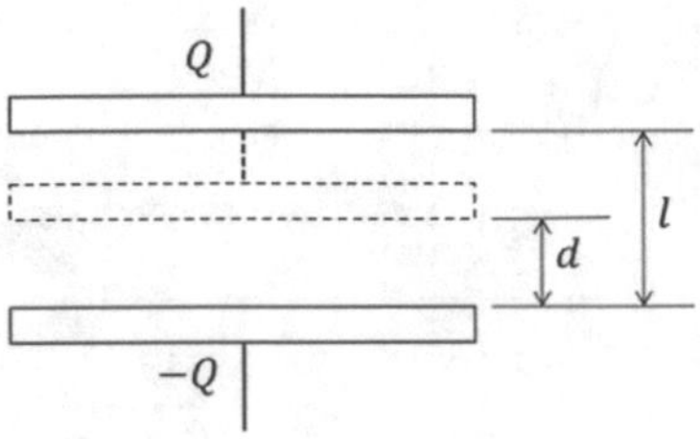

Fig. E3.4 Parallel-plate capacitor. The area of each electrode is S

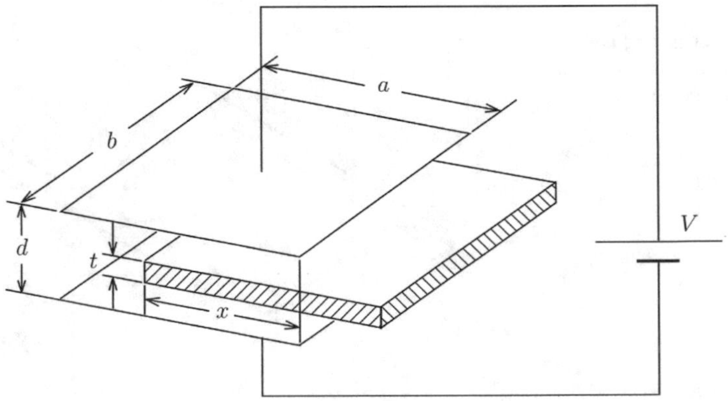

Fig. E3.5 Plate conductor inserted into gap of parallel-plate capacitor

between the electrodes from d to l. Determine also the change in the electrostatic energy during this change.

3.10. A plate conductor is inserted into the gap of a wide parallel-plate capacitor a distance x from the edge, as shown in Fig. E3.5. The capacitor is connected to a power source of output voltage V. Determine the force on the plate conductor.

3.11. The electric power source is removed from the parallel-plate capacitor in Fig. E3.5, when the electric charges in the capacitor are $\pm Q$ under the condition that a plate conductor is inserted into the gap of the capacitor up to the depth x. Determine the force on the plate conductor.

Chapter 4
Dielectric Materials

Abstract This chapter covers electric phenomena in dielectric materials. When an electric field is applied to a dielectric material, the electrons are slightly displaced along the direction opposite to the electric field, even though they are bonded by nuclei. This phenomenon is electric polarization. As a result, electric dipoles are produced and polarization charge appears. Thus, the electric field is produced by true electric charges and also by polarization charges in dielectric materials. We newly define the electric flux density, which describes only the electric field produced by the true electric charge. Electric phenomena are generally described in terms of the electric field and the electric flux density. The refraction of the electric field at an interface between different dielectric materials is also treated.

4.1 Electric Polarization

When two electric charges are separated by a grounded metal sheet, there is no force between the two electric charges. On the other hand, when the two electric charges are separated by a paper or wood, although the force may be weakened slightly, it does not reduce to zero. This shows that the effect of electric charge, i.e., the electric field, penetrates the paper or wood. Such material is called a **dielectric material** or **insulator**.

Here, we discuss the difference in electric behavior between conductors and dielectric materials. When an electric field is applied to a conductor, free electric charges (free electrons) move to the surface of the conductor to shield the interior from the external electric field. When the conductor is grounded, the electric charge on the surface opposite to the applied electric field is zero, and no electric field line appears from this surface. Hence, electric field lines from the electric charge, Q, do not reach a point, A, at which the other charge stays, as shown in Fig. 4.1a. In a dielectric material, electrons cannot freely move to shield because of their bonding to nuclei, which allows the electric field to penetrate, as shown in Fig. 4.1b. Hence, even when an electric field is applied to a dielectric material, no current can flow in it.

T. Matsushita, *Electricity and Magnetism*, Undergraduate Lecture Notes in Physics,
https://doi.org/10.1007/978-3-030-82150-0_4

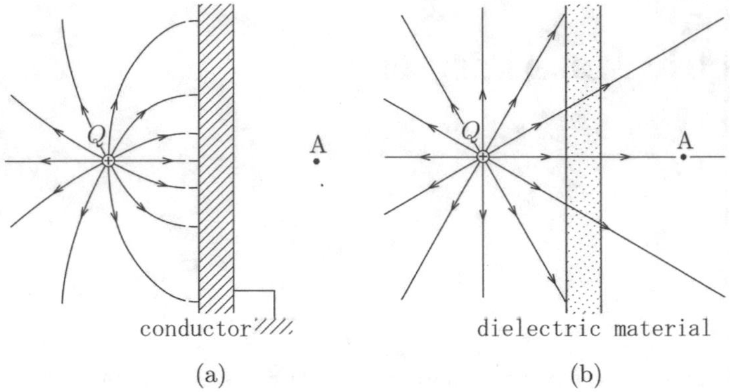

Fig. 4.1 Electric field lines produced by electric charge, Q, for **a** grounded conductor and **b** dielectric material

When no electric field is applied to a dielectric material, positive and negative electric charges are uniformly distributed on the macroscopic scale, resulting in an electrically neutral state. When an electric field is applied the dielectric material, nuclei and electrons bonded by them are slightly displaced opposite to each other, as shown in Fig. 4.2a, and innumerous electric dipoles appear inside the dielectric material. As a result, the electrically neutral state is maintained inside but electric charges appear on the surfaces (see Fig. 4.2b). This phenomenon is called **electric polarization** or **dielectric polarization**, and the electric charge that appears on the surface is called **polarization charge**. Although this phenomenon is similar to the electrostatic induction in conductors learned in Sect. 2.3, the polarization charges cannot be carried outside. This is a difference from true electric charges in conductors. In some dielectric materials, the electric polarization exists even in the absence of an external electric field. Such a material is called a **ferroelectric material** and its electric polarization is called **spontaneous polarization**.

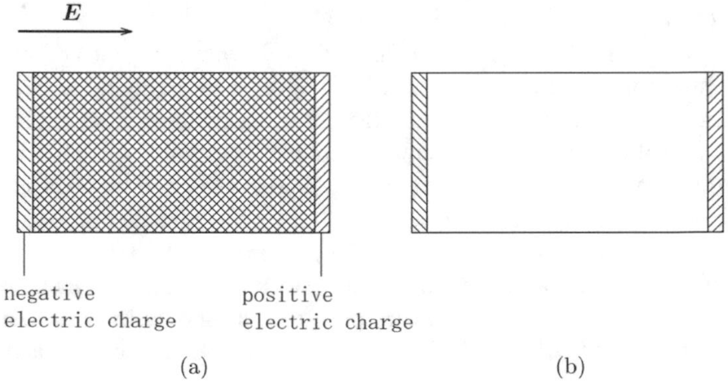

Fig. 4.2 **a** Displacement of positive and negative electric charges in external electric field and **b** resultant polarization charges on the surfaces

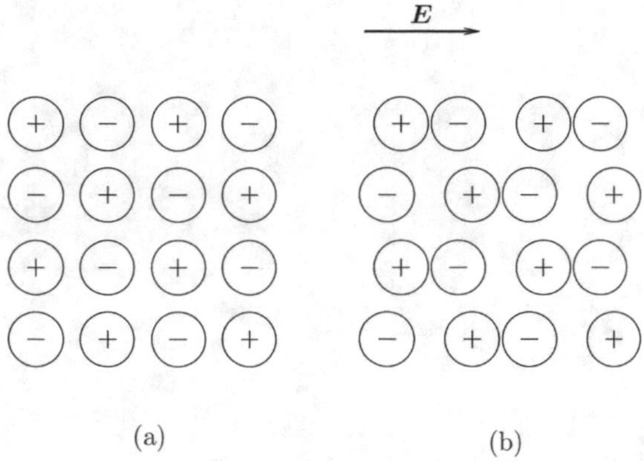

Fig. 4.3 Arrangement of ions in ionic crystal **a** before and **b** after application of electric field

There are also other kinds of electric polarization. The above example is caused by relative displacements of electrons and nuclei under an applied electric field and is called **electronic polarization**. In ionic crystals composed of cations and anions, the relative displacement of positive and negative ions in the electric field brings about the electric polarization called **ionic polarization** (see Fig. 4.3). In the case of water or hydrogen chloride, the molecule itself has an electric dipole moment, as shown in Fig. 4.4. In the usual state of gas or liquid in the absence of external electric field, each electric dipole moment of a polar molecule is directed randomly, resulting in no electric polarization (see Fig. 4.5a). When an electric field is applied, each electric dipole moment tends to incline in the direction of the electric field and the electric polarization appears, as shown in Fig. 4.5b. This is called **orientation polarization**. In many cases, the magnitude of electric polarization is proportional to the electric field strength.

Since electrons are able to respond quickly to a variation in the electric field because of their light mass, they can contribute to the electric polarization even for

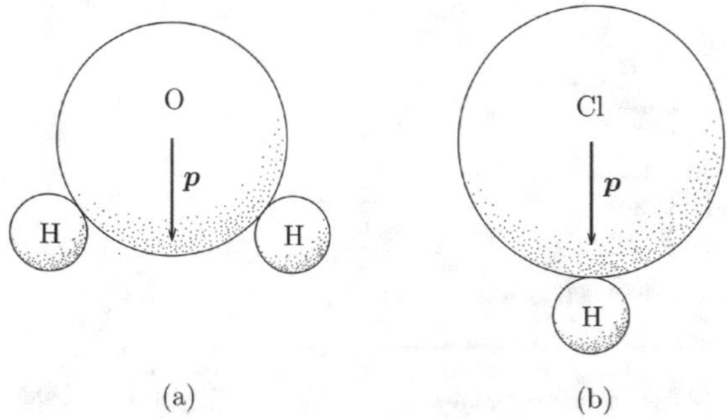

Fig. 4.4 Examples of molecule with electric dipole moment: **a** water (H_2O) and **b** hydrogen chloride (HCl)

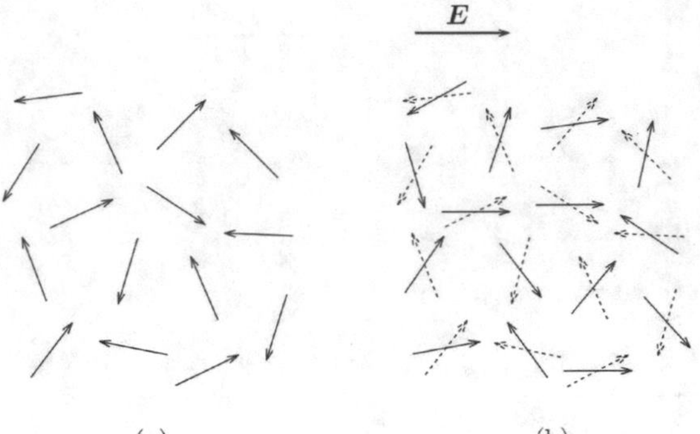

Fig. 4.5 Direction of electric dipole moments of polar molecules **a** before and **b** after application of electric field

an AC (alternating current) electric field with very high frequency. On the other hand, polar molecules cannot quickly rotate themselves to follow the variation in the electric field because of their heavy mass. Hence, those can contribute only to the electric polarization at a fairly low frequency. The ionic polarization is intermediate. Figure 4.6 shows a typical dependence of electric polarization on the frequency of AC electric field. Hence, each contribution can be discriminated by measuring the frequency dependence of electric polarization.

We use $\boldsymbol{P}$ to represent the electric dipole moment appearing in a unit volume of the dielectric material in an electric field of strength $\boldsymbol{E}$. This quantity is also called **electric polarization**. Its unit is [C/m^2]. Usually $\boldsymbol{P}$ is proportional to $\boldsymbol{E}$:

$$\boldsymbol{P} = \epsilon_0 \chi_e \boldsymbol{E}. \tag{4.1}$$

In the above, χ_e is a dimensionless constant of proportionality called **electric susceptibility**.

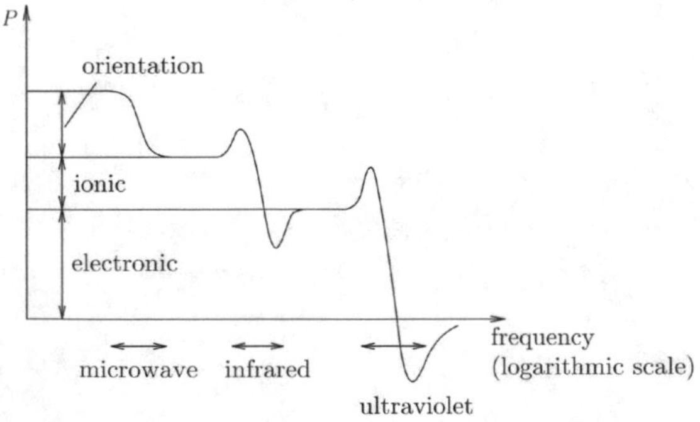

Fig. 4.6 Dependence of electric polarization on the frequency of an AC electric field

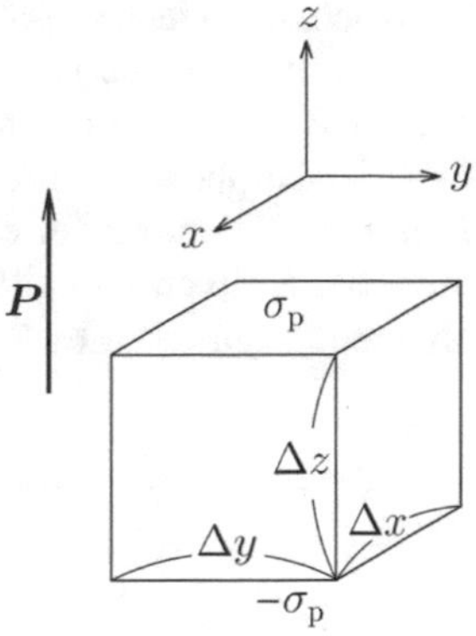

Fig. 4.7 Electric polarization in rectangular parallelepiped

Suppose a small rectangular parallelepiped with each surface perpendicular to a coordinate axis, as shown in Fig. 4.7, in a dielectric material. The extension of this region along each axis is denoted by Δx, Δy, and Δz. We assume that the electric polarization, $\boldsymbol{P}$, is directed along the z-axis. The magnitude of the electric dipole moment in this region of volume $\Delta V = \Delta x \Delta y \Delta z$ is $P\Delta V$. If the surface densities of polarization charge that appears on the top and bottom surfaces of this region are $\pm\sigma_{\rm p}$, polarization charges of $\pm\sigma_{\rm p}\Delta x \Delta y$ appear on the top and bottom surfaces. Hence, the electric dipole moment can also be given by $\sigma_{\rm p}\Delta x \Delta y \Delta z = \sigma_{\rm p}\Delta V$. Thus, we have

$$\sigma_{\rm p} = P. \tag{4.2}$$

That is, the electric polarization is equal to the amount of positive polarization charge that crosses the surface of a unit area perpendicular to the direction of polarization.

Figure 4.8 shows the case where the electric polarization, $\boldsymbol{P}$, is tilted by an angle θ from the unit surface vector, $\boldsymbol{n}$. In this case, the surface density of positive polarization charge that appears on the surface is

$$\sigma_{\rm p} = P\cos\theta = \boldsymbol{P}\cdot\boldsymbol{n}. \tag{4.3}$$

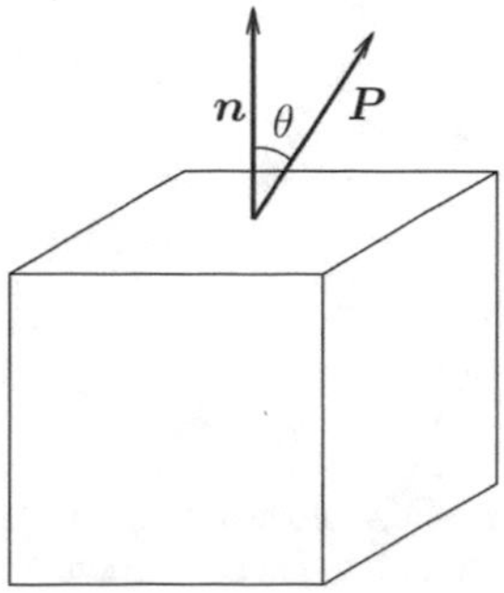

Fig. 4.8 Electric polarization tilted from the normal direction of the surface of a rectangular parallelepiped

When the electric polarization is uniform in space, the polarization charge appears only on the surface as shown above. However, when the electric polarization is not uniform, the polarization charge also appears inside the dielectric material. Suppose a closed surface, S, inside the dielectric material, as shown in Fig. 4.9. The amount of electric charge that goes out of S through a small surface area, dS, is given by $\sigma_{\mathrm{p}} \mathrm{d}S = \boldsymbol{P} \cdot \mathrm{d}\boldsymbol{S}$, using Eq. (4.3). Hence, the total polarization charge that goes out of S is

$$Q'_{\mathrm{p}} = \int_{\mathrm{S}} \boldsymbol{P} \cdot \mathrm{d}\boldsymbol{S} = \int_{\mathrm{V}} \nabla \cdot \boldsymbol{P} \, \mathrm{d}V, \tag{4.4}$$

where V is the interior of S.

The volume density of polarization charge is called **polarization charge density**. The density of polarization charge that appears after the electric polarization is denoted by ρ_{p}. Then, the total polarization charge in V is

$$Q_{\mathrm{p}} = \int_{\mathrm{V}} \rho_{\mathrm{p}} \, \mathrm{d}V. \tag{4.5}$$

The appearance of polarization charge in V is caused by the movement of polarization charge Q'_{p} to the outside. Hence, the principle of conservation of electric charge requires $Q_{\mathrm{p}} + Q'_{\mathrm{p}} = 0$. This is written as

$$\int_{\mathrm{V}} (\nabla \cdot \boldsymbol{P} + \rho_{\mathrm{p}}) \mathrm{d}V = 0. \tag{4.6}$$

Since this relationship holds for arbitrary V, we have

$$\nabla \cdot \boldsymbol{P} = -\rho_{\mathrm{p}}. \tag{4.7}$$

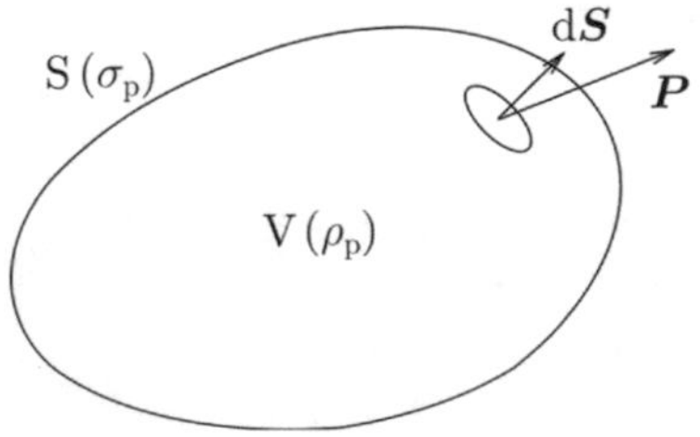

Fig. 4.9 Polarization charge that goes out of closed surface, S, and polarization charge that remains in interior region, V

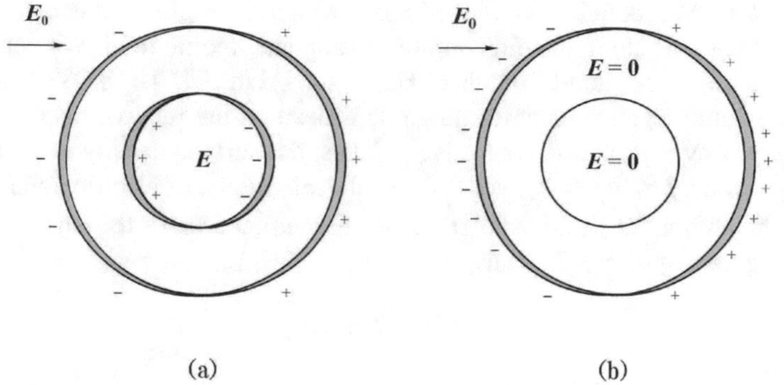

Fig. 4.10 **a** Dielectric hollow cylinder in transverse electric field and induced polarization charge and **b** conducting hollow cylinder in transverse electric field and induced electric charge

In the above, we learned that electric polarization in dielectric materials and electrostatic induction in conductors are similar to each other. However, there is an essential difference between them. When a transverse electric field is applied to a long dielectric hollow cylinder, a similar discussion suggests a distributed polarization charge on the inner surface, as shown in Fig. 4.10a. As a result, the electric field appears also in the interior space. In the case of a hollow cylinder made of a conductor, electric charge does not appear on the inner surface in an external transverse electric field, since this electric charge produces an electric field inside the conductor (see Fig. 4.10b).

Example 4.1. A dielectric sphere of diameter a is placed in a uniform electric field of strength, $\boldsymbol{E}_0$, as shown in Fig. 4.11a. Determine the surface density of polarization charge that appears on the surface of the dielectric sphere. The magnitude of electric polarization, $\boldsymbol{P}$, is P.

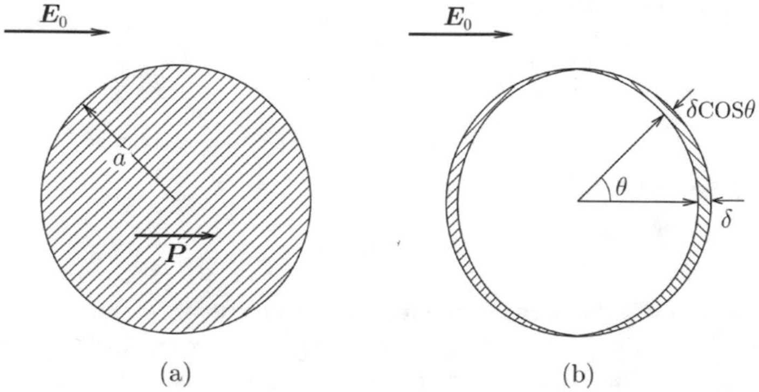

Fig. 4.11 **a** Electric polarization induced in dielectric sphere and **b** polarization charge that appears on the surface

---

Solution 4.1. We define polar coordinates with the origin at the center of the sphere and the axis along the direction of the applied electric field. We denote by θ the zenithal angle measured from this axis, as shown in Fig. 4.11b. We also denote the positive and negative electric charge densities and the relative displacement of these charges by $\pm\rho_{\mathrm{p}}$ and δ, respectively. Thus, the surface density of polarization charge is given by $\sigma_{\mathrm{p}}(\theta) = \rho_{\mathrm{p}}\delta\cos\theta$. Since the magnitude of electric polarization is equal to the amount of positive polarization charge that crosses the surface of a unit area, it is given by $P = \rho_{\mathrm{p}}\delta$. Thus, we have

$$\sigma_{\mathrm{p}}(\theta) = P\cos\theta. \tag{4.8}$$

◇

4.2 Electric Flux Density

We learned that there are two kinds of electric charge, i.e., true electric charge, which can be transferred outside, and polarization charge, which appears on the surface of dielectric materials but cannot be transferred outside. In addition, there are two kinds of electric field in dielectric materials, i.e., the electric field applied from outside and the one caused by electric polarization that occurs inside. Thus, electric phenomena in dielectric materials are rather complicated. To distinguish these contributions, it is convenient to introduce a new concept, electric flux density, defined below.

Suppose a closed surface, S, with inner region V. We denote the densities of true electric charge and polarization charge in V by ρ and ρ_{p}, respectively. Both of them contribute to the electric field and Gauss' law is written as

$$\int_{\mathrm{S}} \boldsymbol{E}\cdot \mathrm{d}\boldsymbol{S} = \frac{1}{\epsilon_0}\int_{\mathrm{V}} (\rho+\rho_{\mathrm{p}})\mathrm{d}V. \tag{4.9}$$

Substituting Eq. (4.7) for ρ_{p} gives

$$\int_{\mathrm{S}} (\epsilon_0\boldsymbol{E}+\boldsymbol{P})\cdot \mathrm{d}\boldsymbol{S} = \int_{\mathrm{V}} \rho\, \mathrm{d}V. \tag{4.10}$$

Here, we define

$$\boldsymbol{D} = \epsilon_0\boldsymbol{E}+\boldsymbol{P} \tag{4.11}$$

and call this **electric flux density** or **electric displacement**. Its unit is the same as for $\boldsymbol{P}$ and is [C/m^2]. In a relatively wide range of electric field strength, $\boldsymbol{P}$ is proportional to $\boldsymbol{E}$, and $\boldsymbol{D}$ is also proportional to $\boldsymbol{E}$. Thus, we can write $\boldsymbol{D}$ as

$$\boldsymbol{D} = \epsilon_0(1 + \chi_e)\boldsymbol{E} = \epsilon\boldsymbol{E}, \tag{4.12}$$

where ϵ is a constant inherent to each material and is called the **dielectric constant**. Its unit is the same as for ϵ_0. If ϵ is written as

$$\epsilon = \epsilon_0\epsilon_r, \tag{4.13}$$

$\epsilon_r = 1 + \chi_e$ is a dimensionless quantity and is called the **relative dielectric constant**. Table 4.1 gives values of ϵ_r for various materials.

Using the electric flux density, Eq. (4.10) gives

$$\int_S \boldsymbol{D} \cdot \mathrm{d}\boldsymbol{S} = \int_V \rho \, \mathrm{d}V. \tag{4.14}$$

This is an extension of Eq. (1.19) and is generally called **Gauss' law**. It should be noted that there is no constant in Eq. (4.14) and only true electric charge is involved. Equation (1.19) describes only phenomena in vacuum or conductors and is Gauss' law in a narrow sense. The surface integral of electric flux density on the left side of Eq. (4.14) is called **electric flux**. Hence, Eq. (4.14) states that the total electric flux going out of a closed surface is equal to the amount of true electric charge in it. Using Gauss' theorem on the left side of Eq. (4.14), it is rewritten as

$$\int_V \nabla \cdot \boldsymbol{D} \, \mathrm{d}V = \int_V \rho \, \mathrm{d}V. \tag{4.15}$$

Since this relationship holds for arbitrary V, we have

$$\nabla \cdot \boldsymbol{D} = \rho. \tag{4.16}$$

This is the general form of **Gauss' divergence law**. It states that the divergence of the electric flux density is caused by true electric charge.

Table 4.1 Values of ϵ_r for various materials at room temperature

Gas (1 atm)		*Solid*	
Oxygen	1.00049	Titanium dioxide	83–183
Nitrogen	1.00055	Quartz glass	3.5–4.5
Carbon dioxide	1.00092	Mica	5–9
Liquid		Ebonite	2.6–5.0
Water	78.54	Bakelite	4.5–9.0
Ethyl alcohol	24.30	Polyethylene	2.3–2.7
Solid		Vinyl chloride	3.3–6.0
Sodium chloride (NaCl)	5.9	*Ferroelectric material*	
Silicon (Si)	10.7–11.8	Barium titanate	1150–4500
Aluminum oxide (Al_2O_3)	8.5–11	Rochelle salt	~4000

Substituting Eqs. (1.24) and (4.12) into Eq. (4.16) gives

$$\Delta\phi = -\frac{\rho}{\epsilon}. \tag{4.17}$$

This equation is an extension of Eq. (1.37) to the case in which a dielectric material is also contained and is the general form of **Poisson's equation**. Its solution is given by Eq. (1.27) by replacing ϵ_0 by ϵ.

Similarly to the definition of an electric field line for $\boldsymbol{E}$, an **electric flux line** can be defined for $\boldsymbol{D}$. Namely, the direction of a tangential line at any point on the electric flux line is the same as the direction of $\boldsymbol{D}$, and its line density is defined to be equal to the magnitude of $\boldsymbol{D}$.

As shown by Eq. (4.16), the true electric charge density changes from the divergence of $\epsilon_0\boldsymbol{E}$ to that of $\boldsymbol{D} = \epsilon\boldsymbol{E}$. Hence, when the space between two electrodes in a capacitor is occupied by a dielectric material with dielectric constant ϵ, its new capacitance is given by the former expression of capacitance by replacing ϵ_0 by ϵ. For example, the capacitance of a parallel-plate capacitor with a material of dielectric constant ϵ is given by

$$C = \frac{\epsilon S}{d} = \frac{\epsilon_0\epsilon_r S}{d} \tag{4.18}$$

[see Eq. (3.22)]. The capacitance increases by factor ϵ_r by filling the space between the two electrodes with a dielectric material.

Example 4.2. Determine the capacitance of the parallel-plate capacitor in Fig. 4.12. The surface area of the electrodes is S, their distance is d, and the thicknesses of dielectric materials 1 and 2 with dielectric constants ϵ_1 and ϵ_2 are t and $d - t$, respectively.

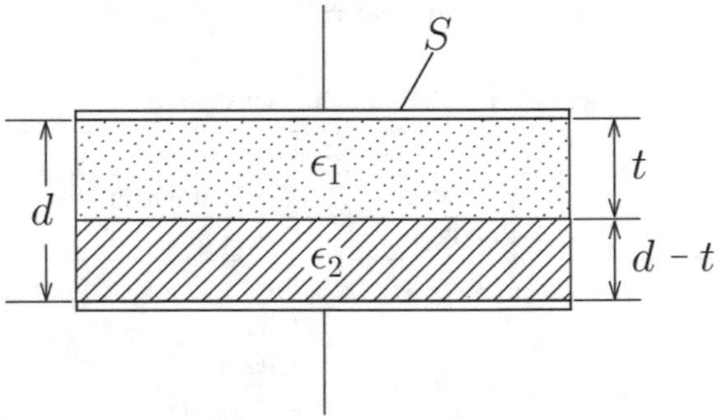

Fig. 4.12 Parallel-plate capacitor with two kinds of dielectric material

Solution 4.2. When a voltage, V, is applied to this capacitor, we assume that electric charges, $\pm Q$, are induced uniformly on the two electrodes. Non-uniformity at the edge can be disregarded if the electrodes are sufficiently wide. The electric field and electric flux density are directed normally to the electrodes and the interface between the two dielectric materials. Since there is no true electric charge on the interface, the magnitude of the electric flux density, D, is continuous there. This value is equal to the surface density of electric charge on the electrode, $\sigma = Q/S$, using Gauss' law. Hence, the electric field strengths in dielectric materials 1 and 2 are $E_1 = D/\epsilon_1$ and $E_2 = D/\epsilon_2$, respectively. The voltage between the two electrodes is

$$V = E_1 t + E_2(d-t) = \frac{Q}{S}\left(\frac{t}{\epsilon_1} + \frac{d-t}{\epsilon_2}\right).$$

This gives a capacitance of

$$C = \frac{Q}{V} = \frac{\epsilon_1 \epsilon_2 S}{\epsilon_2 t + \epsilon_1 (d-t)}.$$

◇

Example 4.3. Determine the capacitance of the parallel-plate capacitor shown in Fig. 4.13 in which dielectric materials 1 and 2 with dielectric constants ϵ_1 and ϵ_2 each occupy half of the space between the two electrodes. The surface area of the electrodes is S and their distance is d.

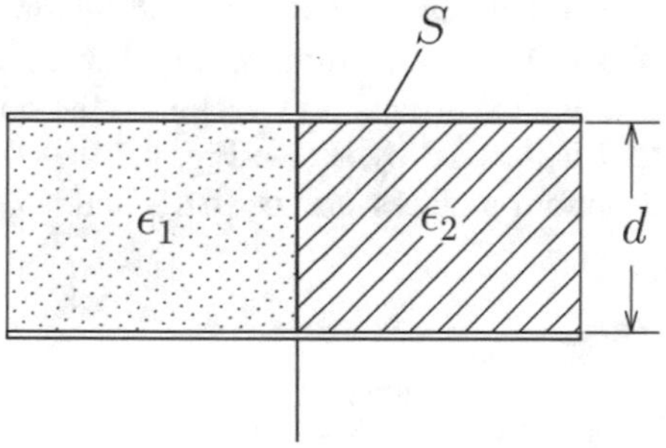

Fig. 4.13 Parallel-plate capacitor with two kinds of dielectric material

Solution 4.3. When a voltage, V, is applied to this capacitor, the electric field strength is $E = V/d$ in the space between the two electrodes. The electric flux densities in dielectric materials 1 and 2 are $D_1 = \epsilon_1 V/d$ and $D_2 = \epsilon_2 V/d$, respectively. Hence, the surface electric charge density on the electrode surface is different between the regions faced to dielectric materials 1 and 2. That is, the density is $\sigma_1 = \epsilon_1 V/d$ and $\sigma_2 = \epsilon_2 V/d$ in respective regions. The total electric charge on the electrode surface is

$$Q = (\sigma_1 + \sigma_2)\frac{S}{2} = \frac{(\epsilon_1 + \epsilon_2)SV}{2d}.$$

The capacitance is

$$C = \frac{Q}{V} = \frac{(\epsilon_1 + \epsilon_2)S}{2d}.$$

◇

4.3 Boundary Conditions

In this section, we investigate the boundary conditions to be fulfilled for the electric field and electric flux density at an interface between two different dielectric materials with dielectric constants ϵ_1 and ϵ_2. Assume that a true electric charge of a surface density, σ, exists on the interface.

First, we discuss the boundary condition for the electric flux density. Assume a closed surface, ΔS, of a small pellet region at the interface with top and bottom surfaces parallel to the interface, as shown in Fig. 4.14a. Assume that the height, Δh, is sufficiently small. We apply Gauss' law, Eq. (4.14), to this region. Since Δh is sufficiently small, the surface integral of electric flux on the side surface can be neglected and only the contributions from the top and bottom surfaces remain. We denote the electric flux density in dielectric materials 1 and 2 near the boundary by $\boldsymbol{D}_1$ and $\boldsymbol{D}_2$, respectively, and the normal unit vector on the boundary directed from dielectric material 1 to 2 is denoted by $\boldsymbol{n}$ (see Fig. 4.14b). The electric flux going out from the top and bottom surfaces are $\boldsymbol{D}_1 \cdot \boldsymbol{n}\Delta S$ and $-\boldsymbol{D}_2 \cdot \boldsymbol{n}\Delta S$, respectively.

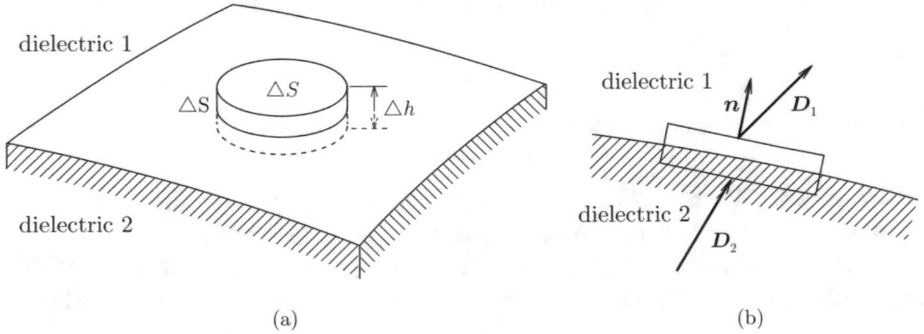

Fig. 4.14 **a** Small closed surface ΔS that contains a part of the interface between two dielectric materials and **b** electric flux density in each dielectric material at the interface

Here, ΔS is the area of the interface in the small region. Since the electric charge inside this small region is $\sigma \Delta S$, Gauss' law gives

$$\boldsymbol{n} \cdot (\boldsymbol{D}_1 - \boldsymbol{D}_2) = \sigma. \tag{4.19}$$

This shows that the difference between the normal components of the electric flux density is equal to the surface electric charge density. Hence, if there is no true electric charge on the interface, the normal component of electric flux density is continuous at the interface. If the electric potentials in dielectric materials 1 and 2 in the vicinity of the interface are ϕ_1 and ϕ_2, Eq. (4.19) is written as

$$\boldsymbol{n} \cdot (\epsilon_1 \nabla \phi_1 - \epsilon_2 \nabla \phi_2) = -\sigma. \tag{4.20}$$

Secondly, we discuss the boundary condition for the electric field. We denote the electric field in dielectric materials 1 and 2 near the boundary by $\boldsymbol{E}_1$ and $\boldsymbol{E}_2$, respectively. One can show that a plane that contains the vectors $\boldsymbol{E}_1$ and $\boldsymbol{E}_2$ is perpendicular to the interface, as illustrated in Fig. 4.15a (see Exercise 4.3). Consider a small rectangle with two sides parallel to the interface on this plane, as shown in Fig. 4.15b. The circular integral of the electric field on the rectangle is zero from Eq. (1.30). When the height, Δh, of the rectangle is sufficiently small, there are only two contributions from the top and bottom sides to the circular integral. We denote the unit vector on the top side along the direction of integral by $\boldsymbol{t}$. Then, the integrals on the top and bottom sides are respectively given by $\boldsymbol{E}_1 \cdot \boldsymbol{t} \Delta s$ and $-\boldsymbol{E}_2 \cdot \boldsymbol{t} \Delta s$, where Δs is the length of these sides. Hence, we have

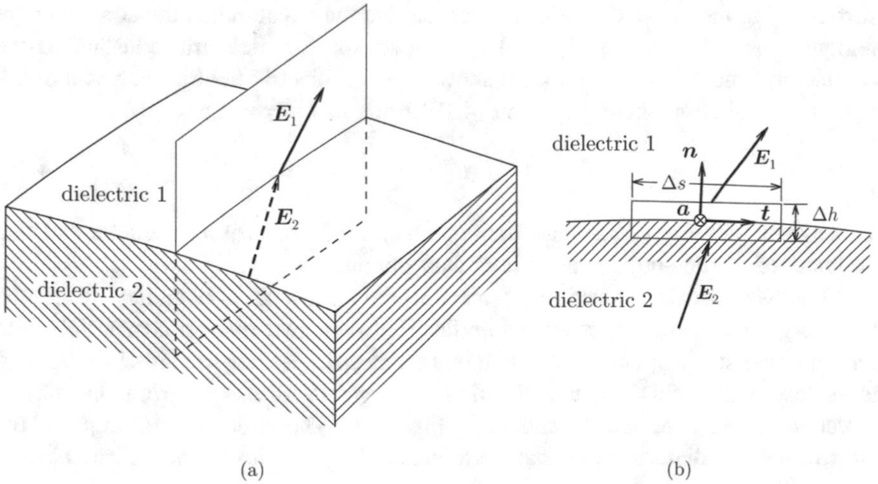

Fig. 4.15 **a** Plane that contains the electric field vectors in each dielectric material at the interface and **b** small rectangle on the plane that contains the interface

$$\boldsymbol{t} \cdot (\boldsymbol{E}_1 - \boldsymbol{E}_2) = 0. \tag{4.21}$$

That is, the component of the electric field parallel to the interface is continuous. However, since $\boldsymbol{t}$ is a unit vector directed along the electric field projected upon the interface, Eq. (4.21) is not a general description. Then, we define a unit vector, $\boldsymbol{a}$, normal to the plane, as shown in Fig. 4.15b. This vector satisfies the relationship $\boldsymbol{a} \times \boldsymbol{n} = \boldsymbol{t}$ and is perpendicular to both the plane that includes the electric field and $\boldsymbol{n}$. Thus, Eq. (4.21) is rewritten as $[\boldsymbol{n} \times (\boldsymbol{E}_1 - \boldsymbol{E}_2)] \cdot \boldsymbol{a} = 0$, and we obtain

$$\boldsymbol{n} \times (\boldsymbol{E}_1 - \boldsymbol{E}_2) = 0. \tag{4.22}$$

This condition can also be described in terms of the electric potential. If the derivative along the direction of $\boldsymbol{t}$ is $\partial/\partial t$, Eq. (4.21) is written as

$$\frac{\partial \phi_1}{\partial t} = \frac{\partial \phi_2}{\partial t}. \tag{4.23}$$

Hence, when integrating along the direction of $\boldsymbol{t}$, we have $\phi_1 - \phi_2 = c$ with c denoting a constant. If there is a finite difference in the electric potential in a very narrow region, it results in an extremely strong electric field in the normal direction at the interface. This is not practical and results in $c = 0$. That is,

$$\phi_1 = \phi_2. \tag{4.24}$$

The electric potential is continuous at the interface.

Confirm that the above boundary conditions are satisfied on the interfaces of two dielectric materials in Examples 4.2 and 4.3.

Here, we discuss the boundary conditions of the electric field at the conductor surface using the above conditions. Suppose that the vacuum and the conductor are regions 1 and 2, respectively, and ϵ_0 is used for the dielectric constant of the conductor. Since $E_2 = 0$, Eq. (4.22) shows that the electric field in the vacuum has only a normal component. Equation (4.19) leads to $D_1 = \sigma$, i.e.,

$$\epsilon_0 E_1 = \sigma. \tag{4.25}$$

This agrees with Eq. (2.5). Thus, the boundary conditions for a conductor can also be included in the above general boundary conditions.

Now, we discuss refraction of electric field lines at a boundary using the boundary conditions. Suppose an interface between two dielectric materials with dielectric constants ϵ_1 and ϵ_2. Assume that an electric field of strength E_1 is applied to dielectric material 1 in the direction of angle θ_1 measured from the normal direction to the interface, as shown in Fig. 4.16. The strength and angle of the electric field in dielectric material 2 are denoted by E_2 and θ_2. Since true electric

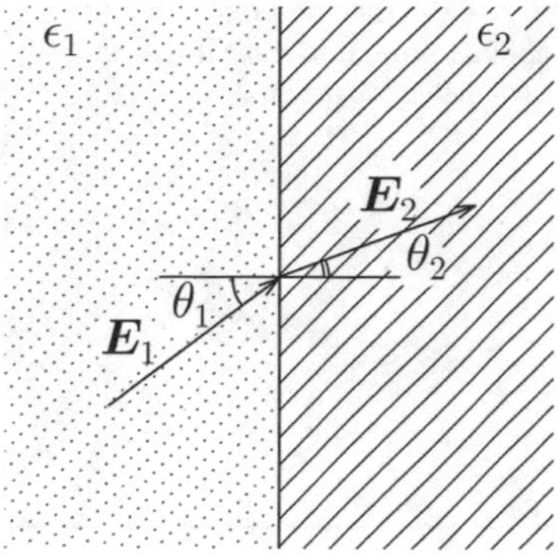

Fig. 4.16 Refraction of electric field lines at interface

charge does not usually exist on the interface, the normal component of the electric flux density is continuous on it:

$$\epsilon_1 E_1 \cos\theta_1 = \epsilon_2 E_2 \cos\theta_2. \tag{4.26}$$

The continuity of the parallel component of the electric field is written as

$$E_1 \sin\theta_1 = E_2 \sin\theta_2. \tag{4.27}$$

These equations give

$$\frac{\tan\theta_1}{\tan\theta_2} = \frac{\epsilon_1}{\epsilon_2}. \tag{4.28}$$

This is the **law of refraction**. We obtain E_2 and θ_2 as

$$E_2 = E_1 \left[\sin^2\theta_1 + \left(\frac{\epsilon_1}{\epsilon_2}\right)^2 \cos^2\theta_1\right]^{1/2}, \tag{4.29}$$

$$\theta_2 = \tan^{-1}\left(\frac{\epsilon_2}{\epsilon_1}\tan\theta_1\right). \tag{4.30}$$

Example 4.4. An electric field of strength E_0 is applied normal to a wide surface of a dielectric material of dielectric constant ϵ, as shown in Fig. 4.17. Determine the electric field strength, electric flux density, and electric polarization inside the dielectric material and the surface density of the polarization charge.

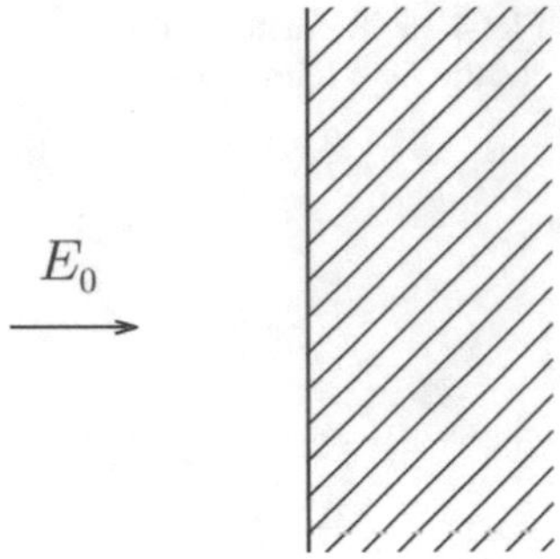

Fig. 4.17 Electric field applied normal to the surface of a dielectric material

Solution 4.4. From the continuity of the normal component of the electric flux density on the surface given by Eq. (4.19), the electric flux density inside the dielectric material is equal to that in the outside, i.e., $D = \epsilon_0 E_0$. Thus, the electric field inside the dielectric material is $E = D/\epsilon = (\epsilon_0/\epsilon)E_0$. The electric polarization is determined to be

$$P = D - \epsilon_0 E = \frac{\epsilon_0(\epsilon - \epsilon_0)}{\epsilon} E_0.$$

Taking account of the direction of the electric field, the surface polarization charge density is

$$\sigma_{\mathrm{p}} = -P = -\frac{\epsilon_0(\epsilon - \epsilon_0)}{\epsilon} E_0.$$

◇

Example 4.5. A dielectric sphere of radius a is placed in a uniform electric field of strength E_0, as shown in Fig. 4.11a. Determine the electric field strength, electric flux density, and electric polarization inside and outside the dielectric sphere and the polarization charge density on the surface.

Solution 4.5. Polar coordinates are defined as in Example 4.1. We can assume that a uniform electric polarization occurs in the dielectric material as in this example. The electric field outside the sphere is given by the sum of the applied electric field and the contribution of the electric dipole placed at the origin after virtual removal of the sphere. The electric field strength inside the sphere is expected to be uniform because of the uniform electric polarization. Let p denote the electric dipole moment directed along the applied electric field. From Eqs. (1.48a) and (1.48b), the radial and zenithal components of the electric field outside the sphere produced by the electric dipole are

$$E_r = \frac{p\cos\theta}{2\pi\epsilon_0 r^3}, \quad E_\theta = \frac{p\sin\theta}{4\pi\epsilon_0 r^3}.$$

We denote the internal electric field strength by E. The continuities of the parallel component of the electric field and the normal component of the electric flux density at the surface ($r = a$) are given by

$$-E_0\sin\theta + \frac{p\sin\theta}{4\pi\epsilon_0 a^3} = -E\sin\theta, \quad \epsilon_0\left(E_0\cos\theta + \frac{p\cos\theta}{2\pi\epsilon_0 a^3}\right) = \epsilon E\cos\theta.$$

From these equations, we have

$$p = \frac{\epsilon - \epsilon_0}{\epsilon + 2\epsilon_0}\, 4\pi\epsilon_0 a^3 E_0, \quad E = \frac{3\epsilon_0}{\epsilon + 2\epsilon_0}\, E_0.$$

We can see that E is smaller than E_0 because of $\epsilon > \epsilon_0$. This means that the dielectric material is shielded by the polarization charge. Using these results, the electric field is

$$\begin{aligned} E_r &= \frac{D_r}{\epsilon_0} = \left(1 + \frac{\epsilon - \epsilon_0}{\epsilon + 2\epsilon_0}\cdot\frac{2a^3}{r^3}\right)E_0\cos\theta, \\ E_\theta &= \frac{D_\theta}{\epsilon_0} = -\left(1 - \frac{\epsilon - \epsilon_0}{\epsilon + 2\epsilon_0}\cdot\frac{a^3}{r^3}\right)E_0\sin\theta \end{aligned}$$

outside the sphere ($r > a$) and

$$E_r = \frac{D_r}{\epsilon} = \frac{3\epsilon_0}{\epsilon + 2\epsilon_0}\, E_0\cos\theta, \quad E_\theta = \frac{D_\theta}{\epsilon} = -\frac{3\epsilon_0}{\epsilon + 2\epsilon_0}\, E_0\sin\theta$$

inside the sphere ($r < a$). These results can also be obtained by solving Eqs. (4.20) and (4.24) for the electric potential. We obtain the electric polarization as

$$P = (\epsilon - \epsilon_0)E = \frac{3\epsilon_0(\epsilon - \epsilon_0)}{\epsilon + 2\epsilon_0}E_0.$$

Here, we suppose a small shell that contains a part of the surface (see Fig. 4.18). We apply Eq. (4.9) to this region. Since there is no true electric charge, the difference in normal component of the electric field is equal to the surface polarization charge density divided by ϵ_0. Thus, we have

$$\sigma_{\rm p}(\theta) = \frac{3\epsilon_0(\epsilon - \epsilon_0)}{\epsilon + 2\epsilon_0}E_0\cos\theta = P\cos\theta.$$

This agrees with Eq. (4.8) in Example 4.1.

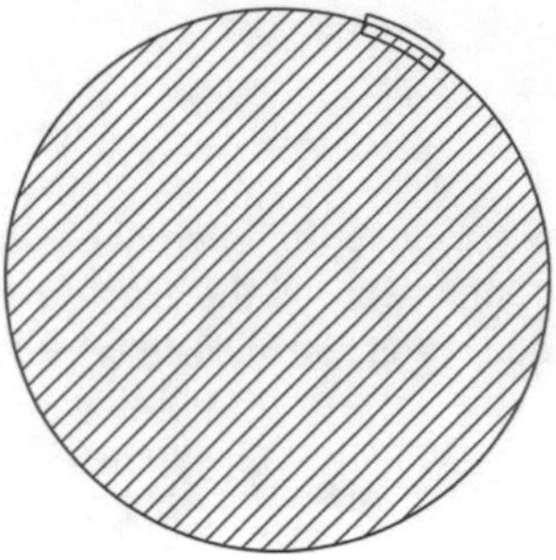

Fig. 4.18 Small shell containing a part of the surface of a dielectric sphere

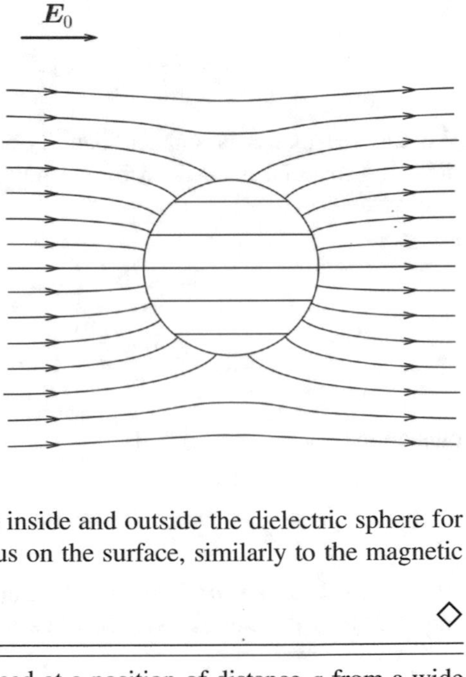

Fig. 4.19 Electric field lines inside and outside the dielectric sphere for $\epsilon = 3\epsilon_0$

Figure 4.19 shows electric field lines inside and outside the dielectric sphere for $\epsilon = 3\epsilon_0$. Electric flux lines are continuous on the surface, similarly to the magnetic flux lines in Fig. 9.17.

◇

Example 4.6. A point charge, q, is placed at a position of distance a from a wide flat surface of a dielectric material of dielectric constant ϵ, as shown in Fig. 4.20. Determine the electric potential in the vacuum and in the dielectric material.

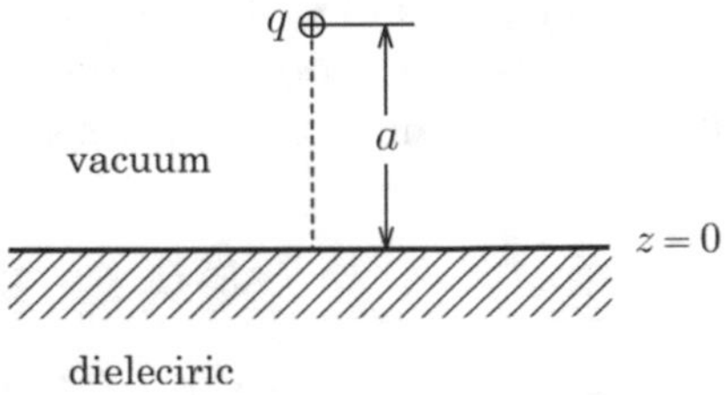

Fig. 4.20 Point charge q placed at distance a from the surface of a dielectric material

Solution 4.6. We draw the z-axis from the point charge in the direction normal to the dielectric material's surface, which is defined to be $z = 0$. We define the x-y plane on the surface with the origin at the foot of a perpendicular line from the point charge. This problem can also be solved using the method of images shown in Sect. 2.2.

We assume that the electric potential in the vacuum region ($z > 0$) is given by the sum of a contribution from q and that from a virtual point charge, q', placed at the symmetric position with respect to the surface with virtual removal of dielectric material, as shown in Fig. 4.21a. Thus, the electric potential at point (x, y, z) is

$$\phi_{\mathrm{v}} = \frac{1}{4\pi\epsilon_0}\left\{\frac{q}{\left[x^2+y^2+(z-a)^2\right]^{1/2}} + \frac{q'}{\left[x^2+y^2+(z+a)^2\right]^{1/2}}\right\}.$$

We assume that the electric potential in the dielectric material ($z < 0$) is equal to that produced by a point charge, q'', placed at the original position with virtual occupation of the vacuum region by the same dielectric material, as shown in Fig. 4.21b. Thus, the electric potential at point (x, y, z) is

$$\phi_{\mathrm{d}} = \frac{1}{4\pi\epsilon}\cdot\frac{q''}{\left[x^2+y^2+(z-a)^2\right]^{1/2}}.$$

From Eq. (4.24), the continuity condition for the parallel component of the electric field is expressed as $\phi_{\mathrm{v}}(z=0) = \phi_{\mathrm{d}}(z=0)$, which gives

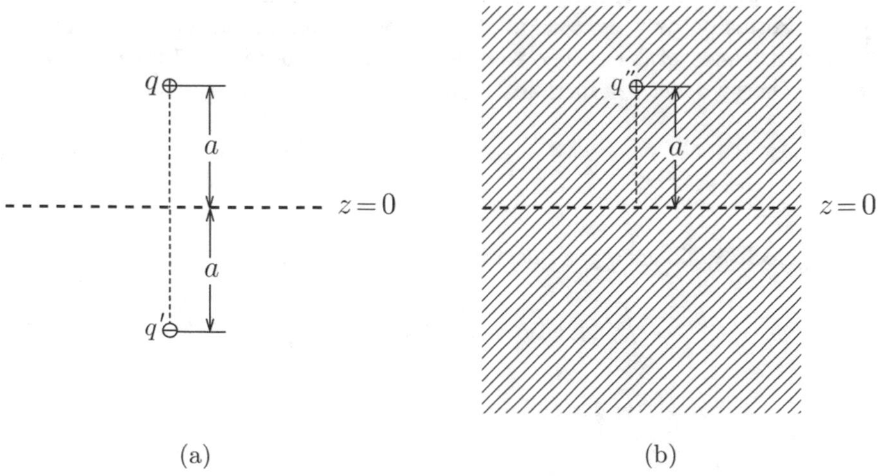

Fig. 4.21 Solution using the method of images: assumed conditions for **a** vacuum and **b** dielectric material

$$\frac{q+q'}{\epsilon_0} = \frac{q''}{\epsilon}.$$

The continuity condition for the normal component of the electric flux density is given by Eq. (4.20) with $\sigma = 0$: $\epsilon_0(\partial\phi_v/\partial z)_{z=0} = \epsilon(\partial\phi_d/\partial z)_{z=0}$, which gives

$$q - q' = q''.$$

From these equations, we have

$$q' = -\frac{\epsilon - \epsilon_0}{\epsilon + \epsilon_0}\, q, \quad q'' = \frac{2\epsilon}{\epsilon + \epsilon_0}\, q.$$

Thus, the electric potential is

$$\begin{aligned}\phi &= \frac{q}{4\pi\epsilon_0(\epsilon+\epsilon_0)}\left\{\frac{\epsilon+\epsilon_0}{[x^2+y^2+(z-a)^2]^{1/2}} - \frac{\epsilon-\epsilon_0}{[x^2+y^2+(z+a)^2]^{1/2}}\right\}; z > 0,\\ &= \frac{q}{2\pi(\epsilon+\epsilon_0)}\cdot\frac{1}{[x^2+y^2+(z-a)^2]^{1/2}}; \qquad z<0.\end{aligned}$$

◇

4.4 Electrostatic Energy in Dielectric Materials

We discussed the electrostatic energy of a conductor system in vacuum in Sect. 3.3. This does not essentially change even when there are dielectric materials in the system. Formally, only the change from ϵ_0 to ϵ occurs in the region of dielectric materials because of the change from Eq. (1.21) to Eq. (4.16). That is, the electrostatic energy density in dielectric materials is

$$u_e = \frac{1}{2}\epsilon \boldsymbol{E}^2 = \frac{1}{2}\boldsymbol{E}\cdot\boldsymbol{D} = \frac{1}{2\epsilon}\boldsymbol{D}^2 \tag{4.31}$$

and the electrostatic energy is given by its volume integral,

$$U_e = \int_V \frac{1}{2}\epsilon \boldsymbol{E}^2\, dV = \int_V \frac{1}{2}\boldsymbol{E}\cdot\boldsymbol{D}\, dV = \int_V \frac{1}{2\epsilon}\boldsymbol{D}^2\, dV. \tag{4.32}$$

Example 4.7. The space between the electrodes in a long cylindrical capacitor is occupied by two dielectric materials with dielectric constants ϵ_1 and ϵ_2, as shown in Fig. 4.22. Determine the electrostatic energy in a unit length of the capacitor, when electric charges $\pm Q'$ are given to the respective electrodes in a unit length.

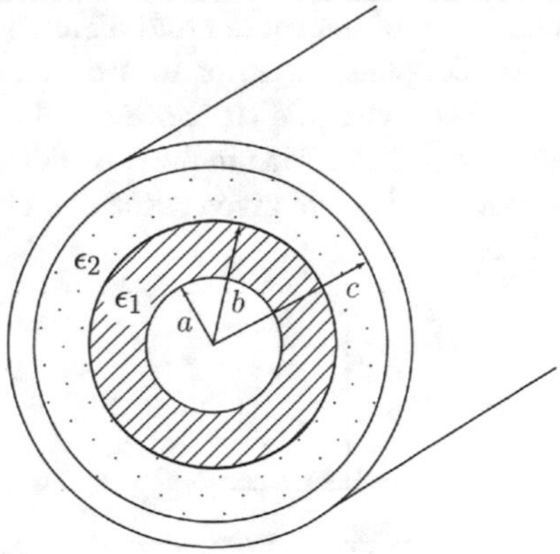

Fig. 4.22 Cylindrical capacitor with two dielectric materials

Solution 4.7. The electric field strength in the region between the electrodes is

$$\begin{aligned} E(R) &= \frac{Q'}{2\pi\epsilon_1 R}; \quad a<R<b, \\ &= \frac{Q'}{2\pi\epsilon_2 R}; \quad b<R<c. \end{aligned}$$

Thus, the electric potential difference between the electrodes is

$$V = \int_a^c E(R)\mathrm{d}R = \frac{Q'}{2\pi}\left(\frac{1}{\epsilon_1}\log\frac{b}{a} + \frac{1}{\epsilon_2}\log\frac{c}{b}\right),$$

and the electrostatic energy in a unit length is given by

$$U'_{\mathrm{e}} = \frac{1}{2}Q'V = \frac{Q'^2}{4\pi}\left(\frac{1}{\epsilon_1}\log\frac{b}{a} + \frac{1}{\epsilon_2}\log\frac{c}{b}\right).$$

The same result is obtained using Eq. (4.32) as

$$U'_{\mathrm{e}} = \int_a^b \frac{Q'^2}{8\pi^2\epsilon_1 R^2} 2\pi R \mathrm{d}R + \int_b^c \frac{Q'^2}{8\pi^2\epsilon_2 R^2} 2\pi R \mathrm{d}R = \frac{Q'^2}{4\pi}\left(\frac{1}{\epsilon_1}\log\frac{b}{a} + \frac{1}{\epsilon_2}\log\frac{c}{b}\right).$$

◇

Example 4.8. A dielectric plate of thickness t and dielectric constant ϵ is inserted into the gap of a wide parallel-plate capacitor to a distance x from the edge, as shown in Fig. 4.23. The surface area and distance of the electrodes are S and d, and the sizes of the dielectric plate and electrode in the direction normal to the sheet are the same. When electric charges, $\pm Q$, are given to the two electrodes, determine the force on the dielectric plate.

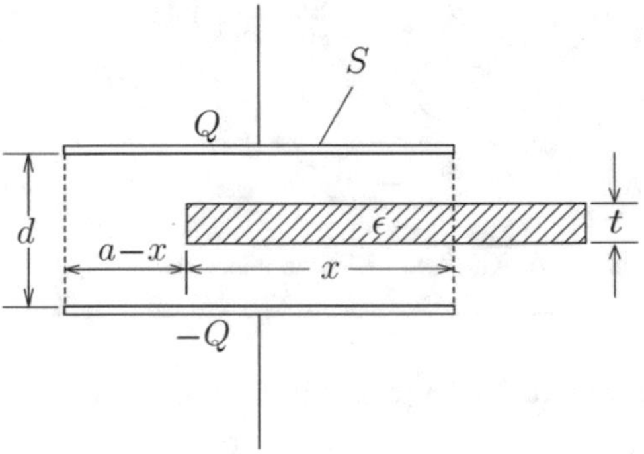

Fig. 4.23 Parallel-plate capacitor with inserted dielectric plate

Solution 4.8. The part in which the dielectric plate is not inserted can be regarded as one capacitor, and from Eq. (3.22), we obtain its capacitance as

$$C_1 = \frac{\epsilon_0 S}{d}\left(1 - \frac{x}{a}\right).$$

The capacitance of the remaining part, in which the dielectric plate is inserted, is similarly given by

$$C_2 = \frac{\epsilon_0 \epsilon S x}{a[\epsilon d - (\epsilon - \epsilon_0)t]}$$

(see Example 4.2). From Eq. (3.27), we obtain the total capacitance as

$$C = C_1 + C_2 = \frac{\epsilon_0 S}{d}\left(1 - \frac{x}{a}\right) + \frac{\epsilon_0 \epsilon S x}{a[\epsilon d - (\epsilon - \epsilon_0)t]}.$$

The variation rate of the electrostatic energy determines the force on the dielectric plate:

$$
\begin{aligned}
F &= -\frac{\mathrm{d}}{\mathrm{d}x}\left(\frac{Q^2}{2C}\right) = \frac{Q^2}{2C^2}\frac{\mathrm{d}C}{\mathrm{d}x} \\
&= \frac{Q^2 atd}{2\epsilon_0 S} \cdot \frac{(\epsilon - \epsilon_0)[\epsilon(d-t) + \epsilon_0 t]}{\{a[\epsilon(d-t) + \epsilon_0 t] + (\epsilon - \epsilon_0)tx\}^2}.
\end{aligned}
$$

This force is positive for increasing x because $\epsilon > \epsilon_0$, showing that it is attractive.

◇

Column: Electric Induction in a Conductor and Electric Polarization in a Dielectric Material

From electric field lines inside and outside the dielectric sphere placed in a uniform electric field shown in Fig. 4.19, one can see that the interior is imperfectly shielded by polarization charge. On the other hand, the electric flux density inside the sphere is higher than the external value because of ϵ larger than ϵ_0, similarly to the magnetic flux lines in Fig. 9.17. The solution for the electric flux density, $\boldsymbol{D}$, has the same form as that for the magnetic flux density, $\boldsymbol{B}$, for a magnetic sphere placed in the uniform magnetic flux density treated in Example 9.5. This similarity comes from the fact that, when there is no electric charge ($\rho = 0$), $\boldsymbol{D}$ obeys the same equation as $\boldsymbol{B}$ ($\nabla \cdot \boldsymbol{D} = \nabla \cdot \boldsymbol{B} = 0$).

If the dielectric constant ϵ is infinitely large in Example 4.5, the solution for the electric field strength coincides with that in the spherical conductor in Sect. 2.3. That is, $E = 0$ inside and Eq. (2.28) holds outside the sphere. In this case, the electric flux density has a finite value. However, its value is meaningless, and we can disregard it. There is also the relationship

$$
\sigma_{\mathrm{p}} \leq \sigma
$$

between the polarization charge density σ_{p} on the dielectric sphere surface in Example 4.5 and the true electric charge density σ on the spherical conductor surface in Sect. 2.3. The equality holds in the limit $\epsilon \longrightarrow \infty$. Hence, there is no large difference between the true electric charge in the conductor and the polarization charge in the dielectric material that are displaced in the electric field ($\sigma_{\mathrm{p}} = (4/7)\sigma$ for $\epsilon = 5\epsilon_0$). Hence, it is reasonable to assume similar models for the two cases. This shows that the electric shielding in conductors and electric polarization in dielectric materials are essentially the same mechanism.

Exercises

4.1. The space between the electrodes in a concentric spherical capacitor is occupied by two dielectric materials with dielectric constants, ϵ_1 and ϵ_2, as shown in Fig. E4.1. Determine the capacitance of the capacitor.

4.2. The space between the electrodes in a concentric spherical capacitor is occupied by two dielectric materials with dielectric constants, ϵ_1 and ϵ_2, as shown in Fig. E4.2. Determine the capacitance of the capacitor.

4.3. The electric fields, $\boldsymbol{E}_1$ and $\boldsymbol{E}_2$, in dielectric materials 1 and 2 are defined in the vicinity of the interface. Prove that these vectors stay in the same plane perpendicular to the interface.

4.4. A uniform electric field of strength E_0 is applied parallel to a thin slit of vacuum in a dielectric material of dielectric constant ϵ, as shown in Fig. E4.3. Determine the electric field and electric flux density inside the slit.

4.5. A uniform electric field of strength E_0 is applied normal to a thin slit of vacuum in a dielectric material of dielectric constant ϵ, as shown in Fig. E4.4. Determine the electric field and electric flux density inside the slit.

4.6. An electric field of strength E_0 is applied parallel to a wide flat surface of a dielectric material of dielectric constant ϵ, as shown in Fig. E4.5. Determine the electric field strength, electric flux density, and electric polarization inside the dielectric material and the surface density of polarization charge.

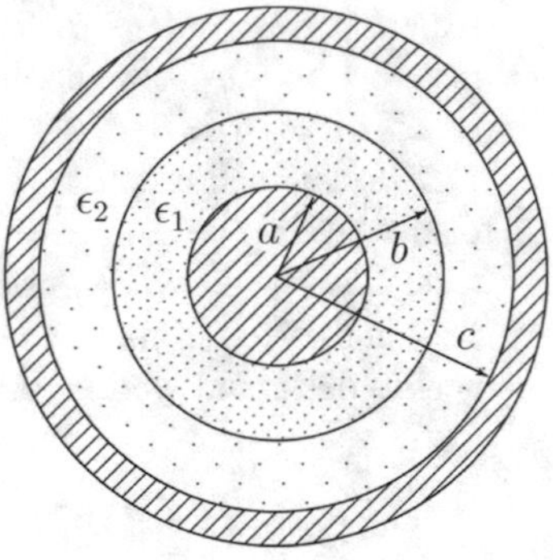

Fig. E4.1 Concentric spherical capacitor with two dielectric materials

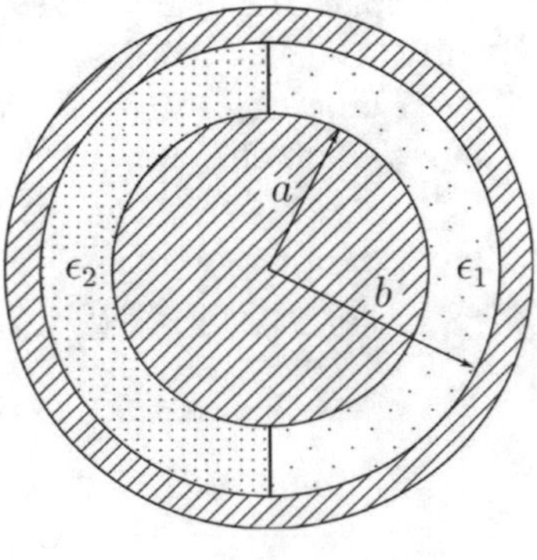

Fig. E4.2 Concentric spherical capacitor with two dielectric materials

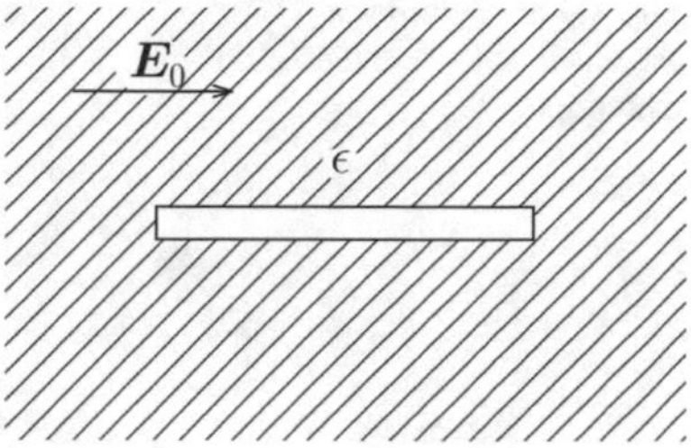

Fig. E4.3 Vacuum slit parallel to the electric field in a dielectric material

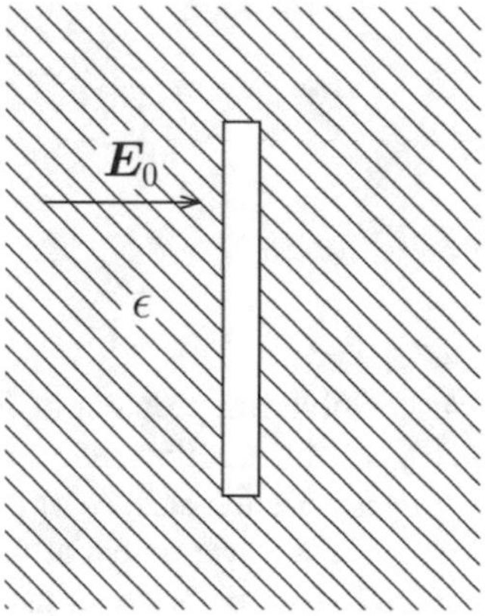

Fig. E4.4 Vacuum slit normal to the electric field in a dielectric material

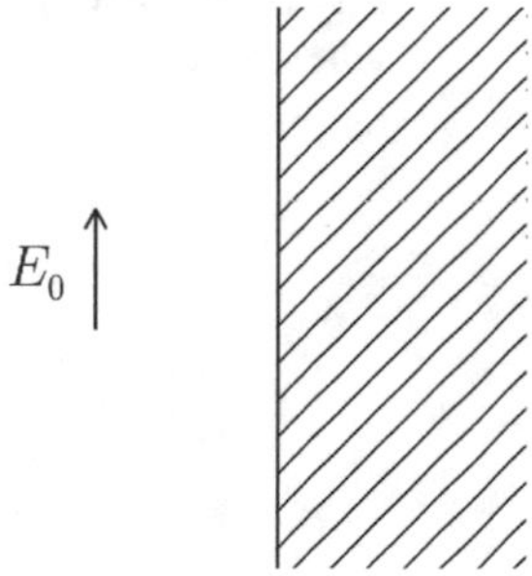

Fig. E4.5 Electric field applied parallel to the surface of a dielectric material

4.7. The internal electric field strength E and surface density of polarization charge σ_p are obtained when a dielectric sphere is placed in a uniform electric field of strength E_0 in Example 4.5. Prove that the obtained value of E coincides with the sum of E_0 and the electric field strength produced by the polarization charge. (Hint: Use Eq. (2.29) for the relationship between the uniform electric field strength and the surface electric charge density.)

4.8. A long dielectric cylinder of radius a and dielectric constant ϵ is placed in a uniform normal electric field of strength E_0. Determine the electric field strength, electric flux density, electric polarization, and surface polarization charge density.

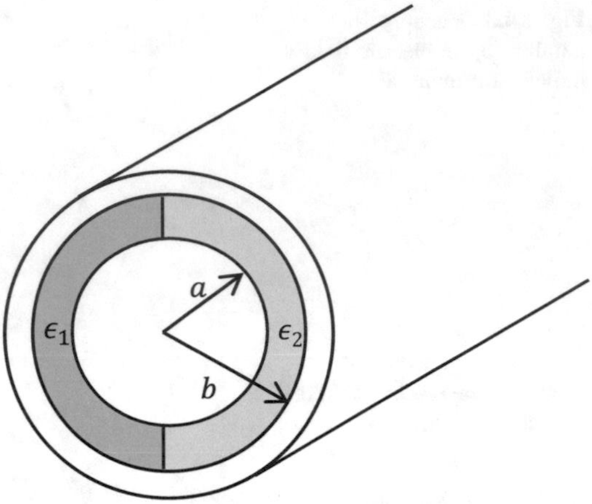

Fig. E4.6 Cylindrical capacitor with two dielectric materials

4.9. Solve the problem of Example 4.5 using the electric potential.

4.10. A uniform line charge of density λ is placed at a position of a distance a from a wide flat surface of a dielectric material of dielectric constant ϵ. Determine the electric potential in the vacuum and the dielectric material.

4.11. The space between the electrodes in a long cylindrical capacitor is occupied by two dielectric materials with dielectric constants ϵ_1 and ϵ_2, as shown in Fig. E4.6. Determine the electrostatic energy in a unit length of the capacitor, when electric charges $\pm Q'$ are given to the respective electrodes in a unit length.

Chapter 5
Steady Current

Abstract This chapter covers electric phenomena when there is steady current flow. Current comes from a movement of electric charge, and the continuity equation of current holds generally based on the principle of conservation of charge. We need to apply an electric potential difference, i.e., a voltage, to a material to transport a current through it. It is well known that Ohm's law holds empirically, which assumes a proportional relationship between the voltage and the current. Its proportional constant, the electric resistance, is proportional to the resistivity, which is a parameter inherent to the material. When there is steady current flow, the energy, which is continuously supplied by an electric power source, is dissipated. Kirchhoff's law, used in direct current electric circuits, is derived.

5.1 Current

A conductor contains freely moving electric charge, and the Coulomb force can move the electric charge when an electric field is directly applied to the conductor. This movement of electric charge is **current**. In this chapter, we discuss electric phenomena when a **steady current** that does not change with time flows. As distinct from magnetization current and displacement current discussed in Chaps. 9 and 11, the current of true electric charge is sometimes called **true current**.

The current is a vector with a magnitude and direction. When electric charge $\mathrm{d}Q$ passes through a cross section within time $\mathrm{d}t$, the current is given by

$$I = \frac{\mathrm{d}Q}{\mathrm{d}t}. \tag{5.1}$$

Its unit is [C/s] and is denoted [A] (**ampere**).

Although the current is an amount of electric charge that passes through a certain cross section in unit time, it is not a quantity representing strength. We define **current density** as a quantity representing the strength of current. The current density $\boldsymbol{i}$ is also a vector. Its direction is the same as that of the current, and its magnitude is given by

T. Matsushita, *Electricity and Magnetism*, Undergraduate Lecture Notes in Physics,
https://doi.org/10.1007/978-3-030-82150-0_5

$$i = \frac{\mathrm{d}I}{\mathrm{d}S}, \tag{5.2}$$

when current dI flows through a small normal cross section of area dS. Its unit is [A/m^2]. When the direction of current is tilted from elementary surface vector d$\boldsymbol{S}$, the current that flows through the elementary surface is

$$\mathrm{d}I = i\mathrm{d}S\cos\theta = \boldsymbol{i}\cdot\mathrm{d}\boldsymbol{S}. \tag{5.3}$$

Since the current is a flow of electric charge, we can describe it using the density and velocity of electric charge. Suppose that particles of electric charge q and density n move with velocity $\boldsymbol{v}$. The current density is then given by

$$\boldsymbol{i} = qn\boldsymbol{v}. \tag{5.4}$$

Since the electric charge density is given by $\rho = qn$, the current density is expressed as

$$\boldsymbol{i} = \rho\boldsymbol{v}. \tag{5.5}$$

The amount of electric charge is conserved similarly to the mass of materials. That is, the algebraic sum of positive and negative charges is conserved. We suppose a region V surrounded by a closed surface S (see Fig. 5.1) and denote the electric charge density inside it by ρ. The total electric charge in V is

$$Q = \int_{\mathrm{V}} \rho\,\mathrm{d}V. \tag{5.6}$$

When current of density $\boldsymbol{i}$ flows across the surface, the electric charge that passes out of V through a small area dS in unit time is $\boldsymbol{i}\cdot\mathrm{d}\boldsymbol{S}$. Hence, the electric charge that goes out of S in unit time is given by

$$\frac{\mathrm{d}Q'}{\mathrm{d}t} = \int_{\mathrm{S}} \boldsymbol{i}\cdot\mathrm{d}\boldsymbol{S}. \tag{5.7}$$

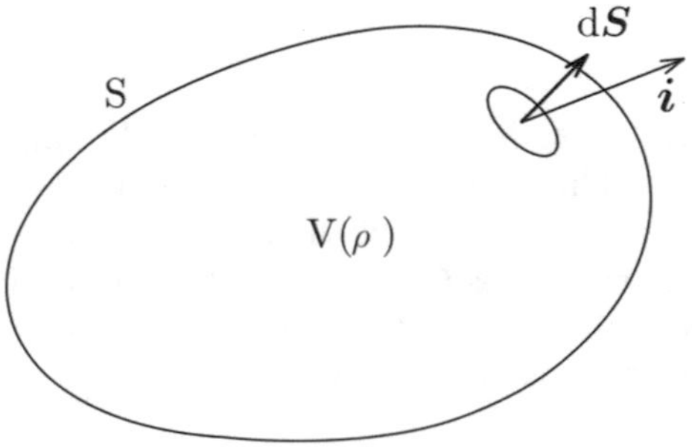

Fig. 5.1 Electric charge in region V and current through surface S

The principle of conservation of electric charge requires that this should be equal to the decrease in electric charge in region V in unit time, $-\mathrm{d}Q/\mathrm{d}t$. This gives

$$\frac{\mathrm{d}}{\mathrm{d}t}\int_{\mathrm{V}} \rho \, \mathrm{d}V + \int_{\mathrm{S}} \boldsymbol{i} \cdot \mathrm{d}\boldsymbol{S} = 0. \tag{5.8}$$

Since region V does not change with time, we can change the order of the time derivative and spatial integral in the first term. Using Gauss' theorem for the second term, Eq. (5.8) is written as

$$\int_{\mathrm{V}} \left(\nabla \cdot \boldsymbol{i} + \frac{\partial \rho}{\partial t} \right) \mathrm{d}V = 0. \tag{5.9}$$

Since this relationship holds for arbitrary V, we have

$$\nabla \cdot \boldsymbol{i} + \frac{\partial \rho}{\partial t} = 0. \tag{5.10}$$

This is called the **continuity equation of current.**

For a steady current that does not change with time Eq. (5.10) reduces to

$$\nabla \cdot \boldsymbol{i} = 0. \tag{5.11}$$

That is, the current density does not diverge and follows the same continuity equation as an incompressible fluid.

5.2 Ohm's Law

It is necessary to apply an electric potential difference to a material such as a metal to get a current. In many cases, it is empirically known that there is a proportional relationship between the electric potential difference V and the current I:

$$V = R_{\mathrm{r}} I. \tag{5.12}$$

The proportional constant R_{r} is called **electric resistance** or simply **resistance**. This constant is determined by the shape and property of the material that carries current. The unit of electric resistance is [V/A] and is denoted [Ω] (**ohm**). Equation (5.12) is called **Ohm's law**. For a material of length l and uniform cross-sectional area S, the electric resistance is given by

$$R_{\mathrm{r}} = \rho_{\mathrm{r}} \frac{l}{S}, \tag{5.13}$$

where ρ_{r} is a constant inherent to material and is called **resistivity** or **specific resistance**. Its unit is [Ω m]. Table 5.1 lists values of the resistivity for various materials. The electric property of a material differs dramatically depending on the

Table 5.1 Resistivity of various materials at 20 °C

Metal	($\times 10^{-8}$ Ω m)	Semiconductor	(Ω m)
Silver (Ag)	1.62	Germanium (Ge)[a]	4.8×10^{-1}
Copper (Cu)	1.72	Silicon (Si)[a]	3.2×10^{3}
Gold (Au)	2.4	Insulator	(Ω m)
Aluminum (Al)	2.75	Epoxy resin	10^{11}–10^{14}
Brass (Cu–Zn)	5–7	Aluminum oxide	10^{12}–10^{13}
Iron (Fe)	9.8	Mica	10^{12}–10^{15}
Platinum (Pt)	10.6	Natural rubber	10^{13}–10^{15}
Constantan	50	Polyethylene	$> 10^{14}$
Mercury (Hg)	95.8	Paraffin	10^{14}–10^{17}
Nichrome	109	Quartz glass	$> 10^{15}$

[a]Values at 27 °C

resistivity: Materials with resistivity less than 10^{-6} Ω m that can easily transport a current are called conductors, and materials with resistivity above 10^{8} Ω m that can hardly transport a current are classified as insulators. Materials with intermediate resistivity are called semiconductors.

The relationship between the current and electric potential difference is also written as

$$I = GV. \tag{5.14}$$

In the above, the proportional constant $G=1/R_r$ is called **conductance**. Its unit is [S] (**siemens**). Using Eq. (5.13), the conductance is written as

$$G = \sigma_c \frac{S}{l}. \tag{5.15}$$

The constant $\sigma_c=1/\rho_r$ is called **electric conductivity**. Its unit is [S/m].

Suppose a small region in which the current flows under electric potential difference (see Fig. 5.2). The length of this region along the current is Δl, and the cross-sectional area normal to the current is ΔS. When the electric field strength is E and the current density is i, the potential difference and the current in this small region are $\Delta V = E\Delta l$ and $\Delta I = i\Delta S$, respectively. Hence, the electric resistance is written as

$$R_r = \frac{\Delta V}{\Delta I} = \frac{\Delta l}{\Delta S} \cdot \frac{E}{i}. \tag{5.16}$$

On the other hand, the electric resistance is defined as

$$R_r = \frac{\Delta l}{\sigma_c \Delta S}. \tag{5.17}$$

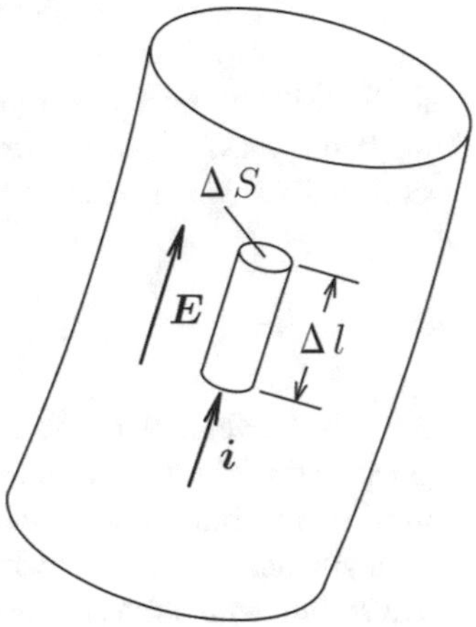

Fig. 5.2 Small region in which current flows under an electric potential difference

Hence, we have $i = \sigma_c E$. Since the current and electric field have the same direction, we can write this relationship in the vector form as

$$\boldsymbol{i} = \sigma_c \boldsymbol{E} \tag{5.18}$$

or

$$\boldsymbol{E} = \rho_r \boldsymbol{i}. \tag{5.19}$$

These are Ohm's law for electromagnetism.

The current is realized by a movement of electric charge, most of which is electrons. Therefore, current is a dynamic phenomenon and is not a static phenomenon, even if it is in a steady state. The reason why some electric field can remain in a conductor is that the phenomenon is in a dynamic state. Even in this case, we can express the electric field using the electric potential as in Eq. (1.24).

5.3 Microscopic Investigation of Electric Resistance

As mentioned in Sect. 5.2, Ohm's law is an empirical law for various kinds of materials, and it should be noted that this law cannot be explained by any physical principle. This is because the resistance is associated with energy dissipation, which cannot be derived theoretically. Hence, Ohm's law is a kind of phenomenological model. On the other hand, in a superconductor, which is introduced as one kind of magnetic material in Chap. 7, Ohm's law does not hold and the current obeys a physical principle. Hence, we can say that a superconductor is a pure material from the viewpoint of physics.

Here we investigate microscopically the occurrence of electric resistance, although it is not rigorously based on a physical principle. When a current flows steadily inside a metal, electrons that compose the current are driven by the electric field. If the mass of an electron is m, the equation of motion of the electron in electric field $\boldsymbol{E}$ is expected to be

$$m\frac{\mathrm{d}\boldsymbol{v}}{\mathrm{d}t} = -e\boldsymbol{E}, \tag{5.20}$$

where $\boldsymbol{v}$ is the velocity of the electron. However, this equation requires the electron to be accelerated by the electric field, which would bring about increasing current with time, resulting in a contradiction with the assumption of a steady state.

In a practical condition, every time an electron is accelerated by the electric field, it will collide with atoms in the metal and lose the energy given by the electric field, as illustrated in Fig. 5.3a. Thus, the velocity of the electron will have some mean value without increasing appreciably (see Fig. 5.3b). The effect of collision can be introduced by assuming a **viscous force** acting on electrons. Although this force cannot be derived from any physical principle, this phenomenological assumption is known to be useful in many examples.

Here we average the motion of an electron within a suitable time scale as in Fig. 5.3b in such a way that we can describe a gradual variation in the average velocity $\boldsymbol{v}$ on a much longer time scale. Then, the viscous force is defined. This force is directed opposite to the motion of the electron, and its magnitude is proportional to the velocity. Hence, when the electron moves with a higher velocity, it is subjected to a stronger force that reduces the velocity. This force is expressed as $-\eta\boldsymbol{v}$, and constant η is called the **coefficient of viscosity**. Hence, the equation of motion of the electron is now given by

$$m\frac{\mathrm{d}\boldsymbol{v}}{\mathrm{d}t} = -e\boldsymbol{E} - \eta\boldsymbol{v}. \tag{5.21}$$

We can easily solve this equation; the solution under the initial condition of $\boldsymbol{v} = 0$ at $t = 0$ is

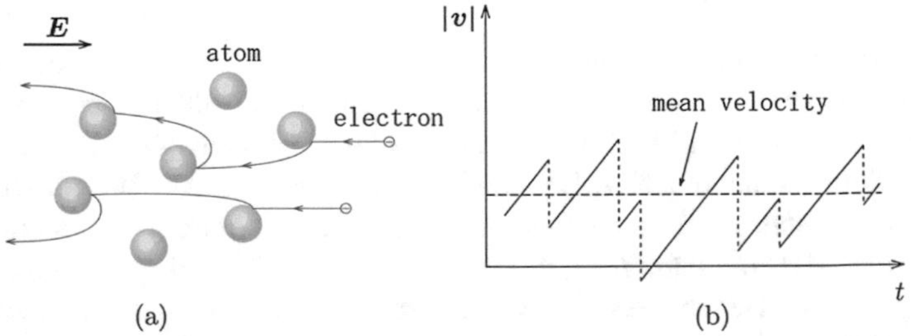

Fig. 5.3 **a** Motion of electrons in a metal and **b** variation in velocity with time

$$\boldsymbol{v} = -\frac{e\boldsymbol{E}}{\eta}\left[1 - \exp\left(-\frac{\eta t}{m}\right)\right]. \tag{5.22}$$

Figure 5.4 shows the variation in the velocity with time. After a sufficiently long time, we can neglect the exponential term and the velocity reaches a constant value,

$$\boldsymbol{v} = -\frac{e\boldsymbol{E}}{\eta}. \tag{5.23}$$

Hence, using Eq. (5.4), the current density in the steady state is given by

$$\boldsymbol{i} = -en_{\mathrm{e}}\boldsymbol{v} = \frac{n_{\mathrm{e}}e^2}{\eta}\boldsymbol{E}, \tag{5.24}$$

where n_{e} is the density of electrons. Thus, we obtain Ohm's law. In the above model, the electric conductivity is given by

$$\sigma_{\mathrm{c}} = \frac{n_{\mathrm{e}}e^2}{\eta}. \tag{5.25}$$

From Eq. (5.22), the time needed for the current to reach the steady-state value is approximately given by

$$t = t_0 = \frac{m}{\eta} = \frac{\sigma_{\mathrm{c}} m}{n_{\mathrm{e}}e^2}. \tag{5.26}$$

Substituting $m \simeq 0.9 \times 10^{-10}$ kg, $e \simeq 1.6 \times 10^{-19}$ C, $n_{\mathrm{e}} \simeq 1 \times 10^{29}$ m^{-3} for usual metals and $\sigma_{\mathrm{c}} \simeq 0.6 \times 10^8$ S/m for copper, we have $t_0 \simeq 2 \times 10^{-14}$ s. Since the time required for observation using measurement instruments is of the order of 10^{-10} s, the above variation in current cannot be observed. This means that Ohm's law is always observed.

When we apply a current to a material with electric resistance, energy dissipation takes place. Suppose that a current I flows in a material under an electric potential difference V given by a power source. Here we estimate the work done by the power

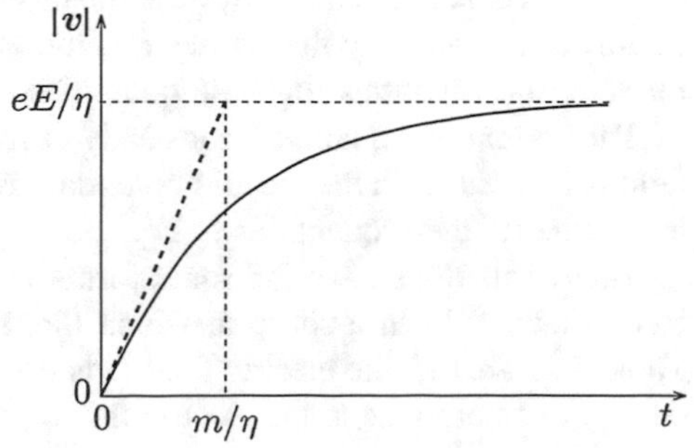

Fig. 5.4 Time variation in velocity of an electron derived from the equation of motion

source during a period Δt. The amount of electric charge that is transferred during this period is $\Delta Q = I\Delta t$. The work done on this electric charge is

$$\Delta W = V\Delta Q = VI\Delta t. \tag{5.27}$$

Hence, the work done in unit time is

$$P = \frac{\Delta W}{\Delta t} = VI. \tag{5.28}$$

This is called **electric power**. Its unit is [VA] and is denoted [W] (**watt**). Using Ohm's law, this is rewritten as

$$P = R_r I^2 = \frac{V^2}{R_r}. \tag{5.29}$$

In this case, the steady state of the current is maintained and the stored electric energy does not change. Hence, the work done by the power source is dissipated to heat.

Suppose a small region of length Δl and cross-sectional area ΔS. We assume that a current, $\Delta I = i\Delta S$, flows through the cross section under an electric potential difference, $\Delta V = E\Delta l$, along the length. The electric power dissipated in this region is

$$\Delta P = \Delta V \Delta I = Ei\Delta l\Delta S. \tag{5.30}$$

Hence, the dissipated electric power in a unit volume (i.e., the electric power density) is

$$\frac{\Delta P}{\Delta l \Delta S} = p = Ei = \sigma_c E^2 = \rho_r i^2. \tag{5.31}$$

5.4 Fundamental Equations for Steady Electric Current

We have discussed the equation describing the phenomena associated with a steady current. The fundamental physical quantity for the steady current is the current density $\boldsymbol{i}$. The quantity that causes it is the electric field, $\boldsymbol{E}$. Here we summarize the fundamental equations for these quantities.

The continuity equation for a steady current is given by Eq. (5.11). The electric field is derived from the electric potential, and hence, Eq. (1.28) holds. Ohm's law that connects these quantities is Eq. (5.18).

These equations have the same forms as fundamental equations describing the electrostatic field in a space in which there is no electric charge, as compared in Table 5.2. That is, the electric field $\boldsymbol{E}$ is common to the two cases, and the current density $\boldsymbol{i}$ corresponds to the electric flux density $\boldsymbol{D}$, and the electric conductivity σ_c

Table 5.2 Comparison of fundamental equations for steady current and electrostatic field in the absence of electric charge

Steady current		Electrostatic field in the absence of electric charge
$\nabla \cdot \boldsymbol{i} = 0$		$\nabla \cdot \boldsymbol{D} = 0$
$\boldsymbol{i} = \sigma_c \boldsymbol{E}$		$\boldsymbol{D} = \epsilon \boldsymbol{E}$
$\nabla \times \boldsymbol{E} = 0$		$\nabla \times \boldsymbol{E} = 0$
$\boldsymbol{i}$	$\longleftrightarrow$	$\boldsymbol{D}$
σ_c	$\longleftrightarrow$	ϵ

corresponds to the dielectric constant ϵ. However, it should be noted that this correspondence is mathematical, and similarity is sometimes broken in real cases as will be shown later.

Here we suppose that electric charges $\pm Q$ are given to the outer and inner electrodes of a concentric spherical capacitor with a dielectric material of dielectric constant ϵ in Fig. 5.5. We have learned the method with which to determine the electric potential difference between the electrodes. The electric flux density is directed outward from the inner electrode, and using Gauss' law, we can determine its magnitude as

$$D = \frac{Q}{4\pi r^2} \tag{5.32}$$

at position $r(a < r < b)$ from the center. Since the electric field is $E = D/\epsilon$, the electric potential difference is given by

$$V = \int_a^b \frac{Q}{4\pi\epsilon r^2} \mathrm{d}r = \frac{Q}{4\pi\epsilon}\left(\frac{1}{a} - \frac{1}{b}\right). \tag{5.33}$$

Thus, the capacitance is

$$C = \frac{Q}{V} = \frac{4\pi\epsilon ab}{b-a}. \tag{5.34}$$

Fig. 5.5 Concentric spherical capacitor

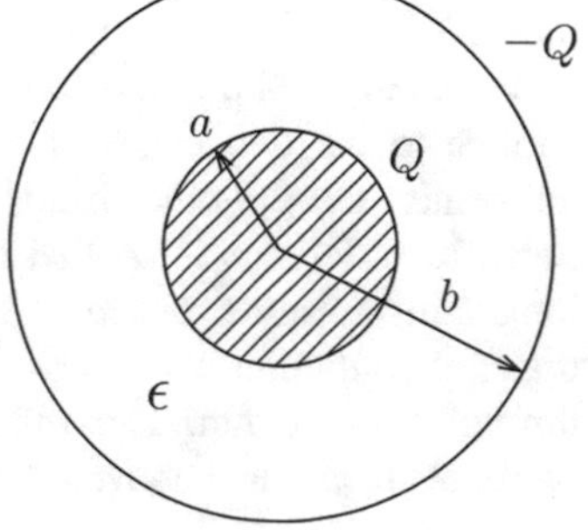

The dielectric material in the concentric spherical capacitor is replaced by a substance with electric conductivity σ_c. Now we determine the electric resistance between the two electrodes under the electric potential difference V. We denote the current by I. Since the current density does not have a divergence similarly to the electric flux density, Gauss' law gives the current density as

$$i = \frac{I}{4\pi r^2}, \tag{5.35}$$

corresponding to Eq. (5.32). Since the electric field is $E = i/\sigma_c$, the electric potential difference is given by

$$V = \int_a^b \frac{I}{4\pi\sigma_c r^2} \mathrm{d}r = \frac{I}{4\pi\sigma_c}\left(\frac{1}{a} - \frac{1}{b}\right). \tag{5.36}$$

The electric resistance is

$$R_r = \frac{V}{I} = \frac{b-a}{4\pi\sigma_c ab}. \tag{5.37}$$

Thus, the above two problems are formally identical. Eliminating V common to each example from Eqs. (5.34) and (5.37), we have

$$CR_r = \frac{\epsilon}{\sigma_c}. \tag{5.38}$$

This quantity—the product of capacitance and electric resistance—does not depend on the shape of capacitor or resistor and is given only by the dielectric constant and electric conductivity. This relationship of Eq. (5.38) generally holds for a capacitor and resistor having electrodes of the same shape. However, this is limited to the case in which we can obtain a rigorous solution for the field.

Similarity is rarely found between phenomena of steady current and an electrostatic field. This is explained by a quite large difference in the electric conductivity between a conductor and vacuum, while the difference in the dielectric constant between a dielectric material and vacuum is a factor of several tens at most.

For example, Fig. 5.6a shows the current when an electric potential difference is applied to the two edges of a long thin resistor. In this case, the current flows uniformly. On the other hand, Fig. 5.6b shows the electric flux when an electric potential difference is applied to the two electrodes of a capacitor composed of a long thin dielectric material. The electric flux lines spread out of the dielectric material, although they pass more easily through the dielectric material than through the vacuum. This difference arises from the large difference in material constants mentioned above. The reason why Eq. (5.38) holds in the case of Fig. 5.5

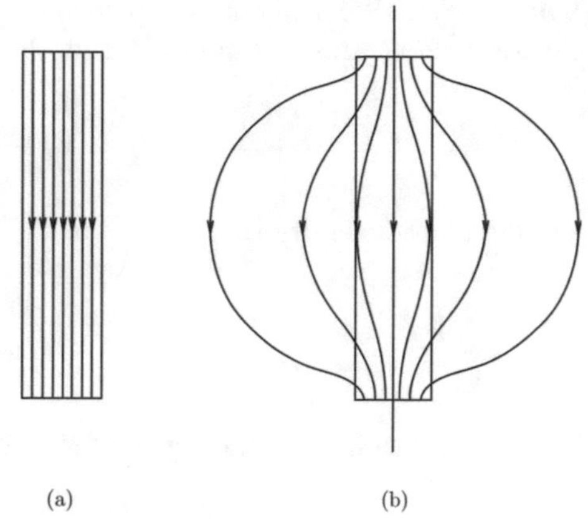

Fig. 5.6 **a** Lines of current in a long thin resistor and **b** electric flux lines in a capacitor with long thin dielectric material

is that the electric field has completely the same form under the rigorous symmetry in both cases.

Here we discuss the boundary condition to be satisfied for the steady current at an interface between substances with different electric resistivities. Since the equation for the current density $\boldsymbol{i}$ is formally the same as that for the electric flux density $\boldsymbol{D}$ in the absence of electric charge, the boundary condition is also the same. That is, from Eq. (4.19), we have

$$\boldsymbol{n} \cdot (\boldsymbol{i}_1 - \boldsymbol{i}_2) = 0, \tag{5.39}$$

where $\boldsymbol{n}$ is a unit vector normal to the interface. This shows that the normal component of the current density is continuous at the interface.

Example 5.1. Determine the electric resistance when current flows along a quarter of a circle of radius R_0 with a rectangular cross-section, as shown in Fig. 5.7a. The electric resistivity is ρ_r.

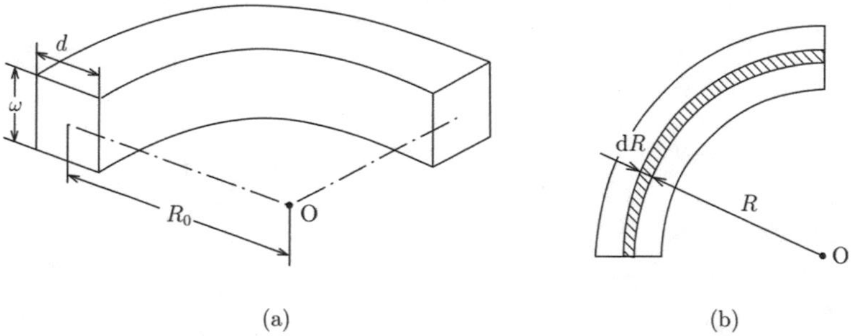

Fig. 5.7 **a** Shape of a quarter circular prism and **b** part of a thin region of radius R to R + dR

Solution 5.1. We apply electric potential difference V between the two edges. The electric field at an arc of radius R from the center in Fig. 5.7b is

$$E(R) = \frac{2V}{\pi R}$$

and the current density is

$$i(R) = \frac{2V}{\pi \rho_r R}.$$

The current that flows in the region of R to $R + \mathrm{d}R$ is $i(R)w\mathrm{d}R$ and the total current is

$$I = \int_{R_0-d/2}^{R_0+d/2} \frac{2wV}{\pi \rho_r R} \mathrm{d}R = \frac{2wV}{\pi \rho_r} \log \frac{R_0 + d/2}{R_0 - d/2}.$$

Thus, the electric resistance is given by

$$R_r = \frac{\pi \rho_r}{2w \log[(R_0 + d/2)/(R_0 - d/2)]}.$$

◇

Example 5.2. Determine the electric resistance along the length of the truncated cone in Fig. 5.8. The electric resistivity is ρ_r.

Fig. 5.8 Long truncated cone

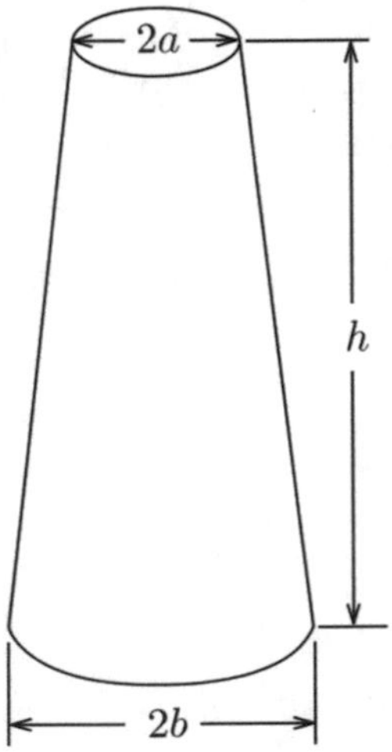

Solution 5.2. The cross-sectional area at position x from the bottom is

$$S(x) = \pi \left(b - \frac{b-a}{h} x \right)^2 .$$

If the applied current is I, the current density at this position is

$$i(x) = \frac{I}{S(x)} .$$

Since the electric field is $E(x) = \rho_r i(x)$, the electric potential difference between the two edges is

$$V = \int_0^h \rho_r \frac{I}{S(x)} \, dx = \frac{I\rho_r h}{\pi ab} .$$

The electric resistance is

$$R_r = \frac{\rho_r h}{\pi ab} .$$

◇

Example 5.3. The space between the electrodes in a concentric spherical resistor is occupied by two materials with resistivity ρ_{r1} and ρ_{r2}, as shown in Fig. 5.9. Determine the electric resistancc bctwcen the two electrodes.

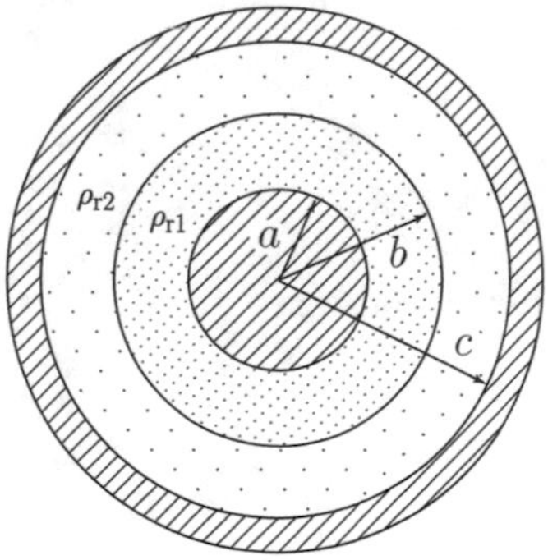

Fig. 5.9 Concentric spherical resistor with two materials

Solution 5.3. When current I is applied between the two electrodes, the current density at a position of radius r from the center is

$$i = \frac{I}{4\pi r^2} .$$

The electric field strength is $E = \rho_{r1} i$ in the region $a < r < b$ and $E = \rho_{r2} i$ in the region $b < r < c$. The electric potential difference between the two electrodes is

$$V = \frac{\rho_{r1} I}{4\pi} \int_a^b \frac{dr}{r^2} + \frac{\rho_{r2} I}{4\pi} \int_b^c \frac{dr}{r^2} = \frac{\rho_{r1}(b-a)I}{4\pi ab} + \frac{\rho_{r2}(c-b)I}{4\pi bc}.$$

Thus, the electric resistance is determined to be

$$R_r = \frac{V}{I} = \frac{\rho_{r1}(b-a)}{4\pi ab} + \frac{\rho_{r2}(c-b)}{4\pi bc}.$$

◇

Example 5.4. Assume that current I flows through the resistor shown in Fig. 5.8 when we apply electric potential difference V between its top and bottom. Prove that the dissipated total electric power is equal to VI.

Solution 5.4. The cross-sectional area of the resistor at a position x from the bottom is denoted by $S(x)$. The density of current flowing at this position is $i(x) = I/S(x)$ and the electric field strength is $E(x) = \rho_r I/S(x)$. Thus, the dissipated electric power density is

$$p(x) = E(x)i(x) = \frac{\rho_r I^2}{S(x)^2}.$$

The volume integral leads to the total dissipated power:

$$P = \int_0^h p(x)S(x)dx = I^2 \int_0^h \frac{\rho_r}{S(x)} dx.$$

In this equation

$$\int_0^h \frac{\rho_r}{S(x)} dx = R_r$$

is the electrical resistance. Thus, we have $P = R_r I^2 = VI$.

◇

The average of a product of two quantities is not generally equal to the product of the average of each quantity. This is the case in which these, P and VI, are equal to each other. Discuss the reason.

Example 5.5. Current of density $\boldsymbol{i}_0$ flows uniformly in a substance of electric conductivity σ_{c0}. When a spherical region of radius a in this substance is replaced by a different substance of electric conductivity σ_c, as shown in Fig. 5.10, determine the current density inside and outside the sphere.

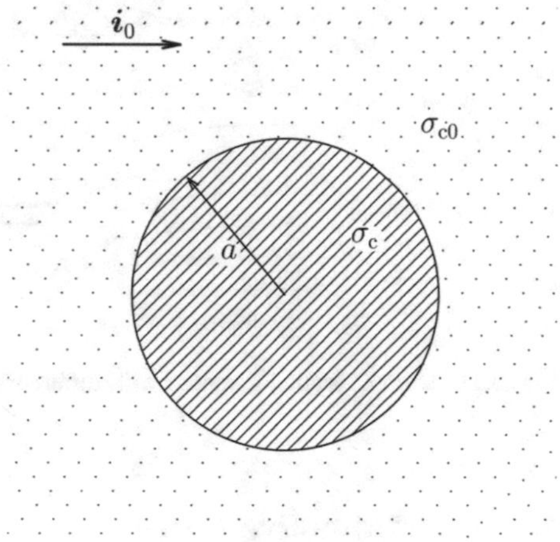

Fig. 5.10 Sphere of radius a with different electrical conductivity from the surrounding uniform substance

Solution 5.5. The equations describing the phenomena have the same form as those describing the electric flux density in the absence of an electric charge. Hence, the solution for the current density and that for the electric flux density are formally the same. That is, we can use the solution for $\boldsymbol{D}$ in Example 4.5 by replacing ϵ_0 and ϵ with σ_{c0} and σ_c. In this case, the uniform electric field E_0 corresponds to i_0/σ_{c0}. We define polar coordinates with the origin at the center of the sphere. We denote by θ the zenithal angle measured from the direction of the applied current. The current density outside the sphere $(r > a)$ is

$$i_r = \left(1 + \frac{\sigma_c - \sigma_{c0}}{\sigma_c + 2\sigma_{c0}} \cdot \frac{2a^3}{r^3}\right) i_0 \cos\theta, \quad i_\theta = -\left(1 - \frac{\sigma_c - \sigma_{c0}}{\sigma_c + 2\sigma_{c0}} \cdot \frac{a^3}{r^3}\right) i_0 \sin\theta$$

and that inside the sphere $(0 \leq r < a)$ is uniform:

$$i_r = \frac{3\sigma_c}{\sigma_c + 2\sigma_{c0}}\, i_0 \cos\theta, \quad i_\theta = -\frac{3\sigma_c}{\sigma_c + 2\sigma_{c0}}\, i_0 \sin\theta.$$

Figure 5.11 shows the current around the sphere.

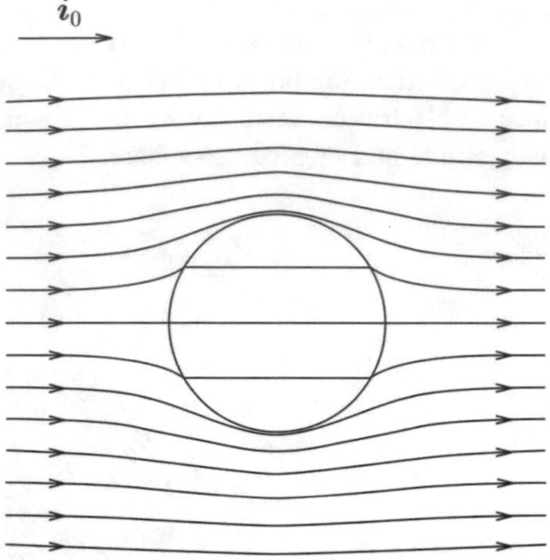

Fig. 5.11 Current around a sphere with different electric conductivity for $\sigma_c = \sigma_{c0}/3$

◇

5.5 Electromotive Force

Suppose that a steady current, $\boldsymbol{I}$, flows in a closed electric circuit, C. When integrating the current along the circuit, we have

$$\oint_C \boldsymbol{I} \cdot \mathrm{d}\boldsymbol{s} = lI, \tag{5.40}$$

where l is the perimeter of the circuit. On the other hand, with the aid of Ohm's law, the left side of this equation can be written as

$$S \oint_C \boldsymbol{i} \cdot \mathrm{d}\boldsymbol{s} = S\sigma_c \oint_C \boldsymbol{E} \cdot \mathrm{d}\boldsymbol{s} = 0, \tag{5.41}$$

where S and σ_c are the cross-sectional area and electric conductivity of the circuit, respectively. Thus, we have $I = 0$, which contradicts the assumption of a steady current. This gives $\boldsymbol{E} = 0$. On the other hand, under the initial condition of $\boldsymbol{v} = \boldsymbol{v}_0$ ($I = I_0 = -en_e|\boldsymbol{v}_0|S$), we obtain the solution to Eq. (5.21) as

$$I = I_0 \exp\left(-\frac{\eta t}{m}\right). \tag{5.42}$$

This also shows that I reduces to zero after a very short period. A steady current can continue to flow only in superconductors with $\rho_r = 0$.

Hence, to get a steady current in the general case, it is necessary to have an **electric power source** that applies an electric potential difference to force the current to flow in a circuit. The electric potential difference that the electric power

Table 5.3 Kinds of electric power source and electromotive force

Electric power source	Kind of electromotive force
Battery	Chemical electromotive force
Generator	Electromagnetic induction
Thermocouple	Thermoelectric power
Photoelectric cell	Photovoltaic effect

source generates is called **electromotive force**. Table 5.3 lists practical electric power sources and the kinds of electromotive force. The unit of the electromotive force is [V]. Except in the case of generator, the electric energy of the electromotive force provided by the sources is transformed from chemical, mechanical, thermal, or optical energy.

Suppose a closed circuit with an electric power source of electromotive force V. We denote the part of electric power source and the remaining part of the electric circuit as $\Delta\mathrm{C}$ and C' $(= \mathrm{C} - \Delta\mathrm{C})$, respectively, as shown in Fig. 5.12. The electric field due to the electromotive force is denoted by $\boldsymbol{E}_{\mathrm{em}}$. We then have

$$V = -\int_{\Delta\mathrm{C}} \boldsymbol{E}_{\mathrm{em}} \cdot \mathrm{d}\boldsymbol{s}. \tag{5.43}$$

In the above, the integral is directed along the current. We define the electric potential ϕ that also includes the electromotive force. This satisfies

$$-\nabla\phi = \rho_{\mathrm{r}}\boldsymbol{i} \tag{5.44}$$

in C' and

$$-\nabla\phi = \boldsymbol{E}_{\mathrm{em}} \tag{5.45}$$

in $\Delta\mathrm{C}$. The condition

$$\oint_{\mathrm{C}} \nabla\phi \cdot \mathrm{d}\boldsymbol{s} = 0, \tag{5.46}$$

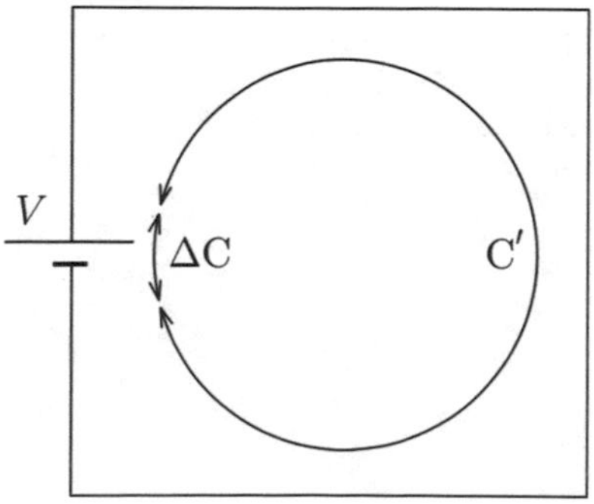

Fig. 5.12 Closed electric circuit with electric power source

which is required for the electric potential, gives

$$\frac{\rho_r}{S}\int_{C'} \boldsymbol{I}\cdot d\boldsymbol{s} = V. \tag{5.47}$$

Thus, the steady current is realized by the electric power source. The electric potential difference on the left side is due to the current that flows in an electric resistor and is called a **voltage drop**. The direction of the electric field is the same as that of the current but is opposite that of the electric field due to the electromotive force.

5.6 Kirchhoff's Law

In an electrical network composed only of resistors and direct current (DC) electric power sources, the current flows in a steady state. An important law that describes the steady current is **Kirchhoff's law**. We derive this law from the principles of electromagnetism.

Kirchhoff's first law states that the algebraic sum of currents passing out of an arbitrary node is zero. Here the currents that pass out and in are considered to be positive and negative, respectively. Applying this law to a node in Fig. 5.13a, we have

$$\sum_n I_n = 0, \tag{5.48}$$

where I_n is the current that passes out of the node through the n-th branch. Suppose a closed surface S that includes the node (see Fig. 5.13b). The surface integral of the current density $\boldsymbol{i}$ on S gives

$$\int_S \boldsymbol{i}\cdot d\boldsymbol{S} = \sum_n I_n. \tag{5.49}$$

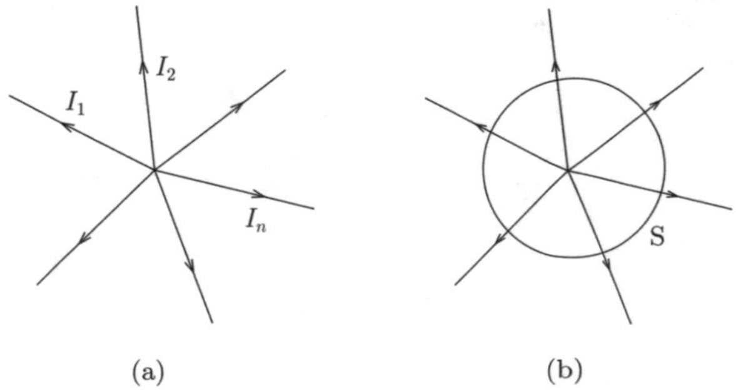

Fig. 5.13 **a** Currents that flow out of and into a node and **b** closed surface that includes the node

Using Gauss' theorem and Eq. (5.11), the left side of Eq. (5.49) is rewritten as

$$\int_{\mathrm{V}} \nabla \cdot \boldsymbol{i}\,\mathrm{d}V = 0, \tag{5.50}$$

where V is the interior of S. Thus, we obtain Kirchhoff's first law, Eq. (5.48).

Kirchhoff's second law states that the sum of electromotive forces is equal to the sum of voltage drops in resistors in an arbitrary closed circuit composed of branches in an electrical network. This is expressed as

$$\sum_m V_m = \sum_m R_m I_m \tag{5.51}$$

for closed circuit C in Fig. 5.14. In the above, V_m, R_m, and I_m are the electromotive force, electric resistance, and current in the m-th branch, respectively. We denote the electric potential including the electromotive force by ϕ. The potential difference between the two edges of the m-th branch is then given by

$$\Delta\phi_m = \int_m \nabla\phi \cdot \mathrm{d}\boldsymbol{s} = V_m - R_m I_m. \tag{5.52}$$

In the above, the relationship $\int_m (\rho_{\mathrm{r}}/S)\mathrm{d}s = R_m$ is used. From uniqueness of the electric potential, we have

$$\oint_{\mathrm{C}} \nabla\phi \cdot \mathrm{d}\boldsymbol{s} = \sum_m \Delta\phi_m = 0. \tag{5.53}$$

Substituting Eq. (5.52) into this equation gives Kirchhoff's second law, Eq. (5.51).

Fig. 5.14 Closed circuit in an electrical network

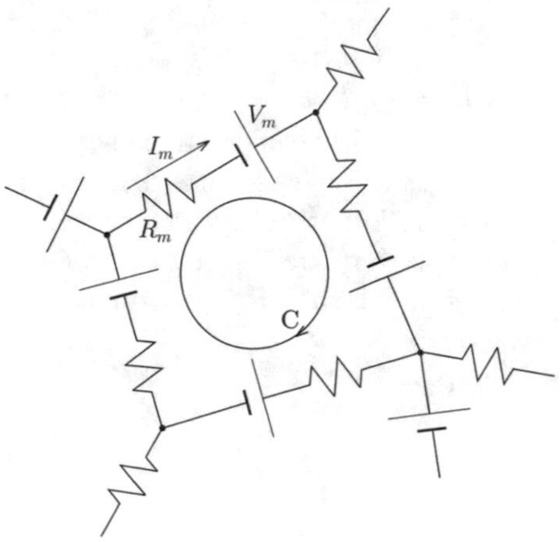

Column: Difference in Electric Resistivity

There are no material constants that differ so much depending on material as electric resistivity. Electric resistivity differs by a factor of 10^{23} between silver and quartz glass. In superconducting materials covered in Chap. 7, the electric resistivity is theoretically predicted to be zero. However, it is impossible to experimentally prove that the resistivity is absolutely zero.

For example, if the electric potential difference is measured to be less than the noise level of 10 nV when current of 1 A is applied to a superconducting wire of cross-sectional area 1 mm^2 and length 1 m, we can prove that the electric resistivity is less than 10^{-14} Ω m. This is close to the limit of sensitivity of measurement instruments. It should be noted that the electric resistivity is of the order of 10^{-11} Ω m even for high-purity copper at extremely low temperatures such as 4.2 K. Observation of the decay of current due to electric resistivity is much more sensitive. If the inductance and electric resistance of a closed circuit are L and R_r, the current decays as

$$I(t) = I(0)\exp\left(-\frac{t}{\tau}\right); \quad \tau = \frac{L}{R_r},$$

as known for an electrical circuit. Hence, we can estimate the electric resistance R_r from the time constant τ. For a closed circuit in which two parallel superconducting wires of length $l = 0.5$ m and diameter $a = 0.5$ mm separated by $d = 10$ mm are connected at both edges with zero resistance, we calculate the inductance as

$$L \simeq \frac{\mu_0 l}{\pi}\log\frac{d}{a} \simeq 6.0 \times 10^{-7}\,\text{H}$$

(see Example 8.1). In the above, μ_0 is the magnetic permeability of vacuum. If no decay is observed with measurement uncertainty less than 0.1% over 3 years (approximately 0.95×10^{8} s), τ is larger than 0.95×10^{11} s. Thus, R_r is less than 6×10^{-18} Ω and we can say that the electric resistivity is less than 6×10^{-24} Ω m.

Hence, the practical difference in electric resistivity reaches the level of 10^{38}. The difference in size between a hydrogen atom and the universe (about 15 billion light years) is of the order of 10^{36}, which gives an idea of the huge range of electric resistivity. Completely different electric properties of different materials are due to such a dramatic difference in the electric resistivity.

Exercises

5.1. Determine the electric resistance along the length of a quarter ring of radius R_0 with a circular cross section of radius a, as shown in Fig. E5.1. The electric resistivity is ρ_r.

5.2. Determine the electric resistance along the length of a substance with electric resistivity ρ_r, as shown in Fig. E5.2.

5.3. The space between the electrodes in a concentric spherical resistor is occupied by two materials with resistivity ρ_{r1} and ρ_{r2}, as shown in Fig. E5.3. Determine the electric resistance between the two electrodes.

5.4. The space between two long coaxial electrodes is occupied by two types of substance with electric resistivities ρ_{r1} and ρ_{r2}, as shown in Fig. E5.4. Determine the electric resistance between the two electrodes.

5.5 The space between two long coaxial electrodes is occupied by two types of substance with electric resistivities ρ_{r1} and ρ_{r2}, as shown in Fig. E5.5. Determine the electric resistance between the two electrodes.

5.6. Suppose that, when we apply electric potential difference V between the two edges of the quarter circular prism shown in Fig. 5.7a, current I flows. Prove that the total electric power dissipated in this resistor is VI. (Hint: Integrate the loss power density given by Eq. (5.31).)

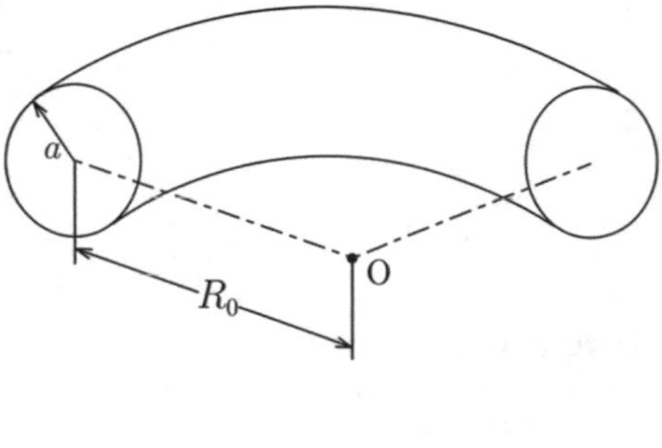

Fig. E5.1 Quarter ring with circular cross-section

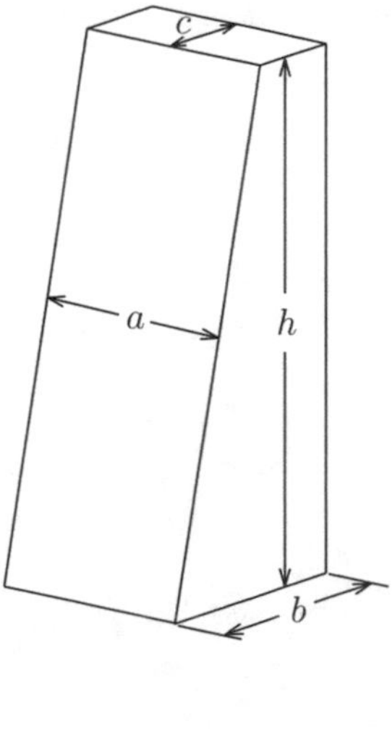

Fig. E5.2 Long substance with rectangular cross-section

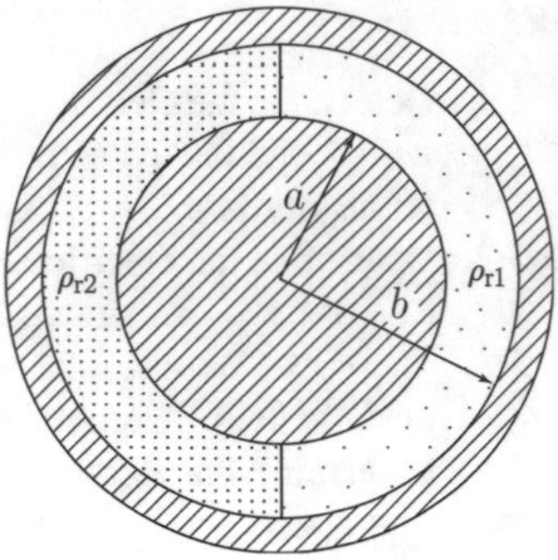

Fig. E5.3 Concentric spherical resistor with two materials

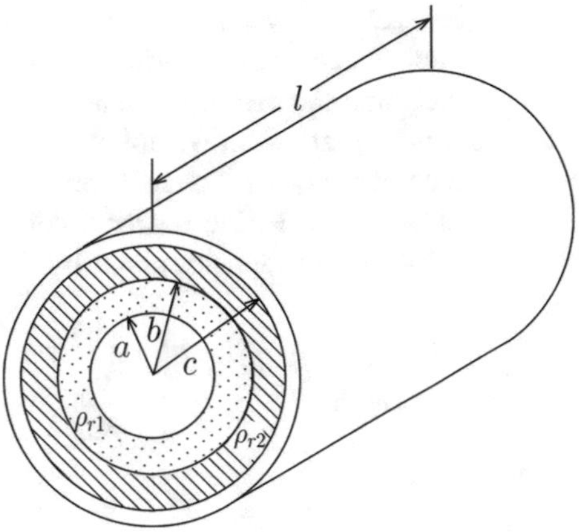

Fig. E5.4 Long coaxial resistor composed of two types of substance with different electric resistivities

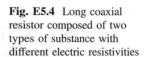

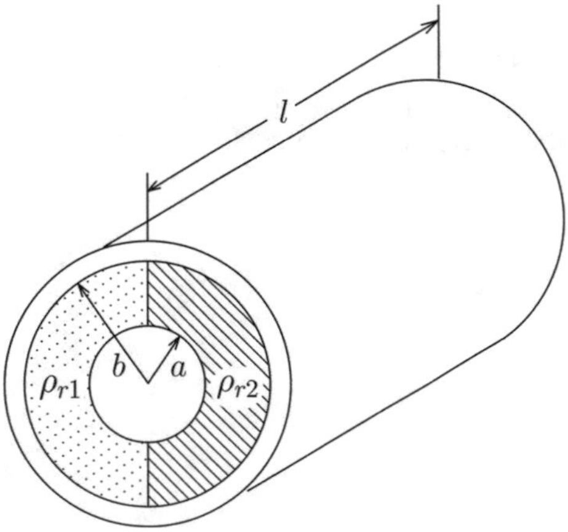

Fig. E5.5 Long coaxial resistor composed of two types of substance with different electric resistivities

5.7. Prove that Eq. (5.38) holds also for two parallel-plate electrodes with area S and distance d.

5.8. A long cylindrical rod of radius a and electric conductivity σ_c is embedded in a substance of electric conductivity σ_{c0}. Determine the current density inside and outside the rod, when we apply a uniform current of density $\boldsymbol{i}_0$ to an infinitely wide region, as shown in Fig. E5.6.

5.9. Two thin tapes A and B of length L are adhered to each other, and current I is applied at the edge of tape A between $x = 0$ and $x = 2L$, as shown in Fig. E5.7a. The electric resistances in unit lengths of tapes A and B are R_A' and R_B', respectively, and the interface conductance between the two tapes in a unit length is g'. Determine the current that flows in tape A. (Hint: Use the distributed constant circuit model shown in Fig. E5.7b.)

5.10. Two parallel long cylindrical conductors of radius a are embedded in a uniform substance of electric conductivity σ_c, as shown in Fig. E5.8. The distance between the central axes of the two conductors is $d(> 2a)$, and the electric resistivity of the two conductors is negligibly small. Determine the

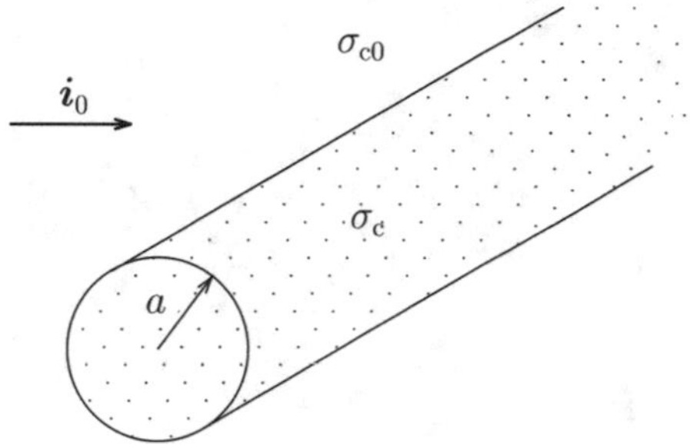

Fig. E5.6 Current applied normal to a long cylindrical rod with a different electric conductivity embedded in a uniform substance

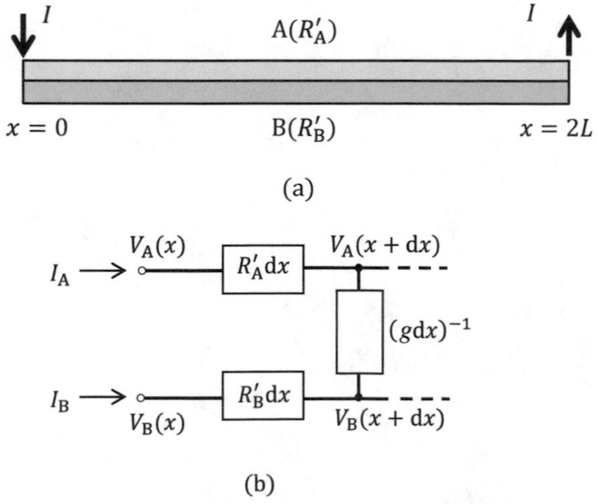

Fig. E5.7 **a** Two thin tapes and **b** equivalent distributed constant circuit

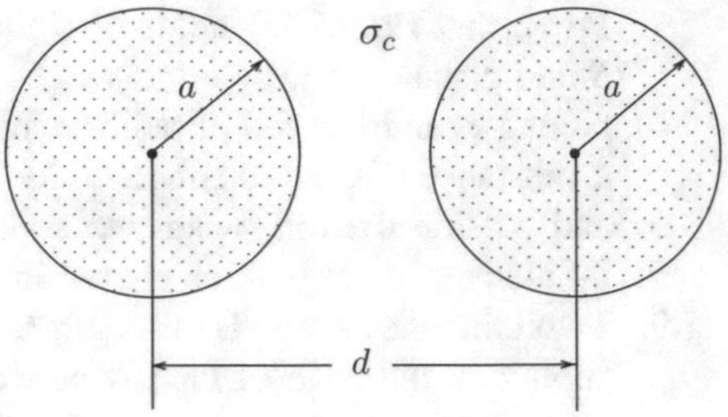

Fig. E5.8 Cross section of two parallel cylindrical conductors embedded in a uniform substance

electric resistance in a unit length between the two conductors. (Hint: Convert this problem to the problem of determining the capacitance between two conductors with replacement of a substance having electric conductivity σ_c with a dielectric material having dielectric constant ϵ. Place virtual positive and negative line charges having densities $\pm\lambda$ at suitable positions so that the electric potential has constant values on the surfaces of the two conductors).

Part II
Static Magnetic Phenomena

Chapter 6
Current and Magnetic Flux Density

Abstract This chapter covers magnetic phenomena caused by currents. The Lorentz force, which works between currents, is caused by a magnetic distortion in space that is produced by the currents. This distortion is the magnetic field and is represented by the magnetic flux density. The magnetic flux density produced by current is described by the Biot-Savart law. On the other hand, Ampere's law describes the global relationship between the current and the magnetic flux density. The magnetic flux density is expressed by the curl of the vector potential and has a nature of a field with no divergence. The virtual surface on which the vector potential is the same is the equivector potential surface. The relationship between the equivector potential surface and the magnetic flux density is discussed. We also learn the magnetic properties of the magnetic moment produced by a small closed current, which appears in magnetic materials.

6.1 Magnetic Flux Density by Current

We know that there exists a force between currents. This phenomenon is similar to the Coulomb force between electric charges. Hence, we can presume that currents also make some field in space similar to the electric field made by electric charges. This field is called the **magnetic field**.

Magnets also make a magnetic field. However, the magnetic field source that we can quantitatively define is current. In fact, the amount of current is defined based on the force between currents. In the case of magnets, the magnetic field strength depends on the substance of the magnet and its condition of magnetization, which cannot be controlled exactly. We discuss the relationship between the current and magnetic field in this chapter. It should be noted, however, that we do not define the magnetic field itself but the **magnetic flux density** to express the magnetic field strength. The relationship between the magnetic field and magnetic flux density will be described in Chap. 9.

T. Matsushita, *Electricity and Magnetism*, Undergraduate Lecture Notes in Physics,
https://doi.org/10.1007/978-3-030-82150-0_6

We suppose that two parallel straight wires separated by distance d carry currents I_1 and I_2, as shown in Fig. 6.1a. In this case, a force of strength

$$F' = -\frac{\mu_0 I_1 I_2}{2\pi d} \tag{6.1}$$

works on each wire of a unit length. It is attractive ($F' < 0$) for currents in the same direction ($I_1 I_2 > 0$) and repulsive ($F' > 0$) for currents in opposite directions ($I_1 I_2 < 0$). The constant

$$\mu_0 = 4\pi \times 10^{-7}\ \mathrm{N/A^2} \tag{6.2}$$

is called the **magnetic permeability of vacuum**. The unit of current, [A], is defined using Eq. (6.1). The magnitude of this force corresponds to the magnitude of the Coulomb force on electric charges. That is, when electric charges of linear densities λ_1 and λ_2 are uniformly distributed on two parallel straight lines separated by d, as shown in Fig. 6.1b, the Coulomb force on each line of a unit length is given by

$$F' = \frac{\lambda_1 \lambda_2}{2\pi\epsilon_0 d}, \tag{6.3}$$

which is of the same form as the magnetic force of Eq. (6.1). Namely, the magnetic force is proportional to the product of two currents and is inversely proportional to their distance. The unique difference is that the magnetic force is attractive for currents in the same direction.

The magnetic flux produced by currents is largely different from the electric field produced by electric charges. The magnetic flux can be visualized using magnetic particles such as iron sand. For example, the magnetic flux lies in a plane normal to a straight current and forms vortices around it, as shown in Fig. 6.2. On the other hand, the electric field radiates from a line charge.

Below are the main differences between the electric interaction between electric charges and the magnetic interaction between currents:

- Electric charges are scalars, and currents are vectors.

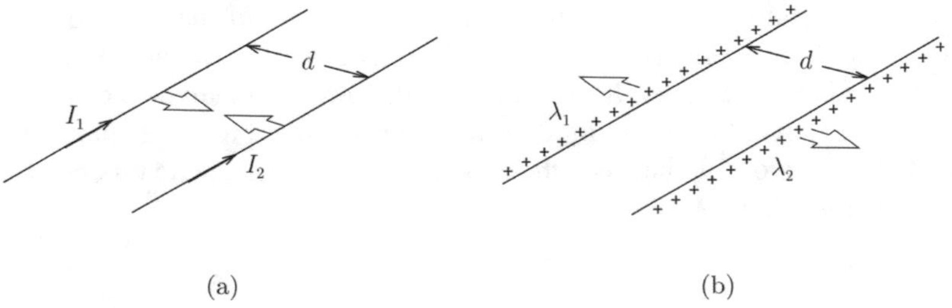

Fig. 6.1 **a** Force between two parallel currents and **b** force between two parallel line charges

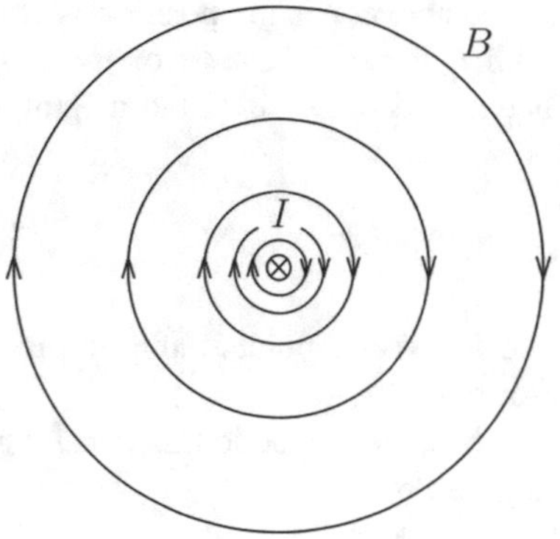

Fig. 6.2 Magnetic flux produced by straight current

- While the electric field is directed parallel to the straight line connecting the electric charge and observation point, the magnetic flux is directed normally to both the current and the straight line connecting a part of the current and the observation point.
- While the electric force between electric charges of the same kind is repulsive, the magnetic force between currents in the same direction is attractive.

6.2 The Biot-Savart Law

The law that expresses the magnetic flux density produced by a current is the **Biot-Savart law**. Suppose that current I flows along line C, as shown in Fig. 6.3. The law states that the magnetic flux density at point P produced by an elementary current $I\mathrm{d}\boldsymbol{s}$ flowing in a small segment $\mathrm{d}\boldsymbol{s}$ is given by

$$\mathrm{d}\boldsymbol{B} = \frac{\mu_0}{4\pi} \cdot \frac{I\mathrm{d}\boldsymbol{s} \times \boldsymbol{i}_r}{r^2} = \frac{\mu_0}{4\pi} \cdot \frac{I\mathrm{d}\boldsymbol{s} \times \boldsymbol{r}}{r^3}. \tag{6.4}$$

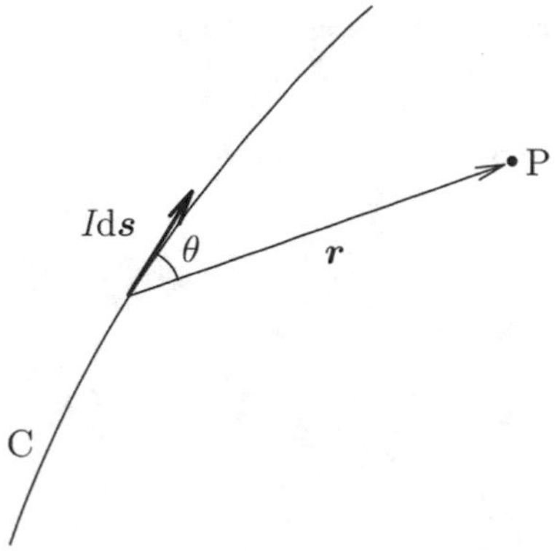

Fig. 6.3 Elementary current along C and observation point P

In the above, $\boldsymbol{r}$ is the position vector from the small segment to point P and $|\boldsymbol{r}| = r$ with $\boldsymbol{i}_r = \boldsymbol{r}/r$. The unit of the magnetic flux density is [T] (**tesla**). If the angle between $\mathrm{d}\boldsymbol{s}$ and $\boldsymbol{r}$ is θ, the magnitude of magnetic flux density is

$$\mathrm{d}B = \frac{\mu_0 I \mathrm{d}s}{4\pi r^2} \sin\theta \tag{6.5}$$

and the vector points along the motion of a screw when a screw driver is rotated from $\mathrm{d}\boldsymbol{s}$ to $\boldsymbol{r}$.

Thus, the magnetic flux density produced at $\boldsymbol{r}$ by current I flowing through line C is given by

$$\boldsymbol{B}(\boldsymbol{r}) = \frac{\mu_0}{4\pi} \int_{\mathrm{C}} \frac{I\mathrm{d}\boldsymbol{r}' \times (\boldsymbol{r} - \boldsymbol{r}')}{|\boldsymbol{r} - \boldsymbol{r}'|^3}. \tag{6.6}$$

When there are many currents, the total magnetic flux density is the superposition of all the individual magnetic flux densities they produce.

Here, we consider a current flowing with density $\boldsymbol{i}$ in space V, as shown in Fig. 6.4. The elementary current that flows in a small region of length $\mathrm{d}s$ and cross-sectional area $\mathrm{d}S$ is $\boldsymbol{i}\mathrm{d}S\mathrm{d}s = \boldsymbol{i}\mathrm{d}V$ with $\mathrm{d}V$ denoting the volume of this region. Thus, the magnetic flux density is given by

$$\boldsymbol{B}(\boldsymbol{r}) = \frac{\mu_0}{4\pi} \int_{\mathrm{V}} \frac{\boldsymbol{i}(\boldsymbol{r}') \times (\boldsymbol{r} - \boldsymbol{r}')}{|\boldsymbol{r} - \boldsymbol{r}'|^3} \mathrm{d}V'. \tag{6.7}$$

One can see that this equation corresponds to Eq. (1.12) for the electric field produced by electric charges. The electric charge density ρ corresponds to the current density $\boldsymbol{i}$, and the vector product of current density and position is necessary for yielding a vector for the magnetic flux density. This explains why the magnetic flux density is perpendicular to the current.

Fig. 6.4 Elementary current in space V

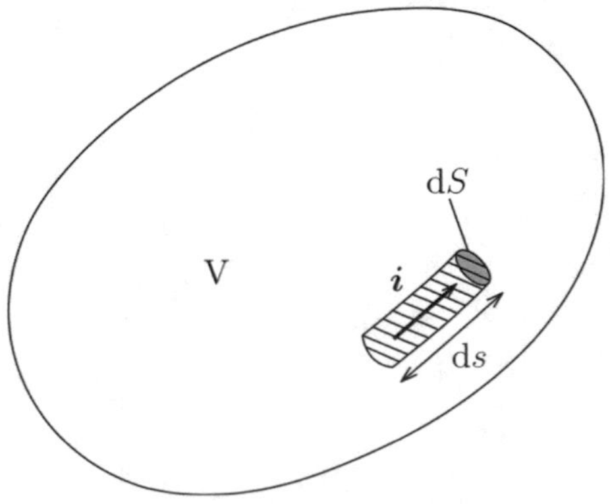

Example 6.1. Current I flows in a circle of radius a, as shown in Fig. 6.5a. Determine the magnetic flux density at point P located at distance z in the normal direction from the center O of the circle.

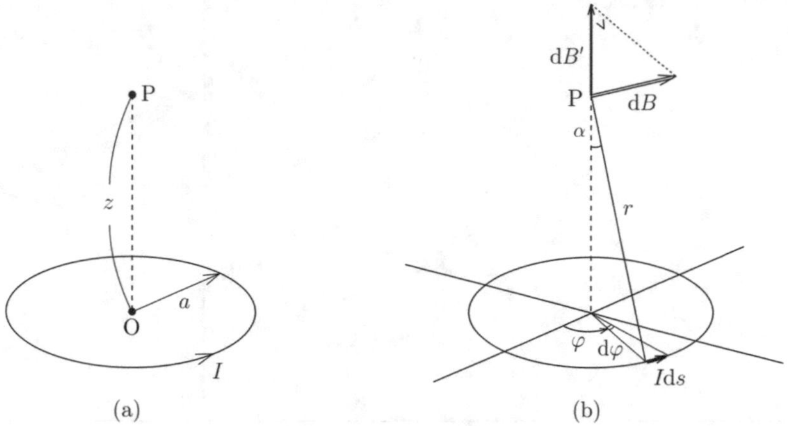

Fig. 6.5 **a** Point P on axis of circular current and **b** magnetic flux density at P produced by elementary current

Solution 6.1. We define angle φ as shown in Fig. 6.5b. The elementary current in the part φ to $\varphi + \mathrm{d}\varphi$ has magnitude $I\mathrm{d}s = Ia\mathrm{d}\varphi$ and is directed normally to the position vector from this segment to point P [$\theta = \pi/2$ in Eq. (6.5)], as shown in the figure. The magnetic flux density at P produced by this elementary current is

$$\mathrm{d}B = \frac{\mu_0 Ia\mathrm{d}\varphi}{4\pi r^2}$$

with $r = (z^2 + a^2)^{1/2}$. From symmetry, only the component along the z-axis remains:

$$\mathrm{d}B' = \frac{\mu_0 Ia\mathrm{d}\varphi}{4\pi r^2}\sin\alpha = \frac{\mu_0 Ia^2\mathrm{d}\varphi}{4\pi r^3}.$$

This will be understood by considering a contribution from the opposite side of the circle at angle $\varphi + \pi$. Integrating with respect to angle φ, we have

$$B = \int_0^{2\pi} \frac{\mu_0 Ia^2\mathrm{d}\varphi}{4\pi r^3} = \frac{\mu_0 Ia^2}{2(z^2 + a^2)^{3/2}}.$$

◇

Example 6.2. Current I flows along a long straight line (see Fig. 6.6). Determine the magnetic flux density at point P at distance a from the line.

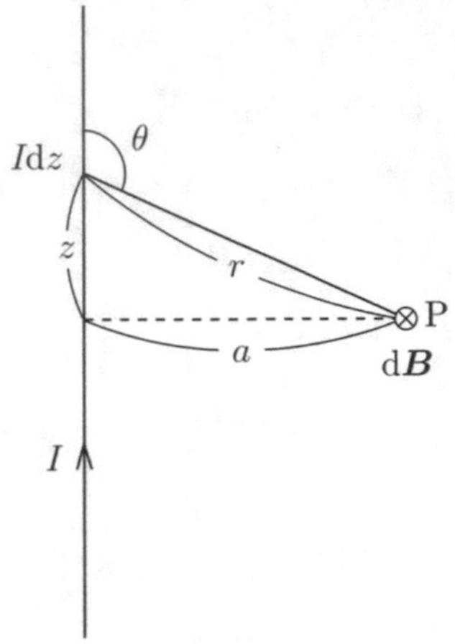

Fig. 6.6 Long straight line with current and observation point P

Solution 6.2. We define coordinates as shown in Fig. 6.6. The magnetic flux density produced at P by an elementary current $I\mathrm{d}z$ in a small region z to $z + \mathrm{d}z$ is

$$\mathrm{d}B = \frac{\mu_0 I \mathrm{d}z}{4\pi r^2}\sin\theta,$$

where angle θ is defined as in the figure and $r = (z^2 + a^2)^{1/2} = a/\sin\theta$. The relationship $z = -a\cot\theta$ gives $\mathrm{d}z = (a/\sin^2\theta)\mathrm{d}\theta$. Thus, the elementary magnetic flux density is transformed to be

$$\mathrm{d}B = \frac{\mu_0 I}{4\pi a}\sin\theta\,\mathrm{d}\theta.$$

Since this vector is directed normally to this sheet, a simple superposition holds for summing the contribution from each small region. We obtain the magnetic flux density as

$$B = \int_0^{\pi} \frac{\mu_0 I}{4\pi a}\sin\theta\,\mathrm{d}\theta = \frac{\mu_0 I}{2\pi a}. \tag{6.8}$$

◇

Example 6.3. Current I is applied to the closed circuit shown in Fig. 6.7. AD is an arc of a circle of radius a. Determine the magnetic flux density at O, located at the intersection of lines extended from BA and CD.

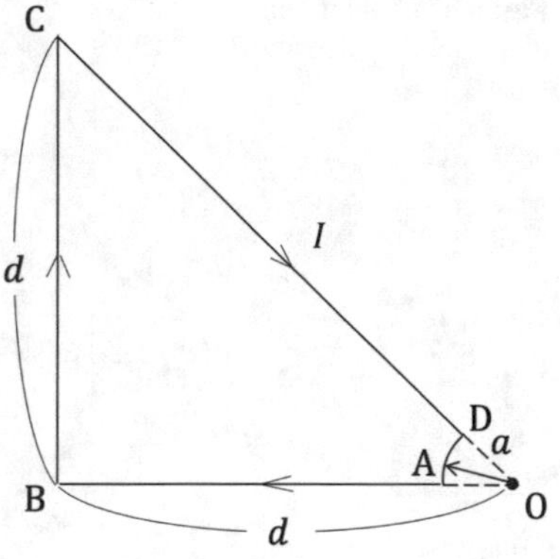

Fig. 6.7 Current on a closed circuit composed of three straight sections and an arc of a circle

Solution 6.3. The angle θ measured from the direction of elementary current $I\mathrm{d}s$ to that of the observation point O is π on the section from A to B, and the magnetic flux density produced by the current flowing in this branch is zero. On the section from B to C, if the distance of $I\mathrm{d}s$ from point B is denoted by y, the distance between $I\mathrm{d}s$ and point O is $r = \left(y^2 + d^2\right)^{1/2} = d/\sin\theta$. Thus, the magnetic flux density produced by the current in this branch is

$$B_{\mathrm{BC}} = \frac{\mu_0 I}{4\pi}\int_0^d \frac{\sin\theta}{y^2+d^2}\mathrm{d}y = \frac{\mu_0 I}{4\pi d}\int_{\pi/2}^{3\pi/4} \sin\theta \mathrm{d}\theta = \frac{\mu_0 I}{4\sqrt{2}\pi d},$$

and is normal to the sheet and is directed backward. On the section from C to D, the angle θ is zero, and there is no contribution from this branch. On the section from D to A, $r = a$ and $\theta = -\pi/2$, so we have

$$B_{\mathrm{DA}} = -\frac{\mu_0 I}{4\pi a^2}\int_{\mathrm{D}}^{\mathrm{A}} \mathrm{d}s = -\frac{\mu_0 I}{16a}.$$

This is normal to the sheet and is directed forward. As a result, the magnetic flux density is directed forward and is

$$B = \frac{\mu_0 I}{4}\left(\frac{1}{4a} - \frac{1}{\sqrt{2}\pi d}\right).$$

◇

6.3 Force on Current

The force on an elementary part $\mathrm{d}s$ of a current I in a magnetic flux density $\boldsymbol{B}$ is given by

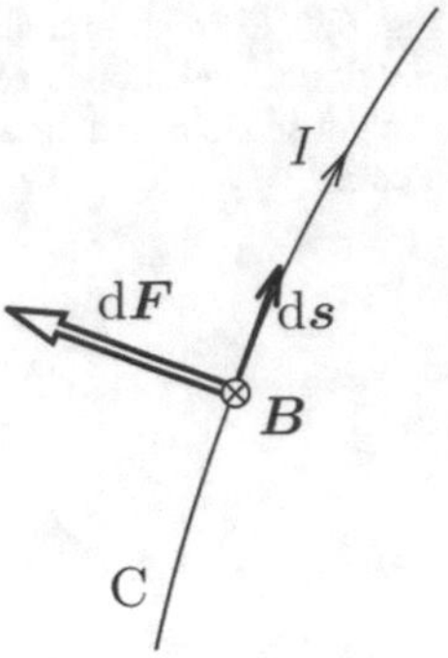

Fig. 6.8 Current I in magnetic flux density $\boldsymbol{B}$

$$\mathrm{d}\boldsymbol{F} = I\mathrm{d}\boldsymbol{s} \times \boldsymbol{B} \tag{6.9}$$

(see Fig. 6.8). This is called the **Lorentz force** in a narrow sense. From the mathematical requirement that the force vector results from the product of two vectors, the vector product appears again. The force that line C with current I experiences in the magnetic flux density $\boldsymbol{B}$ is

$$\boldsymbol{F} = I \int_{\mathrm{C}} \mathrm{d}\boldsymbol{s} \times \boldsymbol{B}. \tag{6.10}$$

We apply this to the case of a force on two parallel currents in Fig. 6.1a. The magnetic flux density that current I_1 produces at the position of current I_2 is $B = \mu_0 I_1/(2\pi d)$ using the result of Example 6.2. It is directed as shown in Fig. 6.9a. Hence, if I_2 is in the same direction as I_1, the force on I_2 is attractive. If I_2 is directed opposite to I_1 as in Fig. 6.9b, the force is repulsive. This results in Eq. (6.1).

Since the current is composed of moving electric charges, the force on the current is written as a force on moving electric charges. The force on the region of length ds along the direction of the current I and cross-sectional area dS in the magnetic flux density $\boldsymbol{B}$ is

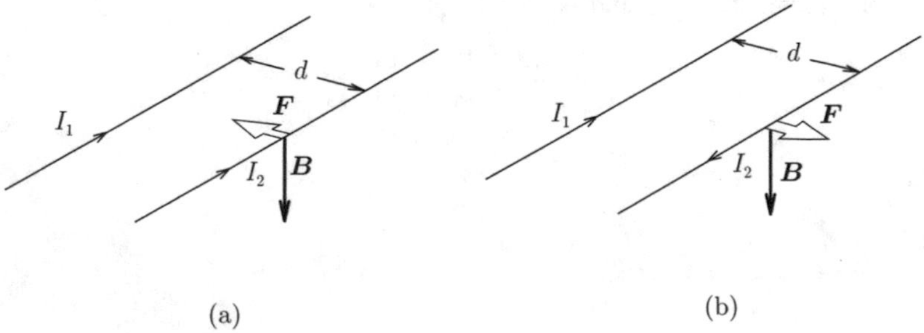

Fig. 6.9 Magnetic flux density produced by current I_1 and resultant force on current I_2 when I_2 has **a** the same and **b** opposite directions to I_1

$$\mathrm{d}\boldsymbol{F} = I\mathrm{d}\boldsymbol{s} \times \boldsymbol{B} = \boldsymbol{i}\mathrm{d}S\mathrm{d}\boldsymbol{s} \times \boldsymbol{B}. \tag{6.11}$$

Since $\mathrm{d}S\mathrm{d}s$ is the volume of this region, $\mathrm{d}V$, we rewrite this force as

$$\mathrm{d}\boldsymbol{F} = (q\boldsymbol{v} \times \boldsymbol{B})n\mathrm{d}V \tag{6.12}$$

using Eq. (5.4). Since $n\mathrm{d}V$ is the number of electric charges in this volume, the force on one electric charge is given by

$$\boldsymbol{F} = q\boldsymbol{v} \times \boldsymbol{B}. \tag{6.13}$$

For the special case where the electric field $\boldsymbol{E}$ and the magnetic flux density $\boldsymbol{B}$ coexist, the force on the electric charge is

$$\boldsymbol{F} = q(\boldsymbol{E} + \boldsymbol{v} \times \boldsymbol{B}). \tag{6.14}$$

This is called the **Lorentz force** in a broad sense. This equation shows that $\boldsymbol{E}$ and $\boldsymbol{B}$ are important variables that connect electromagnetism with dynamics. The ***E-B* analogy** is the standpoint from which electromagnetism is described using these variables. Chapter 9 will include a description based on another standpoint.

We can easily show that the total force on a closed circuit, C, along which current I flows in a uniform magnetic flux density is zero:

$$\boldsymbol{F} = I\left(\oint_{\mathrm{C}} \mathrm{d}\boldsymbol{s}\right) \times \boldsymbol{B} = 0. \tag{6.15}$$

However, it should be noted that the torque, i.e., the moment of force on a closed circuit is not necessarily zero.

For example, we suppose rectangular circuit PQRS in a uniform magnetic flux density, $\boldsymbol{B}$, as shown in Fig. 6.10a. We assume sides PS and QR are normal to $\boldsymbol{B}$. When current I flows in this circuit, the force $aIB\cos\theta$ works on side PQ, where θ is the angle between the unit vector $\boldsymbol{n}$ normal to the circuit and the magnetic flux density $\boldsymbol{B}$. The strength of this force is the same as that on side RS, and these forces are on the same line and directed opposite to each other (see Fig. 6.10b). Hence, these forces completely cancel and do not contribute to the torque. On the other hand, forces of strength bIB work on sides PS and QR in opposite directions to each other, but these forces do not lie on the same line. Hence, a torque appears and rotates the circuit (see Fig. 6.10c). Its magnitude is

$$N = bIBa\sin\theta = BIS\sin\theta, \tag{6.16}$$

where $S = ab$ is the area of the circuit.

The unit vector $\boldsymbol{n}$ specifies the direction of movement of a right thumb when it is rotated along the direction of the current, and the surface vector of the closed circuit is defined by $\boldsymbol{S} = \boldsymbol{n}S$. Then, the torque is expressed in the form of a vector:

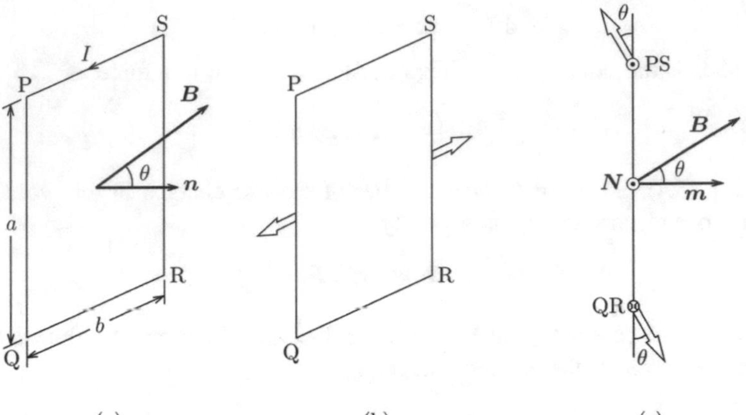

Fig. 6.10 **a** Rectangular circuit with current I in uniform magnetic flux density $\boldsymbol{B}$, **b** two forces that cancel each other, and **c** two forces that cause torque

$\boldsymbol{N} = I\boldsymbol{S} \times \boldsymbol{B}$, where $\boldsymbol{N}$ is a vector with the same magnitude as the torque and is directed along the motion of a screw when a screw driver is rotated along the torque. If we define the **magnetic moment** of the closed circuit as

$$\boldsymbol{m} = I\boldsymbol{S}, \tag{6.17}$$

the torque is given by

$$\boldsymbol{N} = \boldsymbol{m} \times \boldsymbol{B}. \tag{6.18}$$

Equation (6.18) holds for closed current I with an arbitrary shape on a plane in a uniform magnetic flux density. We can explain this as follows. Any given current-carrying closed circuit is expressed as a superposition of small rectangular closed circuits with the same current I, as shown in Fig. 6.11, since two currents in opposite

Fig. 6.11 Division of closed current of arbitrary shape on a plane

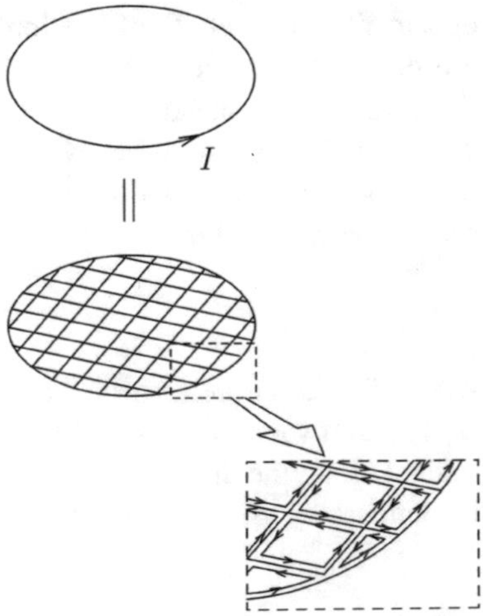

directions in adjacent closed circuits completely cancel out. The above discussion holds for each small rectangular circuit. The magnitude of the total magnetic moment is equal to the product of current I and the total area of the closed circuit. Thus, we can prove that Eqs. (6.17) and (6.18) hold for any closed circuit on a plane.

Example 6.4. Electric charge q of mass m is ejected with velocity v along the x-axis in magnetic flux density B along the z-axis. Discuss the motion of the electric charge after ejection.

Solution 6.4. The Lorentz force given by Eq. (6.13) works on the electric charge along the y-axis just after it is ejected. Since the force is directed perpendicularly to the motion, a circular motion of the electric charge occurs (see Fig. 6.12). We denote the radius of this circular motion as R. The magnitude of the Lorentz force on the electric charge is qvB, and the centrifugal force is mv^2/R. From the condition of balance between these forces, we obtain the radius of the circular motion as

$$R = \frac{mv}{qB}.$$

This circular motion of a charge is called **cyclotron motion**. The angular frequency of the motion, which is called **cyclotron angular frequency**, is given by

$$\omega = \frac{v}{R} = \frac{qB}{m}.$$

Since the Lorentz force is always perpendicular to the direction of motion, the work done by the Lorentz force is zero.

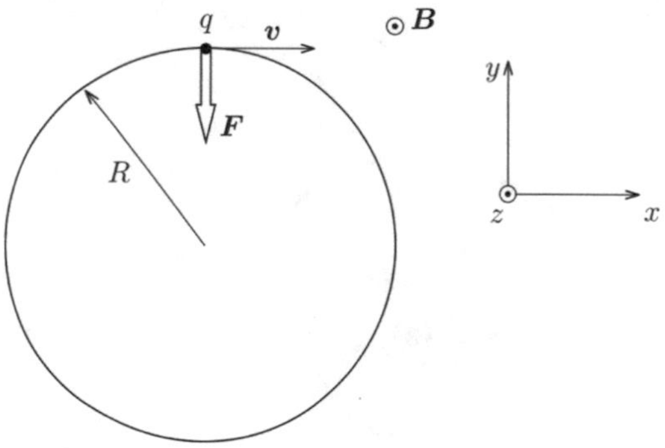

Fig. 6.12 Motion of electric charge in magnetic flux density

◇

6.4 Magnetic Flux Lines

The electric field produced by electric charges can be visualized by electric field lines. We can similarly see the magnetic flux density with **magnetic flux lines**. The magnetic flux line is defined as follows: The direction of a tangential line at any point on the magnetic flux line is the same as the direction of $\boldsymbol{B}$, and its line density is defined as equal to the magnitude of $\boldsymbol{B}$. Figure 6.13 shows examples of magnetic flux lines. We define the **magnetic flux** that passes through arbitrary surface S as

$$\Phi = \int_{\mathrm{S}} \boldsymbol{B} \cdot \mathrm{d}\boldsymbol{S}. \tag{6.19}$$

The unit of magnetic flux is [Tm^2], which is newly defined as [Wb] (**weber**).

From the examples in Fig. 6.13, it seems that magnetic flux lines are closed lines. This is different from electric field lines that start from positive electric charges and terminate at negative electric charges. If this speculation is valid,

$$\int_{\mathrm{S}} \boldsymbol{B} \cdot \mathrm{d}\boldsymbol{S} = 0 \tag{6.20}$$

holds for an arbitrary closed surface, S. Using Gauss' theorem on the left side, Eq. (6.20) gives

$$\int_{\mathrm{V}} \nabla \cdot \boldsymbol{B} \, \mathrm{d}V = 0,$$

where V is the interior of S. Since this holds for arbitrary V, we have

$$\nabla \cdot \boldsymbol{B} = 0. \tag{6.21}$$

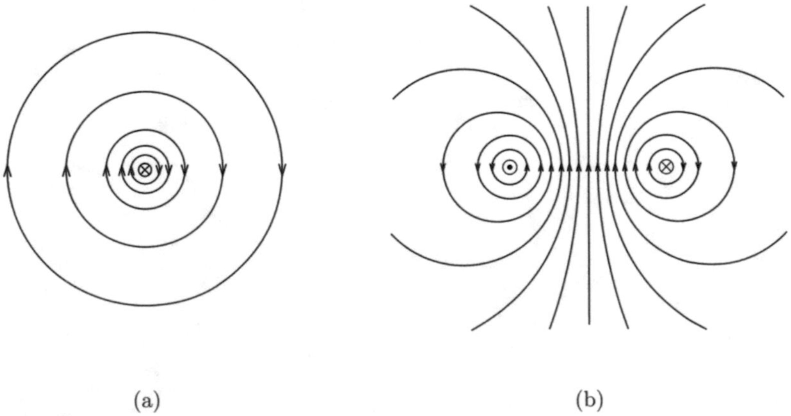

Fig. 6.13 Examples of magnetic flux lines for **a** a straight current and **b** a circular current

In fact, we can prove Eq. (6.21) as described in Sect. A2.2 in the Appendix. That is, the magnetic flux lines are closed lines. Equations (6.20) and (6.21) are called **Gauss' law for magnetic flux** and **Gauss' divergence law for magnetic flux**, respectively.

6.5 Ampere's Law

Consider the circular integral of magnetic flux density $\boldsymbol{B}$ along a closed line, C:

$$\oint_{\mathrm{C}} \boldsymbol{B} \cdot \mathrm{d}\boldsymbol{s}.$$

When current I flows in a straight line and C is a circle of radius R from the current in a plane normal to the current, $\boldsymbol{B}$ is parallel to the line element $\mathrm{d}\boldsymbol{s}$ (see Fig. 6.14), and its magnitude B is a constant, $\mu_0 I/(2\pi R)$, as discussed in Example 6.2. Hence, the above integral gives

$$\oint_{\mathrm{C}} \boldsymbol{B} \cdot \mathrm{d}\mathbf{s} = \frac{\mu_0 I}{2\pi R} \oint_{\mathrm{C}} \mathrm{d}s = \mu_0 I \tag{6.22}$$

and the value is independent of R.

Suppose a closed line, C, on a plane normal to the straight current I that does not encircle the current. For simplicity, we assume that C is composed of two arcs with different radii and two straight segments extending from the current, as shown in Fig. 6.15. Here, we calculate the circular integral of $\boldsymbol{B}$ on C. We denote the angle of arcs KL and MN as α. The integrals along these arcs are

$$\int_{\mathrm{K}}^{\mathrm{L}} \boldsymbol{B} \cdot \mathrm{d}\boldsymbol{s} = \int_{\mathrm{N}}^{\mathrm{M}} \boldsymbol{B} \cdot \mathrm{d}\boldsymbol{s} = \frac{\mu_0 I \alpha}{2\pi}.$$

Fig. 6.14 Circle C around straight current

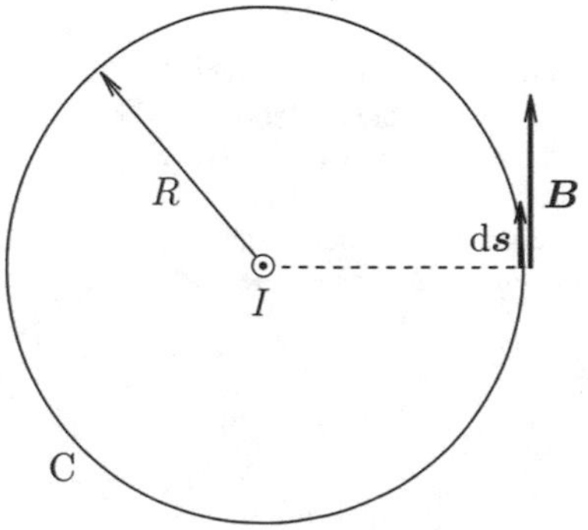

Fig. 6.15 Closed line KLMN on a plane normal to straight current that is composed of two arcs and two straight segments

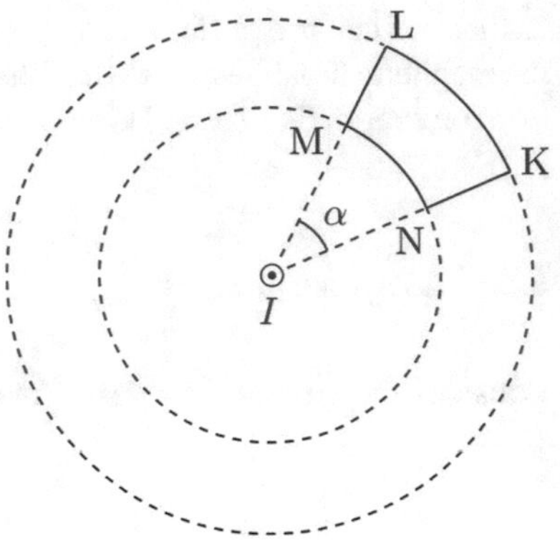

On the two straight segments NK and LM, $\boldsymbol{B}$ is perpendicular to d$\boldsymbol{s}$, and we have

$$\int_{N}^{K} \boldsymbol{B} \cdot \mathrm{d}\boldsymbol{s} = \int_{L}^{M} \boldsymbol{B} \cdot \mathrm{d}\boldsymbol{s} = 0.$$

Thus, the circular integral gives

$$\oint_{C} \boldsymbol{B} \cdot \mathrm{d}\boldsymbol{s} = 0. \tag{6.23}$$

Using the above results, we can show that Eq. (6.22) holds when straight current I penetrates arbitrary closed line C on the normal plane, and Eq. (6.23) holds when this is not so. In the latter case, for example, we can realize an arbitrary closed line with a set of small closed loops composed of two arcs with the center on the current and two straight segments extending from the current, as shown in Fig. 6.16, and Eq. (6.23) holds for each closed loop. Hence, we can show that Eq. (6.23) holds for any closed line.

Fig. 6.16 Closed line divided into a set of closed loops composed of two arcs and two straight segments when straight current does not penetrate the closed line

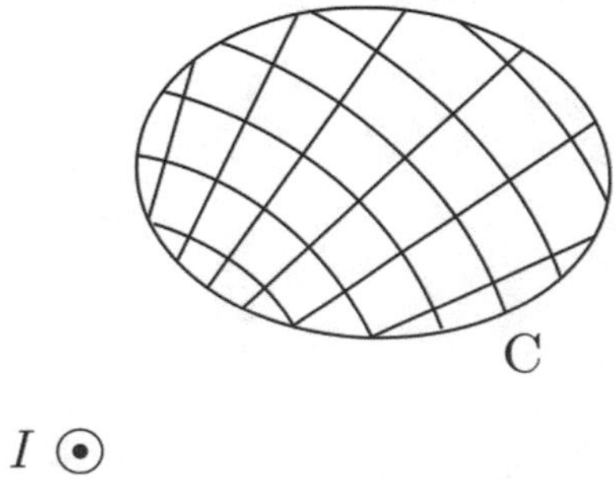

If I is defined as the current that penetrates closed line C in Eq. (6.22), it includes Eq. (6.23). This equation holds also when the closed line is not on a plane normal to the straight current and/or when the current is not straight. Since it is time-consuming to prove each of them, this will be proved indirectly using another method.

We suppose that currents I_1, I_2, $\cdots$ flow separately in space. In this case, Eq. (6.22) is extended to

$$\oint_{\mathrm{C}} \boldsymbol{B} \cdot \mathrm{d}\boldsymbol{s} = \mu_0 \sum_{m=1}^{n} I_m. \tag{6.24}$$

In the above, the right side is the sum of the currents that penetrate C. When current flows with the density $\boldsymbol{i}$, the corresponding equation is

$$\oint_{\mathrm{C}} \boldsymbol{B} \cdot \mathrm{d}\boldsymbol{s} = \mu_0 \int_{\mathrm{S}} \boldsymbol{i} \cdot \mathrm{d}\boldsymbol{S}, \tag{6.25}$$

where S is the surface surrounded by C. Equations (6.22), (6.24), and (6.25) are called **Ampere's law**.

Applying Stokes' theorem to the left side of Eq. (6.25) gives

$$\int_{\mathrm{S}} \nabla \times \boldsymbol{B} \cdot \mathrm{d}\boldsymbol{S} = \mu_0 \int_{\mathrm{S}} \boldsymbol{i} \cdot \mathrm{d}\boldsymbol{S}. \tag{6.26}$$

Since this holds for arbitrary S, we have

$$\nabla \times \boldsymbol{B} = \mu_0 \boldsymbol{i}. \tag{6.27}$$

This is called the **differential form of Ampere's law**. Hence, if Eq. (6.27) holds, then Eq. (6.22) must hold. The proof of Eq. (6.27) is given in Sect. A2.3 in the Appendix.

From Eq. (A1.44) in the Appendix, we can see that Eq. (6.27) satisfies Eq. (5.11). That is, the magnetic flux density $\boldsymbol{B}$ is produced by a steady current. Equation (6.27) shows that the current produces rotation of the magnetic flux density. This is in contrast with Eq. (1.21) that shows that an electric charge produces divergence of the electric field.

Example 6.5. Current I flows uniformly in a long cylinder of radius a. Determine the magnetic flux density inside and outside the cylinder.

Solution 6.5. We can determine the magnetic flux density using the Biot-Savart law. However, it is not easy. On the other hand, this can be easily done using Ampere's law. We apply this law to a circle, C, of radius R from the central axis of the cylinder, as shown in Fig. 6.17. From symmetry, the magnetic flux density $\boldsymbol{B}$ is

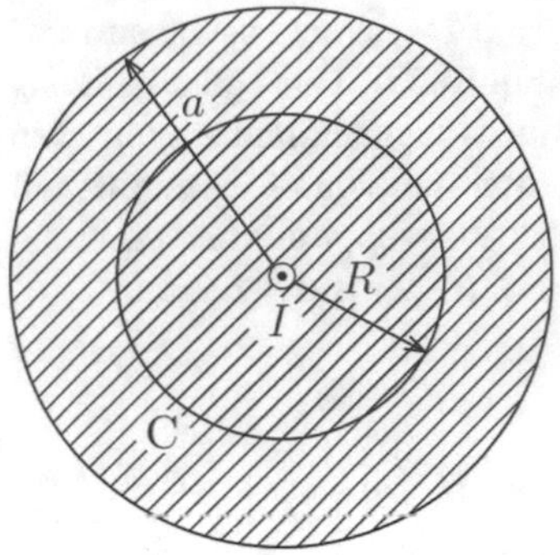

Fig. 6.17 Circle C with the same central axis as cylinder for $R<a$

parallel to a line element, d$\boldsymbol{s}$, and its magnitude B is constant on C. Hence, the left side of Eq. (6.25) gives

$$\oint_{\mathrm{C}} \boldsymbol{B} \cdot \mathrm{d}\boldsymbol{s} = 2\pi RB.$$

For $R > a$, the total current that flows inside C is I, and the right side of Eq. (6.25) is equal to $\mu_0 I$. Thus, we have

$$B = \frac{\mu_0 I}{2\pi R}.$$

This result is the same as for the case where the total current flows along the central axis. For $R < a$, the current inside C is $(R/a)^2 I$, and we have

$$B = \frac{\mu_0 IR}{2\pi a^2}.$$

◇

Example 6.6. Current flows uniformly with surface density τ on a wide plane. Determine the magnetic flux density at a position at distance h from the plane.

Solution 6.6. We can easily obtain the answer using Ampere's law also for this case. Suppose a rectangle, KLMN, normal to the direction of current with two sides (KL and MN) of length w parallel to the plane, as shown in Fig. 6.18. The other two sides (LM and NK) of length $2h$ are normal to the plane. We apply Ampere's law to this rectangle. From symmetry, the magnetic flux density has the same value on sides KL and MN at the same distance from the plane, and its vectors are parallel to these sides but opposite to each other. Hence, the contribution from these sides to the circular integral of the magnetic flux density gives $2wB$. On the other hand, the contribution from other two sides is zero. As a result, we have

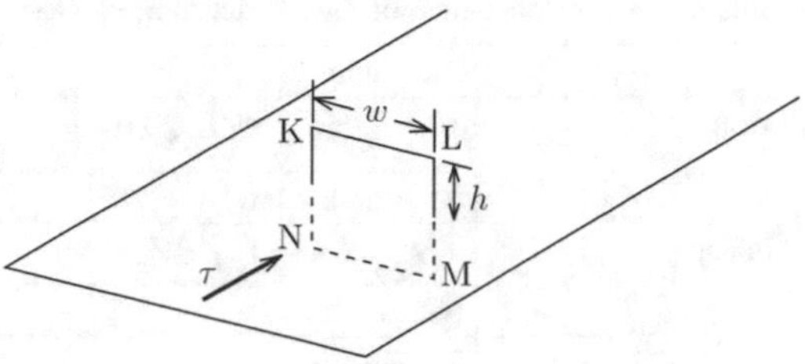

Fig. 6.18 Rectangle normal to the direction of current flowing uniformly on a plane

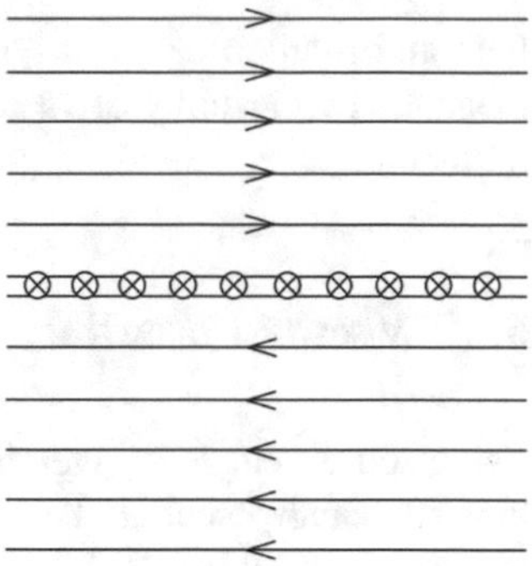

Fig. 6.19 Magnetic flux lines produced by uniform sheet current

$$\oint_C \boldsymbol{B} \cdot \mathrm{d}\boldsymbol{s} = 2wB.$$

The total current inside the rectangle is $w\tau$. Thus, we obtain the magnetic flux density as

$$B = \frac{\mu_0 \tau}{2}. \tag{6.28}$$

It should be noted that this value is independent of the distance h from the plane. Figure 6.19 shows the magnetic flux lines produced by the sheet current. This is similar to the electric field produced by an electric charge distributed uniformly on a wide plane (see Example 1.5).

◇

As shown above, we can say that the Biot-Savart law for the local magnetic flux density produced by a current and Ampere's law for the global relationship between the magnetic flux density and current describes the same magnetic phenomenon from opposite viewpoints. This is similar to the relation between Coulomb's law and Gauss' law describing electric phenomena in Chap. 1. Table 6.1 shows the correspondence between electric and magnetic phenomena.

Although there is a difference between the scalar source (electric charge) and vector source (current) that produce the fields, the correspondence between electric and magnetic phenomena is clear. These sources cause a divergence of the electric

Table 6.1 Correspondence of laws describing electric and magnetic phenomena

	Electricity	Magnetism
Local	$\boldsymbol{E}(\boldsymbol{r}) = \frac{1}{4\pi\epsilon_0}\int_{\mathrm{V}} \frac{\rho(\boldsymbol{r}')(\boldsymbol{r}-\boldsymbol{r}')}{\|\boldsymbol{r}-\boldsymbol{r}'\|^3}\mathrm{d}V'$	$\boldsymbol{B}(\boldsymbol{r}) = \frac{\mu_0}{4\pi}\int_{\mathrm{V}} \frac{\boldsymbol{i}(\boldsymbol{r}')\times(\boldsymbol{r}-\boldsymbol{r}')}{\|\boldsymbol{r}-\boldsymbol{r}'\|^3}\mathrm{d}V'$
	(Coulomb's law)	(The Biot-Savart law)
Global	$\int_{\mathrm{S}} \boldsymbol{E}\cdot \mathrm{d}\boldsymbol{S} = \frac{1}{\epsilon_0}\int_{\mathrm{V}} \rho\, \mathrm{d}V$	$\oint_{\mathrm{C}} \boldsymbol{B}\cdot \mathrm{d}\boldsymbol{s} = \mu_0 \int_{\mathrm{S}} \boldsymbol{i}\cdot \mathrm{d}\boldsymbol{S}$
	(Gauss' law)	(Ampere's law)
Differential	$\nabla \cdot \boldsymbol{E} = \frac{\rho}{\epsilon_0}$	$\nabla \times \boldsymbol{B} = \mu_0 \boldsymbol{i}$

field and rotation of the magnetic flux density. As for the physical constants, the magnetic permeability of vacuum, μ_0, corresponds to the inverse of the permittivity of vacuum, ϵ_0^{-1}.

6.6 Vector Potential

The electric field $\boldsymbol{E}$ is given by Eq. (1.24) in terms of the electric potential, i.e., a kind of scalar potential. This originates from the irrotational nature of the electric field as given by Eq. (1.28) and from the mathematical property that the gradient of a scalar is irrotational.

Magnetic flux density is a solenoidal field with no divergence, as shown by Eq. (6.21). Mathematically, curl of a vector has no divergence, as shown by Eq. (A1.44) in the Appendix. Hence, the magnetic flux density can be mathematically expressed as a curl of some vector:

$$\boldsymbol{B} = \nabla \times \boldsymbol{A}. \tag{6.29}$$

The quantity $\boldsymbol{A}$ is called the **vector potential**. As can be seen from Eq. (A1.45), we could add the gradient of any scalar function to the vector potential, and the vector potential would still correspond to the same magnetic flux density. This arbitrary gradient of some scalar function means the vector potential cannot be uniquely determined without specifying some condition. In the case of static magnetic phenomenon that does not change with time, the condition

$$\nabla \cdot \boldsymbol{A} = 0 \tag{6.30}$$

is usually used. This is called the **Coulomb gauge**. We assume a new vector potential,

$$\boldsymbol{A}' = \boldsymbol{A} + \nabla\alpha, \tag{6.31}$$

under this condition. It yields the same magnetic flux density. Using Eq. (6.30), α satisfies Laplace's equation.

$$\nabla \cdot \nabla \alpha = \Delta \alpha = 0. \tag{6.32}$$

Since α is uniquely determined under a given boundary condition as discussed in Sect. 2.2, the vector potential is uniquely determined using the Coulomb gauge.

The solution of the vector potential is given by

$$\boldsymbol{A}(\boldsymbol{r}) = \frac{\mu_0}{4\pi} \int_{\mathrm{V}} \frac{\boldsymbol{i}(\boldsymbol{r}')}{|\boldsymbol{r} - \boldsymbol{r}'|} \mathrm{d}V'. \tag{6.33}$$

One can show that this proves the Biot-Savart law for the magnetic flux density $\boldsymbol{B}$ (see Sect. A2.4 in the Appendix). In this case, the Coulomb gauge is fulfilled (see Exercise 6.9). This equation corresponds to Eq. (1.27) describing the electric potential produced by electric charge. These are compared in Table 6.2. The result that the potential is scalar or vector depends on whether the source of the field is scalar or vector. The similarity in Table 6.1 is found again in this table. When current I flows along a line circuit, C, Eq. (6.33) reduces to

$$\boldsymbol{A}(\boldsymbol{r}) = \frac{\mu_0 I}{4\pi} \int_{\mathrm{C}} \frac{\mathrm{d}\boldsymbol{r}'}{|\boldsymbol{r} - \boldsymbol{r}'|}. \tag{6.34}$$

Integrating the vector potential along C gives

$$\oint_{\mathrm{C}} \boldsymbol{A} \cdot \mathrm{d}\boldsymbol{s} = \int_{\mathrm{S}} \nabla \times \boldsymbol{A} \cdot \mathrm{d}\boldsymbol{S} = \int_{\mathrm{S}} \boldsymbol{B} \cdot \mathrm{d}\boldsymbol{s} = \Phi, \tag{6.35}$$

where S is a surface surrounded by C, and Φ is a magnetic flux that penetrates C. In the above, we have used Stokes' theorem and Eqs. (6.19) and (6.29).

Substituting Eq. (6.29) into Eq. (6.27) gives

$$\nabla \times (\nabla \times \boldsymbol{A}) = \mu_0 \boldsymbol{i}. \tag{6.36}$$

With Eqs. (A1.46) in the Appendix and (6.30), Eq. (6.36) becomes

$$\Delta \boldsymbol{A} = -\mu_0 \boldsymbol{i}. \tag{6.37}$$

Table 6.2 Comparison between electric potential due to electric charge and vector potential due to current

	Electric potential by charge	Vector potential by current
Potential	$\phi(\boldsymbol{r}) = \frac{1}{4\pi\epsilon_0} \int_{\mathrm{V}} \frac{\rho(\boldsymbol{r}')}{\lvert\boldsymbol{r}-\boldsymbol{r}'\rvert} \mathrm{d}V'$	$\boldsymbol{A}(\boldsymbol{r}) = \frac{\mu_0}{4\pi} \int_{\mathrm{V}} \frac{\boldsymbol{i}(\boldsymbol{r}')}{\lvert\boldsymbol{r}-\boldsymbol{r}'\rvert} \mathrm{d}V'$
Source	ρ	$\boldsymbol{i}$
Constant	ϵ_0	μ_0^{-1}
Equation	$\Delta\phi = -\frac{\rho}{\epsilon_0}$	$\Delta \boldsymbol{A} = -\mu_0 \boldsymbol{i}$

That is, each component of the vector potential satisfies Poisson's equation. In the region where current does not flow ($\boldsymbol{i} = 0$), this reduces to Laplace's equation,

$$\Delta \boldsymbol{A} = 0. \tag{6.38}$$

This is also similar to the electric potential in electric phenomena. When a boundary condition is given, $\boldsymbol{A}$ in Eqs. (6.37) or (6.38) is uniquely determined. One can directly prove that Eq. (6.36) holds for the vector potential $\boldsymbol{A}$ given by Eq. (6.33). This will be apparent from the proof of Eq. (6.33) in Sect. A2.4 and that of Eq. (6.27) in Sect. A2.3 in the Appendix.

Example 6.7. Determine the vector potential for the case discussed in Example 6.5.

Solution 6.7. We use cylindrical coordinates. Since the current flows only along the z-axis, the vector potential has only the z-component A_z, as indicated by Eq. (6.33). In addition, from symmetry, it does not depend on z or the azimuthal angle φ. The magnetic flux density has only the azimuthal component B_φ. Thus, we have

$$B_\varphi = -\frac{\partial A_z}{\partial R}.$$

The vector potential is given by

$$A_z = -\int_{R_0}^{R} B_\varphi \mathrm{d}R,$$

where $R_0(> a)$ is the distance from the central axis to the reference point at which $A_z = 0$. The reason why infinity is not chosen as the reference point is that the vector potential diverges because of the requirement that the current flows over an infinitely long distance. In fact, we find that the vector potential directly estimated from Eq. (6.34) diverges. This corresponds to the divergence of the electric potential for an infinitely long line charge (see Example 1.8).

We determine the vector potential as

$$A_z = \frac{\mu_0 I}{2\pi} \log \frac{R_0}{R}$$

from $B_\varphi = \mu_0 I/(2\pi R)$ for $R > a$ and as

$$A_z = \frac{\mu_0 I}{4\pi a^2}(a^2 - R^2) + \frac{\mu_0 I}{2\pi} \log \frac{R_0}{a}$$

from $B_\varphi = \mu_0 IR/(2\pi a^2)$ for $0 \leq R < a$. In the above, the vector potential takes on a constant value on the cylindrical surface with the radius R. Thus, we can define an

equivector potential surface, which is similar to the equipotential surface. The vector potential is parallel to the equivector potential surface.

◇

Example 6.8. Current I is applied to an infinitely long solenoid coil of radius a with n turns in a unit length. Determine the vector potential.

Solution 6.8. Firstly, we determine the magnetic flux density. Suppose rectangles C_1 and C_2 as shown in Fig. 6.20. Applying Ampere's law to these rectangles, the circular integral of the magnetic flux density is zero in each case. It shows that the magnetic flux density is constant inside and outside the coil. Since the magnetic flux density outside the coil must be uniform up to infinity and the total magnetic flux must be finite, we can show that the magnetic flux density must be zero outside the coil. Then, we apply Ampere's law to rectangle C_3 to determine the magnetic flux density B inside the coil. The left side of Eq. (6.25) is Bl with l denoting the axial length of C_3. Since the total current inside C_3 is nlI, we have

$$B = \mu_0 nI.$$

Then, we can determine the vector potential with Eq. (6.29), but we use Eq. (6.35) here. We apply this equation to a circle, C, of radius R from the central axis of the coil in Fig. 6.21. Since the current flows only in the azimuthal direction, the vector potential has only the azimuthal component A_φ. Hence, this is parallel to C, and its magnitude is constant. We have

$$\oint_C \mathbf{A} \cdot d\mathbf{s} = 2\pi R A_\varphi(R).$$

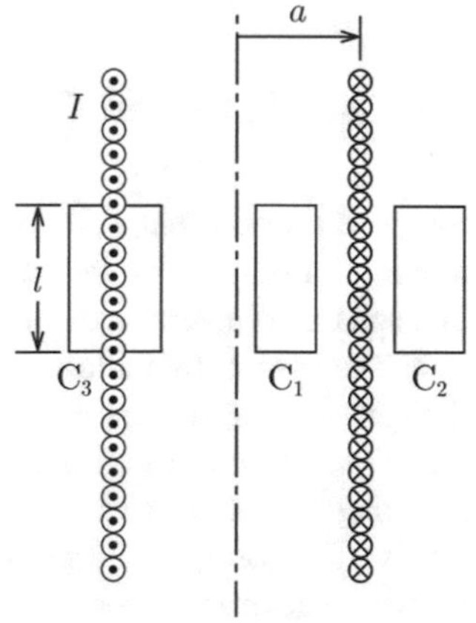

Fig. 6.20 Longitudinal cross section of solenoid coil and rectangles C_1–C_3

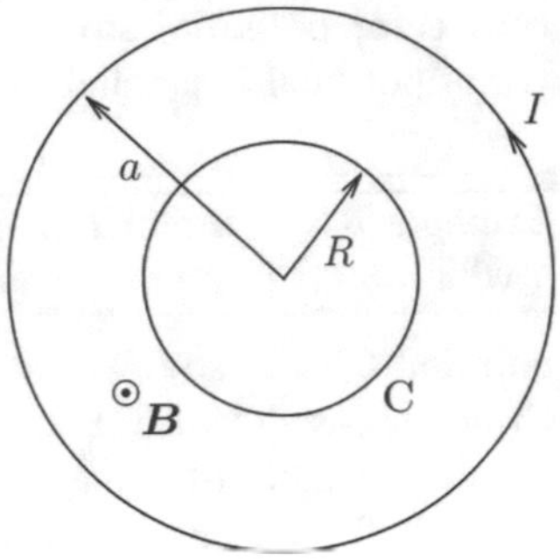

Fig. 6.21 Circle C of radius R from the central axis of the coil (for $R < a$)

On the other hand, the magnetic flux penetrating C is

$$\int_S \boldsymbol{B} \cdot \mathrm{d}\boldsymbol{S} = B\pi R^2 = \pi\mu_0 nIR^2; \qquad 0 \le R < a,$$

$$= B\pi a^2 = \pi\mu_0 nIa^2; \qquad R > a.$$

Thus, we determine the vector potential as

$$A_\varphi(R) = \frac{\mu_0 nIR}{2}; \qquad 0 \le R < a,$$

$$= \frac{\mu_0 nIa^2}{2R}; \qquad R > a.$$

Determine the vector potential with Eq. (6.29) and confirm that it agrees with the above result (see Exercise 6.10).

◇

6.7 Small Closed Current

Suppose that current I flows around a small square of side length d. The vector potential produced by this current is determined at a point, P, sufficiently far from this square. We denote the direction vector from the center of the square to point P as $\boldsymbol{r}$. The assumption allows $|\boldsymbol{r}| = r \gg d$. We define the origin of the coordinates at the center of the square placed on the x-y plane. Their sides are parallel to the x- or y-axis, as shown in Fig. 6.22a. We also assume that P is on the y-z plane, and θ is the angle between $\boldsymbol{r}$ and the z-axis. Hence, the position of P is $\boldsymbol{r} = (0, r\sin\theta, r\cos\theta)$ in Cartesian coordinates. We suppose that a point, Q, with the position vector $\boldsymbol{r}'$ moves on square KLMN (see Fig. 6.22b). When Q is on side KL, $\boldsymbol{r}' = (x, d/2, 0)$ with $-d/2 \le x \le d/2$. A simple calculation gives

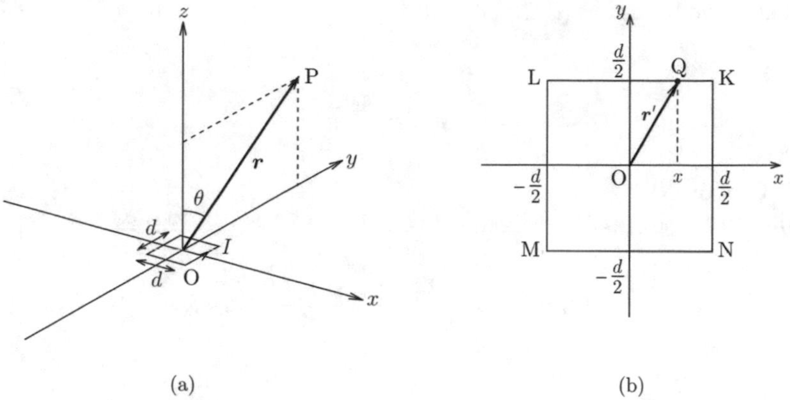

Fig. 6.22 **a** Closed current flowing along a small square with the center at the origin and observation point P and **b** position of Q on the square

$$\frac{1}{|\boldsymbol{r}-\boldsymbol{r}'|}=\frac{1}{[r^2-rd\sin\theta+(d^2/4)+x^2]^{1/2}}\simeq\frac{1}{r}\left(1+\frac{d}{2r}\sin\theta\right).$$

Integrating this from K to L (along the negative x-axis), we have

$$\int_{\mathrm{K}}^{\mathrm{L}}\frac{\mathrm{d}\boldsymbol{r}'}{|\boldsymbol{r}-\boldsymbol{r}'|}=-\frac{d}{r}\left(1+\frac{d}{2r}\sin\theta\right)\boldsymbol{i}_x. \tag{6.39a}$$

When Q is on MN, we similarly have $\boldsymbol{r}'=(x,-d/2,0)$ and

$$\int_{\mathrm{M}}^{\mathrm{N}}\frac{\mathrm{d}\boldsymbol{r}'}{|\boldsymbol{r}-\boldsymbol{r}'|}=\frac{d}{r}\left(1-\frac{d}{2r}\sin\theta\right)\boldsymbol{i}_x. \tag{6.39b}$$

When Q is on LM, substituting $\boldsymbol{r}'=(-d/2,y,0)$ gives

$$\frac{1}{|\boldsymbol{r}-\boldsymbol{r}'|}=\frac{1}{[r^2-2ry\sin\theta+(d^2/4)+y^2]^{1/2}}\simeq\frac{1}{r}\left(1+\frac{y}{r}\sin\theta\right)$$

and

$$\int_{\mathrm{L}}^{\mathrm{M}}\frac{\mathrm{d}\boldsymbol{r}'}{|\boldsymbol{r}-\boldsymbol{r}'|}=-\frac{d}{r}\boldsymbol{i}_y. \tag{6.39c}$$

Similarly, we have $\boldsymbol{r}' = (d/2, y, 0)$ and

$$\int_{\mathrm{N}}^{\mathrm{K}} \frac{\mathrm{d}\boldsymbol{r}'}{|\boldsymbol{r}-\boldsymbol{r}'|} = \frac{d}{r}\boldsymbol{i}_y, \tag{6.39d}$$

when Q is on NK.

The vector potential is determined with Eqs. (6.39a)–(6.39d) as

$$\boldsymbol{A}(\boldsymbol{r}) = \frac{\mu_0 I}{4\pi}\oint_{\mathrm{C}} \frac{\mathrm{d}\boldsymbol{r}'}{|\boldsymbol{r}-\boldsymbol{r}'|} = -\frac{\mu_0 I d^2}{4\pi r^2}\sin\theta\,\boldsymbol{i}_x. \tag{6.40}$$

We rewrite this as

$$\boldsymbol{A}(\boldsymbol{r}) = \frac{\mu_0}{4\pi}\cdot\frac{\boldsymbol{m}\times\boldsymbol{r}}{r^3} \tag{6.41}$$

in terms of the magnetic moment,

$$\boldsymbol{m} = I\boldsymbol{S}, \tag{6.42}$$

with $\boldsymbol{S} = d^2\boldsymbol{i}_z$ denoting the surface vector. Equation (6.41) corresponds to the electric potential, Eq. (1.47), produced by an electric dipole. Practically, Eq. (6.41) reduces to

$$A_\varphi = \frac{\mu_0 m}{4\pi r^2}\sin\theta \tag{6.43}$$

in our polar coordinates. In the above, $m = |\boldsymbol{m}|$ is the magnitude of the magnetic moment. A_φ is constant on a surface on which the following condition holds:

$$\frac{\sin\theta}{r^2} = \text{const.} \tag{6.44}$$

For simplicity, we treated the closed current flowing on the small square in the above. This result is valid for small closed current of arbitrary shape. One can easily prove this, since any closed current can be realized by a superposition of small square currents.

The magnetic flux density produced by the closed current is determined with Eqs. (6.29) and (6.43) as

$$B_r = \frac{1}{r\sin\theta}\cdot\frac{\partial}{\partial\theta}(\sin\theta A_\varphi) = \frac{\mu_0 m\cos\theta}{2\pi r^3}, \tag{6.45a}$$

$$B_\theta = -\frac{1}{r}\cdot\frac{\partial}{\partial r}(rA_\varphi) = \frac{\mu_0 m\sin\theta}{4\pi r^3}, \tag{6.45b}$$

$$B_\varphi = 0. \tag{6.45c}$$

It is found that the obtained magnetic flux density has the same form as the electric field produced by the electric dipole given by Eqs. (1.48a)–(1.48c) under the correspondence of $m \to p$ and $\mu_0 \to \epsilon_0^{-1}$. This is reasonable, since the magnetic flux density is expressed as

$$\boldsymbol{B} = \nabla \times \boldsymbol{A} = -\frac{\mu_0}{4\pi}\nabla\left(\frac{\boldsymbol{m}\cdot\boldsymbol{r}}{r^3}\right) \tag{6.46}$$

as is shown in Sect. A2.5 in the Appendix.

6.8 Magnetic Charge

Magnets are materials that cause magnetic interaction similarly to currents. Magnets have north (N) and south (S) poles. The force between poles of the same kind is repulsive, and the force between poles of the different kinds is attractive. This property is similar to that of electric charges. Hence, one can compare the magnetic interaction between magnetic poles to Coulomb's law for electric charges. A **magnetic charge** is an imaginary source that causes magnetic interaction corresponding to the magnetic pole. In fact, it was assumed in the past that N and S poles had magnetic charges, q_{m} and $-q_{\mathrm{m}}$, respectively, and that a force similar to the Coulomb force worked on magnetic charges. The force exerted by q_{m}' on q_{m} would then be given by

$$\boldsymbol{F}_{\mathrm{m}} = \frac{\mu_0 q_{\mathrm{m}} q_{\mathrm{m}}' \boldsymbol{r}}{4\pi r^3} \tag{6.47}$$

similarly to Eq. (1.3), where $\boldsymbol{r}$ is the position vector from q_{m}' to q_{m} and $r = |\boldsymbol{r}|$.

Since the magnetic force, Eq. (6.47), comes from some magnetic distortion in space, we can define the magnetic flux density $\boldsymbol{B}$ as the magnetic field that quantitatively expresses the strength of magnetic distortion. When we express the magnetic force as

$$\boldsymbol{F}_{\mathrm{m}} = q_{\mathrm{m}}\boldsymbol{B}, \tag{6.48}$$

the magnetic flux density is given by

$$\boldsymbol{B} = \frac{\mu_0 q_{\mathrm{m}}' \boldsymbol{r}}{4\pi r^3}. \tag{6.49}$$

It should be noted that this definition of magnetic charge is different by a factor of μ_0^{-1} from that used in other books, in which the magnetic field $\boldsymbol{H}$ defined in Chap. 9 was used for the magnetic interaction instead of the magnetic flux density

Table 6.3 Formal correspondence between electricity and magnetism

	Electricity	Magnetism
Source	Electric charge (q)	Magnetic charge (q_{m})
Potential	Electric potential (ϕ)	Magnetic potential (ϕ_{m})
Field	Electric field	Magnetic flux density
	($\boldsymbol{E} = -\nabla\phi$)	($\boldsymbol{B} = -\nabla\phi_{\mathrm{m}}$)

$\boldsymbol{B}$. One can also define a scalar potential, ϕ_{m}, called **magnetic potential** similarly to the electric potential. Its relation to $\boldsymbol{B}$ is

$$\boldsymbol{B} = -\nabla\phi_{\mathrm{m}}. \tag{6.50}$$

Table 6.3 compares the charges, scalar potentials, and fields between electricity and magnetism. However, we cannot apply this scheme in a space in which current flows.

Suppose that imaginary magnetic charges $\pm q_{\mathrm{m}}$ are separated by a small distance, d, like an electric dipole, as shown in Fig. 6.23a. This pair of magnetic charges is called a **magnetic dipole**. We define the **magnetic dipole moment** as

$$m = q_{\mathrm{m}}d. \tag{6.51}$$

The magnetic flux density produced by the magnetic dipole is given by Eq. (1.48) with p replaced by m. It was shown above that the magnetic flux density has the same form as Eq. (6.45) produced by a small closed current shown in Fig. 6.23b. Hence, the magnetic dipole and the small closed current are equivalent to each other (see Fig. 6.24). This also supports the formal correspondence between electric and magnetic charges. It is empirically known that even when a magnet is divided into small pieces, it is impossible for any piece to pick up only one type of magnetic pole, as shown in the upper part of Fig. 6.24. This can be explained assuming equivalent closed currents, as shown in the lower figure.

Thus, it may be advantageous in some cases to assume magnetic virtual charges and compare them to electric charges, since we can directly use all knowledge of electric phenomena. However, the obvious problem is that magnetic charge has never been observed. That is, the magnetic flux density empirically satisfies

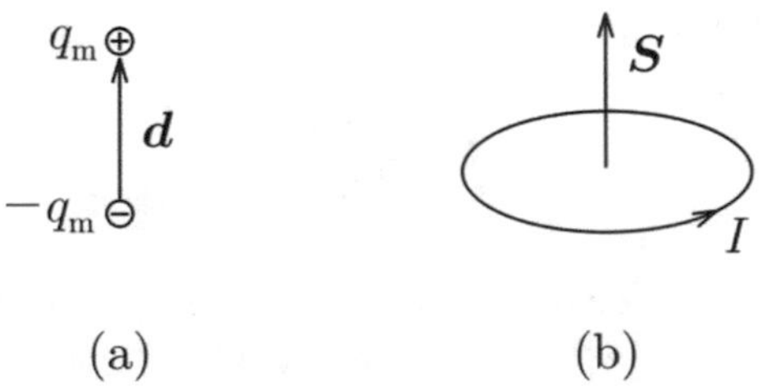

Fig. 6.23 Magnetic moment produced by **a** magnetic dipole and **b** small closed current

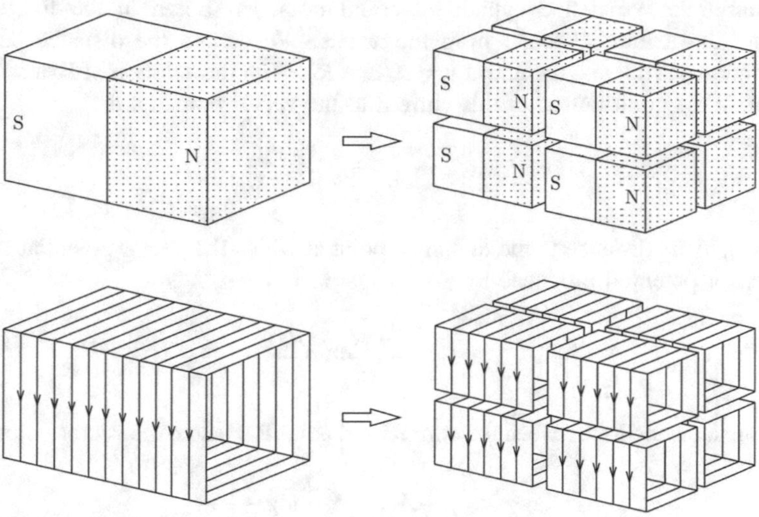

Fig. 6.24 Division of permanent magnet (*upper half*) and corresponding division of closed current (*lower half*)

Eq. (6.21), including the magnetic flux for permanent magnets. On the other hand, the right side should be equal to **magnetic charge density** ρ_m multiplied by μ_0. Hence, this results in $\rho_m = 0$. For this reason, textbooks now base their treatment of magnetic phenomena on current.

Example 6.9. Currents I and $-I$ flow along lines at $y = d/2$ and $y = -d/2$, respectively, on the y-z plane, as shown in Fig. 6.25. Determine the vector potential at point P sufficiently far from the z-axis. This pair of parallel opposite currents is equivalent to a pair of virtual line magnetic charges and is called a **magnetic dipole line**.

Fig. 6.25 Magnetic dipole line composed of pair of parallel opposite currents

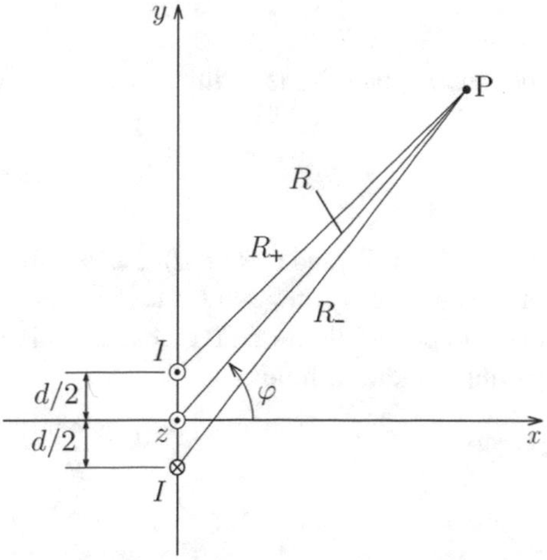

Solution 6.9. We define cylindrical coordinates, as shown in the figure, and measure the azimuthal angle from the x-axis. We denote the distance between observation point P and the line at $y = d/2$ as R_+. With the solution of Example 6.7, we obtain the contribution of this current to the vector potential as

$$A_{z+} = \frac{\mu_0 I}{2\pi} \log \frac{R_0}{R_+},$$

where R_0 is the distance to the reference point at which the vector potential is zero. The vector potential produced by the current at $y = -d/2$ is

$$A_{z-} = -\frac{\mu_0 I}{2\pi} \log \frac{R_0}{R_-}$$

using the distance R_- between the current and point P. Hence, the vector potential is given by

$$A_z = A_{z+} + A_{z-} = \frac{\mu_0 I}{2\pi} \log \frac{R_-}{R_+}.$$

For $R \gg d$, R_+ reduces approximately to

$$R_+ = \left[R^2 + \left(\frac{d}{2} \right)^2 - Rd \sin\varphi \right]^{1/2} \simeq R \left(1 - \frac{d}{2R} \sin\varphi \right).$$

Similarly, we have $R_- \simeq R\{1 + [d/(2R)] \sin\varphi\}$. Thus, the vector potential reduces to

$$A_z(R, \varphi) \simeq \frac{\mu_0 I}{2\pi} \log \frac{1 + [d/(2R)] \sin\varphi}{1 - [d/(2R)] \sin\varphi} \simeq \frac{\mu_0 I d}{2\pi R} \sin\varphi.$$

If we define the **moment of a magnetic dipole line** in a unit length by

$$\hat{m} = Id, \tag{6.52}$$

the vector potential is written as

$$A_z(R, \varphi) \simeq \frac{\mu_0 \hat{m} \sin\varphi}{2\pi R}. \tag{6.53}$$

At sufficient distance ($R \gg d$), the equality holds. In this case, the total vector sum of the current is zero, and hence, the vector potential goes to zero at infinity. Thus, there is no problem of divergence. A_z is constant on a surface on which the following condition holds:

$$\frac{\sin\varphi}{R} = \text{const.} \tag{6.54}$$

If we carry out the same calculation using a virtual magnetic charge, we obtain the magnetic potential $\phi_{\mathrm{m}}(R,\varphi)$ that corresponds to the electric potential in Example 1.9. Assume that the magnetic charges of linear densities $\pm\lambda_{\mathrm{m}}$ stay at $x=\pm d/2$, respectively, and define the moment of the magnetic dipole line by

$$\hat{m}=\lambda_{\mathrm{m}}d. \tag{6.55}$$

Then, the magnetic potential is given by

$$\phi_{\mathrm{m}}(R,\varphi)=\frac{\mu_0\hat{m}\cos\varphi}{2\pi R}. \tag{6.56}$$

This magnetic potential has the same form as the electric potential, Eq. (1.54).

From the above result for vector potential or magnetic potential, we determine the magnetic flux density as

$$B_R=\frac{\mu_0\hat{m}}{2\pi R^2}\cos\varphi, \tag{6.57a}$$

$$B_\varphi=\frac{\mu_0\hat{m}}{2\pi R^2}\sin\varphi, \tag{6.57b}$$

$$B_z=0. \tag{6.57c}$$

This corresponds to the electric field in Eq. (1.56).

◇

6.9 Equivector Potential Surface

It is shown that an equivector potential surface exists in magnetic phenomena, as in Example 6.7. This is a concept that corresponds to the equipotential surface in electrostatic phenomena. Here, the characteristic features of the equivector potential surface are discussed, and analogous points between the two surfaces are shown.

The equivector potential surface shown in Example 6.7 is characterized by the fact that the vector potential has the same value and direction on the surface in Cartesian coordinates. The vector potentials obtained in Examples 6.8 and 6.9 do not have such equivector potential surfaces but have surfaces on which they have a single component that is constant in other coordinates. Namely, the direction of such a vector potential varies, while the value is the same on the surface. Although these are not equivector potential surfaces in the strict sense, we treat them as a kind of equivector potential surface.

Since the electric potential is a scalar, the equipotential surface corresponds to both equivector potential surfaces. Here, we compare the equipotential surface and the equivector potential surface for corresponding electrostatic and magnetic phenomena. The first case is the similarity between the electric potential produced by an electric charge distributed in a long cylinder in Example 1.8 and the vector potential produced by a current flowing in a long cylinder in Example 6.7. These surfaces have the same shape. A similar correspondence can be found between the electric potential produced by a long electric dipole line in Example 1.9 and the vector potential produced by a long magnetic dipole line in Example 6.9.

On the other hand, there is a case where the equivector potential surface has a different structure from the corresponding equipotential surface. The equipotential surface produced by an electric dipole is a surface on which

$$\frac{\cos\theta}{r^2} = \text{const.} \tag{6.58}$$

holds in polar coordinates. The equivector potential surface produced by a magnetic moment is a surface on which

$$\frac{\sin\theta}{r^2} = \text{const.} \tag{6.59}$$

holds. This difference comes from the different nature of the fields. If we change the angle to $\theta \to \theta + \pi/2$, however, $\sin\theta$ in Eq. (6.59) is transformed to $\cos\theta$, and the two surfaces have the same shape. A similar relationship can be found between the equipotential surface given by Eq. (1.55) for the electric dipole line and the equivector potential surface given by Eq. (6.54) for the magnetic dipole line.

Here, we prove two theorems on the equivector potential surface. The first theorem is that the vector potential on an equivector potential surface is parallel to the surface. This is denoted as **Theorem I**. Assume a closed equivector potential surface S. Integration of the vector potential on S leads to

$$\int_{\mathrm{S}} \boldsymbol{A} \cdot \mathrm{d}\boldsymbol{S} = \boldsymbol{A} \cdot \int_{\mathrm{S}} \mathrm{d}\boldsymbol{S} = A_{\mathrm{n}} S, \tag{6.60}$$

where A_{n} is the normal component of the vector potential on S, and S is the surface area of S. The left side of this equation leads to

$$\int_{\mathrm{V}} \nabla \cdot \boldsymbol{A} \mathrm{d}V, \tag{6.61}$$

where V is a region surrounded by S. Using the Coulomb gauge, $\nabla \cdot \boldsymbol{A} = 0$, this quantity is zero. Hence, we have

$$A_{\mathrm{n}} = 0. \tag{6.62}$$

Thus, the vector potential is generally parallel to the equivector potential surface.

The second theorem is that the magnetic flux density on an equivector potential surface is parallel to the surface. This is denoted as **Theorem II**. Assume a closed line C on the equivector potential surface. Integration of the vector potential $\boldsymbol{A}$ on this line gives

$$\oint_{\mathrm{C}} \boldsymbol{A} \cdot \mathrm{d}\boldsymbol{s} = \boldsymbol{A} \cdot \oint_{\mathrm{C}} \mathrm{d}\boldsymbol{s} = 0. \tag{6.63}$$

The left-hand side of this equation leads to

$$\int_{\mathrm{S}} \boldsymbol{B} \cdot \mathrm{d}\boldsymbol{S} = \Phi, \tag{6.64}$$

where S is the surface surrounded by C, and Φ is the magnetic flux that penetrates S. Since this holds for arbitrary S, it can be said that the magnetic flux density is parallel to the equivector potential surface. Thus, if we denote the normal component of the magnetic flux density on the equivector potential surface by B_{n}, we have

$$B_{\mathrm{n}} = 0. \tag{6.65}$$

Theorem I holds generally. On the other hand, Theorem II does not hold for all equivector potential surfaces. For example, the equivector potential surfaces in Examples 6.8 and 6.9 satisfy Theorem II, while that for the magnetic moment does not. It is clear that Theorem II holds for two-dimensional surface structures such as a long cylindrical surface. In the case of three-dimensional surface structures, Eq. (6.63) is not satisfied. We classify the equivector potential surface on which the magnetic flux density is not parallel to it as that of the second kind.

A similar correspondence can be found between the electrostatic phenomena around conductors in Chap. 2 and the magnetic phenomena around superconductors in Chap. 7. The feature of the equivector potential surface that the magnetic flux density is parallel to the surface is in a clear contrast to that of the equipotential surface, that the electric field is perpendicular to the surface.

Example 6.10. Prove that Theorem II holds for the vector potential of the long magnetic dipole line and does not hold for that of the magnetic dipole.

Solution 6.10. The direction of the magnetic flux density produced by the magnetic dipole line is specified by

$$\frac{B_{\varphi}}{B_R} = \tan\varphi.$$

If we change R and φ to $R + \mathrm{d}R$ and $\varphi + \mathrm{d}\varphi$ on the equivector potential surface, the value given by Eq. (6.54) does not change. That is,

$$\frac{\sin\varphi}{R} = \frac{\sin(\varphi + \mathrm{d}\varphi)}{R + \mathrm{d}R}.$$

Since $\sin(\varphi + \mathrm{d}\varphi) \simeq \sin\varphi + \mathrm{d}\varphi\cos\varphi$, the above equation leads to

$$\frac{R\mathrm{d}\varphi}{\mathrm{d}R} = \tan\varphi.$$

Since the direction of the magnetic flux density is the same as that of the extension of the equivector potential surface, the magnetic flux density is parallel to the surface. See Exercise 6.12.

The direction of the magnetic flux density produced by the magnetic dipole is specified by

$$\frac{B_\theta}{B_r} = \frac{1}{2}\tan\theta.$$

If we change r and θ to $r + \mathrm{d}r$ and $\theta + \mathrm{d}\theta$ on the equivector potential surface, the value given by Eq. (6.44) does not change. That is,

$$\frac{\sin\theta}{r^2} = \frac{\sin(\theta + \mathrm{d}\theta)}{(r + \mathrm{d}r)^2}.$$

The above equation leads to

$$\frac{r\mathrm{d}\theta}{\mathrm{d}r} = 2\tan\theta.$$

Hence, the magnetic flux density is not parallel to the surface in this case.

◇

Column: (1) Forces Between Electric Charges and Between Currents
We have said that, since the source of magnetic phenomena is the current, a vector, the resultant vector field of magnetic flux density must be given by the vector product of the current and position vector. As a result, the magnetic flux density is perpendicular to both the current and position vector. The magnetic field is perpendicular to the position vector, while the electric field is parallel to the position vector. So, these phenomena show really contrast. Thus, these results follow the mathematical requirements and the correspondence holds in this sense.

Here, we compare the forces arising from electric charges and currents. To obtain a force vector for electric charges, we need the direct product of the electric charge, a scalar, and the electric field, a vector. Thus, the force between electric charges of the same kind is repulsive. For currents, the vector product must appear again to yield a force vector from the current vector and the magnetic flux density vector. Thus, the magnetic force is perpendicular to both the current and magnetic flux density. This explains why the force is attractive between currents in the same direction. We again show the correspondence between the forces based on the mathematical requirements.

Column: (2) Coulomb Magnetic Field

In a space with no current, as in vacuum, the magnetic flux density obeys

$$\nabla \times \boldsymbol{B} = 0$$

from Eq. (6.27). Hence, we can describe the magnetic flux density in terms of the magnetic potential ϕ_{m} as in Eq. (6.50) from the mathematical viewpoint. Such a magnetic field is called a **Coulomb magnetic field**.

In the past the magnetic phenomena were described using virtual magnetic charges and the Coulomb magnetic field. This is the magnetic field $\boldsymbol{H}$ defined in Chap. 9 and is equal to $\boldsymbol{B}/\mu_0$ in the present description. This method is beneficial because various descriptions of electric phenomena can be used to explain some magnetic phenomena. For example, the static magnetic energy can be simply determined as the work needed to carry magnetic charges in the magnetic field. This is not possible for currents because of electromagnetic induction (see Column (1) in Chap. 8). In addition, from similarity to the electric field, one can easily show that the parallel component of the magnetic field is continuous at an interface using Eq. (4.22). In fact, this coincides with the boundary condition of Eq. (9.24) when there is of no surface current on the interface.

Thus, the magnetic field in the absence of current behaves similarly to the electric field in the absence of a true electric charge. We can see such an analogy between the electric field in a dielectric material in Example 4.5 and the magnetic flux density in a magnetic material in Example 9.5.

Exercises

6.1. Current I flows along a line composed of a semicircle and two straight lines on a common plane, as shown in Fig. E6.1. Determine the magnetic flux density at the center O of curvature of the semicircle.

6.2. Current I flows on a square closed line of side length a, as shown in Fig. E6.2. Determine the magnetic flux density at point P at distance b from the center O of the square.

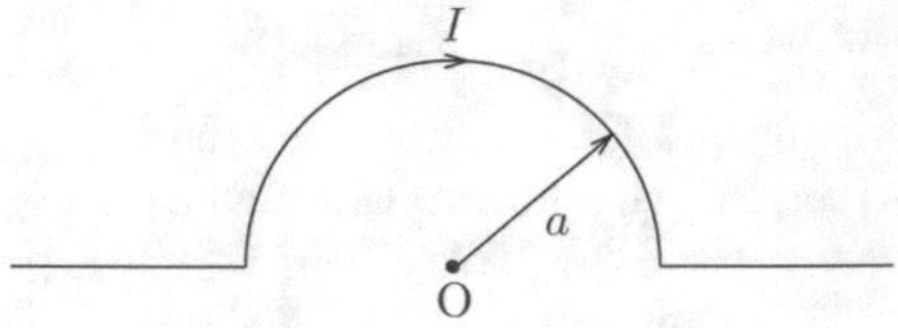

Fig. E6.1 Current on line composed of semicircle and two straight lines

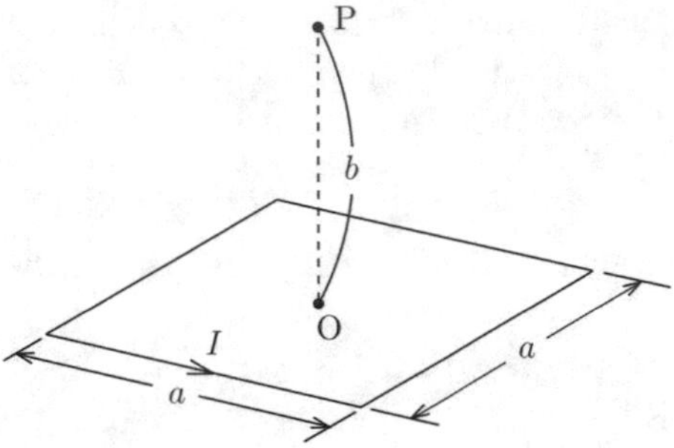

Fig. E6.2 Square current and observation point P

6.3. Current of density $\boldsymbol{i}$ flows in a slab conductor along the x-axis in the normal magnetic flux density $\boldsymbol{B}$ parallel to the z-axis (see Fig. E6.3). Determine the steady electric field produced in the direction normal to both the current and magnetic flux density. The electric charge and number density of particles that carry current are q and n.

6.4. Prove that Eq. (6.22) holds for an arbitrary closed line C that straight current I penetrates (see Fig. E6.4).

6.5. Currents of density i flow uniformly in two parallel wide slab conductors in opposite directions along the y-axis (see Fig. E6.5). Determine the magnetic flux density and vector potential inside and outside of the conductors.

6.6. Current I flows uniformly in a thin planar conductor of width w. Determine the magnetic flux density and vector potential at point P at distance $d(> w/2)$ from the center of the conductor (see Fig. E6.6). The conductor and P are on a common plane.

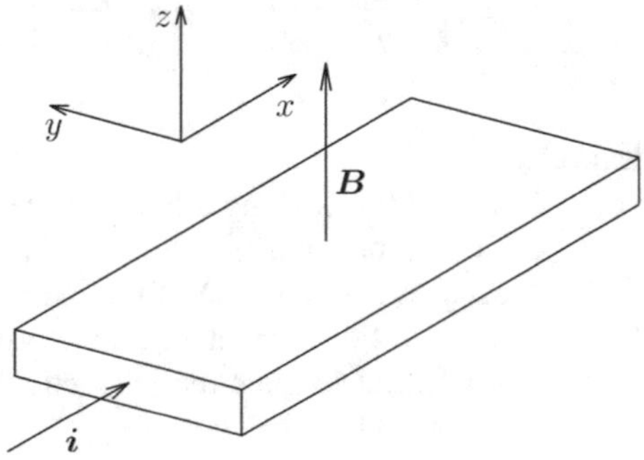

Fig. E6.3 Slab conductor carrying current in normal magnetic flux density

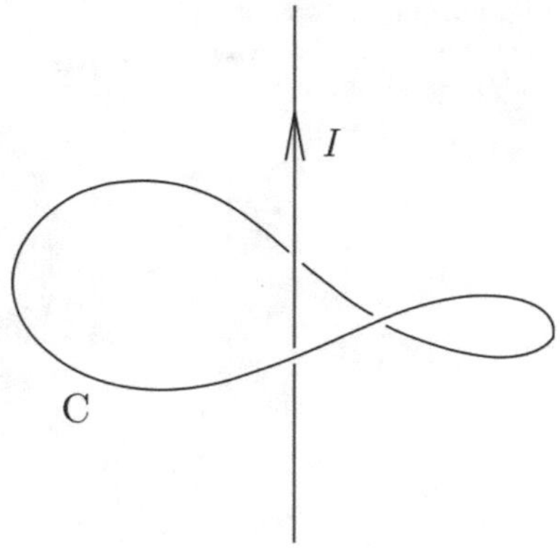

Fig. E6.4 Straight current and surrounding closed line of arbitrary shape

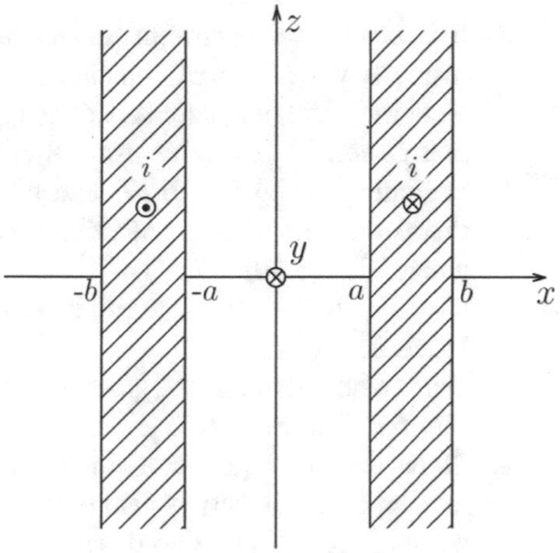

Fig. E6.5 Two parallel slab conductors with uniform currents flowing in opposite directions

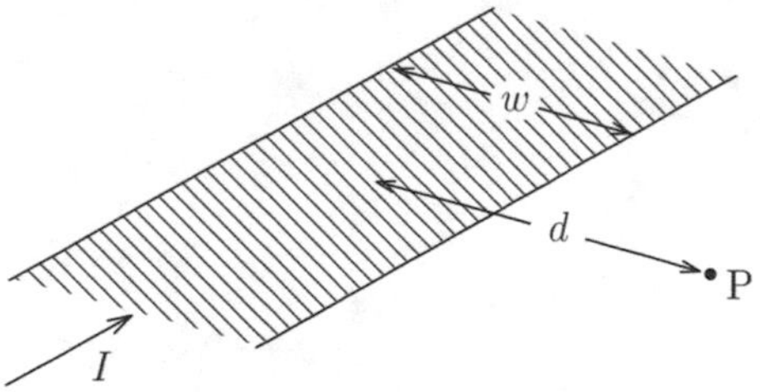

Fig. E6.6 Thin planar conductor with uniform current and observation point P

6.7. Current I flows uniformly in a long cylindrical conductor of radius a that contains a cylindrical hollow of radius b, as shown in Fig. E6.7. The center of the hollow is located at distance d from the center of the conductor, where $a > b + d$. Determine the magnetic flux density at the center of the hollow (point A) and at point B outside the conductor. The distance of point B from the center is $R(> a)$. The central axis O and points A and B are on a common plane.

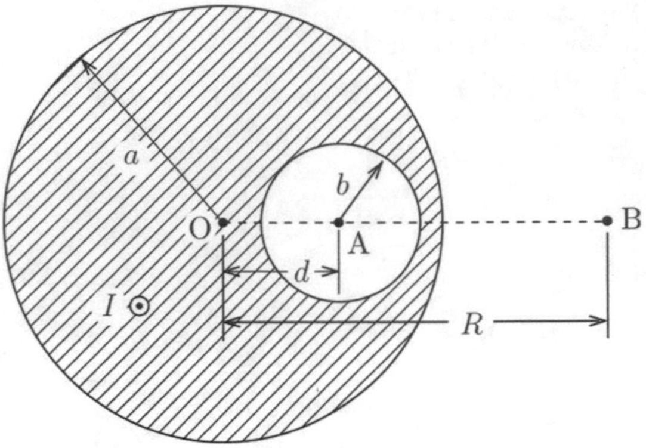

Fig. E6.7 Cross section of long cylindrical conductor with cylindrical hollow

6.8. It is assumed that current flows uniformly along the y-axis with density i_0 inside a wide slab of thickness $2a(-a<x<a)$ parallel to the y-z plane. Determine the magnetic flux density inside and outside the slab. Then, prove that current flows only inside the slab with density i_0, as assumed in the beginning, using Eq. (6.27) and the obtained magnetic flux density.

6.9. Prove that the vector potential given by Eq. (6.33) satisfies the Coulomb gauge, Eq. (6.30).

6.10. Use Eq. (6.29) to determine the vector potential in the solenoid in Example 6.8.

6.11. Determine the magnetic potential produced by the small closed current in Fig. 6.22b in Sect. 6.7.

6.12. Show that the equivector potential surface is a cylindrical surface and the magnetic flux density is also parallel to the cylindrical surface for the case of the magnetic dipole line discussed in Example 6.9.

Chapter 7
Superconductors

Abstract This chapter covers magnetic phenomena inside and outside a superconductor, in which the magnetic flux density is zero. As a result, there is no current inside the superconductor, and the superconductor is at equivector potential. When a current is applied to the superconductor, it flows only on the superconductor surface in such a way that zero magnetic flux density is attained inside. The same thing occurs when an external magnetic flux density is applied to the superconductor. The magnetic flux density is directed parallel to the surface of the superconductor. A special method, i.e., the method of images, is introduced to determine the magnetic flux density outside the superconductor and the distribution of current on the surface in such given conditions. It is theoretically proved that the electrical resistivity of the superconductor is zero.

7.1 Magnetic Properties of Superconductors

A **superconductor** is a material that loses its electric resistance when cooled below a characteristic temperature called the critical temperature. Many elements, alloys, and compounds are superconductors. This state of zero resistivity is called the **superconducting state**. A superconductor has not only this property but also **perfect diamagnetism**. That is, when a magnetic flux density is applied to a superconductor, the interior magnetic flux density is zero:

$$\boldsymbol{B} = 0. \tag{7.1}$$

This state is called the **Meissner state**. When the temperature is above the critical temperature, the superconductor is in a **normal state** with nonzero resistance. In this case, the magnetic flux density penetrates the superconductor. If the superconductor is cooled below the critical temperature, the magnetic flux is expelled from the superconductor and perfect diamagnetism occurs. These diamagnetic phenomena are called the **Meissner-Ochsenfeld effect**.

T. Matsushita, *Electricity and Magnetism*, Undergraduate Lecture Notes in Physics, https://doi.org/10.1007/978-3-030-82150-0_7

The perfect diamagnetic state is realized by a current that flows on the surface of the superconductor, as will be mentioned. This is similar to the electric property of a conductor that the inside is completely shielded by an electric charge induced on the surface when an external electric field is applied. In this chapter we study the magnetic phenomena around a superconductor in the perfect diamagnetic state. The perfect diamagnetic state of $\boldsymbol{B} = 0$ occurs in a type 1 superconductor or type 2 superconductor in a magnetic flux density below the lower critical magnetic flux density (see Sect. A3.2 in the Appendix). Special knowledge is needed to understand the physical mechanism that causes the zero resistivity and perfect diamagnetism. This is not discussed in this textbook. Section A3.1 in the Appendix gives a brief explanation of phenomenological theory. For a more detailed understanding, it is recommended to read technical books.

Materials are roughly classified into conductors and insulators (dielectric materials) with respect to their electric properties. For the magnetic properties, materials are classified into superconductors and **magnetic materials**. The latter will be covered in Chap. 9.

Since Eq. (7.1) holds inside the superconductor, from Eq. (6.27), we have

$$\boldsymbol{i} = 0. \tag{7.2}$$

Hence, current flows only on the surface of the superconductor. Equation (6.29) generally gives

$$\boldsymbol{A} = \nabla\alpha \tag{7.3}$$

with α denoting a scalar function. However, it is no problem to assume as

$$\boldsymbol{A} = \text{const.} \tag{7.4}$$

in most cases. This is a special case of Eq. (7.3). If the superconducting region is not simply connected but a magnetic flux penetrates a space surrounded by the superconductor, the vector potential in the superconductor is not a constant. Equations (7.1), (7.2), and (7.4) correspond to Eqs. (2.1), (2.2), and (2.3) for conductors, respectively.

Here, we discuss the magnetic flux density in the vicinity of the superconductor surface. Suppose a small closed surface of a pellet that includes the interface between the superconductor and vacuum, as shown in Fig. 7.1. We denote the height of the pellet and the area of surface inside the pellet by Δh and ΔS, respectively. Assume that the upper and lower surfaces of the pellet are parallel to the superconductor surface. We apply Gauss' law, Eq. (6.20), to the surface of the pellet, ΔS, and we have

$$\int_{\Delta S} \boldsymbol{B} \cdot \mathrm{d}\boldsymbol{S} = 0. \tag{7.5}$$

Fig. 7.1 Small closed surface that includes a part of superconductor surface

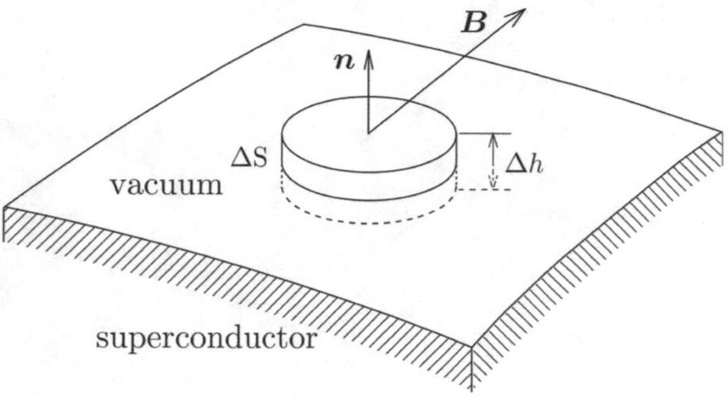

When the height Δh is sufficiently small, we can neglect the magnetic flux going out of the side surface. Since $\boldsymbol{B} = 0$ on the lower surface in the superconductor, there is no magnetic flux going out from this surface. Hence, we can conclude that there is no magnetic flux going out from the upper surface. This means that the magnetic flux density is parallel to the superconductor surface. That is,

$$\boldsymbol{B} \cdot \boldsymbol{n} = 0, \tag{7.6}$$

where $\boldsymbol{n}$ is the unit vector normal to the superconductor surface and directed outward. This is in contrast with the electric field, which is normal to the conductor surface.

Next, we discuss the current that flows on the superconductor surface. Suppose that the magnetic flux density B is parallel to a wide surface of the superconductor, as shown in Fig. 7.2. We define the coordinates as in the figure. We apply Ampere's law, Eq. (6.25), to a small rectangular closed loop, $\Delta\mathrm{C}$, with two sides of length Δl parallel to the z-axis and two sufficiently short sides of length Δt parallel to the x-axis. Since the magnetic flux density inside the superconductor is zero and there is no contribution from the two sides of length Δt, the closed curvilinear integral gives

$$\oint_{\Delta\mathrm{C}} \boldsymbol{B} \cdot \mathrm{d}\boldsymbol{s} = B\Delta l. \tag{7.7}$$

This is equal to the current flowing along the negative y-axis inside $\Delta\mathrm{C}$ multiplied by μ_0, that is, $\mu_0 \tau \Delta l$ with τ denoting the surface current density. Thus, we have

$$B = \mu_0 \tau. \tag{7.8}$$

It should be noted that the directions of the current and magnetic flux density follow the right-hand rule. This relationship is also in contrast with Eq. (2.5) for conductors.

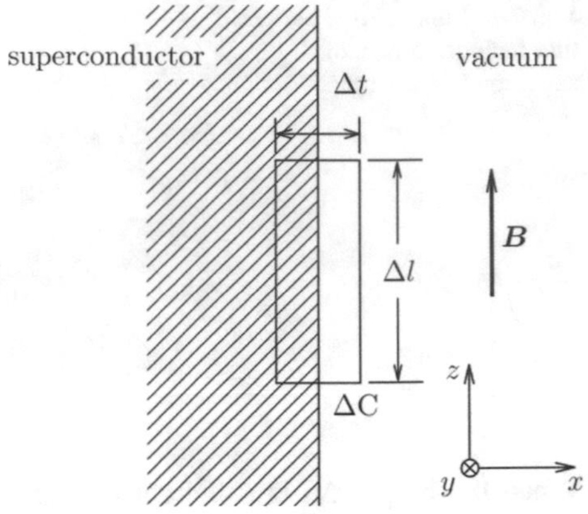

Fig. 7.2 Magnetic flux density parallel to wide superconductor surface

As has been discussed above, magnetic phenomena with perfect diamagnetism ($\boldsymbol{B} = 0$) in superconductors are similar to electric phenomena with perfect electrostatic shielding ($\boldsymbol{E} = 0$) in conductors. This similarity is prominent in the ***E*-*B*** analogy in electromagnetism. In addition, the correspondence of Eqs. (7.2), (7.4), and (7.8) to Eqs. (2.1), (2.3), and (2.5) shows that the similarity between electricity and magnetism is considerably deep. With the ***E*-*B*** analogy, it was possible for someone even in the nineteenth century to predict the existence of a material with perfect diamagnetism, i.e., a superconductor, just after completion of the Maxwell theory. It might be expected that the proof of zero resistivity was easy, since the shielding current must continue to flow to keep the diamagnetic state. It should be noted, however, that this proof was not justified, since the diamagnetism might be caused by the magnetization, i.e., a magnetic property discussed in Chap. 9, for an unknown material in the nineteenth century. In reality, the superconductivity with zero resistivity was discovered independently of the above consideration in 1911, and the perfect diamagnetism was discovered 22 years later in 1933.

Here, we show an example of magnetic phenomena associated with a superconductor. Suppose that current I is applied to a long cylindrical superconductor of radius a. We determine the magnetic flux density and vector potential inside and outside the superconductor. We define cylindrical coordinates with the z-axis on the central axis of the superconductor. The current flows uniformly only on the superconductor surface and the magnetic flux density does not appear inside the superconductor. Thus, the surface current density is $\tau = I/(2\pi a)$.

We define a circle, C, of radius R from the central axis on a plane normal to the axis (see Fig. 7.3). We apply Ampere's law to C. Since the current distribution is cylindrically symmetric, we can assume the magnetic flux density also has cylindrical symmetry. Hence, the magnetic flux density is parallel to C and its magnitude B is constant on C. Thus, the left side of Eq. (6.25) is $2\pi RB(R)$. All the current

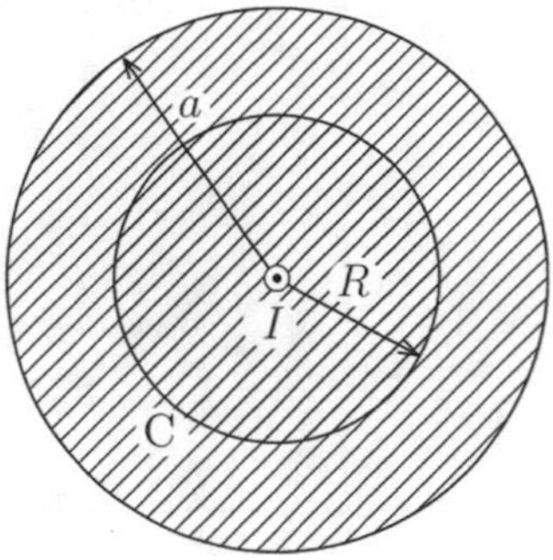

Fig. 7.3 Cross section of cylindrical superconductor and closed circle C (for $R < a$)

I flows inside C and the right side of Eq. (6.25) is $\mu_0 I$ for $R > a$. We obtain the magnetic flux density as

$$B(R) = \frac{\mu_0 I}{2\pi R}; \quad R > a. \tag{7.9}$$

The magnetic flux density outside the superconductor is the same as that when all the current is concentrated along the central axis. For $R < a$, the right side of Eq. (6.25) is zero. This gives

$$B(R) = 0; \qquad 0 \leq R < a. \tag{7.10}$$

Thus, Eq. (7.1) is fulfilled inside the superconductor. We can also show that Eq. (7.9) satisfies Eq. (7.8) on the superconductor surface ($R = a$) with the surface current density $\tau = I/2\pi a$.

Now we determine the vector potential using the above results. From Eq. (6.33) we find that the vector potential has only the z-component, A_z, which is given by

$$A_z(R) = -\int B(R)\mathrm{d}R. \tag{7.11}$$

We choose the reference point for zero vector potential at distance $R_0(> a)$ from the central axis, as done in Example 6.7. Then, the vector potential is determined to be

$$A_z(R) = \frac{\mu_0 I}{2\pi}\log\frac{R_0}{R}; \qquad R > a, \tag{7.12a}$$

$$= \frac{\mu_0 I}{2\pi}\log\frac{R_0}{a}; \qquad 0 \leq R < a. \tag{7.12b}$$

Figure 7.4a, b shows the obtained magnetic flux density and vector potential. The vector potential is constant inside the superconductor and Eq. (7.4) is satisfied.

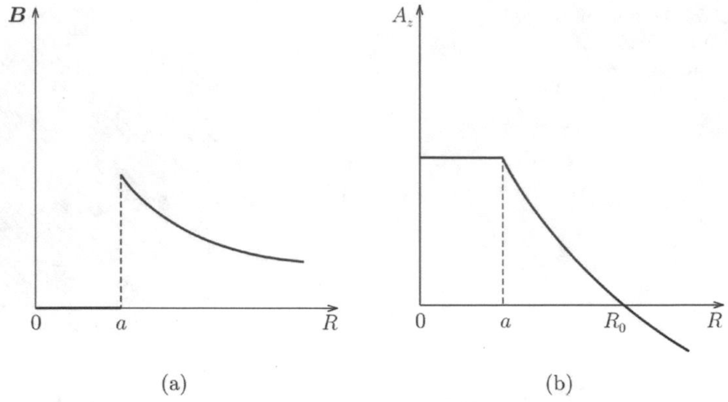

Fig. 7.4 **a** Magnetic flux density and **b** vector potential inside and outside the current-carrying cylindrical superconductor

Example 7.1. Suppose that current flows uniformly with surface density τ in the direction of the y-axis on a thin sheet separated by b from a slab superconductor of thickness a, as shown in Fig. 7.5. Determine the current density on each surface of the superconductor and the magnetic flux density in each region of $x > -b$.

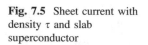

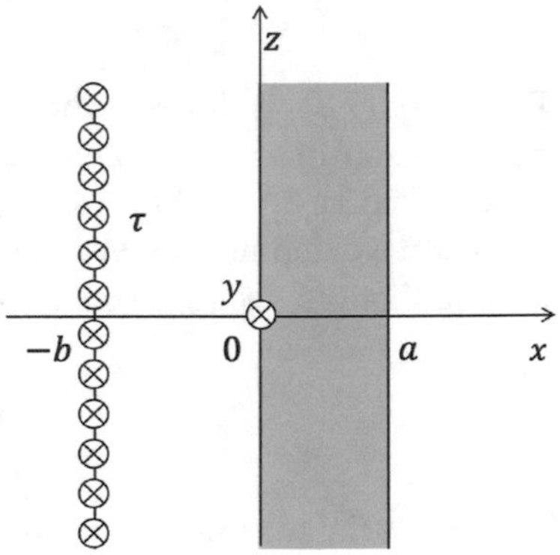

Fig. 7.5 Sheet current with density τ and slab superconductor

Solution 7.1. The magnetic flux density produced by the sheet current on the left side of the superconductor $(-b < x < 0)$ is $-\mu_0\tau/2$ (see Example 6.6) in the direction of the z-axis. A current with surface density $-\tau/2$ is induced in the direction of the y-axis on the left surface $(x = 0)$ of the superconductor from Eq. (6.25) or Eq. (7.8). Then, the current with surface density $\tau/2$ flows on the right surface $(x = a)$ of the superconductor. Thus, the magnetic flux density in each region is

$$B = -\frac{\mu_0 \tau}{2}; \quad -b<x<0,$$
$$= 0; \quad 0<x<a,$$
$$= -\frac{\mu_0 \tau}{2}; \quad x>a.$$

◇

Example 7.2. Suppose a long superconducting coaxial transmission line, as shown in Fig. 7.6. Determine the magnetic flux density and vector potential when we apply current I only to the inner superconductor.

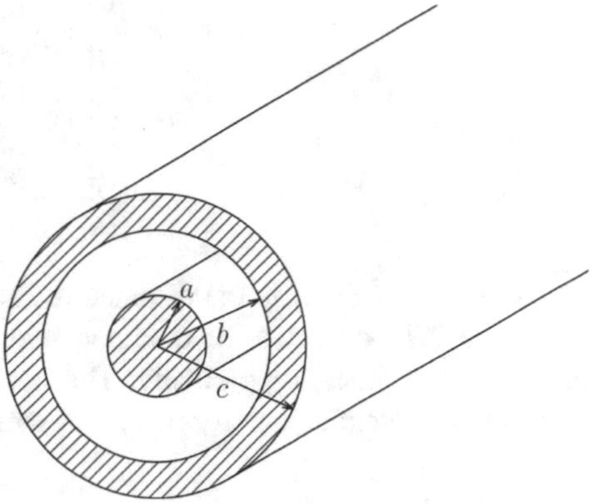

Fig. 7.6 Long superconducting coaxial transmission line

Solution 7.2. We use cylindrical coordinates. From cylindrical symmetry the current I flows uniformly on the surface of the inner superconductor ($R = a$), and the internal magnetic flux density ($R < a$) is zero. The current is induced on the inner surface ($R = b$) of the outer superconductor so that the magnetic flux does not penetrate the outer superconductor. This current is denoted by I_b. We apply Ampere's law to a circle C of radius $R(a < R < b)$ from the central axis on a plane perpendicular to the axis. Then, we have

$$\oint_{\mathrm{C}} \boldsymbol{B} \cdot \mathrm{d}\boldsymbol{s} = \mu_0 (I + I_b).$$

Since $\boldsymbol{B} = 0$ on C, this gives $I_b = -I$. Since the total current is zero in the outer superconductor, an opposite current, I, flows on the outer surface ($R = c$) of the outer superconductor.

If we denote the total current passing through C of radius R as I_R, Ampere's law gives

$$B(R) = \frac{\mu_0 I_R}{2\pi R}.$$

Since I_R is equal to I, 0, and I for $a < R < b$, $b < R < c$, and $R > c$, respectively, we determine the magnetic flux density to be

$$\begin{aligned} B(R) &= 0; && 0 \le R < a, \\ &= \frac{\mu_0 I}{2\pi R}; && a < R < b, \\ &= 0; && b < R < c, \\ &= \frac{\mu_0 I}{2\pi R}; && R > c. \end{aligned}$$

The vector potential has only the z-component and is determined to be.

$$\begin{aligned} A_z(R) &= \frac{\mu_0 I}{2\pi} \log \frac{R_0}{R}; && R > c, \\ &= \frac{\mu_0 I}{2\pi} \log \frac{R_0}{c}; && b < R < c, \\ &= \frac{\mu_0 I}{2\pi} \log \frac{bR_0}{cR}; && a < R < b, \\ &= \frac{\mu_0 I}{2\pi} \log \frac{bR_0}{ac}; && 0 \le R < a. \end{aligned}$$

In the above, $R_0 (> c)$ is the distance to the reference point.

Figure 7.7a–c shows the obtained magnetic flux density, vector potential, and magnetic flux lines, respectively. The above results for the magnetic flux density and vector potential are formally the same as for the electric field and electric

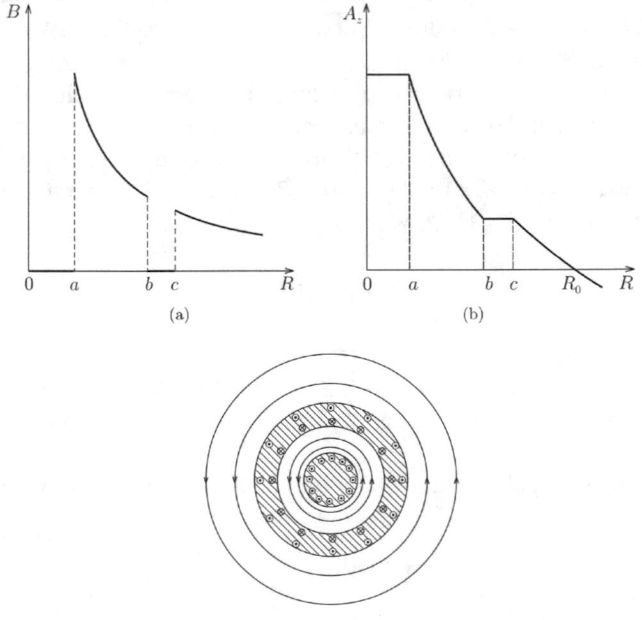

Fig. 7.7 **a** Magnetic flux density, **b** z-component of vector potential, and **c** magnetic flux lines when current is applied to the inner superconductor of a superconducting coaxial transmission line

potential of the coaxial cylindrical conductor when electric charge is given to the inner conductor (see Example 2.3).

◇

There is a treatment called grounding for a conductor. What is the corresponding treatment for a superconductor? It is to connect the superconductor with a current path with infinitely large capacity. In the above case a connection of the outer superconductor to infinity moves the back current on the surface at $R = c$ to infinity. We also call this grounding. Then, the current is I at $R = a$ and $-I$ at $R = b$. The magnetic flux density and vector potential are

$$\begin{aligned} B(R) &= 0; & 0 \le R < a, \\ &= \frac{\mu_0 I}{2\pi R}; & a < R < b, \\ &= 0; & R > b \end{aligned}$$

and

$$\begin{aligned} A_z(R) &= \frac{\mu_0 I}{2\pi} \log \frac{b}{a}; & 0 \le R < a, \\ &= \frac{\mu_0 I}{2\pi} \log \frac{b}{R}; & a < R < b, \\ &= 0; & R > b. \end{aligned}$$

Figure 7.8a, b shows the obtained results.

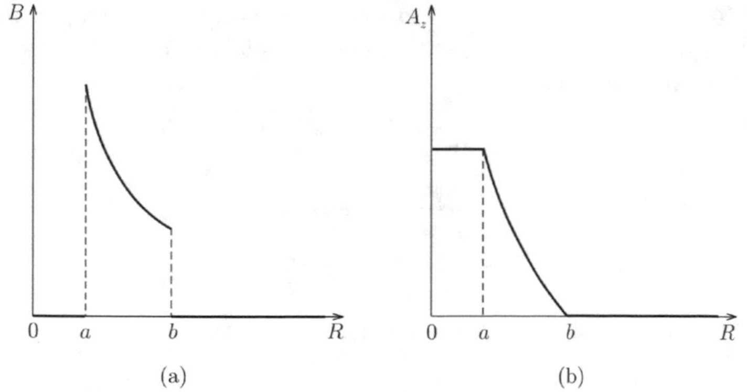

Fig. 7.8 **a** Magnetic flux density and **b** z-component of vector potential when we apply a current to the inner superconductor of superconducting coaxial transmission line and ground the outer superconductor

Example 7.3. Two wide slab superconductors are parallel to each other, as shown in Fig. 7.9, and currents I and $-I$ are applied along the y-axis to the left and right superconductors, respectively. The length along the z-axis of each slab is l. Determine the current that appears on each superconductor surface, the magnetic flux density, and the vector potential inside and outside the superconductors.

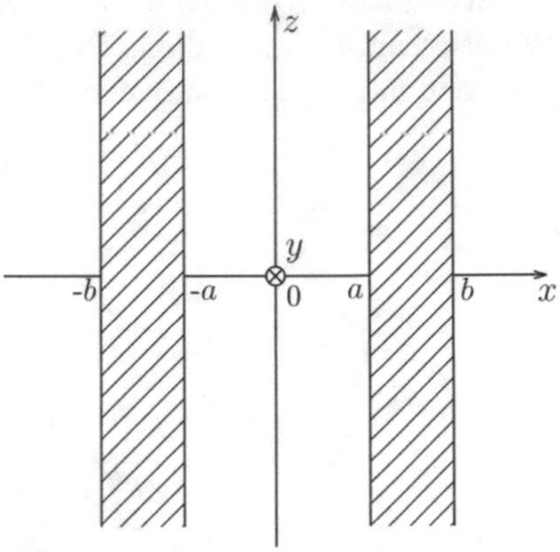

Fig. 7.9 Two parallel slab superconductors

Solution 7.3. We denote the current flowing on the surface at $x = b$ by I_b. Then, the current at $x = a$ is $-I - I_b$. So that the magnetic flux density does not penetrate the right superconductor; the current at $x = b$ (I_b) must be the same as the total current in the region $x \leq a$, i.e., $I - I - I_b = -I_b$. Thus, we have $I_b = 0$ and the current at $x = a$ is $-I$. The total current in the region of $-b<x<b$ must be zero from Eq. (6.25) so that the magnetic flux density is zero in the two superconductors. Hence, the current at $x = -a$ is I, and that at $x = -b$ is 0. Thus, the magnetic flux density along the z-axis is

$$
\begin{aligned}
B &= 0; && x < -a, \\
&= -\frac{\mu_0 I}{l}; && -a<x<a, \\
&= 0; && x > a.
\end{aligned}
$$

The vector potential has a y-component, A_y, and is given by

$$
\begin{aligned}
A_y &= \frac{\mu_0 I a}{l}; && x < -a, \\
&= -\frac{\mu_0 I x}{l}; && -a<x<a, \\
&= -\frac{\mu_0 I a}{l}; && x > a.
\end{aligned}
$$

◇

Example 7.4 Assume a long hollow superconducting cylinder with inner and outer radii, a and b, respectively, as shown in Fig. 7.10. We apply magnetic flux density B_0 along the axis in the normal state of the superconductor, cool down the superconductor below its critical temperature to make the superconductor superconducting, and remove the magnetic flux density. This process is called the field-cooled process, and the magnetic flux density at each stage is shown in Fig. 7.11. Finally, the magnetic flux density B_0 is trapped in the hollow interior, and a current with density $\tau = B_0/\mu_0$ flows on the inner surface of the superconductor. Determine the vector potential inside and outside the superconductor.

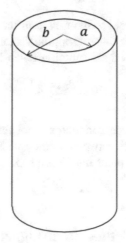

Fig. 7.10 Long hollow superconducting cylinder

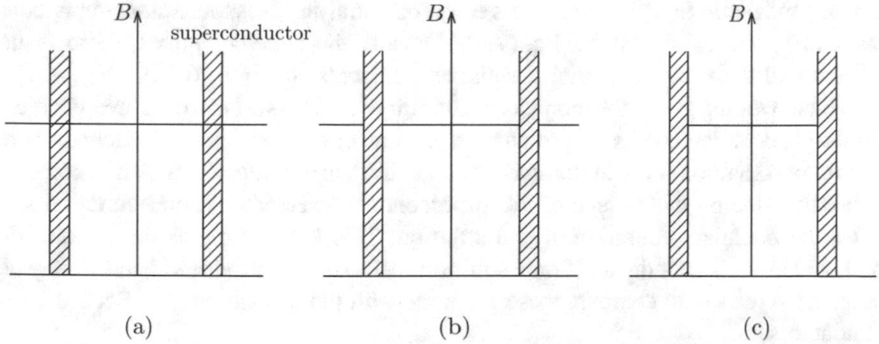

Fig. 7.11 Magnetic flux density in the field-cooled process: **a** after applying magnetic flux density above the critical temperature, **b** after cooling below the critical temperature, and **c** after removing external magnetic flux density

Solution 7.4. The magnetic flux density has a z-component in cylindrical coordinates, and its value is $B_z = B_0$ for $R<a$ and $B_z = 0$ for $R > a$. The vector potential has an azimuthal component, A_φ, and from the relationship,

$$\frac{1}{R} \cdot \frac{\partial R A_\varphi}{\partial R} = B_z,$$

we have

$$A_\varphi = \frac{B_z R}{2}; \quad 0<R<a,$$
$$= \frac{B_z a^2}{2R}; \quad R > a.$$

Thus, the vector potential is not constant even in the superconductor in which the magnetic flux density is zero. In this region, the vector potential is given in the form of Eq. (7.3), and α is a multi-valued function given by

$$\alpha = \frac{B_z a^2}{2} \varphi.$$

This is the case in which the superconductor is not simply connected. That is, if we assume a circle that has a radius between a and b with its center on the central axis, this loop does not shrink to a point inside the superconductor.

◇

7.2 Special Solution Method for Magnetic Flux Density

We determine the current distribution on the superconductor surface or magnetic flux density around a superconductor when the superconductor is placed in an applied magnetic flux density. The vector potential in the superconductor is constant in space, as shown by Eq. (7.4). Outside the superconductor, there is no current and the vector potential $\boldsymbol{A}$ satisfies Laplace's equation (6.38).

When we are given the boundary condition on the surface of a treated area, Laplace's equation can be solved uniquely, as mentioned in Sect. 2.2. Hence, there is only one solution for $\boldsymbol{A}$ in the space outside the superconductor, which becomes a constant value on the surface of the superconductor. Hence, if some function satisfies the boundary condition, it is a solution. This is the same as for the electric potential around a conductor. Some solution methods are introduced here. It may be helpful for readers to compare these methods with those mentioned in Sect. 2.2 for conductors.

First, suppose we apply current I through a thin straight line placed at distance a from a wide flat surface of a superconductor, as shown in Fig. 7.12a. A current of opposite direction appears on the surface of the superconductor to shield it and

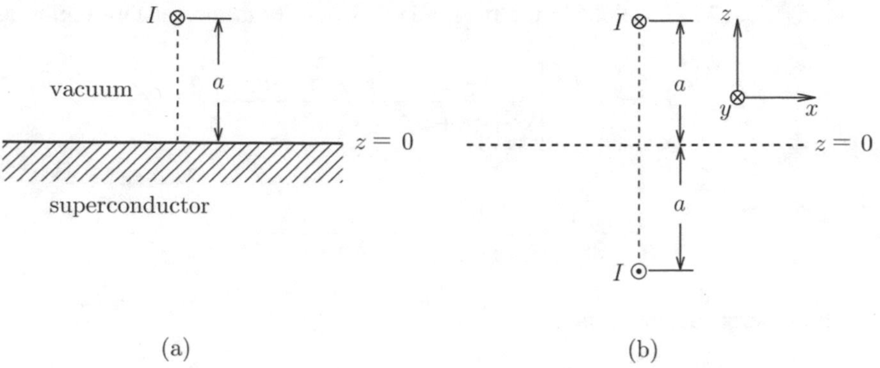

Fig. 7.12 **a** Thin straight current I placed at distance a from wide flat surface of superconductor and **b** imaginary current placed at a mirror position with respect to the superconductor surface

exerts a repulsive force on I. We determine the distribution of the induced current similarly to the electric charge distribution on the conductor surface in Fig. 2.11.

We define the x-y plane on the superconductor surface. Suppose that current I flows on the line at $x = 0$ and $z = a$ along the positive y-axis. The magnetic flux density on the surface ($z = 0$) must be parallel to the superconductor, as shown in Sect. 7.1. We can realize this situation by virtually removing the superconductor and then applying a straight current with the same magnitude and opposite direction at the symmetric position ($x = 0$, $z = -a$) with respect to the superconductor surface. This will be confirmed below. The vector potential in the vacuum region ($z > 0$) produced by the two straight currents has only the y-component parallel to the currents, and we calculate it as

$$\begin{aligned} A_y(x, y, z) &= \frac{\mu_0 I}{2\pi}\left\{\log\frac{R_0}{[x^2 + (z-a)^2]^{1/2}} - \log\frac{R_0}{[x^2 + (z+a)^2]^{1/2}}\right\} \\ &= \frac{\mu_0 I}{4\pi}\log\frac{x^2 + (z+a)^2}{x^2 + (z-a)^2} \end{aligned} \tag{7.13}$$

using Eq. (7.12a). In the above, R_0 is the distance from the current to the reference point. We can easily show that $A_y = 0$ on the superconductor surface ($z = 0$). Thus, the condition, Eq. (7.4), is satisfied. Laplace's equation is satisfied except at the position of the current in the vacuum region ($z > 0$). Hence, we conclude that Eq. (7.13) is the solution for the vector potential. The vector potential inside the superconductor ($z < 0$) is $A_y = 0$. Thus, the method of images is useful also for superconductors. The imaginary current placed at the mirror position is called an **image current**.

Using Eq. (7.13), we obtain the magnetic flux density in the vacuum region as

$$\begin{aligned} B_x &= -\frac{\mu_0 I}{2\pi}\left[\frac{z+a}{x^2+(z+a)^2}-\frac{z-a}{x^2+(z-a)^2}\right], \\ B_y &= 0, \\ B_z &= \frac{\mu_0 I x}{2\pi}\left[\frac{1}{x^2+(z+a)^2}-\frac{1}{x^2+(z-a)^2}\right]. \end{aligned} \tag{7.14}$$

On the superconductor surface it reduces to

$$B_x(x,y,0) = -\frac{\mu_0 I a}{\pi(x^2+a^2)}, \qquad B_y(x,y,0) = B_z(x,y,0) = 0, \tag{7.15}$$

showing that the magnetic flux density is parallel to the surface. Figure 7.13 shows the magnetic flux lines. Then, from Eq. (7.15), we determine the density of current induced on the superconductor surface to be

$$\tau = -\frac{Ia}{\pi(x^2+a^2)}. \tag{7.16}$$

The total current is

$$\int \mathrm{d}x\, \tau = -\frac{Ia}{\pi}\int_{-\infty}^{\infty}\frac{\mathrm{d}x}{x^2+a^2} = -I. \tag{7.17}$$

That is, the total current is equal to the image current. The force on the given current I caused by the current induced on the superconductor surface is equal to the force by the image current, and its magnitude in a unit length is

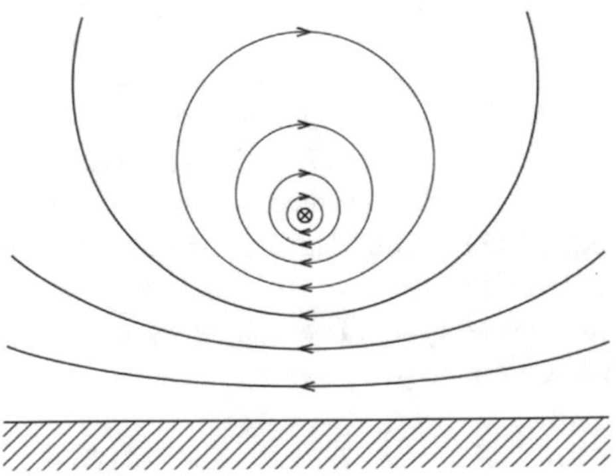

Fig. 7.13 Magnetic flux lines produced by given current and current induced on the superconductor surface

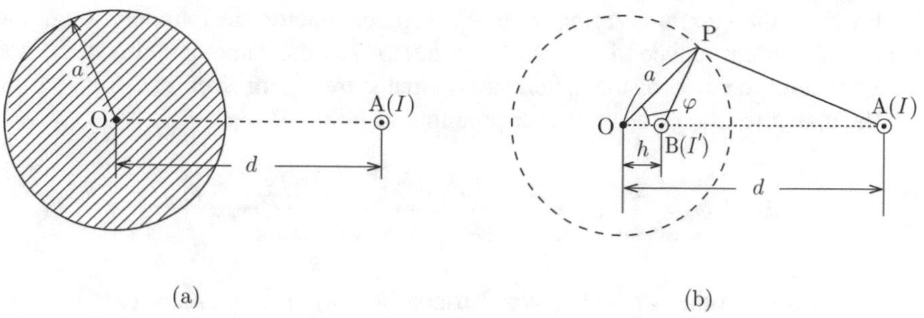

Fig. 7.14 **a** Long superconducting cylinder and straight current parallel to it and **b** image current placed on line B after removing the superconductor

$$F' = \frac{\mu_0 I^2}{4\pi a}. \tag{7.18}$$

This force is repulsive ($F' > 0$). This is called the image force.

One can show that the current induced on the superconductor surface completely cancels the magnetic flux density produced by the given current in the superconductor ($z < 0$).

Suppose we apply current I through a line, A, separated by distance d from the central axis O of a grounded parallel long superconducting cylinder of radius a, as shown in Fig. 7.14a. Now, we determine the vector potential outside the superconductor. We virtually remove the superconductor and apply an image current I' through a line, B, separated by h from the central axis of the superconducting cylinder, as shown in Fig. 7.14b, similarly to the example in Sect. 2.2. The vector potential at point P on the superconductor surface is given by

$$A_z = \frac{\mu_0 I}{2\pi} \log \frac{R_0}{(a^2 + d^2 - 2ad\cos\varphi)^{1/2}} + \frac{\mu_0 I'}{2\pi} \log \frac{R_0'}{(a^2 + h^2 - 2ah\cos\varphi)^{1/2}}, \tag{7.19}$$

where φ is the angle POA. In the above, R_0 and R_0' are distances to a suitable reference point and are not important quantities. For the vector potential to be constant and independent of φ, the following conditions should be satisfied:

$$I' = -I, \tag{7.20}$$

$$\frac{2ad}{a^2 + d^2} = \frac{2ah}{a^2 + h^2}. \tag{7.21}$$

Equation (7.21) reduces to

$$h = \frac{a^2}{d}. \tag{7.22}$$

In this case, the current-carrying wire and superconductor are infinitely long, and hence, the current induced in the superconductor has the same magnitude. This is different from the case of the spherical conductor treated in Sect. 2.2.

The above results give the vector potential at point (R, φ),

$$A_z(R, \varphi) = \frac{\mu_0 I}{2\pi} \log \frac{d[R^2 + (a^2/d)^2 - 2(a^2 R/d) \cos \varphi]^{1/2}}{a(R^2 + d^2 - 2Rd \cos \varphi)^{1/2}}, \tag{7.23}$$

when it is zero on the superconductor surface ($R = a$). That is, $R_0' = (a/d)R_0$. We can calculate the magnetic flux density outside the superconductor using Eq. (7.23) (see Exercise 7.6). Figure 7.15 shows the magnetic flux lines. The current density on the superconductor surface is obtained as

$$\tau(\varphi) = \frac{1}{\mu_0} B_\varphi(R = a) = -\frac{1}{\mu_0} \left(\frac{\partial A_z}{\partial R} \right)_{R=a} = -\frac{I(d^2 - a^2)}{2\pi a(a^2 + d^2 - 2ad \cos \varphi)}. \tag{7.24}$$

The total current is

$$\int_0^{2\pi} \tau(\varphi) \, a\mathrm{d}\varphi = -\frac{I(d^2 - a^2)}{\pi} \int_0^{\pi} \frac{\mathrm{d}\varphi}{a^2 + d^2 - 2ad \cos \varphi} = -I \tag{7.25}$$

and agrees with the image current, Eq. (7.20). In the above, the following formula was used.

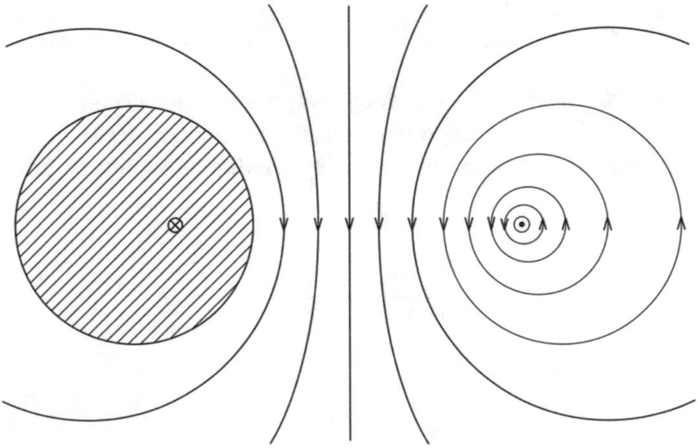

Fig. 7.15 Magnetic flux lines produced by long superconducting cylinder and straight current parallel to it

$$\int_0^{\pi} \frac{d\varphi}{1 - k\cos\varphi} = \frac{\pi}{(1-k^2)^{1/2}}. \tag{7.26}$$

This problem corresponds to estimating electric potential in Exercise 2.9.

Example 7.5. In the above, we discussed the vector potential when straight current I is placed outside the grounded superconducting cylinder, as shown in Fig. 7.14a. Determine the vector potential when the superconductor is isolated.

Solution 7.5. In this situation, the total current flowing in the superconductor must be zero. Hence, we use a superposition. That is, we obtain the current distribution by superposing currents $-I$ and I in a way that makes the vector potential of the superconductor constant. Current $-I$ is distributed according to Eq. (7.24) and I has uniform distribution on the superconductor surface. The former current and the external current give the vector potential, Eq. (7.23), which we denote as $A_{1z}(R, \varphi)$. The latter current gives the vector potential, Eq. (7.12a). Both of them satisfy the requirement to be constant on the superconductor surface. Hence, the sum of these components gives the unique solution. If we determine the vector potential of current I so that it is zero at a reference point, $R = R_0(> a)$, we have.

$$A_z(R, \varphi) = A_{1z}(R, \varphi) + \frac{\mu_0 I}{2\pi}\log\frac{R_0}{R}.$$

◇

7.3 Meissner State

Suppose that a superconducting sphere of radius a is put in a uniform magnetic flux density, B_0 (see Fig. 7.16). The superconducting current flows on the surface to cancel the applied magnetic flux density inside the superconductor. Here, we determine the density of this current and the magnetic flux density outside the superconductor. We use polar coordinates; the origin is at the center of the superconductor and the z-axis is parallel to the direction of the applied magnetic flux density.

When a conducting sphere is in a uniform electric field, the boundary condition is fulfilled by assuming an electric dipole of suitable moment placed at the center of the conductor, as discussed in Sect. 2.3. It seems useful to use a similar method assuming a magnetic moment placed at the center of the superconductor. In fact, the dipole moment of a pair of magnetic charges, which correspond to a pair of electric charges, is equivalent to the magnetic moment of a closed current, as shown in Sect. 6.8. This virtual closed current at the center flows along the azimuthal direction similarly to the real current on the surface.

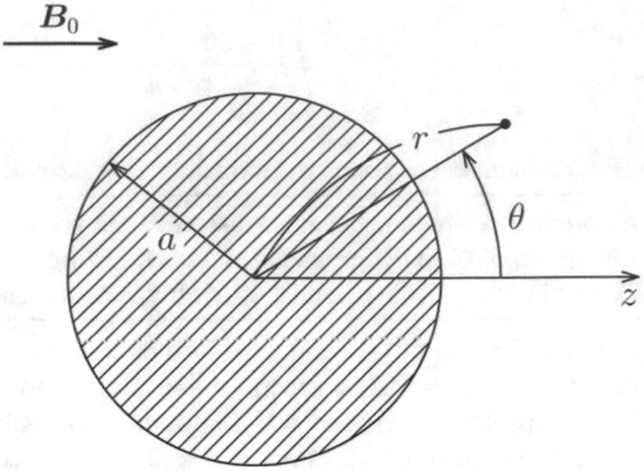

Fig. 7.16 Superconducting sphere in uniform magnetic flux density

The vector potential outside the superconducting sphere has only the azimuthal component, A_φ, corresponding to the current. This component consists of $A_{f\varphi}$ caused by the uniform magnetic flux density B_0 and $A_{d\varphi}$ caused by the magnetic moment m. The former component is given by

$$A_{f\varphi} = \frac{B_0 r}{2} \sin\theta. \tag{7.27}$$

Confirm for yourself that the following conditions are satisfied:

$$\frac{1}{r\sin\theta} \cdot \frac{\partial}{\partial\theta}(\sin\theta A_{f\varphi}) = B_0 \cos\theta, \quad -\frac{1}{r} \cdot \frac{\partial}{\partial r}(r A_{f\varphi}) = -B_0 \sin\theta.$$

The latter component is given by Eq. (6.43),

$$A_{d\varphi} = \frac{\mu_0 m \sin\theta}{4\pi r^2}. \tag{7.28}$$

Thus, the vector potential is

$$A_\varphi = A_{f\varphi} + A_{d\varphi} = \left(\frac{B_0 r}{2} + \frac{\mu_0 m}{4\pi r^2}\right) \sin\theta. \tag{7.29}$$

The requirement that it is zero on the superconductor surface ($r = a$) gives

$$m = -\frac{2\pi a^3 B_0}{\mu_0}. \tag{7.30}$$

Hence, we determine the vector potential to be

$$A_\varphi = \frac{B_0}{2}\left(r - \frac{a^3}{r^2}\right)\sin\theta. \tag{7.31}$$

This satisfies the boundary condition on the superconductor surface and Laplace's equation. Hence, this is the solution. Inside the superconductor, the solution is $A_\varphi = 0$.

We obtain the magnetic flux density outside the superconductor as

$$B_r = \frac{1}{r\sin\theta}\cdot\frac{\partial}{\partial\theta}(\sin\theta A_\varphi) = B_0\left(1 - \frac{a^3}{r^3}\right)\cos\theta, \tag{7.32a}$$

$$B_\theta = -\frac{1}{r}\cdot\frac{\partial}{\partial r}(rA_\varphi) = -B_0\left(1 + \frac{a^3}{2r^3}\right)\sin\theta, \tag{7.32b}$$

$$B_\varphi = 0. \tag{7.32c}$$

Figure 7.17 shows magnetic flux lines on the plane including the z-axis. Equation (7.32a) shows that the magnetic flux lines are parallel to the superconductor surface, $B_r(r = a) = 0$. We find from Eq. (7.32b) that the magnitude of the magnetic flux density takes the maximum value, $3B_0/2$, on the equator ($\theta = \pm\pi/2$). The current density in the azimuthal direction on the superconductor surface is

$$\tau = \frac{B_\theta(r = a)}{\mu_0} = -\frac{3B_0}{2\mu_0}\sin\theta. \tag{7.33}$$

This azimuthal current produces a magnetic flux density opposite to the applied one. Using Eq. (7.30), we have the magnetic moment in a unit volume of the superconductor,

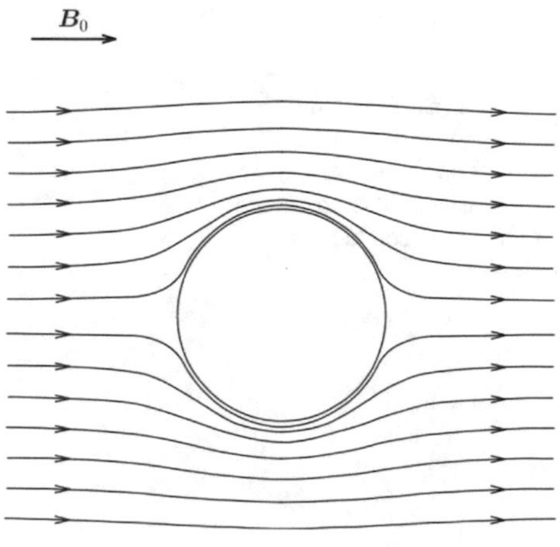

Fig. 7.17 Magnetic flux lines around superconducting sphere in uniform magnetic flux density

$$M = -\frac{3B_0}{2\mu_0}. \tag{7.34}$$

This is called **magnetization**. This corresponds formally to the magnetization of magnetic materials.

The magnetization in a superconductor is given by the magnetic moment in a unit volume of the superconductor as defined above. In the above example, the magnetic flux density that the superconductor experiences on its surface is different from the applied value because of the geometry of the superconductor. Here, we consider a simple case where there is no such geometrical effect. Suppose we apply magnetic flux density B_0 parallel to a long superconducting cylinder along the z-axis of radius a (see Fig. 7.18). The current flows along the negative azimuthal direction and produces a magnetic moment along the negative z-axis. The surface current density is

$$\tau = \frac{B_0}{\mu_0}. \tag{7.35}$$

The magnetic moment in a unit length due to this current is

$$m' = -\pi a^2 \tau = -\frac{\pi a^2 B_0}{\mu_0}. \tag{7.36}$$

Hence, the magnetization is

$$M = -\frac{B_0}{\mu_0}. \tag{7.37}$$

We assume that the superconductor is a type 2 superconductor and is in the mixed state with the internal magnetic flux density B (see Sect. A3.2 in the Appendix). Then, a similar discussion derives $\tau = (B_0 - B)/\mu_0$ and we have

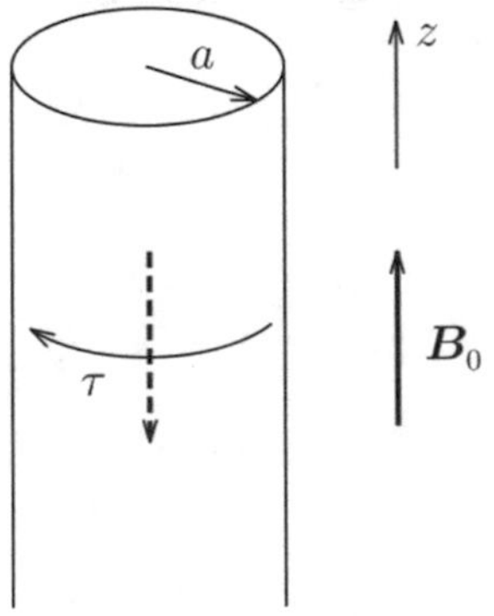

Fig. 7.18 Surface current and resultant magnetic moment (*broken line*) when magnetic flux density is applied parallel to long superconducting cylinder

$$M = -\frac{B_0 - B}{\mu_0}. \tag{7.38}$$

This is the common definition of magnetization of a superconductor when it is not influenced by the geometrical effect.

The magnetic flux density produced by the magnetic moment of a small closed current can be expressed using a magnetic scalar potential:

$$\phi_c = \frac{\mu_0 m}{4\pi r^2}\cos\theta, \tag{7.39}$$

as given by Eq. (6.46). Hence, the magnetic flux density around the spherical superconductor treated in Sect. 7.3 can also be expressed using the magnetic scalar potential. Since the magnetic scalar potential of the applied uniform magnetic flux density is given by

$$\phi_f = -B_0 r\cos\theta, \tag{7.40}$$

the magnetic scalar potential in the space outside the superconductor is

$$\phi_m = \phi_c + \phi_f = -\left(B_0 r - \frac{\mu_0 m}{4\pi r^2}\right)\cos\theta. \tag{7.41}$$

Since the magnetic flux density is parallel to the superconductor surface, the following condition must be satisfied:

$$-\left(\frac{\partial \phi_m}{\partial r}\right)_{r=a} = -\left(B_0 + \frac{\mu_0 m}{2\pi a^3}\right)\cos\theta = 0. \tag{7.42}$$

Thus, we have $m = -2\pi a^3 B_0/\mu_0$ and the magnetic scalar potential is

$$\phi_m = -B_0\left(r + \frac{a^3}{2r^2}\right)\cos\theta. \tag{7.43}$$

The magnetic flux density given by Eq. (7.32) can be derived using the relationship of Eq. (6.50). Thus, this problem can be solved using magnetic scalar potential. However, it cannot give us a correct image of the phenomenon. For example, the magnetic scalar potential on the superconductor surface is not constant but varies as

$$\phi_m(r = a) = -\frac{3a}{2}B_0\cos\theta. \tag{7.44}$$

It takes on a maximum value, $3aB_0/2$, at the left pole ($\theta = \pi$) and a minimum value, $-3aB_0/2$, at the right pole ($\theta = 0$). In addition, the magnetic scalar potential is zero inside the superconductor. The value of the magnetic scalar potential is not continuous on the superconductor surface on which the current flows. This is quite different from the vector potential.

The above result indicates that the electrostatic induction for the spherical conductor treated in Sect. 2.3 can be solved using the vector potential. The obtained result does not give us a correct image of the phenomenon, however.

Example 7.6. An infinitely long superconducting cylinder of radius a is in a uniform perpendicular magnetic flux density of B_0. Determine the vector potential and magnetic flux density outside the superconductor and current density on the superconductor surface.

Solution 7.6. We use cylindrical coordinates with the z-axis at the central axis of the superconductor and the azimuthal angle measured from the direction of the applied magnetic flux density. From a similarity with Example 2.7, we can expect the following treatment to be useful for determining the vector potential: We virtually remove the superconductor and place a magnetic dipole line produced by a pair of anti-parallel straight currents at the axis. We denote the magnitude of the magnetic moment in a unit length along the z-axis by $\hat{m}$. From Eq. (6.53), the vector potential is given by

$$A_z(R, \varphi) = \left(B_0 R + \frac{\mu_0 \hat{m}}{2\pi R} \right) \sin \varphi,$$

where the first and second terms are components of the applied magnetic flux density and magnetic moment, respectively. The requirement $A_z(R = a) = 0$ gives

$$\hat{m} = -\frac{2\pi a^2 B_0}{\mu_0}.$$

Thus, the vector potential outside the superconductor is

$$A_z(R, \varphi) = B_0 \left(R - \frac{a^2}{R} \right) \sin \varphi.$$

We obtain the magnetic flux density as

$$B_R = \frac{1}{R} \cdot \frac{\partial A_z}{\partial \varphi} = B_0 \left(1 - \frac{a^2}{R^2} \right) \cos \varphi,$$

$$B_\varphi = -\frac{\partial A_z}{\partial R} = -B_0 \left(1 + \frac{a^2}{R^2} \right) \sin \varphi,$$

$$B_z = 0.$$

We can show that $B_R(R = a) = 0$. The surface current density is

$$\tau = \frac{B_\varphi(R = a)}{\mu_0} = -\frac{2B_0}{\mu_0} \sin \varphi.$$

The magnetization of the superconducting cylinder is

$$M = -\frac{2B_0}{\mu_0}.$$

◇

From the above examples, we can understand that the interior of a superconductor is completely shielded with zero magnetic flux density. This situation is unchanged even for the case of a hollow superconductor. Hence, if we completely surround a space with a superconductor, the effect of external magnetic flux density can be completely shielded in the space. This is called **magnetic shielding** and corresponds to the electrostatic shielding attained by a conductor.

7.4 Prediction of Superconductivity

Perfect diamagnetism ($\boldsymbol{B} = 0$) is a fundamental property of superconductors. It is well known that superconductors have another characteristic property, i.e., zero resistivity, although the zero resistivity is different from empirical property of usual materials described by Ohm's law.

As stated in Sect. 9.1, it was possible for someone to assume a material that shows perfect diamagnetism based on the ***E-B*** analogy even in the 19th Century just after the completion of Maxwell's theory. He would have been the first person who predicted superconductor with zero resistivity in the world.

Here, the proof of the zero resistivity property is treated. It may be supposed that the zero resistivity can be easily proved, since the current must continue to flow to shield the applied magnetic flux density. We have to consider the point, however, that such shielding can be accomplished by the diamagnetism of the material itself, as will be introduced in Chap. 9. Since this type of material was not discovered in the 19th Century, it was not possible to attribute the diamagnetic response to the shielding by the current or the intrinsic diamagnetism.

Hence, we have to consider the case in which a current is directly applied to a diamagnetic material. In this case, the current must flow on the surface of the material, as shown in Eq. (7.2). As shown in Fig. 7.19, here, we suppose that rectangle C is located inside the material in such a way that one side is placed on the surface and parallel to the current. The current density is integrated on C. The integrated value is not zero on the side on the surface, while the contribution from other three sides is zero. That is, we have

$$\oint_{\mathrm{C}} \boldsymbol{i} \cdot \mathrm{d}\boldsymbol{s} \neq 0. \tag{7.45}$$

If Ohm's law $\boldsymbol{E} = \rho_{\mathrm{r}}\boldsymbol{i}$ holds, Eq. (7.45) leads to

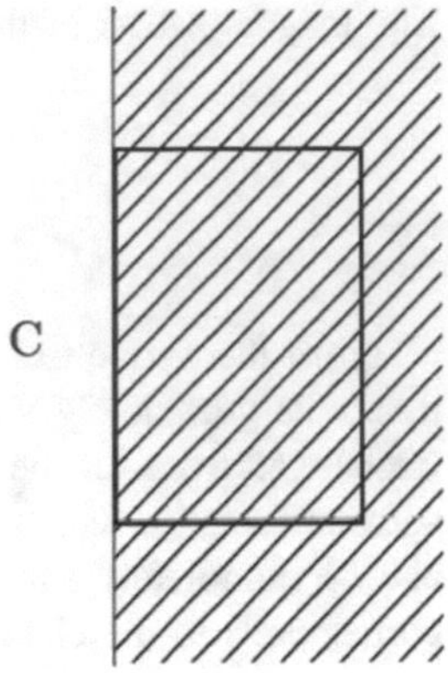

Fig. 7.19 Rectangle C with one side that stays on the surface of the diamagnetic material and is directed parallel to the current

$$\oint_{\mathrm{C}} \boldsymbol{E} \cdot \mathrm{d}\boldsymbol{s} \neq 0. \tag{7.46}$$

This contradicts Eq. (1.30), which must be satisfied under static conditions. Hence, the resistivity must be zero to be consistent with Eq. (1.30). The material that shows $\boldsymbol{B} = 0$ under any conditions is nothing else than a superconductor. Thus, the prediction of a diamagnetic material is equivalent to the prediction of a superconductor.

Column: (1) Penetration of Magnetic Flux into Superconducting Hollow Cylinder with Tilted Slit

Suppose we apply a magnetic flux density parallel to a long superconducting hollow cylinder with a tilted slit as shown in Fig. 7.20. Does the magnetic flux penetrate into the interior of the superconducting hollow cylinder without passing through the superconducting region?

If axial magnetic flux lines move along the radial direction, they surely have to pass through the superconducting region, indicating that the magnetic flux cannot penetrate into the interior. Is this true? In practice, the magnetic flux penetrates into the interior and its density is the same as that of the external one.

Remember here that a superposition holds for electric and magnetic quantities. That is, the resultant magnetic flux density is the sum of the applied one and one produced by the shielding current. The shielding current flows on the surface of the superconductor, as schematically shown in Fig. 7.21a. This is composed of the current flowing on the surface of a virtual hollow cylinder with no slit in Fig. 7.21b and that flowing in the opposite direction only in the region of the slit in Fig. 7.21c. The current in Fig. 7.21b produces a magnetic flux density of the same magnitude and opposite direction to the external one in the superconducting region, resulting in complete shielding there. The important thing is that the current in Fig. 7.21c produces a tilted magnetic flux at the slit. This shows the magnetic flux structure during the penetration. That is, the magnetic flux is tilted when it penetrates into the interior.

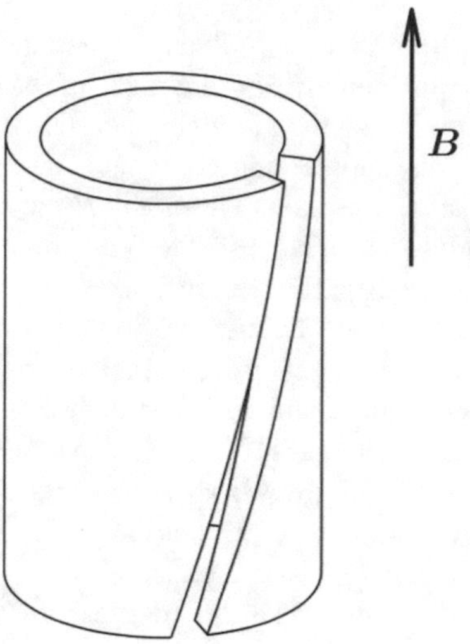

Fig. 7.20 Long superconducting hollow cylinder with a tilted slit in parallel magnetic flux density

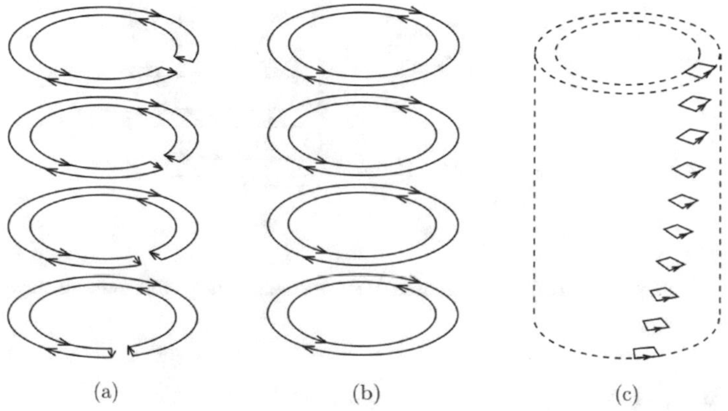

Fig. 7.21 **a** Flow of current on the superconductor surface, **b** that on the surface of virtual superconductor with no slit, and **c** that in the opposite direction in the slit region. The current in (**a**) is obtained by superposing those in (**b**) and (**c**)

Column: (2) Intermediate State

When we apply an external magnetic flux density to a superconducting sphere, the superconductor expels the magnetic flux from its interior. In this case, some part of the superconductor experiences a magnetic flux density higher than the applied value because of the geometrical effect, as shown in Fig. 7.17. Hence, even when the applied magnetic flux density is below the critical value, B_c, the local magnetic flux density can exceed B_c, resulting in a breakdown of the superconductivity. If we assume that the superconductivity is completely broken, the magnetic flux will completely penetrate the superconductor, resulting in a magnetic flux density of the same value as the external one. This will bring about a recovery of the superconductivity. However, this is contradictory. In reality, the superconductor goes into a state in which the superconductivity is partially broken. This is called the **intermediate state**. In this state the superconductor is in the layered structure composed of the superconducting region with perfect extrusion of magnetic flux and normal region with penetration of magnetic flux, as shown in Fig. 7.22. Since the size of these layers is of the order of several 10 μm, the magnetic structure can be regarded as a uniform partially diamagnetic structure on a macroscopic scale. However, we cannot use the condition (7.4), so we need another method to determine the internal magnetic flux density and surface current density. This is the method using the boundary conditions on the magnetic flux density and magnetic field, which will be described in Chap. 9 (see Exercise 9.10). Figure 7.23 shows the magnetization curve of a spherical type 1

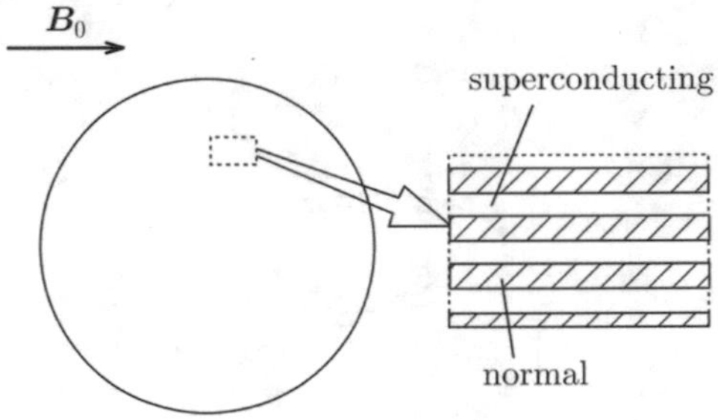

Fig. 7.22 Multilayered structure composed of superconducting and normal layers in the intermediate state

Fig. 7.23 Magnetization curve of spherical type 1 superconductor

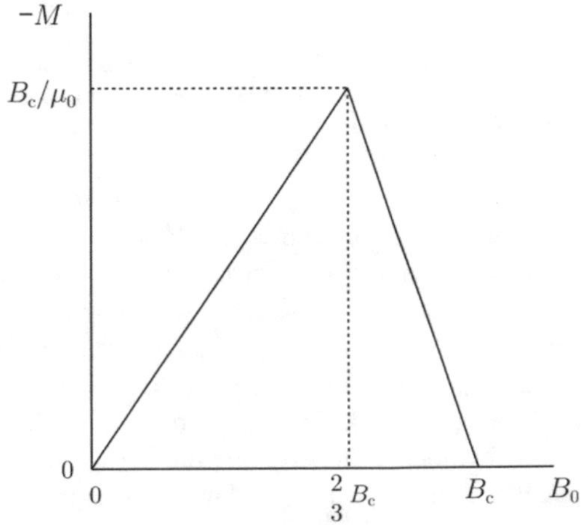

superconductor. The ascending line starting from the origin represents the perfect diamagnetic characteristic given by Eq. (7.34), and the descending line shows the characteristic in the intermediate state.

Exercises

7.1. Determine the magnetic flux density and vector potential when we apply currents I_1 and I_2 to the inner and outer superconductors, respectively, for the coaxial superconductor in Fig. 7.6.

7.2. A current I is given to the left superconductor in Example 7.3. Determine the current that appears on each superconductor surface and the magnetic flux density inside and outside the superconductors.

7.3. When a current flows uniformly with a surface density τ on a thin sheet conductor, the magnetic flux density near the sheet is given by Eq. (6.28). However, Eq. (7.8) yields double this magnetic flux density near the superconductor surface with the same current density. Discuss the reason for the difference.

7.4. When we put straight current I at a distance a from a wide superconductor surface, the current given by Eq. (7.16) is induced on the superconductor surface. Prove that the Lorentz force exerted on I by the induced current is given by Eq. (7.18).

7.5. Straight current I is placed at distances a and b from two flat superconductor surfaces that are perpendicular to each other, as shown in Fig. E7.1. Determine the vector potential and magnetic flux density in the vacuum.

7.6. Determine the magnetic flux density in the space around a superconducting cylinder using the vector potential given by Eq. (7.23).

7.7. The vector potential is given by Eq. (7.13) for the case of a straight current and a wide superconductor surface. Determine the equivector potential surface.

7.8. Prove that the magnetic flux density vector given by Eq. (7.14) is parallel to the equivector potential surface discussed in Exercise 7.7.

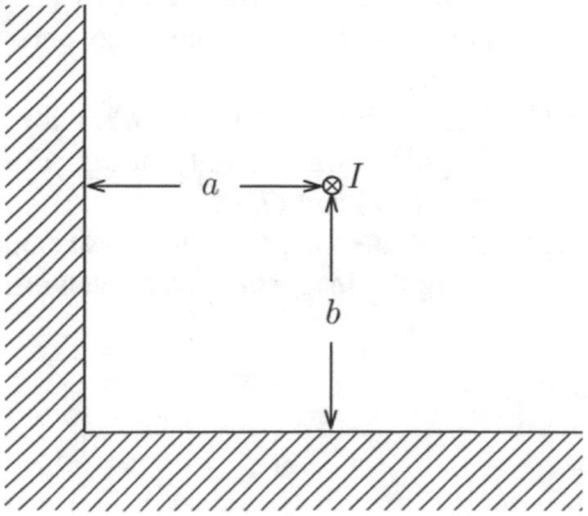

Fig. E7.1 Two perpendicular flat superconductor surfaces and straight current I

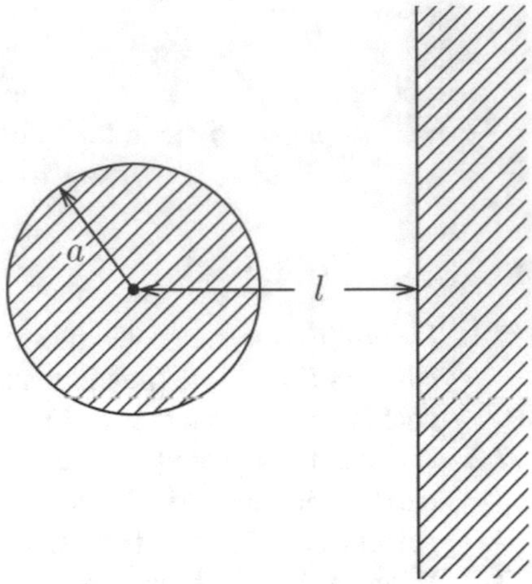

Fig. E7.2 Superconducting cylinder parallel to infinite flat superconductor surface

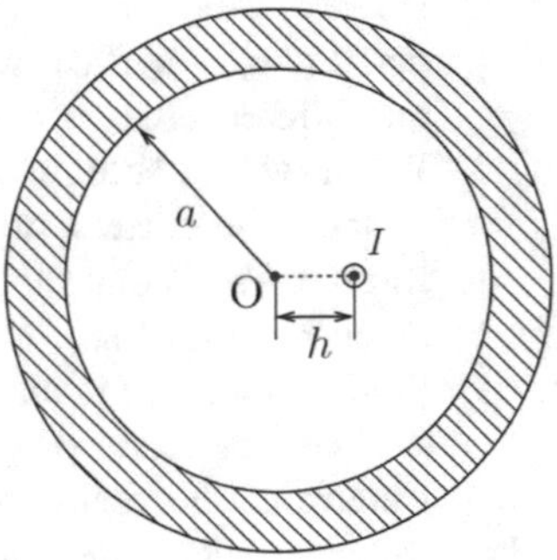

Fig. E7.3 Hollow cylindrical superconductor and straight current inside the superconductor

7.9. A long superconducting cylinder of radius a is placed at distance $l(> a)$ from an infinite flat superconductor surface, as shown in Fig. E7.2, and a current I is applied to the superconducting cylinder. Determine the current density on the surfaces of the two superconductors.

7.10. Straight current I is placed at distance h from the central axis, O, of a hollow cylindrical superconductor, as shown in Fig. E7.3. Determine the vector potential in the vacuum and current density on the inner surface of the superconductor.

7.11. Derive Eqs. (7.30) and (7.33) for a superconducting sphere in a uniform magnetic flux density using Eq. (7.1) with the boundary conditions, Eqs. (7.6) and (7.8).

7.12. Derive the magnetic flux density outside the superconductor in Example 7.6 using the magnetic scalar potential.

Chapter 8
Current Systems

Abstract This chapter covers magnetic phenomena in a current system composed of more than one current loop. The relationship between the current and the magnetic flux in each current loop is generally described in terms of the inductance coefficients. Coils are a typical example of this current system, and the inductance of various coils is treated. We use a current system composed of a superconductor to determine the magnetic energy of the system. The merit of this method is that the magnetic energy can be directly determined from the mechanical work needed against the magnetic force, similarly to the electric energy. The magnetic energy of the current system is described using the magnetic flux and the inductance coefficients.

8.1 Inductance

In Chap. 6, we learned that the magnetic flux is produced around currents. When a current, I, flows in a closed circuit, C, as shown in Fig. 8.1, the magnetic flux penetrating C is proportional to I:

$$\Phi = LI. \tag{8.1}$$

The proportional constant L is called **self-inductance**. The unit of self-inductance is [Wb/A] and is newly defined as [H] (**henry**). The self-inductance is determined only by the shape of C and is defined as a positive quantity. That is, the directions of the current and magnetic flux follow the right-hand rule.

Second, we suppose that there are two closed circuits and current I_1 flows along circuit C_1. The magnetic flux penetrating itself is expressed as

$$\Phi_1 = L_{11} I_1, \tag{8.2}$$

similarly to Eq. (8.1). The magnetic flux also penetrates the other circuit C_2, as illustrated in Fig. 8.2, and it is expressed as

T. Matsushita, *Electricity and Magnetism*, Undergraduate Lecture Notes in Physics,
https://doi.org/10.1007/978-3-030-82150-0_8

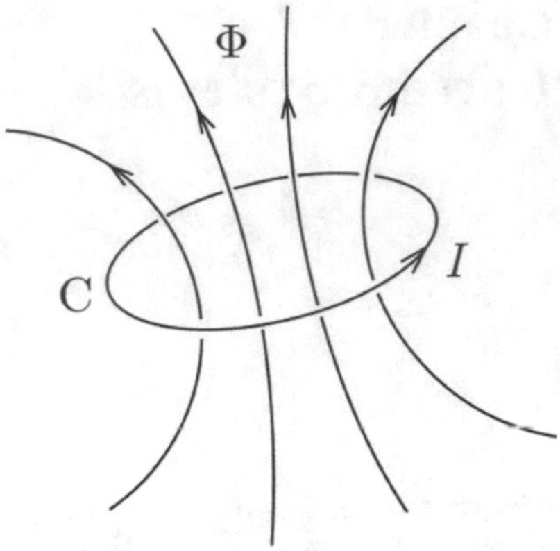

Fig. 8.1 Magnetic flux penetrating closed circuit produced by current flowing along itself

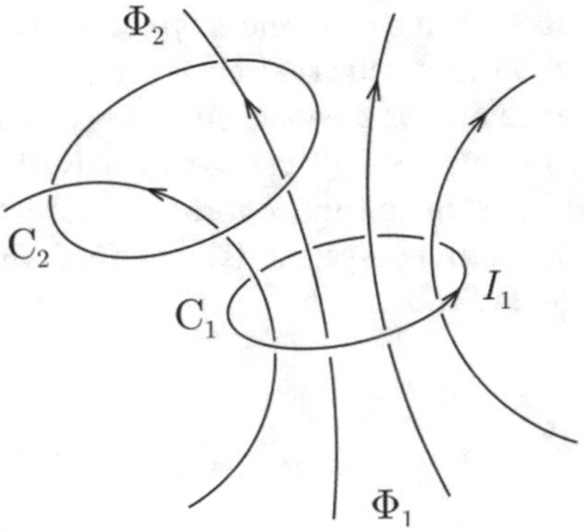

Fig. 8.2 Magnetic flux produced by current flowing along one of two closed circuits

$$\Phi_2 = L_{21} I_1. \tag{8.3}$$

The constant L_{21} is influenced by the geometrical arrangement of C_1 and C_2, while L_{11} is determined only by the shape of C_1. If current I_2 flows along C_2, the resultant magnetic flux penetrates C_2 itself and C_1. Thus, the magnetic fluxes penetrating C_1 and C_2 are formally given by

$$\Phi_1 = L_{11} I_1 + L_{12} I_2, \tag{8.4a}$$

$$\Phi_2 = L_{21} I_1 + L_{22} I_2. \tag{8.4b}$$

The self-inductances L_{11} and L_{22} are positive as mentioned above. The coefficients L_{12} and L_{21} are called **mutual inductances** and have the following relationship,

$$L_{12} = L_{21}. \tag{8.5}$$

The mutual inductance takes a positive or negative value depending on the directions of the current and magnetic flux.

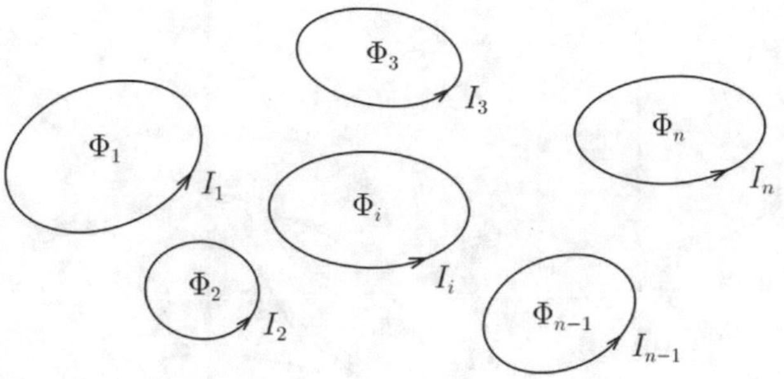

Fig. 8.3 System composed of n closed circuits

Extending the above case of two closed electric circuits, we consider a system composed of n electric circuits in Fig. 8.3, where current I_i flows in the i-th circuit C_i (i = 1, 2, …, n). We express the magnetic flux Φ_i penetrating C_i as

$$\Phi_i = \sum_{j=1}^{n} L_{ij} I_j. \tag{8.6}$$

In the above, the L_{ij}'s are **inductance coefficients**. The L_{ii}'s are self-inductances, and L_{ij}'s ($i \neq j$) are mutual inductances. The reciprocity theorem

$$L_{ij} = L_{ji} \tag{8.7}$$

holds generally.

Now we prove Eq. (8.7). From Eq. (6.34), the vector potential produced by current I_j flowing in the j-th closed circuit C_j is given by

$$\boldsymbol{A}(\boldsymbol{r}) = \frac{\mu_0 I_j}{4\pi} \oint_{\mathrm{C}_j} \frac{\mathrm{d}\boldsymbol{r}_j}{|\boldsymbol{r} - \boldsymbol{r}_j|}. \tag{8.8}$$

With the aid of Eq. (6.35), we rewrite the magnetic flux penetrating C_i as

$$\Phi_{ij} = \oint_{\mathrm{C}_i} \boldsymbol{A}(\boldsymbol{r}_i) \cdot \mathrm{d}\boldsymbol{r}_i = \frac{\mu_0 I_j}{4\pi} \oint_{\mathrm{C}_i} \oint_{\mathrm{C}_j} \frac{\mathrm{d}\boldsymbol{r}_i \cdot \mathrm{d}\boldsymbol{r}_j}{|\boldsymbol{r}_i - \boldsymbol{r}_j|}. \tag{8.9}$$

Hence, the mutual inductance is given by

$$L_{ij} = \frac{\mu_0}{4\pi} \oint_{\mathrm{C}_i} \oint_{\mathrm{C}_j} \frac{\mathrm{d}\boldsymbol{r}_i \cdot \mathrm{d}\boldsymbol{r}_j}{|\boldsymbol{r}_i - \boldsymbol{r}_j|}. \tag{8.10}$$

This is called **Neumann's formula**. The result is the same even if subscripts i and j are exchanged. Thus, Eq. (8.7) is proved.

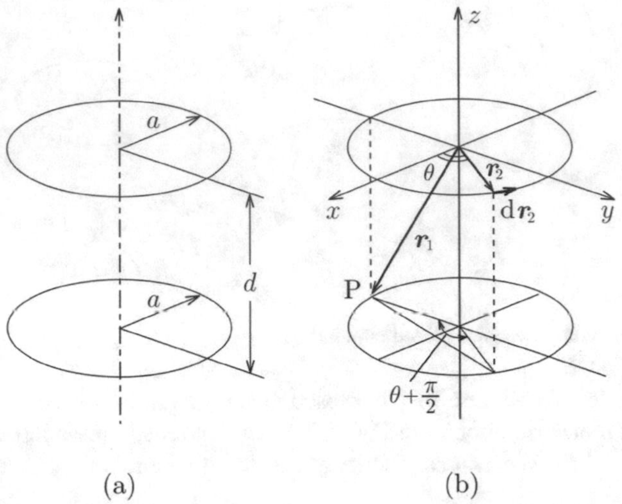

Fig. 8.4 Two circular coils with common axis: **a** arrangement and **b** coordinates

Here, we show an example of calculating a mutual inductance. Two circular coils of radius a are separated by distance d with a common central axis, as shown in Fig. 8.4a. The currents flow in the same direction. We define the coordinates as in Fig. 8.4b and position vector $\boldsymbol{r}_1$ for the lower coil is fixed at point P. Then, the contribution from a small region d$\boldsymbol{r}_2$ in the upper coil to the mutual inductance is written as

$$\frac{\mu_0 \mathrm{d}\boldsymbol{r}_2}{4\pi[2a^2(1+\sin\theta)+d^2]^{1/2}} = \frac{\mu_0 a \mathrm{d}\theta(-\boldsymbol{i}_x \sin\theta + \boldsymbol{i}_y \cos\theta)}{4\pi[2a^2(1+\sin\theta)+d^2]^{1/2}}.$$

Integrating this for the upper coil, the y-component reduces to zero because of symmetry, and only the x-component, i.e., the tangential component at point P, remains. Here we put $\theta = 2\psi + \pi/2$. Then, after a simple calculation, we write the above integration with respect to $\boldsymbol{r}_2$ as

$$-\frac{\mu_0 k}{2\pi}\int_0^{\pi/2} \frac{1-2\sin^2\psi}{(1-k^2\sin^2\psi)^{1/2}}\mathrm{d}\psi,$$

where $k = 2a/(4a^2+d^2)^{1/2}$. Although the integration is not simplified any more, the above calculated vector is directed along d$\boldsymbol{r}_1$ and its value is constant. Hence, integrating with respect to $\boldsymbol{r}_1$ gives simply the factor $2\pi a$. Thus, we obtain the mutual inductance as

$$M = \frac{\mu_0 a}{k}[(2 - k^2)F(k) - 2E(k)], \tag{8.11}$$

where $F(k)$ and $E(k)$ are complete elliptic integrals of the first and second kind, respectively:

$$F(k) = \int_0^{\pi/2} \frac{1}{(1 - k^2 \sin^2 \psi)^{1/2}} d\psi, \tag{8.12}$$

$$E(k) = \int_0^{\pi/2} (1 - k^2 \sin^2 \psi)^{1/2} d\psi. \tag{8.13}$$

When the current path has a finite cross-sectional area, it is not easy to define the magnetic flux penetrating the electric circuit. This may indicate that the inductance cannot be exactly defined. However, the inductance can be exactly determined using the magnetic energy, as will be shown later (see Exercise 8.3).

Example 8.1. Determine the self-inductance of a unit length of the parallel-wire transmission line of radius a separated by distance d in Fig. 8.5. Assume that d is much larger than a and we can neglect the magnetic flux inside the conductors.

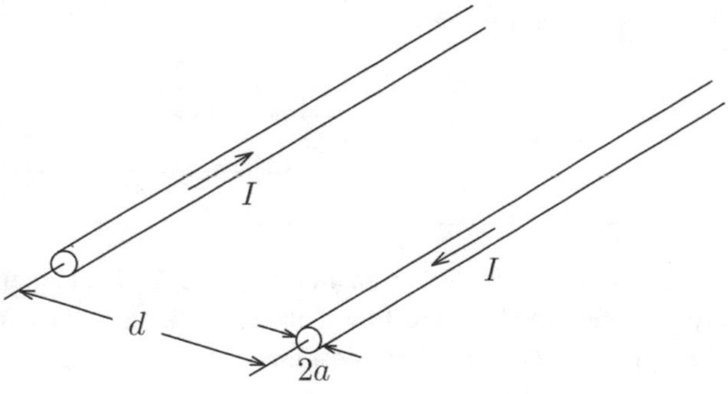

Fig. 8.5 Parallel-wire transmission line

Solution 8.1. We suppose that current I flows as shown in the figure. We assume a plane that includes the axes of the two cylindrical conductors and calculate the magnetic flux penetrating the plane between the two conductors. The magnetic flux density produced by current I flowing along the left conductor at distance x from its central axis is

$$B_1 = \frac{\mu_0 I}{2\pi x}$$

and is directed downwards. Hence, the magnetic flux in a unit length produced by this current is

$$\Phi_1' = \int_a^{d-a} \frac{\mu_0 I}{2\pi x} \mathrm{d}x = \frac{\mu_0 I}{2\pi} \log \frac{d-a}{a}.$$

The magnetic flux produced by the current along the right conductor is the same, and we have the self-inductance of a unit length as

$$L' = \frac{2\Phi_1'}{I} = \frac{\mu_0}{\pi} \log \frac{d-a}{a} \simeq \frac{\mu_0}{\pi} \log \frac{d}{a}.$$

◇

Example 8.2. Determine the mutual inductance in a unit length between the two parallel-wire transmission lines in Fig. 8.6. We define the current directions as in the figure.

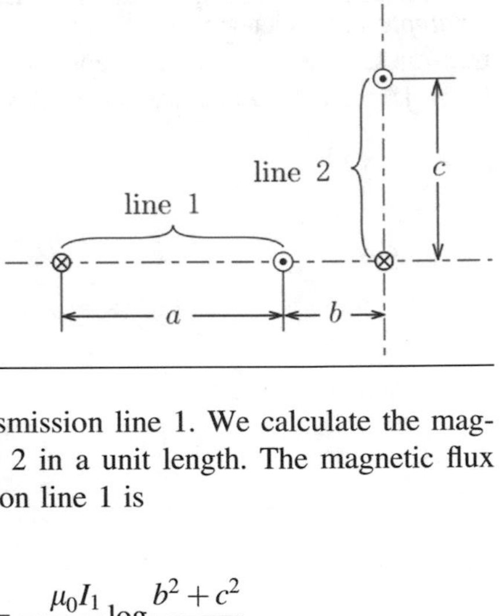

Fig. 8.6 Two parallel-wire transmission lines

Solution 8.2. We apply current I_1 to transmission line 1. We calculate the magnetic flux that penetrates transmission line 2 in a unit length. The magnetic flux produced by the right current of transmission line 1 is

$$\Phi_\mathrm{r}' = -\frac{\mu_0 I_1}{2\pi} \int_b^{\sqrt{b^2+c^2}} \frac{\mathrm{d}r}{r} = -\frac{\mu_0 I_1}{4\pi} \log \frac{b^2+c^2}{b^2}.$$

The magnetic flux produced by the left current of transmission line 1 is

$$\Phi_1' = \frac{\mu_0 I_1}{2\pi} \int_{a+b}^{\sqrt{(a+b)^2+c^2}} \frac{\mathrm{d}r}{r} = \frac{\mu_0 I_1}{4\pi} \log \frac{(a+b)^2+c^2}{(a+b)^2}.$$

The total magnetic flux penetrating transmission line 2 is

$$\Phi_2' = \Phi_r' + \Phi_l' = -\frac{\mu_0 I_1}{4\pi}\log\frac{(b^2+c^2)(a+b)^2}{b^2[(a+b)^2+c^2]},$$

and the mutual inductance is given by

$$L_{21}' = \frac{\Phi_2'}{I_1} = -\frac{\mu_0}{4\pi}\log\frac{(b^2+c^2)(a+b)^2}{b^2[(a+b)^2+c^2]}.$$

Confirm for yourself that the same result is obtained by calculating the magnetic flux produced by transmission line 2 that penetrates transmission line 1.

◇

Example 8.3. A circular coil is placed just between a parallel-wire transmission line separated by d, as shown in Fig. 8.7. These stay on a common plane. Determine the mutual inductance between the transmission line and circular coil.

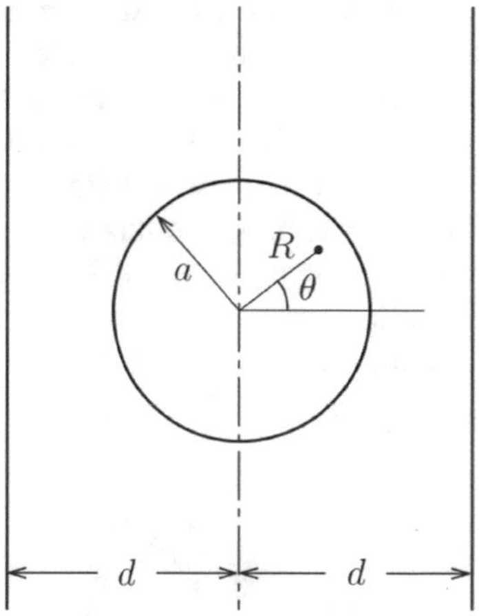

Fig. 8.7 Circular coil placed at the center of the parallel-wire transmission line

Solution 8.3. We calculate the magnetic flux produced by current I flowing along the left line of the transmission line that penetrates the circular coil. Using two-dimensional polar coordinates with the origin at the center of the coil, the magnetic flux density at (R, θ) is $B = \mu_0 I/[2\pi(d + R\cos\theta)]$. Hence, the magnetic flux is

$$\Phi_1 = \frac{\mu_0 I}{2\pi}\int_0^a\int_0^{2\pi}\frac{R\mathrm{d}R\mathrm{d}\theta}{d+R\cos\theta}.$$

The integral with respect to angle θ is carried out using Eq. (7.26):

$$\Phi_1 = \mu_0 I \int_0^a \frac{R\mathrm{d}R}{(d^2 - R^2)^{1/2}} = \mu_0 I[d - (d^2 - a^2)^{1/2}].$$

Since the magnetic flux due to the current flowing along the right line is the same, we obtain the mutual inductance as

$$M = \frac{2\Phi_1}{I} = 2\mu_0[d - (d^2 - a^2)^{1/2}].$$

◇

8.2 Coils

In Sect. 3.2, we learned about the electric property of capacitors used for storing electric charges in electric circuits. The component used to store the magnetic flux is a **coil**. Coils are also used for other purposes such as producing various magnetic flux densities or generating electric power, that will be covered in Chap. 10. Here, we introduce the magnetic property of coils.

When we apply a uniform current, I, to a parallel-plate transmission line in Fig. 8.8a, a magnetic flux with a uniform density is produced in the space between the two plates (see Fig. 8.8b). We assume that the distance d between the two plates is sufficiently small and the magnetic flux density on the outside can be neglected. The magnetic flux in the space is directed normally to the currents, and its density is

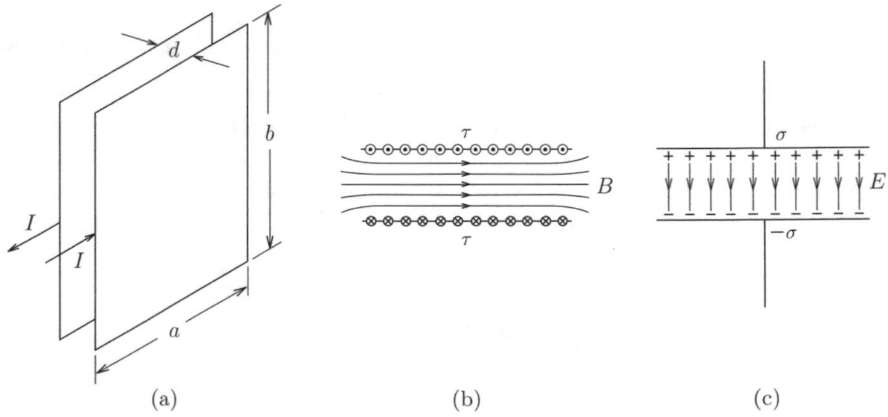

Fig. 8.8 **a** Current in parallel-plate transmission line, **b** magnetic flux lines, and **c** electric field lines in a capacitor of the same geometry

$$B = \frac{\mu_0 I}{b} = \mu_0 \tau, \tag{8.14}$$

where $\tau = I/b$ is the planar current in a unit width. The magnetic flux between the two plates is

$$\Phi = Bad = \frac{\mu_0 Iad}{b}. \tag{8.15}$$

Hence, we obtain the self-inductance as

$$L = \frac{\mu_0 ad}{b}. \tag{8.16}$$

The magnetic flux density produced by parallel planar currents corresponds to the electric field of strength $E = \sigma/\epsilon_0$ produced by planar electric charges in a parallel-plate capacitor (see Fig. 8.8c).

The coil used to produce a uniform magnetic flux density is a **solenoid coil**. For example, when we apply current I to a long solenoid coil with a winding of n turns in a unit length, the interior magnetic flux density is uniform with the value

$$B = \mu_0 nI, \tag{8.17}$$

as shown in Example 6.8. This value does not depend on the radius or length of the coil. For a coil of radius a, the magnetic flux that penetrates one turn of the coil is

$$\phi = \pi\mu_0 na^2 I. \tag{8.18}$$

Thus, the magnetic flux penctrating the coil of a unit length is

$$\Phi' = n\phi = \pi\mu_0 n^2 a^2 I. \tag{8.19}$$

Hence, the self-inductance in a unit length is given by

$$L' = \frac{\Phi'}{I} = \pi\mu_0 n^2 a^2. \tag{8.20}$$

When the length of the solenoid coil is l, its self-inductance is smaller than L' l and is expressed as.

$$L = K\left(\frac{2a}{l}\right)L'l. \tag{8.21}$$

In the above, $K(2a/l)$ is a function only of the ratio $2a/l$ and is called **Nagaoka's coefficient**. Figure 8.9 shows Nagaoka's coefficient.

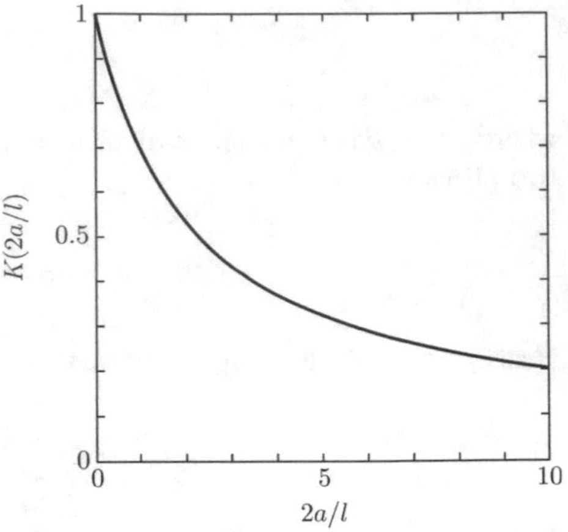

Fig. 8.9 Nagaoka's coefficient

A **spherical coil** is a special coil for producing a uniform magnetic flux density in a limited space. That is, when current flows on the surface of a sphere as given by Eq. (7.33), the interior magnetic flux density is uniform. Assume a spherical coil of radius a with N turns as in Fig. 8.10a. We take the number of turns in a unit zenithal length to be $[N/(2a)]\sin\theta$, where θ is the zenithal angle. When we apply current I, the surface current density on the sphere is

$$\tau = \frac{NI}{2a}\sin\theta \tag{8.22}$$

and from Eq. (7.33) we obtain the interior magnetic flux density as

$$B_0 = \frac{\mu_0 NI}{3a}. \tag{8.23}$$

To realize such a winding, the number of turns in a unit length along the axis is $N/(2a)$.

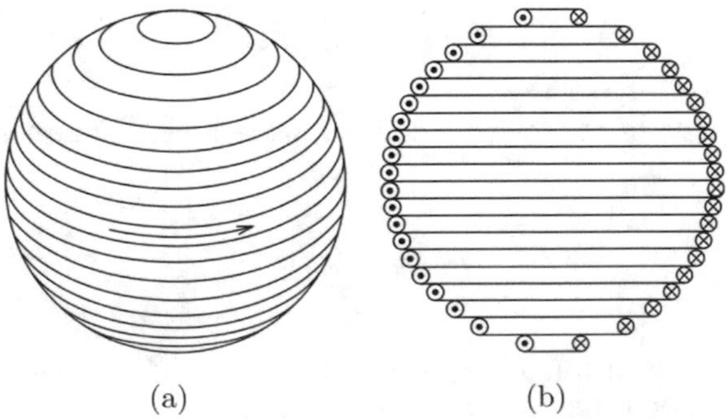

Fig. 8.10 Spherical coil: **a** geometry and **b** windings

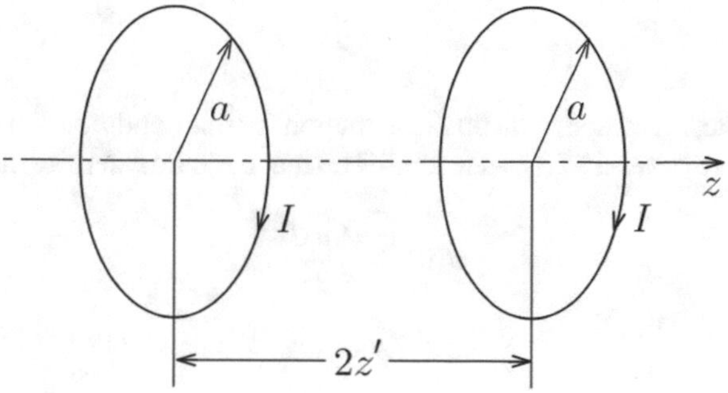

Fig. 8.11 Helmholtz coil

However, fabricating such a spherical coil is not easy, and the **Helmholtz coil** introduced below is commonly used. This coil consists of a pair of circular coils of the same size, as shown in Fig. 8.11. The radius of the circular coils is a, and the distance between the two coils arranged on the common axis is $2z'$. The center is defined to be $z = 0$. We apply current I to the two coils in the same direction. Using the result in Example 6.1, the magnetic flux density on the common axis is given by

$$B(z) = \frac{\mu_0 I a^2}{2}\left\{\frac{1}{[(z-z')^2+a^2]^{3/2}} + \frac{1}{[(z+z')^2+a^2]^{3/2}}\right\}. \tag{8.24}$$

When the distance between the two coils is too large, the magnetic flux density is locally minimum at the center, as shown in Fig. 8.12a. When this distance is too short, the variation in the magnetic flux density around the center is steep (see Fig. 8.12b). No uniform magnetic flux density is achieved in either of these two cases. We obtain the optimum arrangement under the condition $\mathrm{d}^2B/\mathrm{d}z^2 = 0$ at the center, $z = 0$, which gives

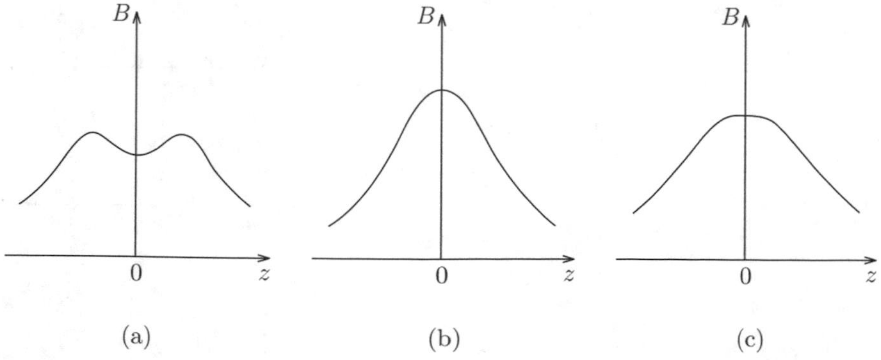

Fig. 8.12 Magnetic flux distribution along the central axis of the Helmholtz coil for the cases where the distance between the two coils is **a** too long, **b** too short, and **c** optimum

$$z' = \frac{a}{2}. \tag{8.25}$$

Figure 8.12c shows the magnetic flux distribution for this condition. The magnetic flux density is uniform over a fairly wide area. The magnetic flux density at the center is

$$B(0) = \frac{8\mu_0 I}{5\sqrt{5}a}. \tag{8.26}$$

◇

Example 8.4. We apply current I to a solenoid coil of radius a, length l, and N total number of turns. Determine the magnetic flux density on the central axis.

Solution 8.4. We define the z-axis on the central axis, as shown in Fig. 8.13, with the origin at the center of the coil. We regard the windings in the region z to z + dz as a one-turn coil. The current flowing there is $\mathrm{d}I = (NI/l)\mathrm{d}z$. Using the result in Example 6.1, the magnetic flux density at $z = z_0$ produced by this current is

$$\mathrm{d}B = \frac{\mu_0 \mathrm{d}I a^2}{2[(z-z_0)^2 + a^2]^{3/2}} = \frac{\mu_0 N I a^2}{2l[(z-z_0)^2 + a^2]^{3/2}}\mathrm{d}z.$$

Hence, the total magnetic flux density at the observation point is

$$B(z_0) = \int_{-l/2}^{l/2} \frac{\mu_0 N I a^2}{2l[(z-z_0)^2 + a^2]^{3/2}}\mathrm{d}z.$$

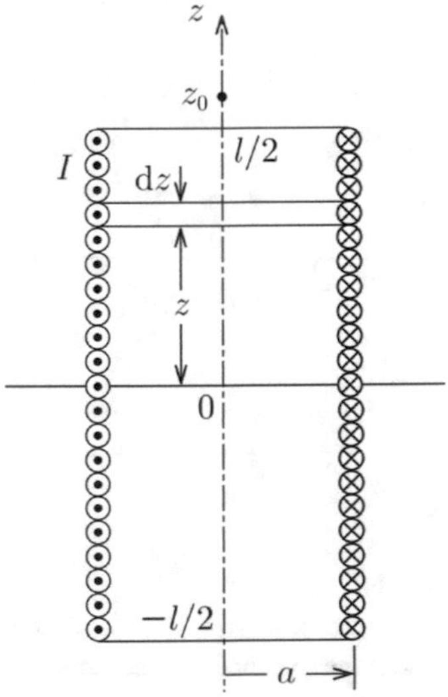

Fig. 8.13 Longitudinal cross section of solenoid coil

Here we define

$$z - z_0 = a \tan \theta$$

with

$$\tan \theta_1 = -\frac{1}{a}\left(\frac{l}{2} + z_0\right), \quad \tan \theta_2 = \frac{1}{a}\left(\frac{l}{2} - z_0\right).$$

Then, we calculate the total magnetic flux density to be

$$\begin{aligned} B(z_0) &= \frac{\mu_0 NI}{2l} \int_{\theta_1}^{\theta_2} \cos\theta \mathrm{d}\theta = \frac{\mu_0 NI}{2l}(\sin \theta_2 - \sin \theta_1) \\ &= \frac{\mu_0 NI}{2l} \left\{ \frac{l + 2z_0}{[(l + 2z_0)^2 + 4a^2]^{1/2}} + \frac{l - 2z_0}{[(l - 2z_0)^2 + 4a^2]^{1/2}} \right\}. \end{aligned}$$

◇

Example 8.5. Determine the self-inductance of the spherical coil of diameter a and total number of windings N shown in Fig. 8.10.

Solution 8.5. When current I is applied to the coil, the magnetic flux that penetrates one turn of the coil at zenithal angle θ is

$$\phi(\theta) = B_0 \pi (a \sin\theta)^2 = \frac{\mu_0}{3} \pi a N I \sin^2\theta,$$

where $B_0 = \mu_0 NI/3a$ is the uniform magnetic flux density in the coil given by Eq. (8.23). Hence, the total magnetic flux penetrating the coil is

$$\Phi = \int_0^{\pi} \frac{1}{2} N \phi(\theta) \sin\theta \mathrm{d}\theta = \frac{\mu_0}{6} \pi a N^2 I \int_0^{\pi} \sin^3\theta \mathrm{d}\theta = \frac{2\mu_0}{9} \pi a N^2 I.$$

The self-inductance is

$$L = \frac{\Phi}{I} = \frac{2\mu_0}{9} \pi a N^2.$$

◇

8.3 Magnetic Energy

The electric field fills the space between two electrodes in a charged capacitor, and we can regard the space as filled with electric energy. For a coil that stores the magnetic flux, we can also regard the interior space as filled with **magnetic energy**.

Here we suppose two superconductors in electrical contact with each other, as shown in Fig. 8.14. One of them is a movable plate. Assume that magnetic flux Φ is trapped within the space surrounded by the superconductors. We can realize this situation by using the field-cooled process shown in Fig. 7.11. A current flows on the inner surface to shield the superconductors from the magnetic flux. Hence, a repulsive force given by Eq. (6.9) is exerted on the movable superconducting plate. If the plate is displaced by distance x, the interior magnetic flux density changes to

$$B = \frac{\Phi}{(a+x)b} \tag{8.27}$$

and the density of current flowing on the inner surface changes to

$$\tau = \frac{B}{\mu_0} = \frac{\Phi}{\mu_0(a+x)b}. \tag{8.28}$$

We assume that a is much smaller than b. Since the magnetic flux density produced by the fixed superconductor is half of the value given by Eq. (8.27), we estimate the force on the movable plate to be

$$F = \frac{1}{2}\tau Bbh = \frac{\Phi^2 h}{2\mu_0(a+x)^2 b}, \tag{8.29}$$

which is directed along increasing x. This is an isolated system, and there is no electromagnetic interaction with the surroundings after the initial condition is established. Thus, this force is attributed to the variation in the magnetic energy U_m of this system. From the relationship

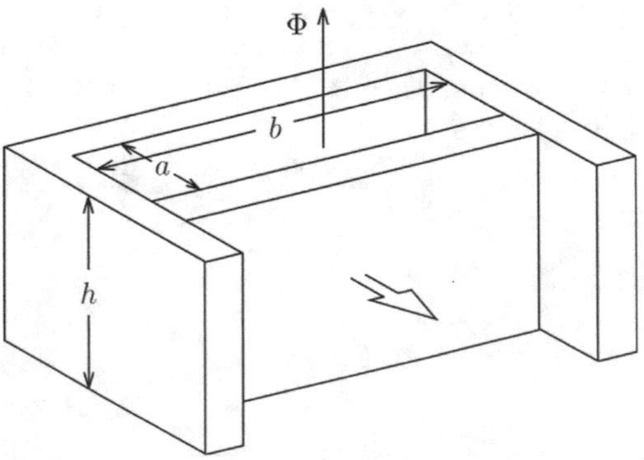

Fig. 8.14 Magnetic flux trapped in closed circuit composed of two superconductors. The superconducting plate is movable as shown by the arrow and in electrical contact with the fixed piece

$$F = -\frac{\partial U_{\mathrm{m}}}{\partial x}, \tag{8.30}$$

we estimate the magnetic energy as

$$U_{\mathrm{m}} = \frac{\Phi^2 h}{2\mu_0 (a+x)b} = \frac{1}{2\mu_0} B^2 (a+x)bh. \tag{8.31}$$

In the above, $(a + x)bh$ is the volume of the space in which the uniform magnetic flux is trapped. Hence, the **magnetic energy density** is given by

$$u_{\mathrm{m}} = \frac{1}{2\mu_0} B^2. \tag{8.32}$$

This is similar to the electric energy density given by Eq. (3.40).

Since the total current flowing in the closed circuit is $I = \tau h$, the self-inductance of the system is

$$L = \frac{\Phi}{I} = \frac{\mu_0 (a+x)b}{h}. \tag{8.33}$$

In terms of the self-inductance, we rewrite the magnetic energy, Eq. (8.31), as

$$U_{\mathrm{m}} = \frac{1}{2} L I^2 = \frac{1}{2} \Phi I = \frac{1}{2L} \Phi^2. \tag{8.34}$$

These expressions are similar to those for electric energy in a capacitor, Eq. (3.38).

We consider a system composed of n closed electric circuits. Suppose that current I_i flows in the i-th circuit and magnetic flux Φ_i penetrates it (i = 1, 2, …, n). Then, extending the result of Eq. (8.34) to this case, the magnetic energy of this system is given by

$$U_{\mathrm{m}} = \frac{1}{2} \sum_{i=1}^{n} \Phi_i I_i = \frac{1}{2} \sum_{i=1}^{n} \sum_{j=1}^{n} L_{ij} I_i I_j. \tag{8.35}$$

The magnetic energy is usually derived using electromagnetic induction, as will be shown in Chap. 10.

Equation (8.32) is the result when the magnetic flux density is uniform in space. Here we determine the magnetic energy density for a non-uniform magnetic flux density. Substituting Eq. (6.35) into Eq. (8.34), the energy of the system is written as

$$U_{\mathrm{m}} = \frac{1}{2} \oint_{\mathrm{C}} I \boldsymbol{A} \cdot \mathrm{d}\boldsymbol{s}, \tag{8.36}$$

where C is the closed circuit with the current I. When the current is not concentrated but flows widely in space, we extend Eq. (8.36) to

Table 8.1 Comparison of electric energy and magnetic energy

	Electric energy	Magnetic energy
Separated system	$\frac{1}{2}\sum_{i=1}^{n}\phi_i Q_i$	$\frac{1}{2}\sum_{i=1}^{n}\Phi_i I_i$
Continuum system	$\frac{1}{2}\int_V \phi\rho \mathrm{d}V$	$\frac{1}{2}\int_V \boldsymbol{A}\cdot\boldsymbol{i}\mathrm{d}V$
Energy density	$\frac{1}{2}\epsilon_0\boldsymbol{E}^2$	$\frac{1}{2\mu_0}\boldsymbol{B}^2$

$$U_{\mathrm{m}} = \frac{1}{2}\int_{\mathrm{V}} \boldsymbol{A}\cdot\boldsymbol{i}\,\mathrm{d}V, \tag{8.37}$$

where V is the region in which the current with density $\boldsymbol{i}$ flows. In terms of Eqs. (6.27) and (A1.41) in the Appendix, the magnetic energy becomes

$$U_{\mathrm{m}} = \frac{1}{2\mu_0}\int_{\mathrm{V}} \boldsymbol{A}\cdot(\nabla\times\boldsymbol{B})\,\mathrm{d}V = \frac{1}{2\mu_0}\int_{\mathrm{V}} [\boldsymbol{B}\cdot(\nabla\times\boldsymbol{A}) - \nabla\cdot(\boldsymbol{A}\times\boldsymbol{B})]\mathrm{d}V. \tag{8.38}$$

Using Gauss' theorem, the second volume integral is transformed to the surface integral

$$-\int_{\mathrm{S}} (\boldsymbol{A}\times\boldsymbol{B})\cdot\mathrm{d}\boldsymbol{S}. \tag{8.39}$$

Assuming a sphere of sufficiently large radius r for V, we have $|\boldsymbol{A}| \propto r^{-1}$, $|\boldsymbol{B}| \propto r^{-2}$, and $\int \mathrm{d}S \propto r^2$ on its surface, and the surface integral is proportional to r^{-1}. Hence, taking the limit $r \to \infty$, the integral reduces to zero. Neglecting this integral, the magnetic energy reduces to

$$U_{\mathrm{m}} = \frac{1}{2\mu_0}\int_{\mathrm{V}} \boldsymbol{B}^2\,\mathrm{d}V, \tag{8.40}$$

where we have used Eq. (6.29). Thus, we can prove that the magnetic energy density is given by Eq. (8.32) even when the magnetic flux density is not uniform in space.

We compare the magnetic energy obtained here and the electric energy obtained in Chap. 3 in Table 8.1. These are quite analogous to each other.

Example 8.6. Current I is applied to the spherical coil of diameter a and total number of windings N shown in Fig. 8.10. Determine the total magnetic energy using Eq. (8.40).

Solution 8.6. Using the uniform magnetic flux density $B_0 = \mu_0 NI/3a$, the magnetic energy inside the coil is

$$U_{\mathrm{m1}} = \frac{B_0^2}{2\mu_0} \cdot \frac{4\pi}{3} a^3 = \frac{2\pi}{27} \mu_0 a N^2 I^2.$$

Since the magnetic flux density outside the coil is equal to that given by the magnetic moment placed on the coil center, it is given by Eq. (7.32) subtracted by each component of the applied magnetic flux density B_0. Thus, the magnetic energy density outside the coil is

$$\begin{aligned} u_{\mathrm{m2}} &= \frac{1}{2\mu_0}\left(B_r^2 + B_\theta^2\right) = \frac{B_0^2 a^6}{2\mu_0 r^6}\left(\cos^2\theta + \frac{\sin^2\theta}{4}\right) \\ &= \frac{\mu_0 a^4 N^2 I^2}{18 r^6}\left(\cos^2\theta + \frac{\sin^2\theta}{4}\right). \end{aligned}$$

Integrating this in the space outside the coil, the magnetic energy in this region is obtained as

$$\begin{aligned} U_{\mathrm{m2}} &= \frac{\mu_0 a^4 N^2 I^2}{18} \int_a^\infty \frac{1}{r^6} 2\pi r^2 \mathrm{d}r \int_0^\pi \left(\cos^2\theta + \frac{\sin^2\theta}{4}\right) \sin\theta \mathrm{d}\theta \\ &= \frac{\pi \mu_0 a^4 N^2 I^2}{9} \cdot \frac{1}{3a^3} \cdot \left(\frac{2}{3} + \frac{1}{3}\right) = \frac{\pi}{27} \mu_0 a N^2 I^2. \end{aligned}$$

The total magnetic energy is determined to be

$$U_{\mathrm{m}} = U_{\mathrm{m1}} + U_{\mathrm{m2}} = \frac{\pi}{9} \mu_0 a N^2 I^2.$$

◇

Example 8.7. We apply current I to the superconducting coaxial transmission line in Fig. 7.6. Determine the magnetic energy stored in a unit length of the transmission line and derive the self-inductance with this result.

Solution 8.7. Currents flow only on the surfaces $R = a$ and $R = b$ so that the magnetic flux does not penetrate the superconductors. The magnetic flux density is $B = \mu_0 I/(2\pi R)$ only in the region $a < R < b$ and is zero in other regions. Hence, the magnetic energy is non-zero only in the region $a < R < b$ and its density is

$$\frac{B^2}{2\mu_0} = \frac{\mu_0 I^2}{8\pi^2 R^2}.$$

Integrating this over the volume in a unit length, we have

$$U'_{\mathrm{m}} = \frac{\mu_0 I^2}{8\pi^2} \int_a^b \frac{1}{R^2} \cdot 2\pi R \mathrm{d}R = \frac{\mu_0 I^2}{4\pi} \log \frac{b}{a}.$$

We can also obtain the magnetic energy from Eq. (8.36). The vector potential has only the axial component A_z similarly to the current. From the relationship $\partial A_z / \partial R = -B$ with $A_z(b) = 0$, we obtain the vector potential as

$$A_z(R) = \frac{\mu_0 I}{2\pi} \log \frac{b}{R}$$

in the region $a < R < b$. Thus, the magnetic energy is

$$U'_{\mathrm{m}} = \frac{1}{2} A_z(a) I = \frac{\mu_0 I^2}{4\pi} \log \frac{b}{a},$$

which agrees with the above result.

Using this result, the self-inductance in a unit length is

$$L' = \frac{2U'_{\mathrm{m}}}{I^2} = \frac{\mu_0}{2\pi} \log \frac{b}{a}.$$

◇

Example 8.8. We apply current I to a sufficiently long solenoid coil of radius a with n turns in a unit length. Calculate the magnetic energy in a unit length using either Eq. (8.37) or (8.40).

Solution 8.8. The magnetic flux density inside the coil is $B = \mu_0 n I$ (see Example 6.8). Using Eq. (8.40), the magnetic energy in a unit length of the coil is

$$U'_{\mathrm{m}} = \frac{1}{2\mu_0} (\mu_0 n I)^2 \pi a^2 = \frac{\pi \mu_0}{2} (n a I)^2.$$

On the other hand, the current flows only on the coil surface ($R = a$). The vector potential on this surface is $A_\varphi(a) = \mu_0 n I a / 2$, and the surface current density is $\tau = nI$. Thus, using Eq. (8.37), the magnetic energy is

$$U'_{\mathrm{m}} = \frac{1}{2} \int_{\mathrm{S}} A_\varphi(a) \tau \mathrm{d}S = \frac{1}{2} \cdot \frac{\mu_0 n I a}{2} n I \cdot 2\pi a = \frac{\pi \mu_0}{2} (n a I)^2.$$

This agrees with the above result.

◇

Example 8.9. We apply a current I to a coaxial transmission line with the cross section shown in Fig. 8.15. Determine the magnetic energy in a unit length using Eq. (8.37) and the self-inductance. The thickness of the outer conductor is neglected.

Fig. 8.15 Cross section of coaxial transmission line

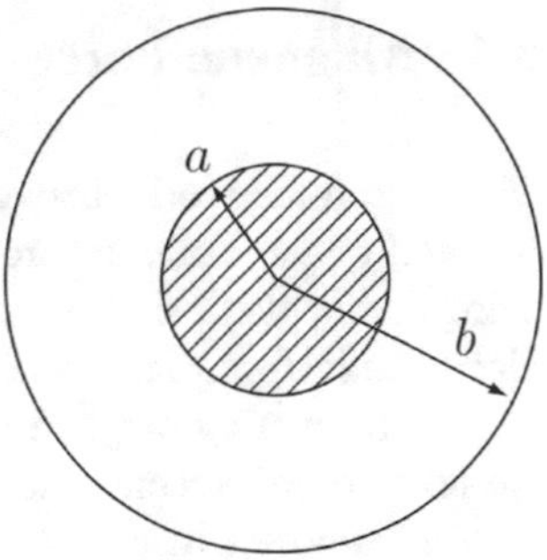

Solution 8.9. The magnetic flux density is

$$B = \frac{\mu_0 IR}{2\pi a^2}; \quad 0 < R < a,$$
$$= \frac{\mu_0 I}{2\pi R}; \quad a < R < b,$$
$$= 0; \quad R > b.$$

Since $B = 0$ for $R > b$, we obtain to be $A(R = b) = 0$. Thus, the vector potential is determined to be

$$A(R) = \int_R^b B(R)\mathrm{d}R = \frac{\mu_0 I}{2\pi} \log\frac{b}{R}; \qquad a < R < b,$$
$$= \frac{\mu_0 I}{2\pi} \log\frac{b}{a} + \frac{\mu_0 I}{4\pi a^2}(a^2 - R^2); \quad 0 < R < a.$$

Thus, the magnetic energy in a unit length is

$$U'_{\mathrm{m}} = \frac{1}{2}\int_0^a A(R)\frac{I}{\pi a^2} 2\pi R\mathrm{d}R = \frac{\mu_0 I^2}{4\pi}\left(\log\frac{b}{a} + \frac{1}{4}\right),$$

and the self-inductance in a unit length is

$$L' = \frac{\mu_0}{2\pi}\left(\log\frac{b}{a} + \frac{1}{4}\right).$$

The difference from the result of Example 8.7 is due to the contribution from the additional magnetic flux in the inner conductor.

◇

8.4 Magnetic Force

Magnetic force works between current-carrying conductors. This force is equal to the sum of the Lorentz force on each current. We can expect to derive this force using the magnetic energy and the principle of virtual displacement, similarly to the electrostatic force learned in Sect. 3.4. In fact, we learned the reverse process, namely estimating magnetic energy from the Lorentz force, in Sect. 8.3. However, we need to pay special attention when applying such a method to general cases.

We suppose that current I_1 flows in a straight line and current I_2 flows along a rectangular circuit with two sides parallel to the straight line, as shown in Fig. 8.16. These are placed on a common plane. Now we determine the force on the rectangular circuit using the Lorentz force. An attractive force works on the closer side and a repulsive force works on the opposite side. If the distance between the straight line and the closer side is x, the force on the rectangular circuit is

$$F' = -\frac{\mu_0 I_1 I_2}{2\pi x} b + \frac{\mu_0 I_1 I_2}{2\pi(x+a)} b = -\frac{\mu_0 ab I_1 I_2}{2\pi x(x+a)}, \tag{8.41}$$

where we define the force in the direction of increasing x to be positive. This force is negative, i.e., attractive.

Next, we calculate the force with the magnetic energy. The magnetic energy due to current I_1 only and that due to current I_2 only are expressed in terms of the self-inductances of each circuit. These energies are independent of the displacement of the rectangular circuit, since the self-inductances do not change under the relative displacement between the two circuits. From Eq. (8.35) for n = 2, the associated energy is the interaction energy between the two currents, i.e., the energy from

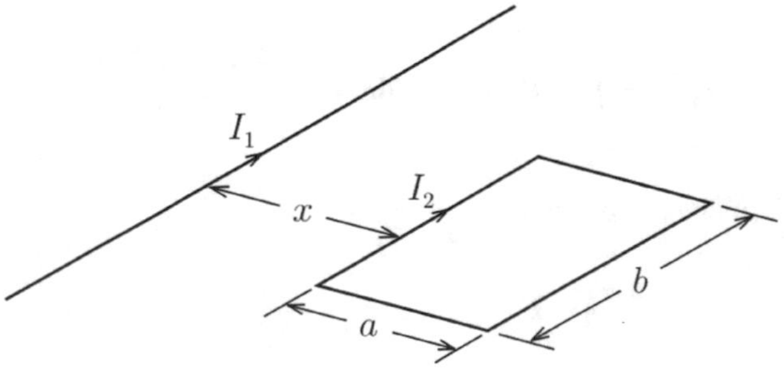

Fig. 8.16 Long straight line and rectangular circuit

mutual induction. The magnetic flux produced by current I_1 that penetrates the rectangular circuit is

$$\Phi = L_{21}I_1 = \frac{\mu_0 I_1 b}{2\pi}\int_x^{x+a}\frac{dr}{r} = \frac{\mu_0 I_1 b}{2\pi}\log\frac{x+a}{x}. \tag{8.42}$$

Hence, the associated energy is

$$U_m = \frac{1}{2}(L_{12}+L_{21})I_1I_2 = \Phi I_2 = \frac{\mu_0 b I_1 I_2}{2\pi}\log\frac{x+a}{x}, \tag{8.43}$$

using Eq. (8.5). Thus, the magnetic force seems to be

$$F = -\frac{\partial U_m}{\partial x} = \frac{\mu_0 ab I_1 I_2}{2\pi x(x+a)}. \tag{8.44}$$

However, this disagrees with F' in Eq. (8.41) which shows that there is a problem with the above procedure.

What does this mean? The above procedure seems to indicate that the magnetic energy decreases by $F\Delta x$ during the displacement of the rectangular circuit from x to $x+\Delta x$. However, the total magnetic energy must additionally increase by $\Delta U_m = 2F\Delta x$ to reach the correct result. From the reciprocity theorem in Eq. (8.5), half of this increment comes from the increase in the magnetic energy in the rectangular circuit. We expect this to be caused by the **induced electromotive force** V in the circuit. Namely, when the rectangular circuit is displaced by Δx within time Δt, the work done by the induced electromotive force is $VI_2\Delta t$. We can rewrite this as $-\Delta\Phi I_2$ in terms of the change in the magnetic flux that penetrates the circuit,

$$\Delta\Phi = -\frac{\mu_0 I_1 ab}{2\pi x(x+a)}\Delta x. \tag{8.45}$$

Taking the limit $\Delta t \to 0$, we have

$$V = -\frac{\Delta\Phi}{\Delta t} \to -\frac{d\Phi}{dt}. \tag{8.46}$$

This is the induced electromotive force that will be learned in Chap. 10.

What we can say from the above example is that estimating the magnetic force from the magnetic energy is valid only when the circuit is isolated and the magnetic flux is conserved as treated in Sect. 8.3. In most cases, electromagnetic induction is involved and such an estimation is not correct.

Column: (1) Method of Deriving Static Magnetic Energy

This chapter shows that there is a formal similarity between the electric energy and magnetic energy. However, the method of deriving the energy is completely different between the two cases. For example, the electric energy is estimated from the mechanical work needed to carry electric charges from infinity until the final distribution of electric charge is attained, whereas the magnetic energy cannot be estimated from the mechanical work to carry currents from infinity. That is, there is a problem of divergence of the energy. In addition, the more severe problem is that the final energy must be negative, since the work to carry a current is negative because of the attractive force between currents of the same direction. Thus, we conclude that the magnetic energy cannot be derived with this method. This is because of the induced electromotive force mentioned in Sect. 8.4.

For this reason, we estimate the magnetic energy in a virtual experiment using a superconducting circuit in this chapter. The merits of this method are that it enables us to construct an isolated system from surroundings such as a power source, and that it is free from electromagnetic induction since it conserves the magnetic flux because of the perfect diamagnetic property (see Exercise 8.11).

Consider possible ways to derive the magnetic energy other than the method introduced here.

Column: (2) Is Magnetic Flux a Magnetic Potential?

There is a famous analogy between the electric energy and magnetic energy, as shown in Table 8.1. The electric energy is given by the product of the electric source, i.e., the electric charge Q_i and the resultant electric potential ϕ_i divided by 2. On the other hand, the magnetic energy is given by the product of the magnetic source, i.e., the current I_i and the resultant magnetic flux Φ_i divided by 2. Does this analogy mean that the magnetic flux is a magnetic potential, i.e., the vector potential? It is clear that the magnetic flux is not the vector potential. How can we explain such a disagreement between electricity and magnetism?

The answer is that this difference is caused by the difference in the dimension of the sources. Originally the electric charge density corresponds to the current density, as can be seen in Table 8.1. For a conductor system, the electric potential is constant in each conductor. In the i-th conductor, the electric energy reduces to

$$\frac{1}{2}\int_{\mathrm{V}_i} \phi\rho \mathrm{d}V = \frac{1}{2}\phi_i \int_{\mathrm{V}_i} \rho \mathrm{d}V = \frac{1}{2}\phi_i Q_i.$$

On the other hand, we obtain the current by integrating the current density in the cross-sectional area. In the i-th circuit, the magnetic energy reduces to

$$\frac{1}{2}\int_{\mathrm{V}_i} \boldsymbol{A}\cdot\boldsymbol{i}\,\mathrm{d}V = \frac{1}{2}\int_{\mathrm{S}_i} i\,\mathrm{d}S \oint_{\mathrm{C}_i} \boldsymbol{A}\cdot\mathrm{d}\boldsymbol{s} = \frac{1}{2}\Phi_i I_i.$$

In the above, the volume integral was divided into the cross-sectional integral and the integral along the current path as $\mathrm{d}V = \mathrm{d}S\mathrm{d}s$, and we used the relationship $\boldsymbol{i}\mathrm{d}s = i\mathrm{d}\boldsymbol{s}$. That is, the magnetic flux is not the vector potential but is the vector potential integrated along the circuit. This may be simply understood from dimensions. In this sense, the magnetic flux is a kind of magnetic potential.

Exercises

8.1 Determine the mutual inductance between a parallel-wire transmission line and a triangular circuit on the common plane in Fig. E8.1.

8.2 Calculate the self-inductance directly from the penetrating magnetic flux for the superconducting coaxial transmission line in Example 8.7.

8.3 Suppose that the coaxial transmission line in Example 8.7 is not made of a superconductor but of a usual conductor. Determine the self-inductance. (Hint: Determine the self-inductance using the magnetic energy for a current.)

8.4 Determine the mutual inductance between a parallel-wire transmission line and a rectangular circuit placed at distance b from the transmission line (see Fig. E8.2).

8.5 Determine the mutual inductance between two coaxial solenoid coils in Fig. E8.3. The inner and outer coils have n_a and n_b turns in a unit length, respectively.

8.6 We apply current I through the parallel-plate transmission line shown in Fig. E8.4. Determine the magnetic energy and self-inductance in a unit length. Discuss the difference between conducting and superconducting transmission lines.

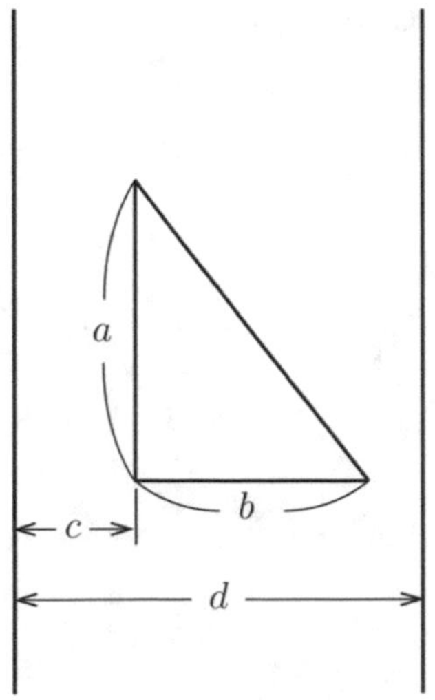

Fig. E8.1 Parallel-wire transmission line and triangular circuit

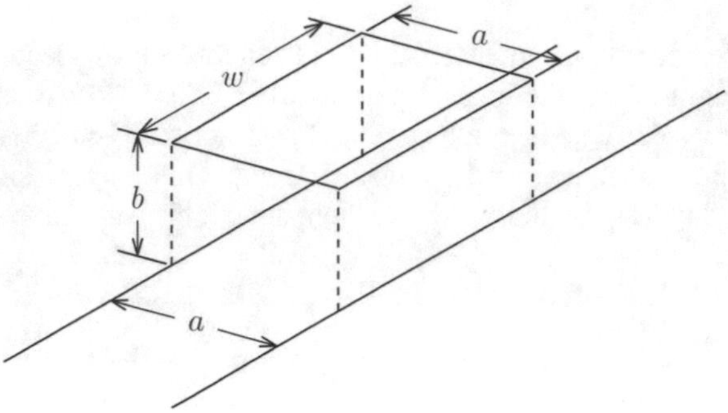

Fig. E8.2 Parallel-wire transmission line and rectangular circuit

Fig. E8.3 Two coaxial solenoid coils

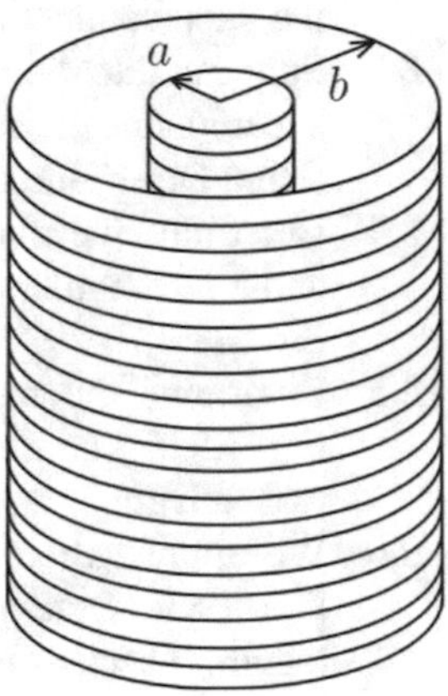

Fig. E8.4 Parallel-plate transmission line

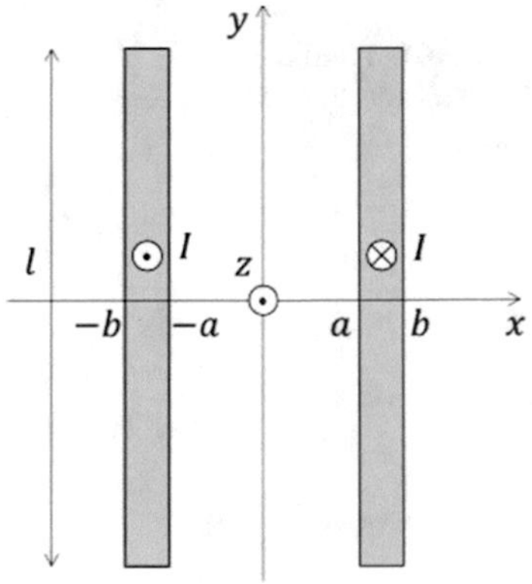

8.7 Solve the same problem in Example 8.6 using Eq. (8.37).

8.8 Determine the interior magnetic flux density, magnetic energy and self-inductance when we apply current I to the toroidal coil in Fig. E8.5. The radius of the central axis is d, the radius of the winding region is a and the total number of turns is N.

8.9 We denote the inner, middle, and outer long coaxial superconducting cylinders in Fig. E8.6 as superconductors 1–3. (a) Determine the inductance coefficients assuming the reference point for zero vector potential at $R = R_\infty > R_4$, and (b) determine the magnetic energy in a unit length using the inductance coefficients when we apply currents I_1, I_2, and I_3 to superconductors 1, 2, and 3, respectively.

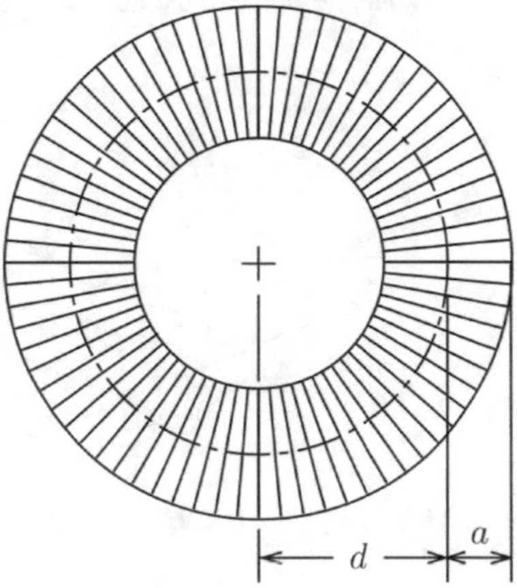

Fig. E8.5 Toroidal coil

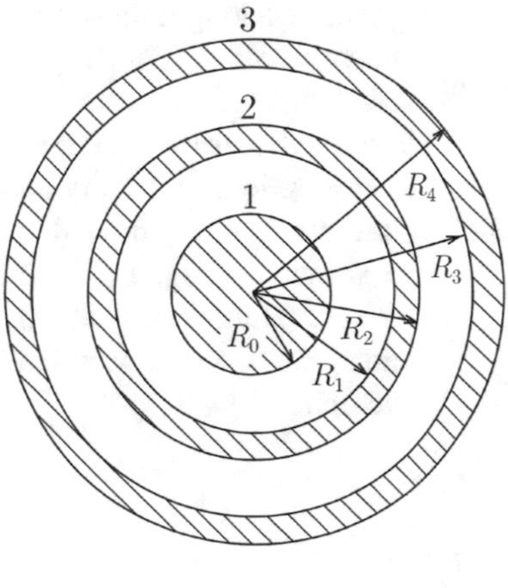

Fig. E8.6 Cross section of coaxial superconducting cylinders

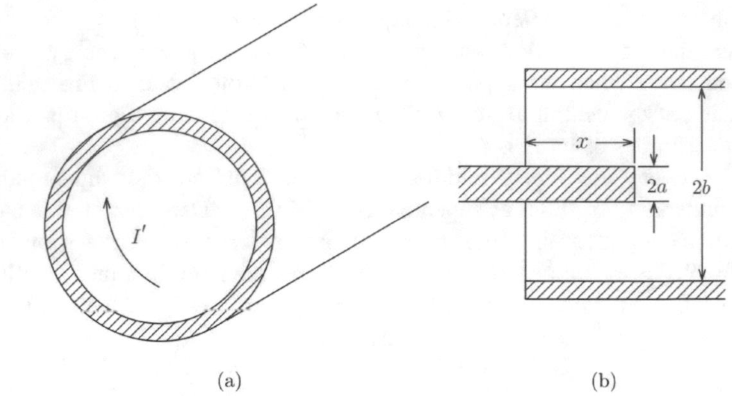

Fig. E8.7 **a** Long hollow superconducting cylinder with azimuthal current and **b** penetration of superconducting rod into the hollow superconducting cylinder

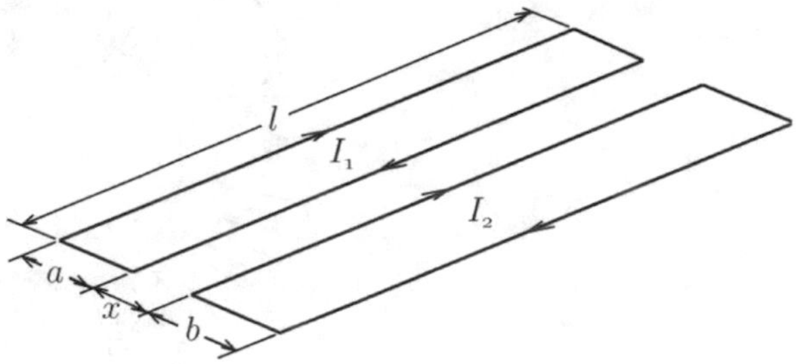

Fig. E8.8 Two current-carrying long rectangular superconducting circuits placed on a common plane

8.10 Suppose that current I' is flowing along the azimuthal direction in a unit length of a long hollow superconducting cylinder with inner diameter $2b$, as shown in Fig. E8.7a. Determine the force on the cylindrical superconducting rod of diameter $2a$ when we insert the rod into the superconducting cylinder to a depth x from the edge, as shown in Fig. E8.7b. Neglect the disturbance of magnetic flux density around the edge of the rod.

8.11 Currents I_1 and I_2 flow along two long rectangular superconducting circuits, as shown in Fig. E8.8. Calculate the magnetic force using the magnetic energy. (Hint: Note that when the distance x changes to $x + \Delta x$, currents I_1 and I_2 change in a way that keeps the penetrating magnetic flux constant in each circuit.)

Chapter 9
Magnetic Materials

Abstract This chapter covers magnetic phenomena in magnetic materials. When a magnetic flux density is applied to a magnetic material, a magnetic moment due to electron spins appears. This phenomenon is magnetization, and the magnetic moment is assumed to be produced by virtual small magnetizing current loops. Thus, the magnetic flux density is produced by currents and magnetizing currents in magnetic materials. We newly define the magnetic field that describes only the magnetic flux density produced by the current. Magnetic phenomena are generally described in terms of the magnetic flux density and the magnetic field. The refraction of the magnetic flux density at an interface between different magnetic materials is treated. Finally, static electric phenomena and magnetic phenomena are systematically compared in detail.

9.1 Magnetization

Some materials possess a magnetic moment that causes magnetic phenomena, when an external magnetic flux density is applied to those materials. These materials are classified as **magnetic materials**. Commonly used permanent magnets are made of magnetic materials and possess a magnetic moment even without an external magnetic flux density.

The appearance of the magnetic moment in magnetic materials is analogous to the way the electric polarization arises because of a relative displacement between electric charges of different signs in an external electric field. In fact, it looks as if the magnetic moment appears because of a relative displacement of magnetic charges of different signs in an external magnetic flux density, as shown in Eq. (6.51). However, magnetic charges do not exist, as discussed in Sect. 6.8. On the other hand, we learned that the magnetic moment can be equivalently expressed by closed currents. This equivalent current that produces the magnetic moment is called **magnetizing current**. However, true currents produce the static magnetic moment only in superconductors. The magnetizing current in a magnetic material cannot be taken outside. This is also similar to the polarization charge, which

T. Matsushita, *Electricity and Magnetism*, Undergraduate Lecture Notes in Physics,
https://doi.org/10.1007/978-3-030-82150-0_9

cannot be taken out of a dielectric material. Thus, the magnetizing current is a virtual substance like the magnetic charge. The magnetic moment is caused by electron spins, electron orbital motion, nuclear magnetic moment, etc.

The resultant magnetic moment in a unit volume of a material is called **magnetization** and is represented by $\boldsymbol{M}$. Its unit is [A/m]. This definition of magnetization is the same as that for superconductors (see Sect. 7.3). The magnetization $\boldsymbol{M}$ in a magnetic material is usually directed parallel to an applied magnetic flux density, $\boldsymbol{B}$, and is proportional to $B = |\boldsymbol{B}|$. Thus, it can be expressed as

$$\boldsymbol{M} = \frac{\chi_{\mathrm{m}}}{\mu_0}\boldsymbol{B}, \tag{9.1}$$

where the dimensionless proportional constant χ_{m} is called the **magnetic susceptibility**. Magnetic materials are classified into various kinds depending on the value of χ_{m}. It is recommended to read technical books to learn more about different kinds and characteristics of magnetic materials.

1. **Non-magnetic materials**: This group includes **diamagnetic materials** with negative χ_{m} and **paramagnetic materials** with χ_{m} that is positive but smaller than 1. The reason why the magnetization is proportional to the applied magnetic flux density is that magnetic moments directed randomly tend to incline in the direction of the applied magnetic flux density, as illustrated in Fig. 9.1. Table 9.1 gives the magnetic susceptibility for non-magnetic materials.

   If we apply Eq. (9.1) to superconductors, those with $\chi_{\mathrm{m}} = -1$ may be classified as diamagnetic material. However, the usual definition in electromagnetism is different, as will be shown later.

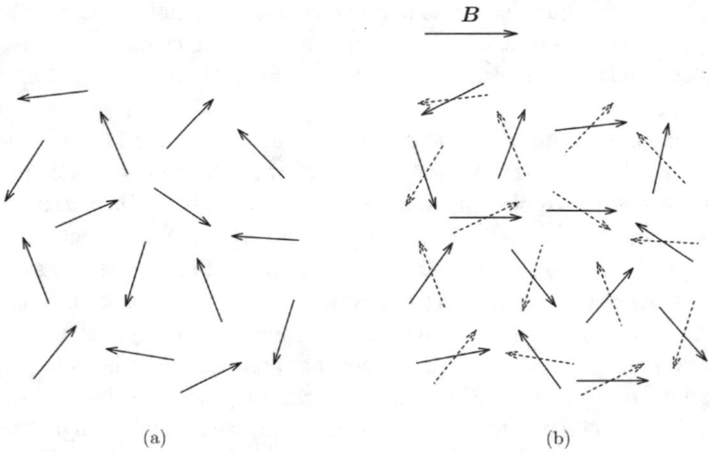

Fig. 9.1 Arrangement of individual magnetic moments in paramagnetic material **a** in the absence of a magnetic flux density and **b** in an applied magnetic flux density

Table 9.1 Magnetic susceptibility per kg/m^3 of non-magnetic materials

Material	χ_m ($\times 10^{-3}$)	Material	χ_m ($\times 10^{-3}$)
Diamond	−0.49	Oxygen	106.2
Graphite	−6 to −7	Air	24.1
Gold	−0.139	Nitrogen	−0.43
Copper	−0.086	Hydrogen	−1.97
Zinc	−0.157	Pure water	−0.720
Germanium	−0.12	Benzene	−0.712
Aluminum	0.62	Quartz glass	−0.5
Manganese	9.6	Alumina	−0.34
Chromium	3.17	Iron dioxide	20.6

2. **Magnetic materials**: Materials with positive χ_m of 10^2–10^4 are classified into this group. The origin of this type of magnetism is spin of electrons. One group of magnetic materials is **ferromagnetic materials**, with magnetic moments of atoms aligned in the same direction, as shown in Fig. 9.2a. This alignment is favored because the positive exchange interaction reduces the free energy. **Anti-ferromagnetic materials** have magnetic moments aligned alternately because of the negative exchange interaction, and **ferrimagnetic materials** with magnetic moments of different magnitudes aligned alternately because of the negative exchange interaction (see Fig. 9.2b, c). Thus, ferromagnetic materials and ferrimagnetic materials have nonzero magnetization without an applied magnetic flux density. This magnetization is called **spontaneous magnetization**. Even for materials with spontaneous magnetization, the total magnetization in a specimen is sometimes zero. This occurs because the specimen is divided into finite regions called magnetic domains. In each magnetic domain, the magnetic moment is aligned in thc same direction, and the magnetic moments of these domains are directed randomly, as illustrated in Fig. 9.3.

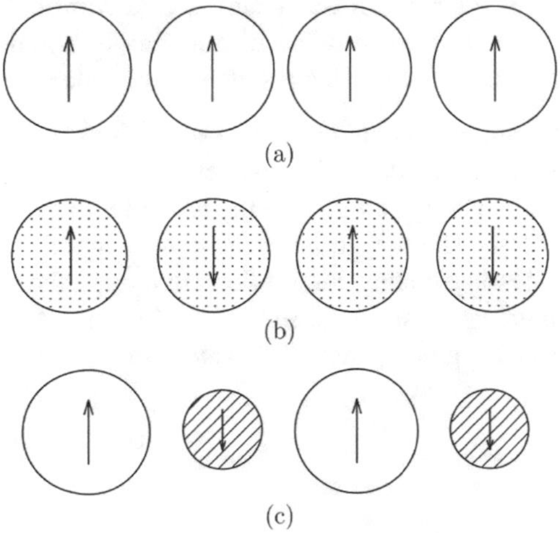

Fig. 9.2 Alignment of magnetic moments in **a** ferromagnetic, **b** anti-ferromagnetic, and **c** ferrimagnetic materials

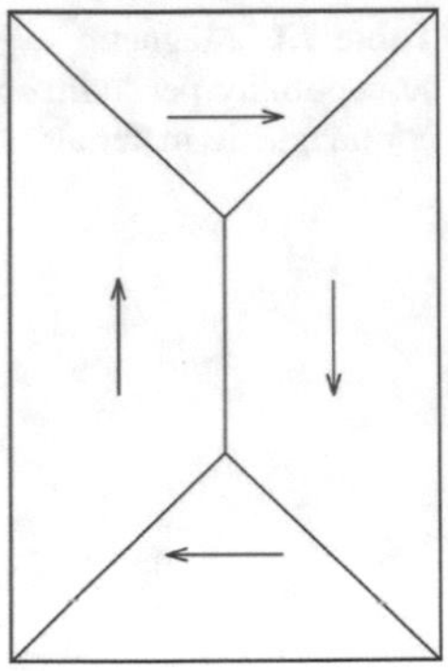

Fig. 9.3 Structure of magnetic domain with magnetic moment for zero magnetization

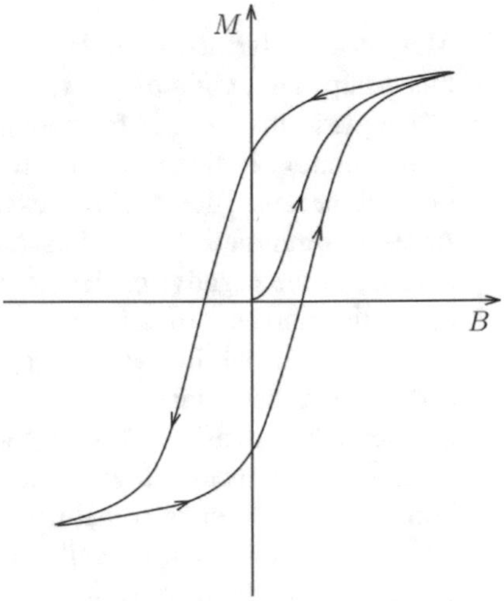

Fig. 9.4 Magnetization curve of ferromagnetic material. The magnetization changes with a change in applied magnetic flux density as indicated by the *arrows*

When we apply a magnetic flux density to such a specimen, the structure of its magnetic domains changes in a complex manner, including movement of the wall between adjacent domains or rotation of the magnetic moment in each domain, etc. This results in a complex magnetization as shown in Fig. 9.4.

Suppose a thin magnetic slab of thickness Δh that is magnetized uniformly in the normal direction. The magnetization of a small region in this slab is expressed by magnetizing current flowing around it (see Fig. 9.5). We denote the area of the top surface and the surface density of magnetizing current by ΔS and τ_{m}, respectively. The magnetizing current flowing around the small region is $\tau_{\mathrm{m}}\Delta h$, and the magnetic moment is $\Delta m = \tau_{\mathrm{m}}\Delta h\Delta S = \tau_{\mathrm{m}}\Delta V$ with $\Delta V = \Delta S\Delta h$ denoting the volume of this small region. Hence, the magnetization is given by

$$M = \frac{\Delta m}{\Delta V} = \tau_{\mathrm{m}}. \tag{9.2}$$

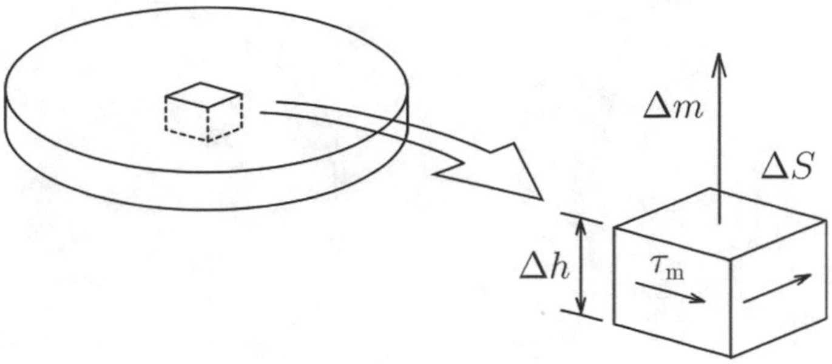

Fig. 9.5 Thin magnetic slab magnetized in the normal direction and a small part

Namely the magnetization is equal to the surface density of the magnetizing current. This relationship is similar to that of Eq. (4.2) stating that the electric polarization is equal to the surface density of the polarization charge in dielectric materials.

When this slab is divided into small regions as shown in Fig. 9.6, the magnetic moment in each region can be expressed by the magnetizing current flowing around it. These currents cancel out between adjacent regions, leaving only the magnetizing current flowing on the periphery of the slab. This is similar to the fact that the polarization charge remains only on the surface of a uniformly polarized dielectric material. When the magnetization is not uniform, the magnetizing current remains inside the magnetic material.

This result may suggest that the magnetizing current in magnetic materials and the magnetizing current in superconductors are similar to each other. However, there is an essential difference between them. Suppose we apply a magnetic flux density along a long hollow magnetic material and a long hollow superconductor. When we use reasoning similar to that for Fig. 9.6 for a magnetic material, the magnetizing current remains not only on the outer surface but also on the inner surface, as shown in Fig. 9.7. In this case, we can easily show that a magnetic flux

Fig. 9.6 Magnetizing current in each divided region (*upper half*) and resultant magnetizing current flowing on the periphery of the slab (*lower half*)

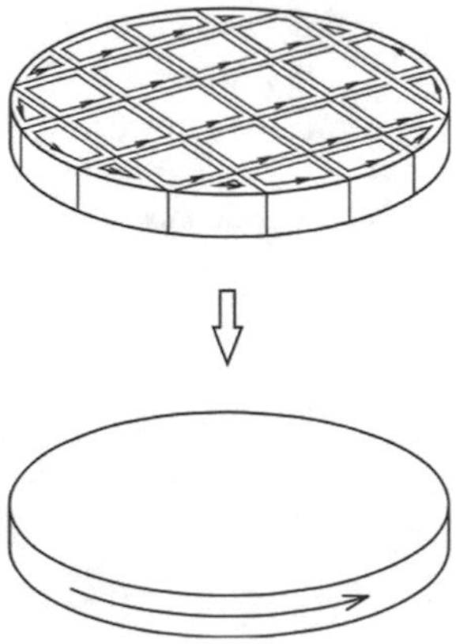

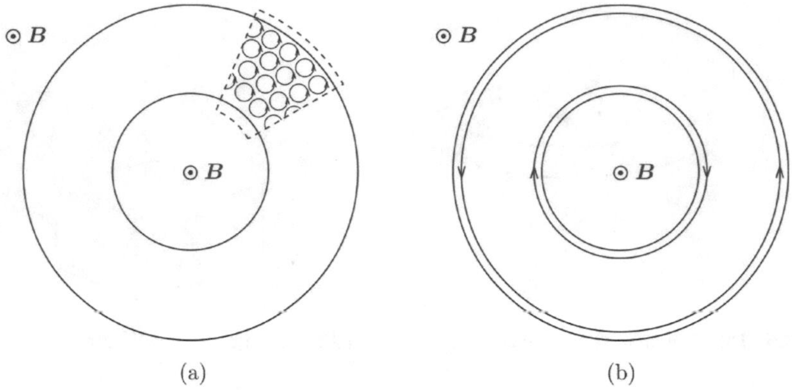

Fig. 9.7 **a** Small closed loops of magnetizing current and **b** resultant magnetizing current on the surfaces in hollow magnetic material in parallel magnetic flux density

density of the same value as the external one appears in the hollow. On the other hand, the current flows only on the outer surface of the superconductor, as shown in Fig. 9.8, and the interior is completely shielded including the hollow. It should be noted that the direction of the current is opposite to the magnetizing current because of the diamagnetic nature of the superconductor. This difference comes from the origin of the magnetic moment: The magnetic moment in magnetic materials originates from the electron spins, which are equivalent to the magnetizing current, while the magnetic moment in superconductors originates from the true current. The currents cannot be necessarily expressed by a superposition of small closed currents like the magnetizing currents.

A similar difference can be found between a hollow dielectric material and a hollow conductor in a transverse electric field, as shown in Fig. 4.10. This difference comes from the difference between the polarization charge and electric charge.

We assume that the magnetization is not uniform in region V occupied by a magnetic material. The magnetic moment of a small region of volume dV' at point

Fig. 9.8 Current that shields a hollow superconductor in parallel magnetic flux density

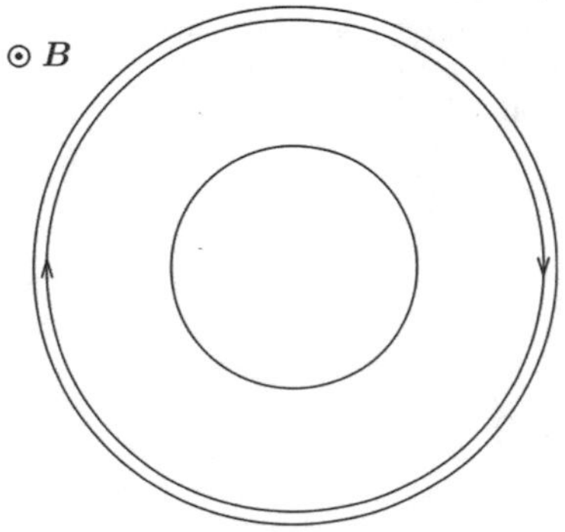

A positioned at $\boldsymbol{r}'$ is $\boldsymbol{m}(\boldsymbol{r}') = \boldsymbol{M}(\boldsymbol{r}')\mathrm{d}V'$ (see Fig. 9.9). Using Eq. (6.41), the vector potential produced by this magnetic moment at observation point P at $\boldsymbol{r}$ is

$$\mathrm{d}\boldsymbol{A} = \frac{\mu_0}{4\pi} \cdot \frac{\boldsymbol{M}(\boldsymbol{r}') \times (\boldsymbol{r} - \boldsymbol{r}')}{|\boldsymbol{r} - \boldsymbol{r}'|^3} \mathrm{d}V'. \tag{9.3}$$

Hence, the total vector potential is given by

$$\boldsymbol{A}(\boldsymbol{r}) = \frac{\mu_0}{4\pi} \int_{\mathrm{V}} \frac{\boldsymbol{M}(\boldsymbol{r}') \times (\boldsymbol{r} - \boldsymbol{r}')}{|\boldsymbol{r} - \boldsymbol{r}'|^3} \mathrm{d}V'. \tag{9.4}$$

As shown in Sect. A2.6 in the Appendix, this transforms to

$$\boldsymbol{A}(\boldsymbol{r}) = \frac{\mu_0}{4\pi} \int_{\mathrm{V}} \frac{\nabla' \times \boldsymbol{M}(\boldsymbol{r}')}{|\boldsymbol{r} - \boldsymbol{r}'|} \mathrm{d}V', \tag{9.5}$$

where the operation $\nabla' \times$ is curl with respect to $\boldsymbol{r}'$. Comparing this and Eq. (6.33), we understand that this vector potential is produced by the **magnetizing current density**,

$$\nabla \times \boldsymbol{M} = \boldsymbol{i}_{\mathrm{m}}. \tag{9.6}$$

In addition, $\mu_0 \boldsymbol{M}$ represents the magnetic flux density produced by the magnetizing current. Thus, the following equation holds:

$$\nabla \cdot \boldsymbol{M} = 0. \tag{9.7}$$

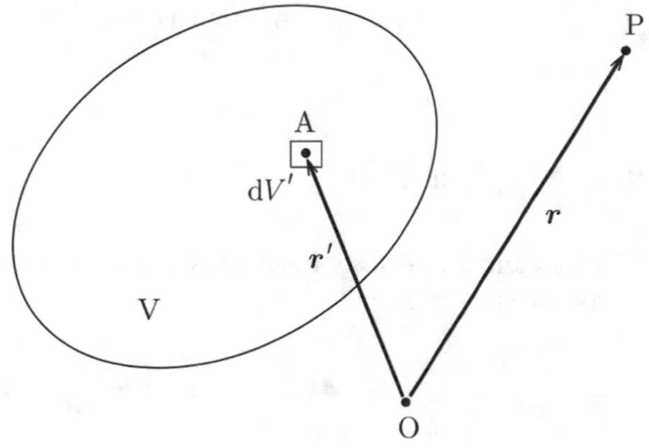

Fig. 9.9 Small region of volume dV' at point A in magnetic material and observation point P

Example 9.1 A spherical magnetic material of radius a is in a uniform magnetic flux density B_0, as shown in Fig. 9.10a. Determine the magnetizing current density on the surface. The magnitude of magnetization $\boldsymbol{M}$ is M.

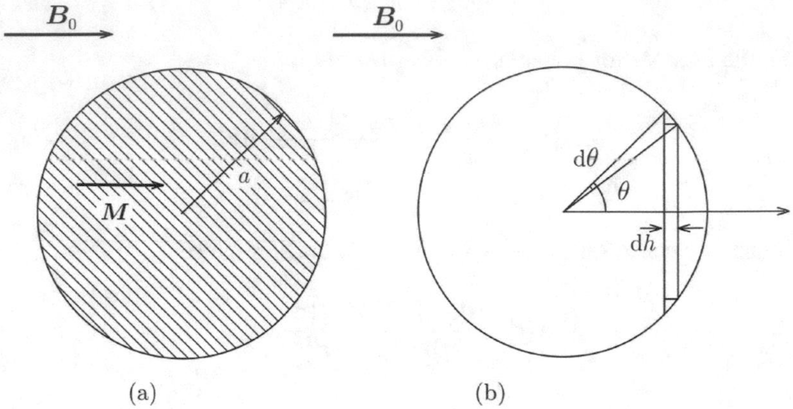

Fig. 9.10 **a** Magnetization of spherical magnetic material in magnetic flux density and **b** thin plate of magnetic material

Solution 9.1 We define the origin at the center of the sphere and the z-axis along the direction of the applied magnetic flux density. We measure the zenithal angle θ from this axis. We slice the spherical magnetic material into thin plates as shown in Fig. 9.10b, and denote the surface magnetizing current density on the edge of a plate by τ. Equation (9.2) gives $\tau = M$. If we denote the thickness of the thin plate and the corresponding zenithal angle interval by $\mathrm{d}h$ and $\mathrm{d}\theta$, we have $\mathrm{d}h = a\sin\theta\mathrm{d}\theta$. The magnetizing current that flows on the edge of this plate is $\tau\mathrm{d}h = \tau_{\mathrm{m}}a\mathrm{d}\theta$. Thus, the surface magnetizing current density is given by

$$\tau_{\mathrm{m}} = \tau \sin\theta = M \sin\theta. \tag{9.8}$$

◇

9.2 Magnetic Field

When a current of density $\boldsymbol{i}$ and a magnetizing current of density $\boldsymbol{i}_{\mathrm{m}}$ coexists, the vector potential is given by

$$\boldsymbol{A}(\boldsymbol{r}) = \frac{\mu_0}{4\pi}\int_{\mathrm{V}} \frac{\boldsymbol{i}(\boldsymbol{r}') + \boldsymbol{i}_{\mathrm{m}}(\boldsymbol{r}')}{|\boldsymbol{r} - \boldsymbol{r}'|}\mathrm{d}V'. \tag{9.9}$$

The curl of the magnetic flux density is

$$\nabla \times \boldsymbol{B} = \mu_0(\boldsymbol{i} + \boldsymbol{i}_{\mathrm{m}}). \tag{9.10}$$

Here we define a new physical quantity

$$\boldsymbol{H} = \frac{1}{\mu_0}\boldsymbol{B} - \boldsymbol{M}. \tag{9.11}$$

Then, we have

$$\nabla \times \boldsymbol{H} = \boldsymbol{i}. \tag{9.12}$$

That is, $\boldsymbol{H}$ is a variable that corresponds only to the current and is called the **magnetic field** or **magnetic field strength**. The unit of the magnetic field is [A/m] and is the same as for the magnetization. Equation (5.11) is satisfied also in this case as shown in Chap. 6, and hence, this is the magnetic field produced by a steady current. The definition of the magnetic field is similar to the definition of electric flux density, which corresponds only to electric charges in dielectric materials.

The word "field" represents a fundamental property of space that exerts force at a distance such as the electric field. The corresponding field for magnetic interaction is the magnetic flux density as described in Eqs. (6.10) or (6.14). In this sense, $\boldsymbol{B}$ can be called the magnetic field. However, $\boldsymbol{H}$ defined by Eq. (9.11) has been called the magnetic field. As is customary, we will call $\boldsymbol{H}$ and $\boldsymbol{B}$ the magnetic field and magnetic flux density, respectively.

As stated above, the magnetic field $\boldsymbol{H}$ is the variable associated only with a current. Hence, the Biot-Savart law and Ampere's law are the laws for $\boldsymbol{H}$. Namely the general form of the **Biot-Savart law** is given by

$$\boldsymbol{H}(\boldsymbol{r}) = \frac{1}{4\pi}\int_{\mathrm{C}} \frac{I\mathrm{d}\boldsymbol{r}' \times (\boldsymbol{r} - \boldsymbol{r}')}{|\boldsymbol{r} - \boldsymbol{r}'|^3} \tag{9.13}$$

instead of Eq. (6.6). The general form of **Ampere's law** is now written as

$$\oint_{\mathrm{C}} \boldsymbol{H} \cdot \mathrm{d}\boldsymbol{s} = \int_{\mathrm{S}} \boldsymbol{i} \cdot \mathrm{d}\boldsymbol{S} \tag{9.14}$$

instead of Eq. (6.25). Equation (9.12) is the general **differential form of Ampere's law**.

Here we explain the definition of the magnetic susceptibility in Eq. (9.1). In this definition, $\boldsymbol{M}$ is described in terms of $\boldsymbol{B}$. This corresponds to the description of the electric polarization in Eq. (4.1) and follows the $\boldsymbol{E}$-$\boldsymbol{B}$ analogy. Hence, the dimensionless magnetic susceptibility χ_{m} in Eq. (9.1) is consistent with the dimensionless electric susceptibility χ_e in Eq. (4.1). However, electromagnetism currently uses the definition

$$\boldsymbol{M} = \chi_m \boldsymbol{H}. \tag{9.15}$$

In practical measurement of magnetization, the magnetic flux density $\boldsymbol{B}$ applied to a specimen is nothing else than $\mu_0 \boldsymbol{H}$, which can be clearly defined using current. In this sense, there is no contradiction between Eqs. (9.1) and (9.15). Substituting Eq. (9.15) into Eq. (9.11) gives

$$\boldsymbol{H} = \frac{1}{\mu} \boldsymbol{B}. \tag{9.16}$$

In the above

$$\mu = \mu_0 (1 + \chi_m) \tag{9.17}$$

is a material constant called the **magnetic permeability** and has a unit of [N/A^2]. The **relative magnetic permeability** μ_r is a dimensionless parameter defined by

$$\mu = \mu_0 \mu_r. \tag{9.18}$$

It is related to the magnetic susceptibility as

$$\mu_r = 1 + \chi_m. \tag{9.19}$$

We can define the **magnetic field line** for the magnetic field $\boldsymbol{H}$ similarly to the magnetic flux line for the magnetic flux density $\boldsymbol{B}$. That is, the direction of a tangential line at any point on the magnetic field line is the same as the direction of $\boldsymbol{H}$, and its line density is equal to the magnitude of $\boldsymbol{H}$.

Substituting Eqs. (9.12) and (9.16) into Eq. (6.29), we have

$$\nabla \times (\nabla \times \boldsymbol{A}) = \mu \boldsymbol{i}. \tag{9.20}$$

Using the Coulomb gage, Eq. (6.30), for static magnetism, the above equation gives Poisson's equation

$$\Delta \boldsymbol{A} = -\mu \boldsymbol{i}. \tag{9.21}$$

This is simply obtained by replacing μ_0 with μ in Eq. (6.37). Hence, its solution is given by Eq. (6.33) when μ_0 is replaced by μ.

It should be noted that the magnetic moment produced by the current is not included in the magnetization $\boldsymbol{M}$, since the current affects only the magnetic field $\boldsymbol{H}$ in Eq. (9.11). That is, Eq. (9.11) is applicable only to magnetic materials and not to superconductors. In fact, we have $\boldsymbol{B} = \mu_0 \boldsymbol{H}$ and $\boldsymbol{M} = 0$ in superconductors. Thus, the magnetization in the superconductor cannot be expressed using Eq. (9.11). However, the same term "magnetization" is used to express the magnetic moment in a unit volume for both magnetic materials and superconductors. The magnetization in a superconductor is proportional to the difference between the mean magnetic flux density in the superconductor and the applied magnetic flux density, as shown in Eq. (7.38). In other words, the magnetization is a local variable in magnetic materials but is a variable averaged over a specimen for superconductors.

Example 9.2 Two magnetic materials have magnetic permeabilities μ_1 and μ_2. Each occupies half of the inner space of a long solenoid coil of radius a and n turns in a unit length, as shown in Fig. 9.11. Determine the self-inductance in a unit length of this coil.

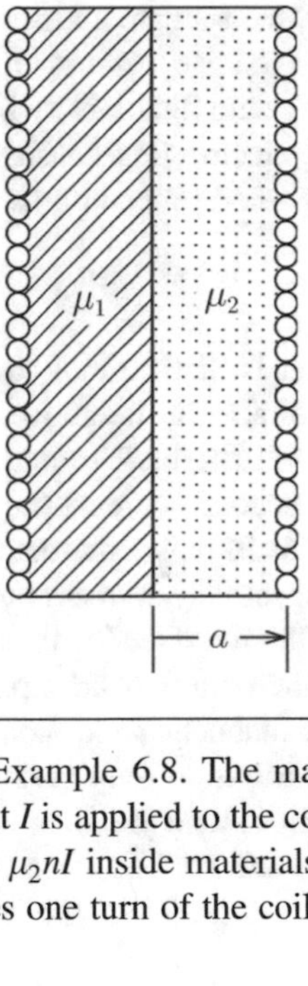

Fig. 9.11 Solenoid coil occupied by two magnetic materials with different magnetic permeabilities

Solution 9.2 We apply Ampere's law, Eq. (9.14), as in Example 6.8. The magnetic field is $H = nI$ in both magnetic materials when current I is applied to the coil. The magnetic flux density is given by $B_1 = \mu_1 nI$ and $B_2 = \mu_2 nI$ inside materials 1 and 2, respectively. Hence, the magnetic flux that penetrates one turn of the coil is

$$\phi = \frac{\pi}{2}(\mu_1 + \mu_2)na^2 I.$$

Since the magnetic flux that penetrates the coil of a unit length is

$$\Phi' = n\phi = \frac{\pi}{2}(\mu_1 + \mu_2)n^2 a^2 I,$$

the self-inductance in a unit length is given by

$$L' = \frac{\Phi'}{I} = \frac{\pi}{2}(\mu_1 + \mu_2)n^2a^2.$$

◇

9.3 Boundary Conditions

In Sect. 4.3, we discussed the conditions for the electric field and electric flux density to be fulfilled at an interface between different dielectric materials. Here, we discuss the conditions for the magnetic flux density and magnetic field to be fulfilled at an interface between different magnetic materials. Suppose an interface between magnetic materials 1 and 2 with magnetic permeabilities μ_1 and μ_2.

First we discuss the boundary condition for the magnetic flux density. The magnetic flux density $\boldsymbol{B}$ satisfies Eq. (6.21). This is the same form as Eq. (4.16) for the electric flux density $\boldsymbol{D}$ when the electric charge density ρ is zero. Hence, using the same procedure as that for deriving Eq. (4.19), we have

$$\boldsymbol{n} \cdot (\boldsymbol{B}_1 - \boldsymbol{B}_2) = 0. \tag{9.22}$$

That is, the normal component of the magnetic flux density is continuous at the interface. In the above, $\boldsymbol{n}$ is the unit vector normal to the interface and is directed from material 2 to material 1.

Second we treat the condition for the magnetic fields. We denote the magnetic fields in magnetic materials 1 and 2 in the vicinity of the interface by $\boldsymbol{H}_1$ and $\boldsymbol{H}_2$, respectively. We assume that the plane containing the vectors $\boldsymbol{H}_1$ and $\boldsymbol{H}_2$ is normal to the interface, as illustrated in Fig. 9.12a. Suppose a small rectangle, ΔC, on the plane with two sides parallel to the interface, as shown in Fig. 9.12b. We integrate the magnetic field along ΔC. If the height Δh is sufficiently small, we obtain the contribution only from the two sides parallel to the interface. We denote the unit vector along the direction of integration on the upper side and the length of this side by $\boldsymbol{t}$ and Δs. The integrations on the upper and lower sides are given by $\boldsymbol{H}_1 \cdot \boldsymbol{t}\Delta s$ and $-\boldsymbol{H}_2 \cdot \boldsymbol{t}\Delta s$. Thus, the integration gives

$$\oint_{\Delta \mathrm{C}} \boldsymbol{H} \cdot \mathrm{d}\boldsymbol{s} = (\boldsymbol{H}_1 - \boldsymbol{H}_2) \cdot \boldsymbol{t}\Delta s. \tag{9.23}$$

From Eq. (9.14), this is equal to the total current flowing through the region surrounded by ΔC. Since the height Δh is infinitesimal, this is zero when the current flows with a finite density. This takes a finite value only when a planar current flows on the interface. Here, we use $\boldsymbol{\tau}$ and $\boldsymbol{a}$ to denote the planar current density and unit vector normal to ΔC, respectively. These vectors are defined to point along the right

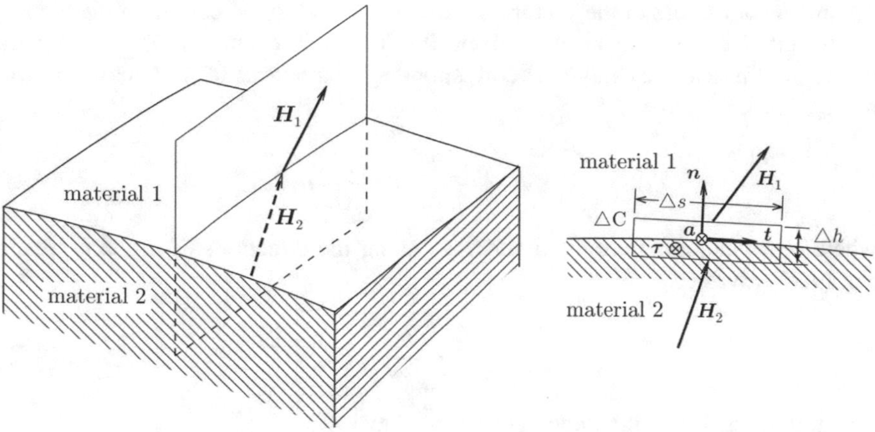

(a) (b)

Fig. 9.12 **a** Plane that contains the magnetic field vectors in each magnetic material at the interface and **b** small rectangle on the plane that contains the interface

thumb when it is rotated along the direction of the integration. The total current that flows through ΔC is $\boldsymbol{\tau} \cdot \boldsymbol{a}\Delta s$. Using the relationship $\boldsymbol{a} \times \boldsymbol{n} = \boldsymbol{t}$, Eq. (9.23) gives

$$[\boldsymbol{n} \times (\boldsymbol{H}_1 - \boldsymbol{H}_2)] \cdot \boldsymbol{a} = \boldsymbol{\tau} \cdot \boldsymbol{a},$$

and finally, we obtain

$$\boldsymbol{n} \times (\boldsymbol{H}_1 - \boldsymbol{H}_2) = \boldsymbol{\tau}. \tag{9.24}$$

In the above, we assumed that the plane containing $\boldsymbol{H}_1$ and $\boldsymbol{H}_2$ is normal to the interface. You can prove this assumption in Exercise 9.3. Equation (9.24) states that a planar current flows on the interface and its density is equal to the difference in the parallel component of the magnetic field. When there is no planar current, the parallel component of the magnetic field is continuous. This condition is satisfied in Example 9.2.

Now, we describe the above boundary conditions using the vector potential. We denote the vector potentials in magnetic materials 1 and 2 in the vicinity of the interface by $\boldsymbol{A}_1$ and $\boldsymbol{A}_2$, respectively. Then, Eqs. (9.22) and (9.24) are rewritten as

$$\boldsymbol{n} \cdot (\nabla \times \boldsymbol{A}_1 - \nabla \times \boldsymbol{A}_2) = 0, \tag{9.25}$$

$$\boldsymbol{n} \times \left(\frac{1}{\mu_1} \nabla \times \boldsymbol{A}_1 - \frac{1}{\mu_2} \nabla \times \boldsymbol{A}_2 \right) = \boldsymbol{\tau}. \tag{9.26}$$

The main component of the vector potential is parallel to the current in the vicinity of the interface, as can be seen from Eq. (6.33). The current flows along the direction of $\boldsymbol{a}$, and we denote the corresponding component of the vector potential by A_a. Thus, we have

$$\nabla \times \boldsymbol{A} = -\frac{\partial A_a}{\partial n}\boldsymbol{t} + \frac{\partial A_a}{\partial t}\boldsymbol{n}, \tag{9.27}$$

where $\partial/\partial n$ and $\partial/\partial t$ are the derivatives along the directions of $\boldsymbol{n}$ and $\boldsymbol{t}$. Hence, Eq. (9.25) gives

$$\frac{\partial A_{a1}}{\partial t} = \frac{\partial A_{a2}}{\partial t}. \tag{9.28}$$

Integrating this along the direction of $\boldsymbol{t}$, we have

$$A_{a1} = A_{a2}. \tag{9.29}$$

Equation (9.26) gives

$$\frac{1}{\mu_1} \cdot \frac{\partial A_{a1}}{\partial n} - \frac{1}{\mu_2} \cdot \frac{\partial A_{a2}}{\partial n} = -\tau. \tag{9.30}$$

Example 9.3 Magnetic materials 1 and 2 of magnetic permeabilities μ_1 and μ_2 each occupy half of the space between two long parallel superconducting plates, as shown in Fig. 9.13. We apply current I to each superconducting plate in opposite directions. Determine the magnetic flux density and magnetic field in each magnetic material and self-inductance in a unit length of superconducting plates. Assume that the width of the plate, w, is sufficiently long in comparison with the distance, d, between the two plates.

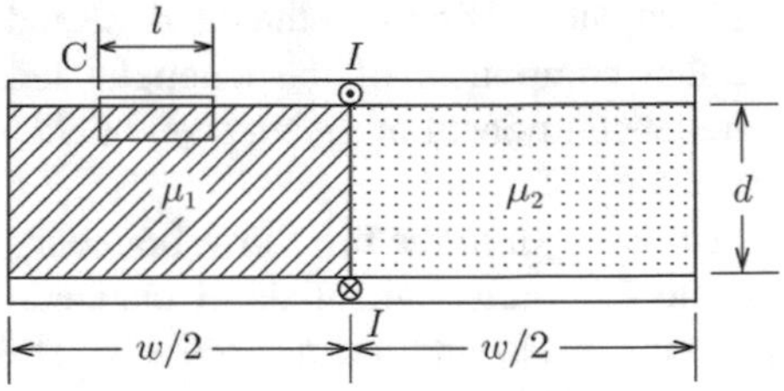

Fig. 9.13 Two superconducting plates with magnetic materials with different magnetic permeabilities

---

Solution 9.3 The magnetic flux density and magnetic field are directed parallel to the superconducting plates in the space between them. We denote these values in magnetic materials 1 and 2 by B_1, H_1, B_2, and H_2. From Eq. (9.22), we have

$$B_1 = B_2,$$

which gives

$$H_2 = \frac{\mu_1}{\mu_2} H_1.$$

We apply Eq. (9.14) to rectangle C in Fig. 9.13. If the surface current density on the superconducting plate in this region is τ_1, we have $H_1 l = \tau_1 l$. This reduces to

$$H_1 = \tau_1.$$

The surface current density on the superconducting plate in contact with magnetic material 2 is

$$\tau_2 = H_2 = \frac{\mu_1}{\mu_2} \tau_1.$$

Since the total current I is equal to $(w/2)(\tau_1 + \tau_2)$, we obtain

$$H_1 = \frac{2\mu_2 I}{w(\mu_1 + \mu_2)}, \qquad H_2 = \frac{2\mu_1 I}{w(\mu_1 + \mu_2)},$$

and

$$B_1 = B_2 = \frac{2\mu_1 \mu_2 I}{w(\mu_1 + \mu_2)}.$$

The magnetic flux that penetrates the superconducting plates in a unit length is

$$\Phi' = \frac{2\mu_1 \mu_2 d I}{w(\mu_1 + \mu_2)},$$

and we obtain the self-inductance as

$$L' = \frac{2\mu_1 \mu_2 d}{w(\mu_1 + \mu_2)}.$$

It should be noted that the situation is not simple if the superconducting plates are replaced with usual conducting plates. This situation is similar to that in Example 4.3.

◇

Here we derive the boundary conditions on the surface of a superconductor from the above general boundary conditions. We denote the vacuum and superconductor as regions 1 and 2, respectively. We use μ_0 for the magnetic permeability of the superconductor. Since $B_2 = 0$, we can say from Eq. (9.22) that the magnetic flux density is parallel to the surface. Equation (9.24) derives $H_1 = \tau$, which gives the same result as Eq. (7.8),

$$B_1 = \mu_0 \tau. \tag{9.31}$$

Here we discuss the refraction of magnetic flux lines and magnetic field lines using the boundary conditions. Suppose an interface between magnetic materials with magnetic permeabilities μ_1 and μ_2 (see Fig. 9.14). We assume that magnetic flux density B_1 is applied in material 1 in the direction of angle θ_1 from the normal direction to the interface. We denote by B_2 and θ_2 the magnitude and angle of magnetic flux density in material 2. The continuity of the normal component of the magnetic flux density gives

$$B_1 \cos\theta_1 = B_2 \cos\theta_2. \tag{9.32}$$

Since no surface current flows usually on the interface, the parallel component of the magnetic field is continuous at the interface:

$$\frac{1}{\mu_1} B_1 \sin\theta_1 = \frac{1}{\mu_2} B_2 \sin\theta_2. \tag{9.33}$$

These equations give

$$\frac{\tan\theta_1}{\tan\theta_2} = \frac{\mu_1}{\mu_2}. \tag{9.34}$$

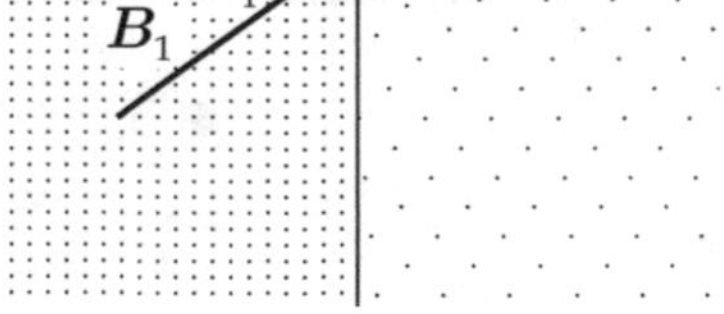

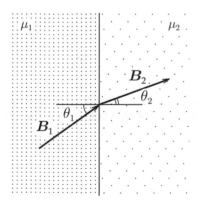

Fig. 9.14 Refraction of magnetic flux lines at interface

This is the **law of refraction**. We obtain B_2 and θ_2 as

$$B_2 = B_1\left[\left(\frac{\mu_2}{\mu_1}\right)^2 \sin^2\theta_1 + \cos^2\theta_1\right]^{1/2}, \tag{9.35}$$

$$\theta_2 = \tan^{-1}\left(\frac{\mu_2}{\mu_1}\tan\theta_1\right). \tag{9.36}$$

Example 9.4 A magnetic flux density of B_0 is applied parallel to a wide surface of magnetic material of magnetic permeability μ, as shown in Fig. 9.15. Determine the magnetic flux density, magnetic field, magnetization inside the magnetic material, and the surface magnetizing current density.

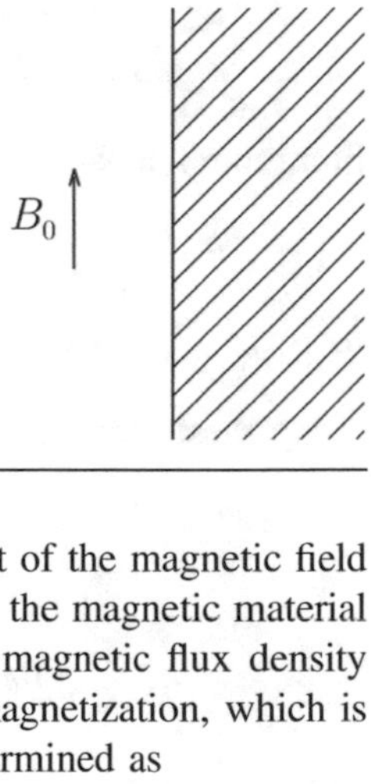

Fig. 9.15 Magnetic flux density applied parallel to the surface of a magnetic material

Solution 9.4 From the continuity of the parallel component of the magnetic field on the surface given by Eq. (9.24), the magnetic field inside the magnetic material is equal to that on the outside, i.e., $H = B_0/\mu_0$. Thus, the magnetic flux density inside the magnetic material is $B = \mu H = (\mu/\mu_0)B_0$. The magnetization, which is equal to the surface magnetizing current density, τ_m, is determined as

$$M = \tau_m = \frac{1}{\mu_0}B - H = \frac{\mu - \mu_0}{\mu_0^2}B_0.$$

The magnetizing current is directed normal forward.

◇

Example 9.5 A magnetic sphere of radius a is in a uniform magnetic flux density of B_0, as shown in Fig. 9.10a. Determine the magnetic flux density, magnetic field, and surface magnetizing current density.

Solution 9.5 We define polar coordinates as in Example 9.1. We can assume that magnetization is uniform in the magnetic material. Hence, the magnetic flux density outside the sphere is given by the sum of the applied magnetic flux density and the contribution of the magnetic moment placed at the origin after removal of the sphere. We expect a uniform magnetic flux density inside the sphere due to the uniform magnetization. We denote the magnetic moment directed to the z-axis by m. Equations (6.45a and 6.45b) give the radial and zenithal components of the magnetic flux density outside the sphere due to the magnetic moment

$$B_r = \frac{\mu_0 m \cos\theta}{2\pi r^3}, \quad B_\theta = \frac{\mu_0 m \sin\theta}{4\pi r^3}.$$

We use B to denote the internal magnetic flux density along the z-axis. Then, the continuities of the normal component of the magnetic flux density and the parallel component of the magnetic field on the surface ($r = a$) are given by

$$\begin{aligned} B_0 \cos\theta + \frac{\mu_0 m \cos\theta}{2\pi a^3} &= B\cos\theta, \\ \frac{1}{\mu_0}\left(-B_0 \sin\theta + \frac{\mu_0 m \sin\theta}{4\pi a^3}\right) &= -\frac{1}{\mu} B \sin\theta. \end{aligned}$$

From these equations, we have

$$m = \frac{\mu - \mu_0}{\mu + 2\mu_0} \cdot \frac{4\pi a^3 B_0}{\mu_0}, \quad B = \frac{3\mu}{\mu + 2\mu_0} B_0. \tag{9.37}$$

Using these results, the magnetic flux density outside the sphere ($r > a$) is

$$\begin{aligned} B_r = \mu_0 H_r &= \left(1 + \frac{\mu - \mu_0}{\mu + 2\mu_0} \cdot \frac{2a^3}{r^3}\right) B_0 \cos\theta, \\ B_\theta = \mu_0 H_\theta &= -\left(1 - \frac{\mu - \mu_0}{\mu + 2\mu_0} \cdot \frac{a^3}{r^3}\right) B_0 \sin\theta, \end{aligned}$$

and that inside the sphere ($r < a$) is

$$\begin{aligned} B_r = \mu H_r &= \frac{3\mu}{\mu + 2\mu_0} B_0 \cos\theta, \\ B_\theta = \mu H_\theta &= -\frac{3\mu}{\mu + 2\mu_0} B_0 \sin\theta. \end{aligned}$$

We can also obtain these results by solving Eqs. (9.29) and (9.30) for the vector potential. We determine the magnetization to be

$$M = \left(\frac{1}{\mu_0} - \frac{1}{\mu}\right) B = \frac{3(\mu - \mu_0)}{\mu_0(\mu + 2\mu_0)} B_0.$$

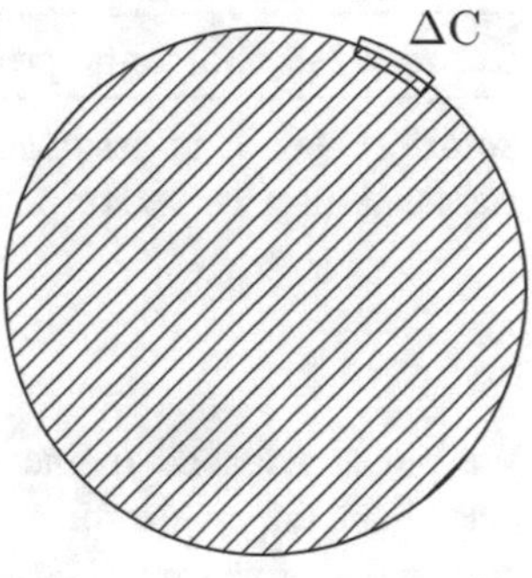

Fig. 9.16 Small closed loop ΔC on a plane that contains a part of the surface of magnetic sphere

Here we suppose small closed loop ΔC that contains a part of the interface on a plane including the *z*-axis with an arbitrary azimuthal angle (see Fig. 9.16). We apply the integrated form of Eq. (9.10)

$$\oint_{\Delta C} \boldsymbol{B} \cdot \mathrm{d}\boldsymbol{s} = \mu_0 \int_{\Delta S} (\boldsymbol{i} + \boldsymbol{i}_{\mathrm{m}}) \cdot \mathrm{d}\boldsymbol{S},$$

to this region, where ΔS is the surface surrounded by ΔC. Since there is no true current, the difference in the parallel component of the magnetic flux density divided by μ_0 is equal to the surface magnetizing current density. Thus, we have

$$\tau_{\mathrm{m}}(\theta) = \frac{3(\mu - \mu_0)}{\mu_0(\mu + 2\mu_0)} B_0 \sin\theta = M \sin\theta.$$

This agrees with Eq. (9.8) in Example 9.1.

Figure 9.17 shows the magnetic flux lines inside and outside the magnetic sphere for $\mu = 3\mu_0$.

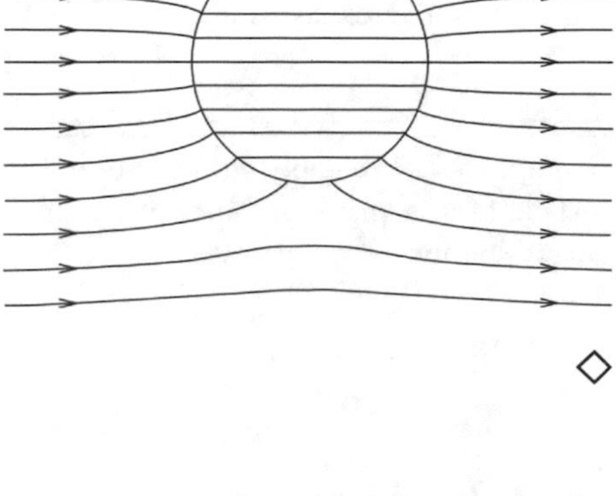

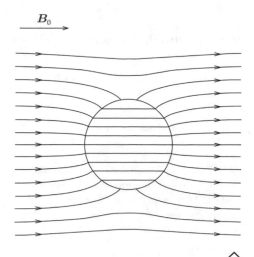

Fig. 9.17 Magnetic flux lines inside and outside the magnetic sphere for $\mu = 3\mu_0$

◇

Example 9.6 Solve the problem of Example 9.5 using the vector potential.

Solution 9.6 It is assumed that the magnetic flux density is applied along the z-axis. Hence, its vector potential has an azimuthal component and is given by

$$A_{f\varphi} = \frac{B_0 r}{2}\sin\theta.$$

We put a magnetic moment m directed along the applied magnetic flux density placed on the origin after the magnetic material is virtually removed. Its vector potential is given by

$$A_{d\varphi} = \frac{\mu_0 m}{4\pi r^2}\sin\theta.$$

Thus, the vector potential outside the magnetic material is

$$A_{1\varphi} = A_{f\varphi} + A_{d\varphi} = \left(\frac{B_0 r}{2} + \frac{\mu_0 m}{4\pi r^2}\right)\sin\theta.$$

The magnetic flux density inside the spherical magnetic material is uniform along the z-axis, and its value is denoted by B. Thus, the vector potential inside is

$$A_{2\varphi} = \frac{Br}{2}\sin\theta.$$

Equation (9.29) leads to

$$B_0 + \frac{\mu_0 m}{2\pi a^3} = B.$$

In Eq. (9.30), $\partial A_\varphi / \partial n$ means $-(1/r)\partial(rA_\varphi)/\partial r$, and we have

$$\frac{1}{\mu_0}\left(B_0 - \frac{\mu_0 m}{4\pi a^3}\right) = \frac{1}{\mu}B.$$

Thus, the same results as in Example 9.5 are obtained.

◇

Example 9.7 We apply current I to a thin straight line at distance a above the flat surface of a magnetic material of magnetic permeability μ, as shown in Fig. 9.18. Determine the vector potential.

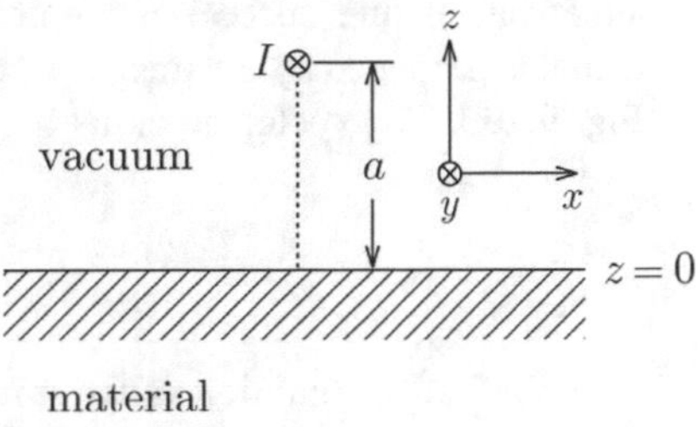

Fig. 9.18 Straight current and flat surface of magnetic material

Solution 9.7 We use the method of images similar to that in Example 4.6. We define the y-axis along the direction of the current and the x-y plane (z = 0) on the surface of the magnetic material with $x = 0$ as the position of the current.

To determine the vector potential in the vacuum region ($z > 0$), we virtually assume that all the space is vacuum and the vector potential is given by the sum of the component caused by I and that caused by the image current I' placed at the mirror position with respect to the surface of the magnetic material, as shown in Fig. 9.19a. Hence, the vector potential has only the y-component, A_y, and depends only on x and z. The vector potential at point (x, z) in the vacuum is given by

$$A_{vy}(x, z) = \frac{\mu_0}{2\pi}\left\{ I \log \frac{R_0}{[x^2 + (z - a)^2]^{1/2}} + I' \log \frac{R_0}{[x^2 + (z + a)^2]^{1/2}} \right\},$$

where R_0 is the distance to the reference point of the vector potential. On the other hand, to determine the vector potential in the magnetic material ($z < 0$), we virtually

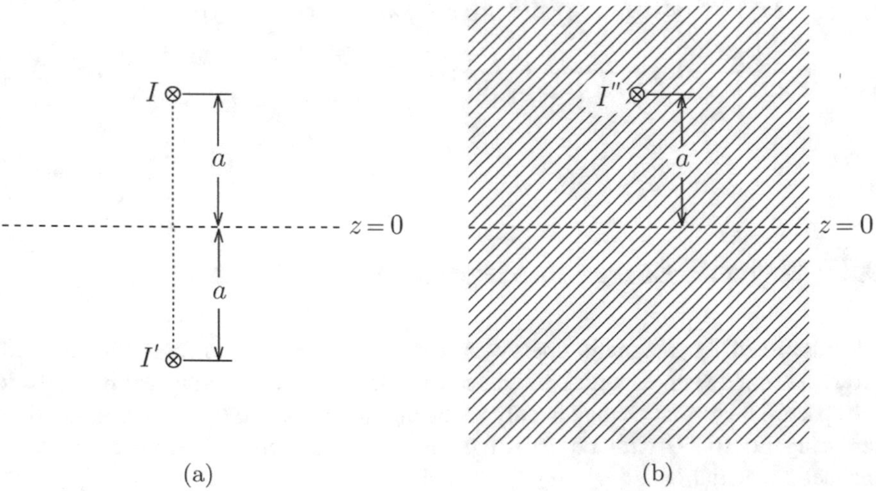

Fig. 9.19 Solution using method of images: assumed conditions for **a** vacuum and **b** magnetic material

assume that all the space is occupied by the magnetic material, and the vector potential is produced by the image current I'' placed at the original position, as shown in Fig. 9.19b. The vector potential at point (x, z) in the magnetic material is

$$A_{my}(x,z) = \frac{\mu}{2\pi} I'' \log \frac{R_0}{[x^2 + (z-a)^2]^{1/2}}.$$

The continuity condition of the normal component of the magnetic flux density on the surface is given by Eq. (9.29), i.e., $A_{vy}(z=0) = A_{my}(z=0)$. This condition gives

$$\mu_0(I + I') = \mu I''.$$

The continuity condition of the parallel component of the magnetic field on the surface is given by Eq. (9.30), i.e., $(1/\mu_0)(\partial A_{vy}/\partial z)_{z=0} = (1/\mu)(\partial A_{my}/\partial z)_{z=0}$. This condition gives

$$I - I' = I''.$$

Thus, we have

$$I' = \frac{\mu - \mu_0}{\mu + \mu_0} I, \quad I'' = \frac{2\mu_0}{\mu + \mu_0} I.$$

We obtain the vector potential as

$$\begin{aligned} A_y &= \frac{\mu_0 I}{2\pi} \left\{ \log \frac{R_0}{[x^2 + (z-a)^2]^{1/2}} + \frac{(\mu - \mu_0)}{(\mu + \mu_0)} \log \frac{R_0}{[x^2 + (z+a)^2]^{1/2}} \right\}; \quad && z > 0, \\ &= \frac{\mu_0 \mu I}{\pi(\mu + \mu_0)} \log \frac{R_0}{[x^2 + (z-a)^2]^{1/2}}; && z < 0. \end{aligned}$$

◇

9.4 Magnetic Energy in Magnetic Material

We discussed the magnetic energy in vacuum caused by currents in Sect. 8.3. However, there is no essential difference even if magnetic materials are involved in the space. The formal change is only replacing the permeability μ_0 with μ based on replacing Eq. (6.27) with Eq. (9.12). That is, the magnetic energy density in the magnetic material is

$$u_{\mathrm{m}} = \frac{1}{2\mu}\boldsymbol{B}^2 = \frac{1}{2}\boldsymbol{B}\cdot\boldsymbol{H} = \frac{1}{2}\mu\boldsymbol{H}^2, \tag{9.38}$$

and the magnetic energy is given by its volume integral:

$$U_{\mathrm{m}} = \int\limits_{\mathrm{V}} \frac{1}{2\mu}\boldsymbol{B}^2\,\mathrm{d}V = \int\limits_{\mathrm{V}} \frac{1}{2}\boldsymbol{B}\cdot\boldsymbol{H}\,\mathrm{d}V = \int\limits_{\mathrm{V}} \frac{1}{2}\mu\boldsymbol{H}^2\,\mathrm{d}V. \tag{9.39}$$

Example 9.8 The space between the superconductors in a superconducting coaxial transmission line is occupied by two magnetic materials with magnetic permeabilities μ_1 and μ_2, as shown in Fig. 9.20. Determine the magnetic energy in a unit length of the transmission line, when current I is applied.

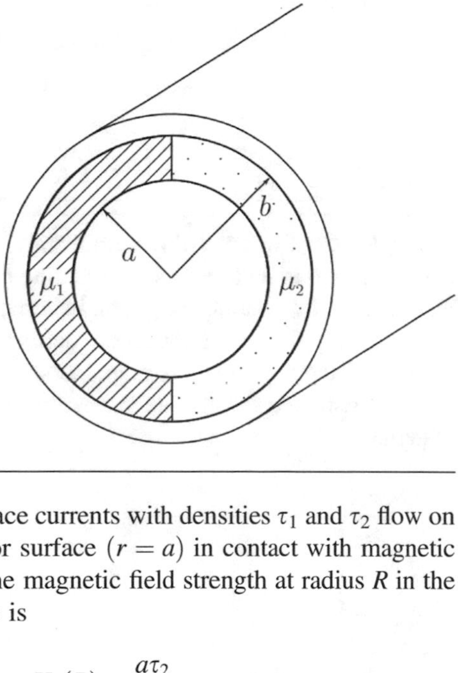

Fig. 9.20 Coaxial superconducting transmission line with two magnetic materials that have different magnetic permeabilities

Solution 9.8 It is assumed that the surface currents with densities τ_1 and τ_2 flow on the interface of the inner superconductor surface ($r = a$) in contact with magnetic materials 1 and 2, respectively. Then, the magnetic field strength at radius R in the space between the two superconductors is

$$H_1(R) = \frac{a\tau_1}{R}, \quad H_2(R) = \frac{a\tau_2}{R},$$

and the corresponding magnetic flux density is

$$B_1(R) = \frac{\mu_1 a\tau_1}{R}, \quad B_2(R) = \frac{\mu_2 a\tau_2}{R}.$$

These quantities must be equal to each other from the continuity of the magnetic flux density, and we have $\mu_1\tau_1 = \mu_2\tau_2$. From the condition of the total current density, $I = \pi a(\tau_1 + \tau_2)$, the surface current densities are obtained, and the magnetic flux density is determined:

$$B(R) = \frac{\mu_1\mu_2 I}{\pi(\mu_1 + \mu_2)R}.$$

The magnetic flux in a unit length is

$$\Phi' = \frac{\mu_1\mu_2 I}{\pi(\mu_1 + \mu_2)} \int_a^b \frac{\mathrm{d}R}{R} = \frac{\mu_1\mu_2 I}{\pi(\mu_1 + \mu_2)} \log\frac{b}{a},$$

and we determine the magnetic energy in a unit length to be

$$U'_\mathrm{m} = \frac{1}{2}\Phi' I = \frac{\mu_1\mu_2 I^2}{2\pi(\mu_1 + \mu_2)} \log\frac{b}{a}.$$

◇

Example 9.9 Current I flows uniformly in the azimuthal direction on the inner surface of a long hollow superconducting cylinder, as shown in Fig. 9.21a. Assume that the length l is sufficiently long in comparison with the radius a. Then, we insert a magnetic cylinder of magnetic permeability μ and the same radius into the hollow to depth x from the edge, as shown in Fig. 9.21b. How does the current change? Estimate the magnetic energy and determine the force on the magnetic material.

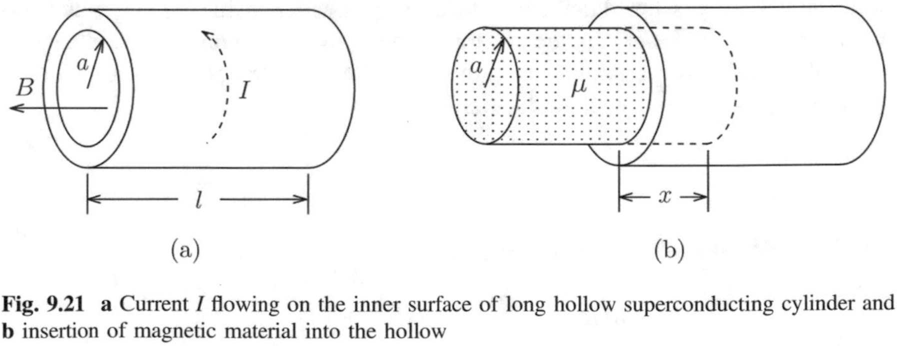

Fig. 9.21 **a** Current I flowing on the inner surface of long hollow superconducting cylinder and **b** insertion of magnetic material into the hollow

Solution 9.9 The magnetic flux density in the hollow before insertion of the magnetic material is $B = \mu_0 I/l$. This does not change even after the insertion. This is because the magnetic flux does not appear or disappear, and the magnetic flux lines must go out of or into the hollow for the change to occur. However, this is prohibited by the superconductor.

Hence, the surface current density is $\tau = I/l$ in the region where the magnetic material is not inserted but changes to $\tau = \mu_0 I/(\mu l)$ in the region where the magnetic material is inserted. Thus, the total current changes to

$$I' = I\left[1 - \left(1 - \frac{\mu_0}{\mu}\right)\frac{x}{l}\right].$$

Hence, the magnetic energy density in the hollow is

$$u_{\mathrm{m}} = \frac{1}{2\mu_0}B^2 = \frac{\mu_0 I^2}{2l^2},$$

and that in the magnetic material is

$$u_{\mathrm{m}} = \frac{1}{2\mu}B^2 = \frac{\mu_0^2 I^2}{2\mu l^2}.$$

We have the total magnetic energy as

$$U_{\mathrm{m}} = \frac{\pi\mu_0 a^2 I^2}{2l^2}\left[l - \left(1 - \frac{\mu_0}{\mu}\right)x\right]$$

and determine the force on the magnetic material to be

$$F = -\frac{\partial U_{\mathrm{m}}}{\partial x} = \frac{\pi\mu_0 a^2 I^2}{2l^2}\left(1 - \frac{\mu_0}{\mu}\right).$$

Since $\mu > \mu_0$ in normal magnetic materials, we have $F > 0$. That is, the force is directed along the positive x-axis and attractive. In this problem, the magnetic flux does not change, and there is no influence of the electromagnetic induction.

◇

9.5 Analogy Between Electric and Magnetic Phenomena

In Part II, we learned static magnetic phenomena through similarities to and differences from static electric phenomena in Part I. To understand these phenomena further, we summarize the corresponding relationships here. Tables 9.2, 9.3, and 9.4 give the electromagnetic variables, equations, and boundary conditions, respectively.

It is also useful to refer to Table 6.1, which summarizes integral equations corresponding to the differential equations in Table 9.3, and Table 6.2, which

Table 9.2 Fundamental variables, supplementary variables, and sources in static electric and magnetic phenomena

	Electric phenomena	Magnetic phenomena
Fundamental variable 1	Electric field $\boldsymbol{E}$	Magnetic flux density $\boldsymbol{B}$
Fundamental variable 2	Electric flux density $\boldsymbol{D}$	Magnetic field $\boldsymbol{H}$
Supplementary variable	Electric polarization $\boldsymbol{P}$	Magnetization $\boldsymbol{M}$
Source 1	Electric charge (density) ρ	Current (density) $\boldsymbol{i}$
Source 2	Polarization charge (density) ρ_{p}	Magnetizing current (density) $\boldsymbol{i}_{\mathrm{m}}$

Table 9.3 Equations of fundamental variables describing static electric and magnetic phenomena

	Electric phenomena	Magnetic phenomena
Property of field	$\nabla \times \boldsymbol{E} = 0$	$\nabla \cdot \boldsymbol{B} = 0$
Potential	$\phi(\boldsymbol{E} = -\nabla \phi)$	$\boldsymbol{A}(\boldsymbol{B} = \nabla \times \boldsymbol{A})$
Relation to source	$\nabla \cdot \boldsymbol{D} = \rho$	$\nabla \times \boldsymbol{H} = \boldsymbol{i}$
Fundamental variable 2	$\boldsymbol{D} = \epsilon_0 \boldsymbol{E} + \boldsymbol{P}$	$\boldsymbol{H} = \frac{1}{\mu_0}\boldsymbol{B} - \boldsymbol{M}$
Constant	ϵ_0	$\frac{1}{\mu_0}$
Supplementary variable	$-\boldsymbol{P}(-\nabla \cdot \boldsymbol{P} = \rho_{\mathrm{p}})$	$\boldsymbol{M}(\nabla \times \boldsymbol{M} = \boldsymbol{i}_{\mathrm{m}})$

Table 9.4 Boundary conditions for fundamental variables describing static electric and magnetic phenomena

	Electric phenomena	Magnetic phenomena
Continuity 1	Parallel component of $\boldsymbol{E}$	Normal component of $\boldsymbol{B}$
Continuity 2	Normal component of $\boldsymbol{D}$	Parallel component of $\boldsymbol{H}$
Source for discontinuity 2	Electric charge (density) σ	Current (density) $\boldsymbol{\tau}$

summarizes the potentials. In these tables, it is necessary to replace ϵ_0 and μ_0 with ϵ and μ, respectively, for generalization.

The contents in Tables 9.2, 9.3, and 9.4 are explained as follows:

1. The most fundamental variables are the electric field $\boldsymbol{E}$ and the magnetic flux density $\boldsymbol{B}$, which are directly connected to the force. That is, these variables describe the fields in electric and magnetic phenomena. The electric field is a conservative field with no rotation and is given by a gradient of the scalar potential, the electrostatic potential. On the other hand, the magnetic flux density is a field with no divergence and is given by a curl of the vector potential.
2. The secondary fundamental variables are the electric flux density $\boldsymbol{D}$ and the magnetic field $\boldsymbol{H}$, which are associated with source-1 quantities such as electric charges and currents. The electric charge causes a divergence of the electric flux density and the current causes a curl of the magnetic field.
3. The supplementary variables are the electric polarization $\boldsymbol{P}$ and the magnetization $\boldsymbol{M}$, and these are associated with source-2 quantities, which cannot be taken out of materials. That is, the polarization charge causes a divergence of the electric polarization and the magnetizing current causes a curl of the magnetization. The signs are opposite between electric polarization and magnetization

because only magnetic materials do not shield themselves from an applied field. This is mathematically expressed by the fact that the relative dielectric constant ϵ_r and relative magnetic permeability μ_r are larger than 1.

4. Table 9.4 summarizes the conditions for the fundamental variables on a boundary. For the primary fundamental variables, the parallel component of the electric field is continuous because there is no rotation, and the normal component of the magnetic flux density is continuous because there is no divergence. For secondary fundamental variables, the corresponding components can be discontinuous at a boundary because of the existence of the source-1 quantities. That is, the normal component of the electric flux density is discontinuous by an amount equal to the surface charge density, and the parallel component of the magnetic field is discontinuous by an amount equal to the surface current density. When there is no such source, these components are continuous.
5. We compare special phenomena in electricity and magnetism in conductors and superconductors in Table 9.5. The primary fundamental variables, the electric field and magnetic flux density, are zero inside the respective materials, and the source-1 quantities, the charge and current, exist only on the surface. There are no source-2 quantities or resultant supplementary variables. To satisfy the boundary conditions in Table 9.4, the electric field is normal, and the magnetic flux density is parallel to the surface. The values of these variables on the surface are proportional to the surface densities of the source-1 quantities.

Table 9.6 summarizes electric and magnetic phenomena in the ***E*-*H* analogy**, in which $\boldsymbol{H}$ is treated as a primary fundamental variable. In the past, Coulomb's law was used for assumed magnetic charges. This method is beneficial as a mathematical analogy. In this case, the magnetization $\boldsymbol{M}$ and the magnetic charge q_m correspond to $\mu_0\boldsymbol{M}$ and $\mu_0 q_m$ in this textbook, respectively.

Table 9.5 Static electric phenomena in conductor and magnetic phenomena in superconductor

	Electric phenomena in conductor	Magnetic phenomena in superconductor
Fundamental variable	$\boldsymbol{E}=0$	$\boldsymbol{B}=0$
Source on the surface	Electric charge (density) σ	Current (density) τ
Supplementary variable	$\boldsymbol{P}=0$	$\boldsymbol{M}=0$
Condition on the surface	Normal $\boldsymbol{E}$ $(E=\sigma/\epsilon_0)$	Parallel $\boldsymbol{B}$ $(B=\mu_0\tau)$

Table 9.6 Fundamental variables in the ***E*-*H*** analogy

	Electric phenomena	Magnetic phenomena
Property of fundamental variable 1	$\nabla\times\boldsymbol{E}=0$	$\nabla\times\boldsymbol{H}=\boldsymbol{i}$
Property of fundamental variable 2	$\nabla\cdot\boldsymbol{D}=\rho$	$\nabla\cdot\boldsymbol{B}=0$
Definition of fundamental variable 2	$\boldsymbol{D}=\epsilon_0\boldsymbol{E}+\boldsymbol{P}$	$\boldsymbol{B}=\mu_0\boldsymbol{H}+\boldsymbol{M}$
Constant	ϵ_0	μ_0
Supplementary variable	$-\boldsymbol{P}\ (-\nabla\cdot\boldsymbol{P}=\rho_p)$	$\boldsymbol{M}\ (\nabla\times\boldsymbol{M}=\mu_0\boldsymbol{i}_m)$

Column: (1) Magnetic Shielding by a Superconductor and That by a Magnetic Material

In the Column in Chap. 4, we showed that electric shielding by a conductor and that by a dielectric material are similar phenomena in spite of the quantitative difference. How about the magnetic shielding by superconductor and that by magnetic material?

From correspondence with Chap. 4, we can derive Eq. (7.32) for the magnetic flux density around a superconducting sphere by taking the magnetic flux density around the magnetic sphere in Example 9.5 in the limit $\mu \to 0$. In this case, although a magnetic field of finite strength remains inside the superconducting sphere, we can disregard it because it has no meaning. This is similar to electric phenomena as shown in the Column in Chap. 4.

However, the essential difference is that the magnetization in a superconductor is negative, while the magnetization in a magnetic material is positive. The internal magnetic flux density is weakened in a superconductor from the outside, and the shielding is really similar to electric shielding. However, the situation is completely different for a magnetic material. Namely the internal magnetic flux density is strengthened in comparison with the external value because of the positive magnetization (see Eq. (9.37) and Fig. 9.17). This characteristic property of magnetic material comes from the origin of the magnetic moment. Is such a magnetic material useful for magnetic shielding?

The answer is yes. When we make a slit with the structure in Fig. E9.3 in Exercise 9.4, the magnetic flux density inside the slit decreases by a factor of μ_0/μ from the value in the surrounding magnetic material. Namely magnetic flux lines are likely to pass through the region with higher magnetic permeability. It is possible to weaken the magnetic flux density in a target region by introducing the magnetic flux to other regions using this property.

Column: (2) Magnetization and Electric Polarization

The substance in a material that causes the magnetic action is the magnetic moment, and the current in a superconductor and the virtual magnetizing current in a magnetic material are the origins of the magnetic moment. However, there is a problem in the fact that the same term "magnetization" is used to express the magnetic moment in a unit volume even for different origins. The mathematical description is different. For example, according to the definition of magnetization used for magnetic materials, the magnetization in superconductors is zero as mentioned in Sect. 9.2. Hence, we need a new term such as **magnetic moment density** instead of magnetization to define the magnetic moment in a unit volume of superconductor.

Electric polarization in dielectric materials is comparable to the magnetization. The electric action is caused by a relative displacement of polarization charges of different signs that cannot move freely in a given electric field and partially contributes to the shielding of the material from the external electric field. This electric phenomenon is quantitatively characterized by the electric polarization P, the electric dipole moment in a unit volume. A similar electric phenomenon in conductors is the perfect electrostatic shielding by electric charges that can move freely. This phenomenon can also be explained by a relative displacement of electric charges of different signs as discussed in Sect. 2.3. That is, appearance of electric dipoles is common for both electric phenomena (see, for example, Eq. (2.30) for a conductor). This situation is similar to the production of a magnetic moment by two kinds of origin mentioned above. Since the electric dipole moment in a unit volume had not been discussed for a conductor, there was no confusion as with the magnetization in a magnetic material and superconductor. However, it is meaningful to study the similarity between the electrostatic shielding in a conductor and the electric polarization in a dielectric material. In this sense, we need a new term such as **electric moment density** to define the electric dipole moment in a unit volume of conductor.

Exercises

9.1. We apply current I to two long parallel superconducting plates with magnetic materials of magnetic permeabilities μ_1 and μ_2 between them, as shown in Fig. E9.1. Assume that the width of the plate, w, is sufficiently larger than the distance, d, between the plates. Determine the magnetic flux density and magnetic field in each magnetic material and the self-inductance in a unit length.

9.2. Determine the self-inductance of a unit length for the superconducting coaxial transmission line with two magnetic materials of different magnetic permeabilities in Fig. E9.2.

9.3. We define the magnetic fields, $\boldsymbol{H}_1$ and $\boldsymbol{H}_2$, in magnetic materials 1 and 2 in the vicinity of an interface. Prove that these vectors stay in the same plane perpendicular to the interface.

9.4. We apply a magnetic flux density B_0 parallel to a thin slit of vacuum in a magnetic material of magnetic permeability μ, as shown in Fig. E9.3. Determine the magnetic flux density and magnetic field inside the slit.

9.5. We apply magnetic flux density B_0 normal to a thin slit of vacuum in a magnetic material of magnetic permeability μ, as shown in Fig. E9.4. Determine the magnetic flux density and magnetic field inside the slit.

9.6. We apply magnetic flux density B_0 normal to a wide flat surface of a magnetic material of magnetic permeability μ, as shown in Fig. E9.5.

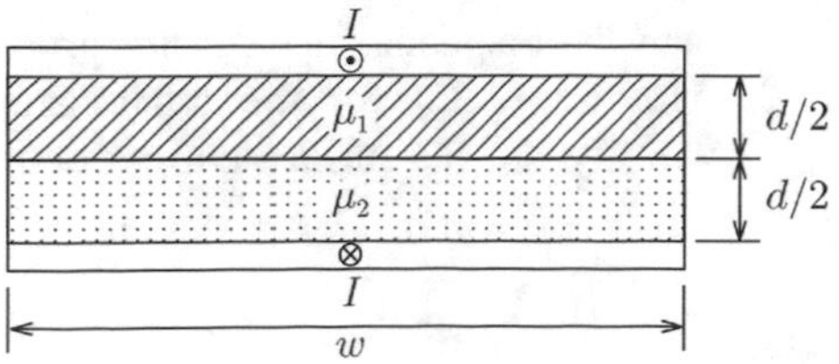

Fig. E9.1 Two parallel superconducting plates with two magnetic materials of different magnetic permeabilities

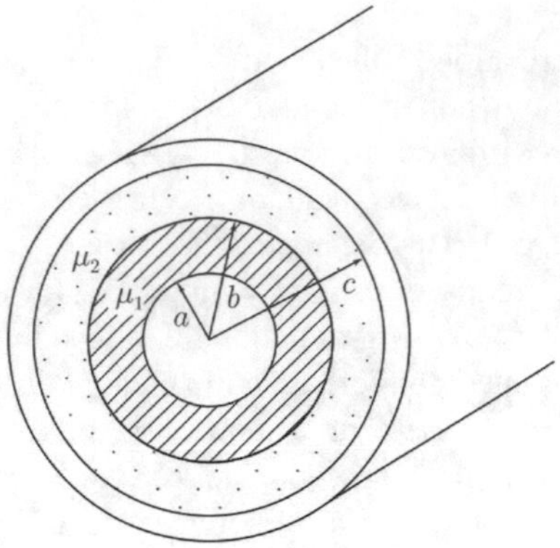

Fig. E9.2 Coaxial superconducting transmission line with two magnetic materials that have different magnetic permeabilities

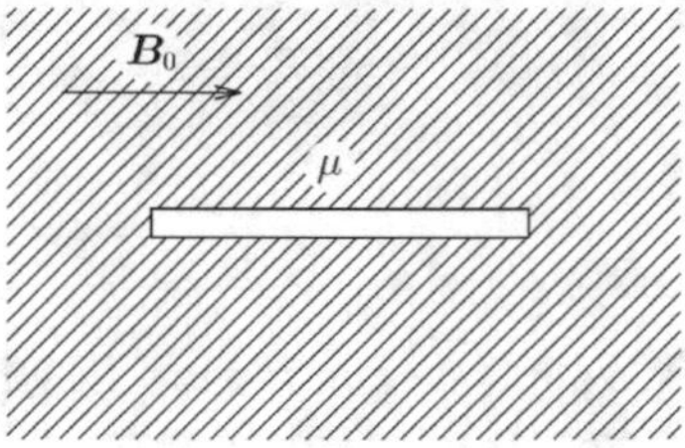

Fig. E9.3 Vacuum slit parallel to the magnetic flux density in magnetic material

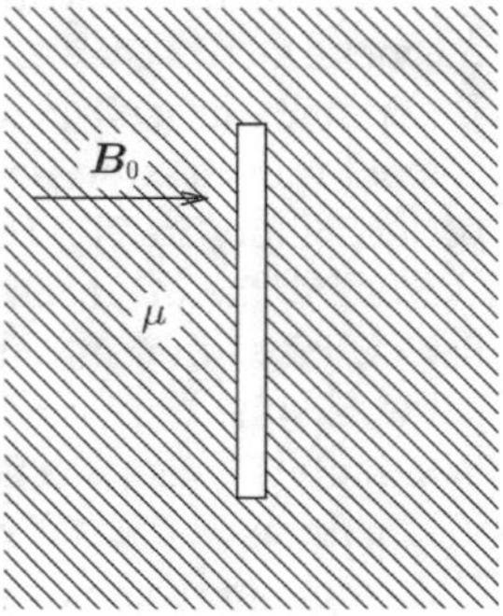

Fig. E9.4 Vacuum slit normal to the magnetic flux density in magnetic material

Determine the magnetic flux density, magnetic field, magnetization inside the magnetic material, and the surface magnetizing current density.

9.7. When a magnetic sphere is in a uniform magnetic flux density B_0, we obtain the magnetic flux density B and the surface magnetizing current density τ_{m} (see Example 9.5). Prove that the obtained B coincides with the sum of B_0 and the magnetic flux density produced by the magnetizing current. (Hint: Use Eq. (7.33) for the relationship between the uniform magnetic flux density and the surface current density).

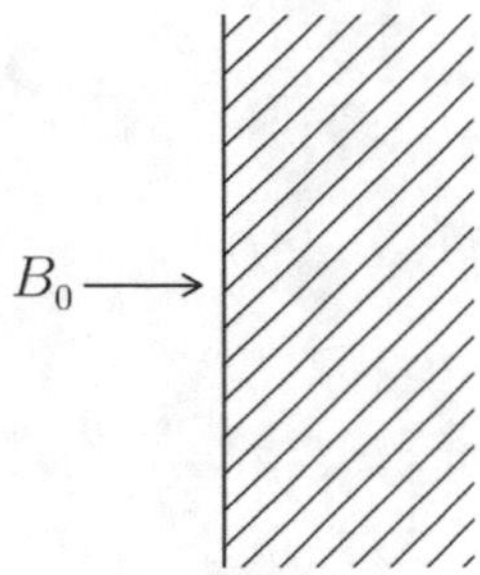

Fig. E9.5 Magnetic flux density applied normal to the surface of a magnetic material

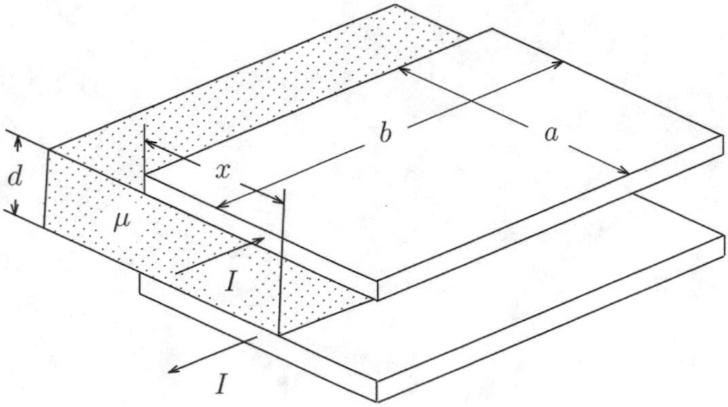

Fig. E9.6 Two parallel superconducting plates with applied current with inserted magnetic material

9.8. A long magnetic cylinder of radius a and magnetic permeability μ is in a uniform normal magnetic flux density B_0. Determine the magnetic flux density, magnetic field, magnetization, and surface magnetizing current density.

9.9. The vector potential is used to determine the magnetic flux density outside the spherical magnetic material in Example 9.6. Solve this problem using the magnetic scalar potential, and discuss the difference between the two solution methods.

9.10. Determine the magnetic flux density and magnetization in a superconducting sphere of radius a in the intermediate state in an applied magnetic flux density B_0. Confirm that the obtained magnetization agrees with the characteristic shown in Fig. 7.23 in column (2) of Chap. 7. Use the boundary conditions for $\boldsymbol{B}$ and $\boldsymbol{H}$ with the critical condition that the maximum magnetic flux density that the superconductor suffers is the critical value, B_c.

9.11. We apply current I to two long parallel superconducting plates and insert a magnetic material of magnetic permeability μ into the space between the two superconducting plates by distance x from the edge, as shown in Fig. E9.6. Determine the magnetic energy and the force on the magnetic material.

Part III
Time-Dependent Electromagnetic Phenomena

Chapter 10
Electromagnetic Induction

Abstract This chapter covers the electromagnetic induction, which is one of most important phenomena that describe electric and magnetic fields varying with time. The electromotive force appears to induce a current in a closed circuit in such a way as to reduce any change in the magnetic flux penetrating it. This is Faraday's law. As a result, the electric field is no longer irrotational and has a vortex caused by a variation in the magnetic flux density with time. Faraday's law is rewritten for the local relationship called the motional law, $\boldsymbol{E} = \boldsymbol{v} \times \boldsymbol{B}$, in terms of the velocity $\boldsymbol{v}$ of a material in a space with a magnetic flux density $\boldsymbol{B}$. The magnetic energy, which is covered in Chap. 8, is estimated again from the electric energy that the electric power source supplies under the electromagnetic induction. Finally, the skin effect is discussed in association with the electromagnetic induction.

10.1 Induction Law

Magnetic flux density is produced by current as described in Chap. 6. Faraday conducted experiments based on the idea that current might be produced by magnetic flux density. Although he could not produce a steady current by using a static magnetic flux density, he produced a non-steady current by varying a magnetic flux density with time. The results are summarized as follows. Suppose two coils. When a current is applied to coil 1 as shown in Fig. 10.1, a magnetic flux density is produced around it. A current flows in coil 2 in the following cases:

1. the current in coil 1 changes as shown in Fig. 10.1a;
2. coil 1 is moved as shown in Fig. 10.1b.

This phenomenon is called **electromagnetic induction**, and the electromotive force that induces the current in coil 2 is called **induced electromotive force**. One can observe that the current flows in coil 2 in such a way as to reduce any change in the magnetic flux penetrating coil 2. This shows a conservative property in nature similarly to the law of inertia for the matter.

T. Matsushita, *Electricity and Magnetism*, Undergraduate Lecture Notes in Physics,
https://doi.org/10.1007/978-3-030-82150-0_10

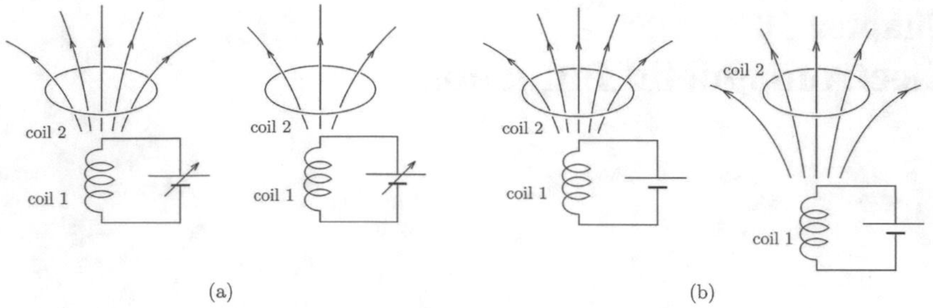

Fig. 10.1 Current changes in coil 2, when **a** current in coil 1 changes or **b** coil 1 moves

From the above results, if the magnetic flux that penetrates the coil is Φ, the induced electromotive force in the coil is given by

$$V_{\mathrm{em}} = -\frac{\mathrm{d}\Phi}{\mathrm{d}t}, \tag{10.1}$$

where the directions of magnetic flux and electromotive force follow the right-hand rule. This is exactly the result predicted in Sect. 8.4 and is called **Faraday's law**. This shows that the variation in magnetic flux with time causes the electromotive force. In this sense, this is also called the **magnetic flux law** or **transformer law**. If the number of turns of the coil is N and the magnetic flux that penetrates one turn of the coil is Φ, the electromotive force is

$$V_{\mathrm{em}} = -N\frac{\mathrm{d}\Phi}{\mathrm{d}t}. \tag{10.2}$$

Here we calculate the electromotive force induced in a solenoid coil of radius a and number of turns N that is rotating with angular frequency ω in a uniform magnetic flux density, B, as shown in Fig. 10.2. If the angle between the coil axis

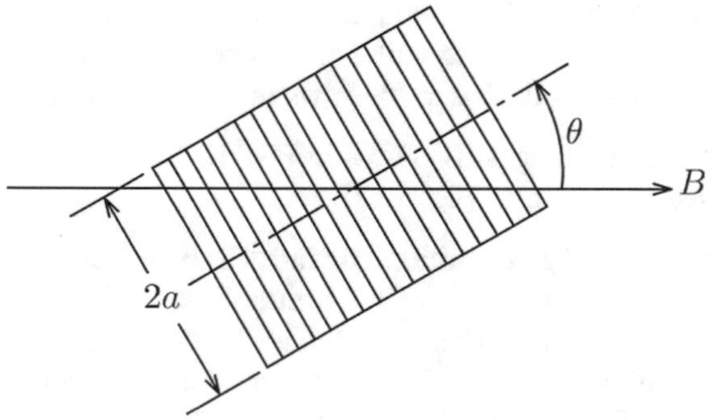

Fig. 10.2 Solenoid coil rotating in uniform magnetic flux density

and magnetic flux density is $\theta = \omega t$, the magnetic flux that penetrates one turn of the coil is $\Phi = \pi a^2 B \cos \omega t$, and the electromotive force is

$$V_{\mathrm{em}} = \pi N a^2 B \omega \sin \omega t. \tag{10.3}$$

Here, we derive the differential expression of Faraday's law for the above case (1). Using the electric field, the electromotive force induced in closed coil 2 is written as

$$V_{\mathrm{em}} = \oint_{\mathrm{C}} \boldsymbol{E} \cdot \mathrm{d}\boldsymbol{s} = \int_{\mathrm{S}} \nabla \times \boldsymbol{E} \cdot \mathrm{d}\boldsymbol{S}, \tag{10.4}$$

where S is the surface surrounded by C, and we have used Stokes' theorem. On the other hand, from Eq. (6.19), the magnetic flux is

$$\Phi = \int_{\mathrm{S}} \boldsymbol{B} \cdot \mathrm{d}\boldsymbol{S}. \tag{10.5}$$

Thus, Eq. (10.1) gives

$$\int_{\mathrm{S}} \nabla \times \boldsymbol{E} \cdot \mathrm{d}\boldsymbol{S} = -\frac{\mathrm{d}}{\mathrm{d}t} \int_{\mathrm{S}} \boldsymbol{B} \cdot \mathrm{d}\boldsymbol{S} = -\int_{\mathrm{S}} \frac{\partial \boldsymbol{B}}{\partial t} \cdot \mathrm{d}\boldsymbol{S}. \tag{10.6}$$

In the above, we changed the order of the surface integral and differentiation with respect to time, since surface S does not change with time. In this process, we change the total differentiation with time is changed to partial differentiation, since we are treating a stationary system. The relationship, Eq. (10.6), holds for arbitrary S, and we have

$$\nabla \times \boldsymbol{E} = -\frac{\partial \boldsymbol{B}}{\partial t}. \tag{10.7}$$

This is the **differential form of the induction law**. When the magnetic flux density does not change with time, Eq. (10.7) reduces to the equation for electrostatic field, Eq. (1.28). Hence, we can conclude that the electric field given by Eq. (10.7) is a general electric field that includes the induced and static components.

Second, we consider case (2) where coil 2 of closed loop C moves with the velocity $\boldsymbol{v}$ in a magnetic flux density $\boldsymbol{B}$ that does not change with time. The area of the hatched region in Fig. 10.3 that a small segment of the coil, $\mathrm{d}\boldsymbol{s}$, sweeps in short period Δt is $|\boldsymbol{v}\Delta t \times \mathrm{d}\boldsymbol{s}|$, and the magnetic flux that enters the coil through this region is

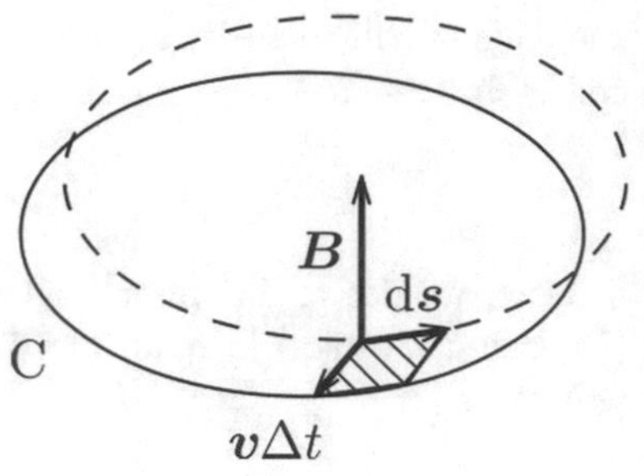

Fig. 10.3 Coil C moving in magnetic flux density $\boldsymbol{B}$. It moves from the position shown by the *dotted line* to that by the *solid line* during short period Δt

$$(\boldsymbol{v}\Delta t \times \mathrm{d}\boldsymbol{s}) \cdot \boldsymbol{B} = (\boldsymbol{B} \times \boldsymbol{v}) \cdot \mathrm{d}\boldsymbol{s}\Delta t. \tag{10.8}$$

It should be noted that the directions of $\mathrm{d}\boldsymbol{s}$ and $\boldsymbol{B}$ follow the right-hand rule. Hence, the total magnetic flux that enters the coil during Δt is

$$\Delta\Phi = \Delta t \oint_{\mathrm{C}} (\boldsymbol{B} \times \boldsymbol{v}) \cdot \mathrm{d}\boldsymbol{s}. \tag{10.9}$$

In the limit $\Delta t \to 0$, we have

$$\frac{\Delta\Phi}{\Delta t} \to \frac{\mathrm{d}\Phi}{\mathrm{d}t} = \oint_{\mathrm{C}} (\boldsymbol{B} \times \boldsymbol{v}) \cdot \mathrm{d}\boldsymbol{s}. \tag{10.10}$$

Hence, from Eqs. (10.4) and (10.10), we obtain the relationship describing the induced electric field,

$$\boldsymbol{E} = \boldsymbol{v} \times \boldsymbol{B}. \tag{10.11}$$

This is called the **motional law**. Mathematically, the induced electric field should be given by $\boldsymbol{E} = \boldsymbol{v} \times \boldsymbol{B} - \nabla\phi$ with ϕ being an arbitrary scalar function. However, it is empirically known that $\nabla\phi$ is zero in usual cases. Electrons in a conductor of a coil suffer the Lorentz force, $-e\boldsymbol{v} \times \boldsymbol{B}$, when the coil moves with the velocity $\boldsymbol{v}$. We can think of this force as a force caused by the electric field, Eq. (10.11), induced in the coil.

Thus, we can conclude that the electromotive force induced in the coil is caused by the change in the magnetic flux that penetrates the coil in both cases (1) and (2). The induced electromotive force can be summarized as

$$V_{\mathrm{em}} = \oint_{\mathrm{C}} (\boldsymbol{E} + \boldsymbol{v} \times \boldsymbol{B}) \cdot \mathrm{d}\boldsymbol{s} = -\int_{\mathrm{S}} \frac{\mathrm{d}\boldsymbol{B}}{\mathrm{d}t} \cdot \mathrm{d}\boldsymbol{S}. \tag{10.12}$$

Using Eq. (A1.43), the condition $\nabla \cdot \boldsymbol{B} = 0$, and the condition that the coil is not deformed during the movement, $\nabla \cdot \boldsymbol{v} = 0$, we have

$$\nabla \times (\boldsymbol{v} \times \boldsymbol{B}) = (\boldsymbol{B} \cdot \nabla)\boldsymbol{v} - (\boldsymbol{v} \cdot \nabla)\boldsymbol{B}. \tag{10.13}$$

When the velocity is constant, the first term on the right side is zero. The total differentiation with respect to time is written as

$$\frac{\mathrm{d}\boldsymbol{B}}{\mathrm{d}t} = \frac{\partial \boldsymbol{B}}{\partial t} + (\boldsymbol{v} \cdot \nabla)\boldsymbol{B}. \tag{10.14}$$

Hence, rewriting Eq. (10.12) with Stokes' theorem, we have

$$\int_{\mathrm{S}} \nabla \times \boldsymbol{E} \cdot \mathrm{d}\boldsymbol{S} = -\int_{\mathrm{S}} \frac{\partial \boldsymbol{B}}{\partial t} \cdot \mathrm{d}\boldsymbol{S}. \tag{10.15}$$

Thus, we have derived Eq. (10.7), indicating that the above conclusion is valid.

When a conductor carries current $\boldsymbol{I}$ in magnetic flux density $\boldsymbol{B}$, the Lorentz force acts on the conductor. If this force forces the conductor to move with velocity $\boldsymbol{v}$, the power in a unit length of the conductor given by the Lorentz force is

$$\boldsymbol{F}' \cdot \boldsymbol{v} = (\boldsymbol{I} \times \boldsymbol{B}) \cdot \boldsymbol{v}. \tag{10.16}$$

This seems to contradict the fact that the Lorentz force does not do any work on electric charges (see Example 6.4). In a practical case, the induced electric field directed opposite to the current works to reduce the current. To have the same current continue to flow, the electric power source must supply additional electric power,

$$-\boldsymbol{I} \cdot (\boldsymbol{v} \times \boldsymbol{B}) = (\boldsymbol{I} \times \boldsymbol{B}) \cdot \boldsymbol{v}, \tag{10.17}$$

which is equal to the power given by the Lorentz force. That is, the power by the Lorentz force is nothing other than the electric power by the electric source (see Exercise 10.9).

Example 10.1. Coil A with number of turns N_{A} and coil B with number of turns N_{B} are wound in directions on a cylindrical magnetic material with cross-sectional area S and magnetic permeability μ, as shown in Fig. 10.4. Determine the current induced in coil B that is short-circuited when current $I_{\mathrm{A}} = I\sin\omega t$ is applied to coil A.

Fig. 10.4 Coils A and B wound on a magnetic material

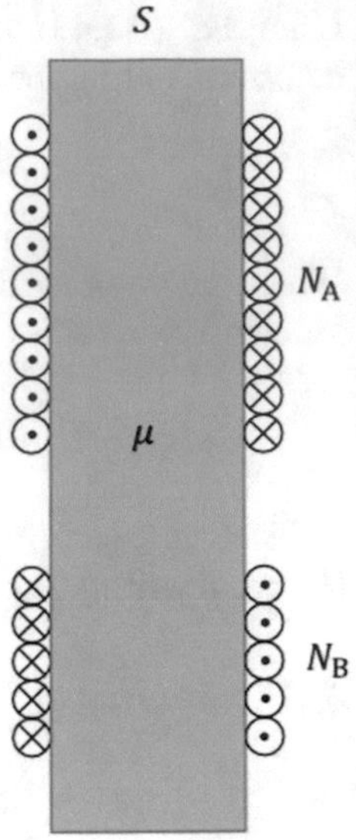

Solution 10.1. The magnetic field induced in the cylindrical magnetic material by the current flowing in coil A is $H = N_\mathrm{A} I \sin\omega t$, and the magnetic flux that penetrates the magnetic material is

$$\Phi = \mu S H = \mu S N_\mathrm{A} I \sin\omega t.$$

The electromotive force induced in Coil B is

$$V_\mathrm{em} = -N_\mathrm{B}\frac{\mathrm{d}\Phi}{\mathrm{d}t} = -\mu S N_\mathrm{A} N_\mathrm{B} \omega I \cos\omega t.$$

We denote the current induced in coil B by I_B, and from Eq. (8.1), this electromotive force is

$$V_\mathrm{em} = -L_\mathrm{B}\frac{\mathrm{d}I_\mathrm{B}}{\mathrm{d}t},$$

where L_B is the self-inductance of coil B wound on a magnetic material;

$$L_\mathrm{B} = \mu S N_\mathrm{B}^2.$$

Hence, we have

$$I_\mathrm{B} = \frac{N_\mathrm{A}}{N_\mathrm{B}} I \sin\omega t = \frac{N_\mathrm{A}}{N_\mathrm{B}} I_\mathrm{A}.$$

Thus, the relationship, $I_B/I_A = N_A/N_B$, holds. The voltages induced in coils A and B are denoted by V_A and V_B, respectively. Then, the relationship is also written as $V_B/V_A = N_B/N_A$. This is the principle of the transformer.

◇

Example 10.2. A triangular closed circuit and a straight current, I, are placed on a common plane. The closed circuit is moving away with velocity v from the current, as shown in Fig. 10.5. The distance between the closed circuit and the current is $r = r_0$ in the initial condition ($t = 0$). Determine the electromotive force induced in the closed circuit with the magnetic flux law. We define the electromotive force to be positive along the direction of ABC.

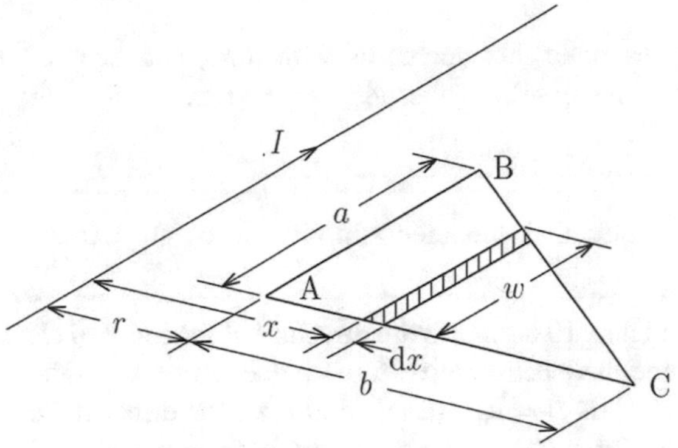

Fig. 10.5 Straight current and triangular closed circuit moving away with constant velocity

Solution 10.2. At time t the distance between the circuit and current is $r(t) = r_0 + vt$. The width of the triangle at distance x ($r \leq x \leq r + b$) from the current is

$$w(x) = a - \frac{a(x-r)}{b}.$$

The direction of the magnetic flux produced by current I inside the circuit is the same as that of the magnetic flux produced by the current flowing along ABC. Hence, the magnetic flux produced by current I is positive. The magnetic flux density at distance x from the current is $B(x) = \mu_0 I/(2\pi x)$. The magnetic flux penetrating the narrow region x to $x + \mathrm{d}x$ in the circuit is

$$\mathrm{d}\Phi = B(x)w(x)\mathrm{d}x = \frac{\mu_0 aI}{2\pi b}\left(\frac{r+b}{x} - 1\right)\mathrm{d}x.$$

Thus, the total magnetic flux penetrating the circuit is

$$\Phi = \frac{\mu_0 Ia}{2\pi b}\int_r^{r+b}\left(\frac{r+b}{x} - 1\right)\mathrm{d}x = \frac{\mu_0 Ia}{2\pi b}\left[(r+b)\log\frac{r+b}{r} - b\right].$$

The electromotive force induced in the circuit is

$$V_{\mathrm{em}} = -\frac{\mathrm{d}\Phi}{\mathrm{d}t} = -\frac{\partial\Phi}{\partial r}\cdot\frac{\partial r}{\partial t} = \frac{\mu_0 Iav}{2\pi b}\left[\frac{b}{r_0+vt} - \log\left(1+\frac{b}{r_0+vt}\right)\right].$$

The penetrating magnetic flux decreases with time, and hence, the electromotive force is induced to increase it. Thus, V_{em} is positive.

◇

Example 10.3. Determine the electromotive force in Example 10.2 with the motional law.

Solution 10.3. Figure 10.6a shows the direction of the induced electric field on each side. We determine the electromotive force induced on each side. On side AB, the magnetic flux density is $B = \mu_0 I/(2\pi r)$, and $\boldsymbol{v}\times\boldsymbol{B}$ is directed from A to B and its magnitude is $\mu_0 Iv/(2\pi r)$. Hence, the contribution from this side to the electromotive force is

$$\int_{\mathrm{A}}^{\mathrm{B}}(\boldsymbol{v}\times\boldsymbol{B})\cdot\mathrm{d}\boldsymbol{s} = \frac{\mu_0 Iv}{2\pi r}\,a.$$

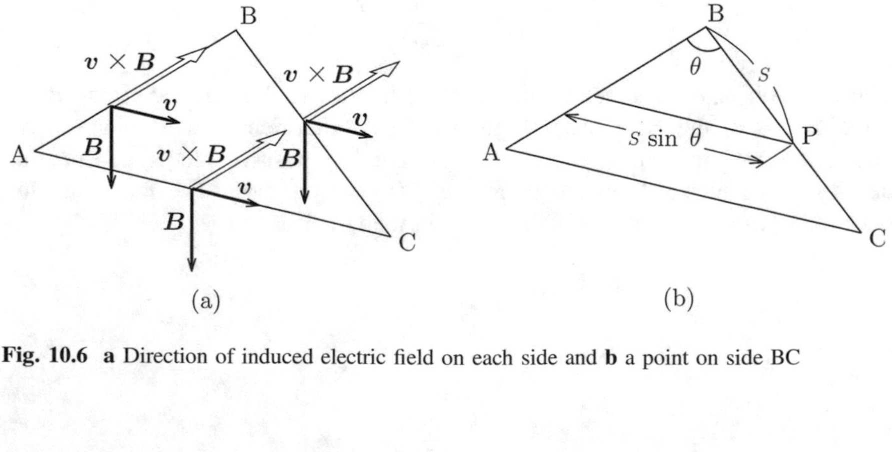

Fig. 10.6 **a** Direction of induced electric field on each side and **b** a point on side BC

Next, the magnetic flux density at point P at distance s from B is $\mu_0 I/[2\pi(r+s\sin\theta)]$ with θ denoting the angle of B (see Fig. 10.6b). The induced electric field has magnitude $\mu_0 Iv/[2\pi(r+s\sin\theta)]$, and its direction is tilted by $\pi-\theta$ from the direction of integration, ds. Thus, we have

$$(\boldsymbol{v}\times\boldsymbol{B})\cdot \mathrm{d}\boldsymbol{s} = \frac{\mu_0 Iv}{2\pi(r+s\sin\theta)}\cos(\pi-\theta)\,\mathrm{d}s = -\frac{\mu_0 Iv\cos\theta}{2\pi(r+s\sin\theta)}\,\mathrm{d}s,$$

and the contribution from side BC is

$$\int_{\mathrm{B}}^{\mathrm{C}}(\boldsymbol{v}\times\boldsymbol{B})\cdot \mathrm{d}\boldsymbol{s} = -\frac{\mu_0 Iv\cos\theta}{2\pi}\int_0^{\overline{\mathrm{BC}}}\frac{\mathrm{d}s}{r+s\sin\theta} = -\frac{\mu_0 Iv\cot\theta}{2\pi}\log\frac{r+a\tan\theta}{r}$$
$$= -\frac{\mu_0 Iav}{2\pi b}\log\frac{r+b}{r}.$$

Finally, the induced electric field is perpendicular to the direction of integration on side CA. Hence, there is no contribution from this side. As a result, the induced electromotive force is

$$V_{\mathrm{em}} = \oint(\boldsymbol{v}\times\boldsymbol{B})\cdot \mathrm{d}\boldsymbol{s} = \frac{\mu_0 Iav}{2\pi b}\left[\frac{b}{r_0+vt} - \log\left(1+\frac{b}{r_0+vt}\right)\right],$$

which agrees with the result obtained in Example 10.2.

◇

Here we discuss a phenomenon that is usually explained using only the motional law. Suppose that a conducting circular plate of radius a is rotated with angular frequency ω around its axis in uniform magnetic flux density B, as shown in Fig. 10.7a. We determine the electromotive force induced between the center O of the plate and point P on the edge. Since the magnetic flux penetrating the closed loop composed of the straight line connecting O and P and the line C outside the plate does

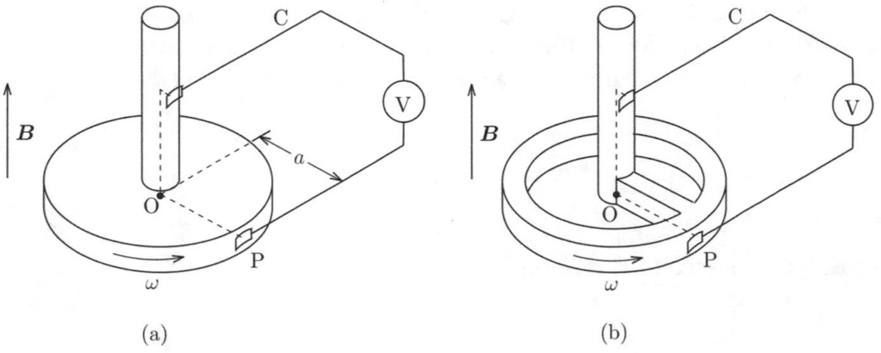

Fig. 10.7 Unipolar induction: **a** usual system and **b** new system

not change with time, it seems that no electromotive force is induced in it. However, an electromotive force is induced in reality. This is called **unipolar induction**.

According to the motional law, since the magnetic flux crosses line OP, the electromotive force is induced there. In the arrangement in Fig. 10.7a, the induced electric field is directed from O to P. At a point at distance R from the central axis, the velocity of rotation, v, is equal to $R\omega$. Hence, the induced electric field is $E = R\omega B$. Integrating this from O to P, the electromotive force is

$$V_{\mathrm{em}} = \int_0^a R\omega B \, \mathrm{d}R = \frac{1}{2}\omega B a^2. \tag{10.18}$$

The unipolar induction is really known. However, since the electromotive force is not stable, it is not practically used. If the circular plate is magnetized, the same thing occurs even if no magnetic field is applied. In addition, when a hollow dielectric plate is rotated in a magnetic flux density, an electric polarization occurs because of the electromotive force.

The same unipolar induction can be observed even for the system shown in Fig. 10.7b, in which bar OP rotates. In this case, the magnetic flux penetrating the circuit changes with time, and the result can also be explained with the magnetic flux law. To explain the result for the system in Fig. 10.7a with the magnetic flux law, we can suppose that part of the circuit rotates with the plate as the bar in Fig. 10.7b. Another part of the circuit on the edge of the plate is equipotential. Hence, it is difficult to distinguish the two mechanisms.

In the above, we learned the magnetic flux law and motional law to describe the induced electromotive force for simple cases. If we discuss the case where a conductor is forced to move in a magnetic flux density varying with time, we have two contributions to the electromotive force, and it is necessary to calculate each contribution using the two laws. Here we propose a general law that combines the two laws.

We assume that the external magnetic field is increasing with time. In this case the magnetic flux density inside a material also increases because of the penetrating magnetic flux. Thus, we can define the velocity of the magnetic flux lines and denote it as $\boldsymbol{V}$. If the coil in Fig. 10.3 stays stationary, but the magnetic flux lines move with velocity $\boldsymbol{V} = -\boldsymbol{v}$, the same amount of magnetic flux penetrates into the coil through a segment d$\boldsymbol{s}$ within period Δt. Hence, repeating a similar argument up to Eq. (10.10), the time variation in magnetic flux Φ that penetrates the coil is given by

$$\frac{\mathrm{d}\Phi}{\mathrm{d}t} = -\oint_{\mathrm{C}} (\boldsymbol{B} \times \boldsymbol{V}) \cdot \mathrm{d}\boldsymbol{s}. \tag{10.19}$$

Thus, we obtain the local relationship,

$$\nabla \times (\boldsymbol{B} \times \boldsymbol{V}) = -\frac{\partial \boldsymbol{B}}{\partial t}. \tag{10.20}$$

This is called the **continuity equation of magnetic flux** and is frequently used for analyzing electromagnetic phenomena in superconductors. Comparing this equation with Eq. (10.7), we have

$$\boldsymbol{E} = \boldsymbol{B} \times \boldsymbol{V}. \tag{10.21}$$

This is called **Josephson's relation**. It should be noted that this relation expresses the magnetic flux law. Using the above two velocities, the relative velocity of magnetic flux lines from coil C is

$$\boldsymbol{V}' = \boldsymbol{V} - \boldsymbol{v}, \tag{10.22}$$

and combining the above equation with Eq. (10.11) gives the **general law for the induced electric field**,

$$\boldsymbol{E} = \boldsymbol{B} \times \boldsymbol{V}'. \tag{10.23}$$

The Column in this chapter shows an example of calculating induced electromotive force.

Example 10.4. Suppose that we increase by ΔB the magnetic flux density B_0 applied parallel to a long cylindrical conductor of radius a during short period, Δt. Determine with Eq. (10.21) the electromotive force measured with potential leads with the different arrangements shown in Fig. 10.8a, b.

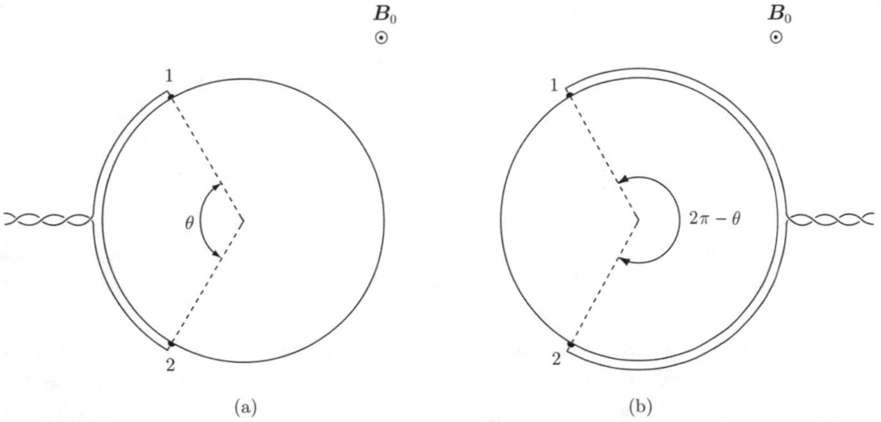

Fig. 10.8 Different arrangements of potential leads for measuring electromotive force induced in cylindrical conductor

Solution 10.4. We denote the velocity of the magnetic flux as

$$\boldsymbol{V} = -\boldsymbol{i}_R V.$$

Then, the continuity equation of magnetic flux shown earlier reduces to

$$-\frac{1}{R} \cdot \frac{\partial}{\partial R}(RB_0 V) = -\frac{\Delta B}{\Delta t}$$

and we have

$$E = B_0 V = \frac{1}{2} \cdot \frac{\Delta B}{\Delta t} a$$

on the surface ($R = a$). The induced electric field is directed counterclockwise. One can easily show that the induced electric field integrated along the circumference in this direction, $2\pi aE$, is equal to the total electromotive force, $\Delta\Phi/\Delta t = \pi a^2 \Delta B/\Delta t$.

Here we consider the case shown in Fig. 10.8a. The azimuthal angle between potential terminals 1 and 2 is θ, and we place the potential leads on the surface of the conductor and twist to eliminate the electromotive force outside the conductor. The measured electromotive force is

$$V_{\mathrm{em}} = a\theta B_0 V = \frac{1}{2} \cdot \frac{\Delta B}{\Delta t} a^2 \theta,$$

where we set a reference point on terminal 1.

Second, we determine the electromotive force for the arrangement shown in Fig. 10.8b. We assume an integral path of the induced electric field on the right conductor surface. In this case, we can neglect the magnetic flux penetrating the closed loop composed of this path and the potential leads and similarly determine the electromotive force only by integrating Eq. (10.21) along the path. Thus, we have

$$V'_{\mathrm{em}} = -(2\pi - \theta)B_0 V = -\frac{1}{2} \cdot \frac{\Delta B}{\Delta t} a^2 (2\pi - \theta)$$

for the same reference point.

On the other hand, it is possible to choose the left conductor surface for the integral path. The integral gives $a^2\theta(\Delta B/\Delta t)/2$. In this case, the electromotive force due to the magnetic flux penetrating the integral path should be taken into account. This additional component is $-\pi a^2(\Delta B/\Delta t)$, and we obtain the same result by adding it.

Thus, there is a freedom in choosing the integral path. For example, it is also possible to choose the path shown by the dotted line in Fig. 10.8b. In this case, the induced electric field is perpendicular to the path, resulting in a zero line integral. From the magnetic flux penetrating the area surrounded by the path and the potential leads (denoted by the azimuthal angle $2\pi - \theta$), we directly obtain the same result.

◇

10.2 Potential

As described in Sect. 10.1, the electric field $\boldsymbol{E}$ contains not only the electrostatic field but also the induced electric field. Hence, such a general electric field cannot be described only by the electric potential. Here we note that the right side of Eq. (10.7) is written in terms of the vector potential $\boldsymbol{A}$ as

$$-\frac{\partial \boldsymbol{B}}{\partial t} = -\nabla \times \left(\frac{\partial \boldsymbol{A}}{\partial t} \right), \tag{10.24}$$

where we have changed the order of the time differentiation and spatial differentiation. Comparing this with the left side of Eq. (10.7), it is obvious that the induced electric field is given by $-\partial \boldsymbol{A}/\partial t$. Hence, with the electrostatic field, the general electric field is given by

$$\boldsymbol{E} = -\nabla \phi - \frac{\partial \boldsymbol{A}}{\partial t}. \tag{10.25}$$

This satisfies Eq. (10.7) and reduces to Eq. (1.24) in the static condition. The vector potential satisfies Eq. (9.21) under the Coulomb gauge even in this case.

To discuss the general electromagnetic fields, it is necessary to extend Ampere's law, Eq. (9.12), for the static magnetic field produced by a steady current to a general law including the magnetic field by a non-steady current. This will be covered in Chap. 11 in which we complete the set of Maxwell's equations.

10.3 Boundary Conditions

Here we investigate whether the boundary condition for the electric field $\boldsymbol{E}$ changes in going from Eqs. (1.28) to (10.7).

We denote the electric fields in materials 1 and 2 near the boundary as $\boldsymbol{E}_1$ and $\boldsymbol{E}_2$, respectively. Consider a plane normal to the boundary that includes vectors $\boldsymbol{E}_1$ and $\boldsymbol{E}_2$ (see Fig. 4.15a, b). Integrating the electric field around a small rectangle, ΔC, with two sides parallel to the boundary, with Eq. (10.7) and Stokes' theorem, we have

$$\oint_{\Delta C} \boldsymbol{E} \cdot \mathrm{d}\boldsymbol{s} = \int_{\Delta S} \nabla \times \boldsymbol{E} \cdot \mathrm{d}\boldsymbol{S} = -\frac{\partial}{\partial t} \int_{\Delta S} \boldsymbol{B} \cdot \mathrm{d}\boldsymbol{S}, \tag{10.26}$$

where ΔS is the surface surrounded by ΔC, and we have changed the order of the spatial integration and time differentiation. When the height of the small rectangle, Δh, is sufficiently small, the amount of magnetic flux that penetrates ΔS is negligible. Thus, the circular integral of the electric field reduces to zero similarly to the case in Sect. 4.3, and we obtain the same result as Eq. (4.22),

$$\boldsymbol{n} \times (\boldsymbol{E}_1 - \boldsymbol{E}_2) = 0. \tag{10.27}$$

That is, the parallel component of the electric field is continuous across the boundary even when electromagnetic induction occurs.

10.4 Magnetic Energy

We already learned the magnetic energy in Sect. 8.3. In the system shown in Fig. 8.14, the magnetic flux that links the circuit made of a superconductor does not change even when the plate moves. This means that there is no electromagnetic induction. This is why the magnetic energy can be determined from the mechanical work needed against the magnetic force for this system. Most textbooks on electromagnetism explain magnetic energy after electromagnetic induction. Hence, this textbook gives the same explanation. An explanation from a different viewpoint is helpful for a deep understanding of the phenomenon. Since there is no new phenomenon in this section, however, readers who do not feel a need for this explanation can skip it.

Suppose a coil of self-inductance L. When we apply current I' to this coil, the magnetic flux penetrating this coil is

$$\Phi' = LI'. \tag{10.28}$$

When the current is increased by a small amount $\mathrm{d}I'$ in short period $\mathrm{d}t$, the induced electromotive force is

$$V_{\mathrm{em}} = -L\frac{\mathrm{d}I'}{\mathrm{d}t}. \tag{10.29}$$

This acts to restrict the increase in the current. This phenomenon is called **self-induction**. Thus, the electric power source must work against this electromotive force to increase the current, and the electric power in this period is

$$P = -V_{\mathrm{em}} I' = L I' \frac{\mathrm{d}I'}{\mathrm{d}t}. \tag{10.30}$$

The energy stored in the coil when we apply the current I to the coil is equal to the energy supplied by the electric power source until the current increases from 0 to I and is given by

$$U_{\mathrm{m}} = W = \int L I' \frac{\mathrm{d}I'}{\mathrm{d}t}\, \mathrm{d}t = \int_0^I L I' \mathrm{d}I' = \frac{1}{2} L I^2. \tag{10.31}$$

Using the magnetic flux $\Phi = LI$, this is also written as

$$U_{\mathrm{m}} = \frac{1}{2} L I^2 = \frac{1}{2} \Phi I = \frac{1}{2L} \Phi^2. \tag{10.32}$$

This result agrees with Eq. (8.34), and we can understand that this energy is exactly the magnetic energy.

Following a similar discussion, we can show that the magnetic energy of a system composed of n coils is given by Eq. (8.35). In this case, the electromotive force in the i-th coil induced by current I_j flowing in the j-th coil is given by

$$V_{\mathrm{em}i} = -L_{ij} \frac{\mathrm{d}I_j}{\mathrm{d}t}, \tag{10.33}$$

using the mutual inductance. This is called **mutual induction**.

Example 10.5. Prove Eq. (8.35) for a system composed of two coils and prove the reciprocity, $L_{12} = L_{21}$.

Solution 10.5. Currents I_1 and I_2 flow in coils 1 and 2, and the resultant penetrating magnetic fluxes in these coils are Φ_1 and Φ_2, respectively. Assume that this final situation is reached after we apply the currents to coil 1 and then to coil 2. We suppose an intermediate situation where coil 2 has no current and coil 1 has current I_1' (see Fig. 10.9a). With the inductance coefficient the magnetic flux that penetrates coil 1 is $L_{11} I_1'$. Hence, the work done to apply current I_1 to coil 1 is

$$W_1 = \int_0^{I_1} L_{11} I_1' \mathrm{d}I_1' = \frac{1}{2} L_{11} I_1^2.$$

Next, we consider the situation where coils 1 and 2 have currents I_1 and I_2', respectively (see Fig. 10.9b). The magnetic flux penetrating coil 2 is $L_{21} I_1 + L_{22} I_2'$. Hence, the work done to apply current I_2 to coil 2 is

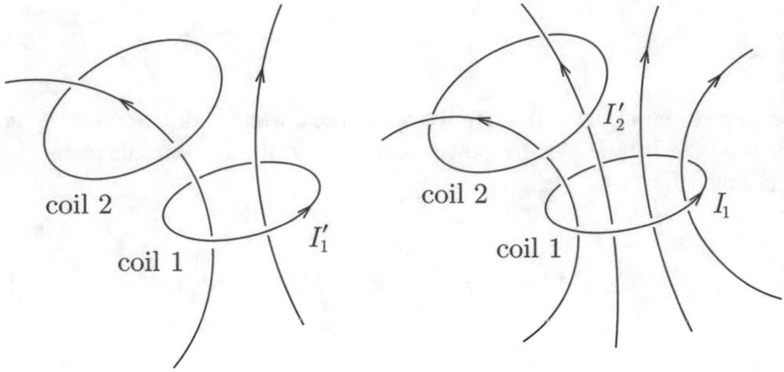

Fig. 10.9 **a** Situation in which current I_1' is flowing in coil 1 but no current in coil 2 and **b** that in which currents I_1 and I_2' are flowing in coils 1 and 2, respectively

$$W_2 = \int_0^{I_2} (L_{21}I_1 + L_{22}I_2')\mathrm{d}I_2' = L_{21}I_1I_2 + \frac{1}{2}L_{22}I_2^2.$$

Thus, the magnetic energy of this system is

$$U_\mathrm{m} = W_1 + W_2 = \frac{1}{2}L_{11}I_1^2 + L_{21}I_1I_2 + \frac{1}{2}L_{22}I_2^2. \tag{10.34}$$

If we apply the currents in reversed order, the magnetic energy is

$$U_\mathrm{m} = \frac{1}{2}L_{11}I_1^2 + L_{12}I_1I_2 + \frac{1}{2}L_{22}I_2^2. \tag{10.35}$$

Since Eqs. (10.34) and (10.35) must be the same, we can prove the reciprocity, Eq. (8.5),

$$L_{12} = L_{21}.$$

Thus, if we write $L_{21} = (L_{12} + L_{21})/2$ in Eq. (10.34), the magnetic energy is

$$U_\mathrm{m} = \frac{1}{2}I_1(L_{11}I_1 + L_{12}I_2) + \frac{1}{2}I_2(L_{21}I_1 + L_{22}I_2) = \frac{1}{2}(I_1\Phi_1 + I_2\Phi_2).$$

Thus, Eq. (8.35) holds for n = 2.

◇

Suppose we change magnetic flux densities in coils with electric power sources. We denote the induced electric field by $\boldsymbol{E}$. From Eq. (10.30), the input power into the system is given by

$$P = -\int_V \boldsymbol{E} \cdot \boldsymbol{i} \, \mathrm{d}V. \tag{10.36}$$

Using Eqs. (6.27) and (A1.41) in the Appendix, this leads to

$$\begin{aligned} P &= -\frac{1}{\mu_0} \int_V \boldsymbol{E} \cdot (\nabla \times \boldsymbol{B}) \mathrm{d}V \\ &= \frac{1}{\mu_0} \int_V [\nabla \cdot (\boldsymbol{E} \times \boldsymbol{B}) - \boldsymbol{B} \cdot (\nabla \times \boldsymbol{E})] \mathrm{d}V. \end{aligned} \tag{10.37}$$

Using Gauss' theorem, we rewrite the first integral as the surface integral

$$\frac{1}{\mu_0} \int_S (\boldsymbol{E} \times \boldsymbol{B}) \cdot \mathrm{d}\boldsymbol{S},$$

where S in the surface of V. We assume a sphere of sufficiently large radius r for V. Since $|\boldsymbol{B}| \propto r^{-2}$, $|\boldsymbol{E}| \propto r^{-1}$, and $\int \mathrm{d}S \propto r^2$, the integral is proportional to r^{-1}. Hence, if we assume a very large sphere, the integral approaches zero and can be disregarded. Substituting Eq. (10.7) for the second integral, we have

$$P = \frac{1}{2\mu_0} \int_V \frac{\partial \boldsymbol{B}^2}{\partial t} \mathrm{d}V. \tag{10.38}$$

From the above result, we can generally show that the magnetic power density is given by $[1/(2\mu_0)]\partial \boldsymbol{B}^2/\partial t$ and the magnetic energy density by Eq. (8.32).

10.5 Skin Effect

The fundamental equations that we have learned up to now are

$$\nabla \times \boldsymbol{E} = -\frac{\partial \boldsymbol{B}}{\partial t}, \tag{10.39}$$

$$\boldsymbol{i} = \nabla \times \boldsymbol{H}, \tag{10.40}$$

$$\boldsymbol{B} = \mu \boldsymbol{H}, \tag{10.41}$$

$$\boldsymbol{i} = \sigma_c \boldsymbol{E}. \tag{10.42}$$

The unknown variables, $\boldsymbol{E}$, $\boldsymbol{B}$, $\boldsymbol{H}$, and $\boldsymbol{i}$ can be obtained by solving the set of these equations. Here we rewrite the above equations in terms of $\boldsymbol{E}$ and $\boldsymbol{B}$. Then, Eqs. (10.40)—(10.42) are summarized as

$$\nabla \times \boldsymbol{B} = \mu \sigma_c \boldsymbol{E}. \tag{10.43}$$

We can solve this equation with Eq. (10.39).

Now we learn the **skin effect** for the example of dynamic phenomena. Supposed that we apply an AC magnetic flux density of amplitude B_0 and angular frequency ω along the z-axis parallel to the surface of a semi-infinite conductor that occupies $x \geq 0$ (see Fig. 10.10). We can assume that no physical variable changes along the y- and z-axes:

$$\frac{\partial}{\partial y}, \frac{\partial}{\partial z} \rightarrow 0. \tag{10.44}$$

The variation with time can be expressed with the factor $e^{i\omega t}$, and we replace the time differentiation with

$$\frac{\partial}{\partial t} \rightarrow i\omega. \tag{10.45}$$

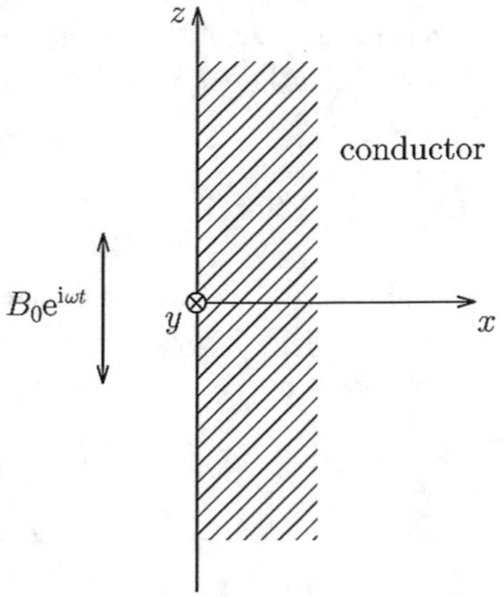

Fig. 10.10 Magnetic flux density applied parallel to the surface of semi-infinite conductor

In addition, since the external magnetic flux density is directed along the z-axis, we can assume that the internal magnetic flux density has only a z-component. Hence, we have

$$\nabla \times \boldsymbol{B} = \begin{vmatrix} \boldsymbol{i}_x & \boldsymbol{i}_y & \boldsymbol{i}_z \\ \mathrm{d}/\mathrm{d}x & 0 & 0 \\ 0 & 0 & B_z \end{vmatrix} = -\boldsymbol{i}_y \frac{\mathrm{d}B_z}{\mathrm{d}x}.$$

Thus, we find that the electric field has only a y-component, E_y, and Eq. (10.43) reduces to

$$\frac{\mathrm{d}B_z}{\mathrm{d}x} = -\mu\sigma_{\mathrm{c}} E_y. \tag{10.46}$$

The left side of Eq. (10.39) is

$$\nabla \times \boldsymbol{E} = \begin{vmatrix} \boldsymbol{i}_x & \boldsymbol{i}_y & \boldsymbol{i}_z \\ \mathrm{d}/\mathrm{d}x & 0 & 0 \\ 0 & E_y & 0 \end{vmatrix} = \boldsymbol{i}_z \frac{\mathrm{d}E_y}{\mathrm{d}x},$$

leading to

$$\frac{\mathrm{d}E_y}{\mathrm{d}x} = -\mathrm{i}\omega B_z. \tag{10.47}$$

This is consistent with the initial assumption that the magnetic flux density has only a z-component. Eliminating E_y in Eqs. (10.46) and (10.47), we have

$$\frac{\mathrm{d}^2 B_z}{\mathrm{d}x^2} - \mathrm{i}\omega\mu\sigma_{\mathrm{c}} B_z = 0. \tag{10.48}$$

The equation for E_y has the same form as this.

We can derive this equation generally. Taking a curl of Eq. (10.43) and substituting Eq. (10.39), we have

$$\nabla \times (\nabla \times \boldsymbol{B}) = -\mu\sigma_{\mathrm{c}} \frac{\partial \boldsymbol{B}}{\partial t}. \tag{10.49}$$

Using Eqs. (A1.46) in the Appendix and (6.21), this equation becomes

$$\Delta \boldsymbol{B} - \mu\sigma_{\mathrm{c}} \frac{\partial \boldsymbol{B}}{\partial t} = 0. \tag{10.50}$$

This is a differential equation of the second-order called a diffusion equation. Substituting the spatial symmetry, Eq. (10.44), and the time variation, Eq. (10.45), derives Eq. (10.48).

Assume a solution of type $B_z(x) \sim e^{\alpha x}$. Substituting this into Eq. (10.48), we have $\alpha^2 = i\omega\mu\sigma_c$. That is,

$$\alpha = \pm(1+i)\left(\frac{\omega\mu\sigma_c}{2}\right)^{1/2}. \tag{10.51}$$

From the condition that the magnetic flux density must be finite in the limit $x \to \infty$, α with the negative real part is the solution. The boundary condition is

$$B_z(x=0) = B_0. \tag{10.52}$$

Thus, we obtain the solution of the magnetic flux density as

$$B_z(x,t) = B_0 e^{-x/\delta} \exp\left[i\left(\omega t - \frac{x}{\delta}\right)\right] \to B_0 e^{-x/\delta} \cos\left(\omega t - \frac{x}{\delta}\right). \tag{10.53}$$

In the above

$$\delta = \left(\frac{2}{\omega\mu\sigma_c}\right)^{1/2} \tag{10.54}$$

is a quantity with the dimension of length and is called the **skin depth**. In Eq. (10.53), we have adopted the real part for the solution. Figure 10.11 shows the spatial variation in the magnetic flux density given by Eq. (10.53). The magnetic flux density propagates along the x-axis while decaying. The depth of penetration is roughly equal to δ, which is the reason for the name. For larger ω and/or larger σ_c the shielding current density is higher, resulting in shorter δ. The position of a plane on which the phase of the propagating wave is constant is given by the condition

$$\omega t - \frac{x}{\delta} = c, \tag{10.55}$$

where c is a constant. Thus, the velocity of propagation is $dx/dt = \omega\delta$, and the wave length λ is $2\pi\delta$. Substituting this solution into Eq. (10.46) gives the solution of the electric field,

$$\begin{aligned} E_y(x,t) &= B_0\left(\frac{\omega}{\mu\sigma_c}\right)^{1/2} e^{-x/\delta} \exp\left[i\left(\omega t - \frac{x}{\delta} + \frac{\pi}{4}\right)\right] \\ &\to B_0\left(\frac{\omega}{\mu\sigma_c}\right)^{1/2} e^{-x/\delta} \cos\left(\omega t - \frac{x}{\delta} + \frac{\pi}{4}\right). \end{aligned} \tag{10.56}$$

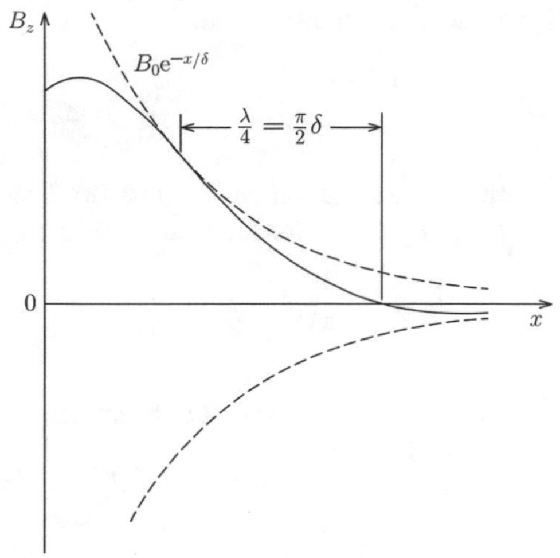

Fig. 10.11 Spatial variation in magnetic flux density

We find that, although this solution is similar to that of the magnetic flux density, the phase is ahead by $\pi/4$. The magnetic field and current density are obtained from $H_z(x,t) = B_z(x,t)/\mu$ and $i_y(x,t) = \sigma_c E_y(x,t)$ with the above results.

We estimate the skin depth of copper at 60 Hz at room temperature. Substituting typical values, $\sigma_c = 0.58 \times 10^8$ S/m and $\mu \simeq \mu_0 = 4\pi \times 10^{-7}$N/A^2, into Eq. (10.54) gives $\delta = 0.85 \times 10^{-2}$ m.

Example 10.6. We apply an AC electric field of amplitude E_0 and angular frequency ω along the z-axis parallel to an infinitely wide slab conductor that occupies $-d \leq x \leq d$. Determine the electric field in the conductor.

Solution 10.6. We can assume that $\partial/\partial y$ and $\partial/\partial z$ are zero as in Eq. (10.44) and the electric field has only a z-component. The time differentiation is replaced by the operator to multiply iω. Hence, Eq. (10.39) reduces to

$$\frac{\mathrm{d}E_z}{\mathrm{d}x} = \mathrm{i}\omega B_y,$$

showing that the magnetic flux density has only a y-component. Thus, Eq. (10.43) becomes

$$\frac{\mathrm{d}B_y}{\mathrm{d}x} = \mu\sigma_c E_z.$$

From these equations, we have

$$\frac{\mathrm{d}^2 E_z}{\mathrm{d}x^2} - \mathrm{i}\omega\mu\sigma_\mathrm{c} E_z = 0,$$

which has the same form as Eq. (10.48). Using Eqs. (10.51) and (10.54), we obtain the general solution for the electric field as

$$E_z(x) = K_1 \exp\left[(1+\mathrm{i})\frac{x}{\delta}\right] + K_2 \exp\left[-(1+\mathrm{i})\frac{x}{\delta}\right].$$

The coefficients K_1 and K_2 are determined by the boundary conditions:

$$E_z(x=-d) = E_z(x=d) = E_0.$$

Thus, the electric field is given by

$$E_z(x,t) = E_0 \frac{\cosh[(1+\mathrm{i})x/\delta]}{\cosh[(1+\mathrm{i})d/\delta]} \mathrm{e}^{\mathrm{i}\omega t}$$

and its real part is

$$\begin{aligned} E_z(x,t) = & \frac{E_0}{\cosh(2d/\delta) + \cos(2d/\delta)} \\ & \times \left\{ \left[\cosh\left(\frac{x+d}{\delta}\right) \cos\left(\frac{x-d}{\delta}\right) + \cosh\left(\frac{x-d}{\delta}\right) \cos\left(\frac{x+d}{\delta}\right) \right] \cos\omega t \right. \\ & \left. - \left[\sinh\left(\frac{x-d}{\delta}\right) \sin\left(\frac{x+d}{\delta}\right) + \sinh\left(\frac{x+d}{\delta}\right) \sin\left(\frac{x-d}{\delta}\right) \right] \sin\omega t \right\}. \end{aligned}$$

◇

Column: General Law of Electromagnetic Induction
Suppose that the plate in Fig. 10.7a is rotating with angular frequency ω in magnetic flux density, B, that increases with time as $B = B_0 + \gamma t$. For simplicity, we assume that the potential leads are arranged as in Fig. 10.12 to eliminate the magnetic flux that interlinks the circuit outside the plate. That is, one potential lead is aligned on the plate edge between P and A′ and twisted with another one up to a voltmeter. Now, we determine the induced electromotive force in the direction of OPA′AO. In this case, the electromotive force appears on line OP from the motional law and appears also from the magnetic flux law because of the penetrating magnetic flux. We now use Eq. (10.23) for the determination.

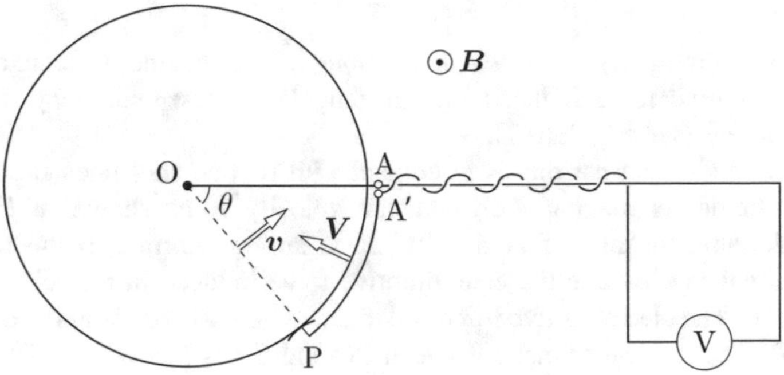

Fig. 10.12 Arrangement of plate and potential leads for measurement

The velocity of the plate is

$$\boldsymbol{v} = R\omega \boldsymbol{i}_\theta$$

and from the answer to the problem in Example 10.4, the velocity of the magnetic flux lines is

$$\boldsymbol{V} = -\frac{\gamma R}{2B}\boldsymbol{i}_R.$$

The contribution from line OP to the electromotive force is

$$\int_0^a [\boldsymbol{B} \times (\boldsymbol{V} - \boldsymbol{v})] \cdot \mathrm{d}\boldsymbol{R} = \int_0^a vB\mathrm{d}R = \frac{1}{2}\omega Ba^2.$$

That from arc PA′ is

$$\int_{\mathrm{P}}^{\mathrm{A}'} [\boldsymbol{B} \times (\boldsymbol{V} - \boldsymbol{v})] \cdot \mathrm{d}\boldsymbol{s} = \int_0^\theta BV(R = a)a\mathrm{d}\theta = -\frac{1}{2}\gamma a^2\theta.$$

On line AO outside the plate, $\boldsymbol{v} = 0$, and the induced electric field, $\boldsymbol{B} \times \boldsymbol{V}$, is perpendicular to the integration path. Thus, there is no contribution from this line to the electromotive force. Finally, we have

$$V_{\mathrm{em}} = \frac{1}{2}(\omega B - \gamma\theta)a^2.$$

Exercises

10.1. AC current $I(t) = I_m \sin\omega t$ flows along a straight line. Calculate the electromotive force induced in a rectangular coil separated by d from the current (see Fig. E10.1).

10.2. A conducting bar that is in contact with two parallel lines shunted at the terminal is moving with constant velocity v, as shown in Fig. E10.2. A static magnetic flux density, B, is applied normal to the rectangular circuit. Determine the electromotive force induced in the rectangular circuit. The electromotive force is defined to be positive along the direction of PQRS, and the distance between PQ and SR is $b + vt$.

10.3. Determine the electromotive force induced in the rectangular circuit in Exercise 10.2 when the magnetic flux density changes with time as $B(t) = B_0 + \alpha t$. Use the general law, Eq. (10.23).

10.4. A rectangular coil is moving with constant velocity v on a horizontal plane of distance R_0 from a straight line carrying constant current I, as shown in Fig. E10.3. Calculate the electromotive force induced in this coil with the motional law. The electromotive force is defined to be positive along the direction of PQRS and $d = d_0 + vt$.

10.5. A rectangular closed circuit starts to fall down from the distance d_0 at $t = 0$ below a straight line carrying current I, as shown in Fig. E10.4. Determine the electromotive force induced in the direction of ABCD of the rectangular circuit. The gravity acceleration is g.

10.6. A constant current, I, is applied to a straight line and a rectangular coil is rotating with angular frequency ω ($\theta = \omega t$) around side RS parallel to the straight line, as shown in Fig. E10.5. Calculate the electromotive force induced in the coil with the magnetic flux law. The distance d is larger than a, and the electromotive force is defined to be positive along the direction of PQRS.

10.7. Solve Exercise 10.6 with the motional law.

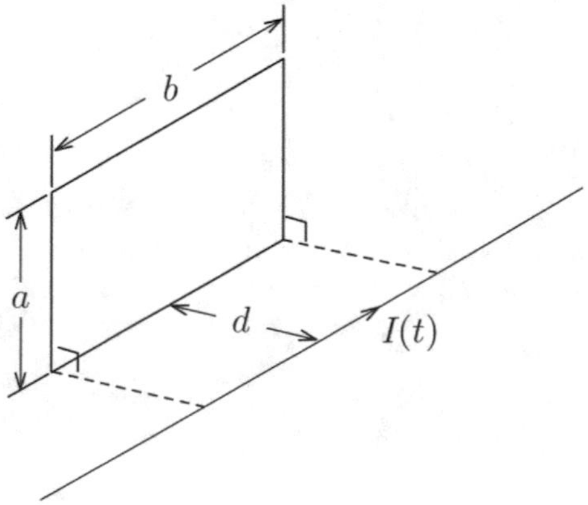

Fig. E10.1 Straight line carrying AC current and rectangular coil

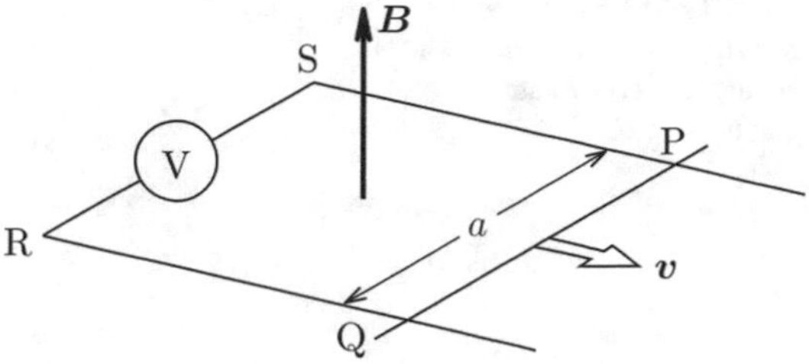

Fig. E10.2 Conducting bar moving with constant velocity on two parallel lines shunted at the terminal

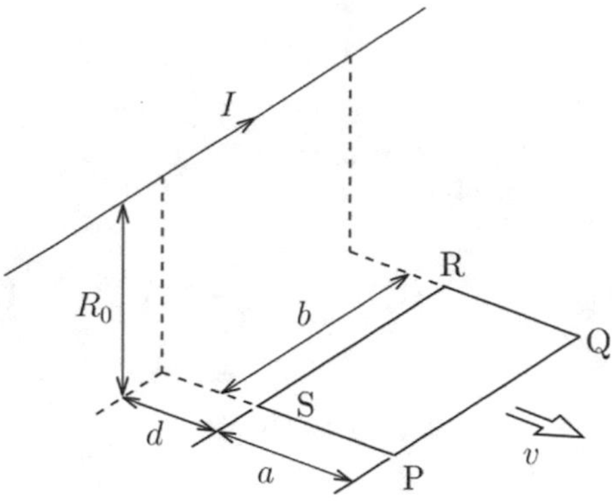

Fig. E10.3 Straight line carrying constant current and rectangular coil moving with constant velocity

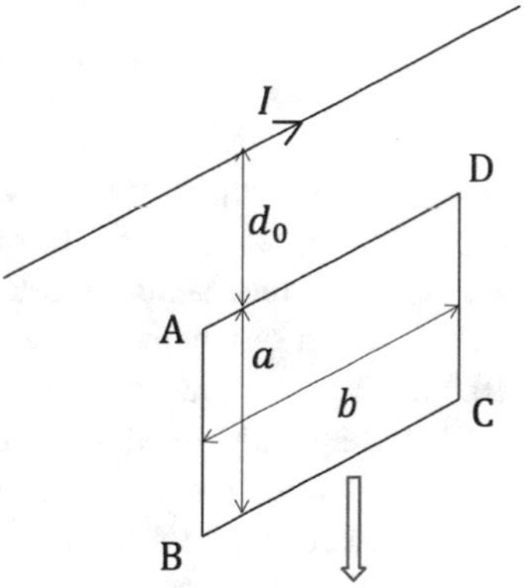

Fig. E10.4 Rectangular circuit that starts to fall down below the straight current

10.8. Voltage V is applied to an electric circuit composed of a resistor of electric resistance R_r and a coil of inductance L at $t \geq 0$, as shown in Fig. E10.6. Derive the equation for the circuit and determine the current.

10.9. Suppose that a conductor carrying current $\boldsymbol{I}$ in magnetic flux density $\boldsymbol{B}$ is forced to move with velocity $\boldsymbol{v}$ by the Lorentz force. Thus, we may say that the Lorentz force does mechanical work. In this case, the work in unit time done on electric charge by the induced electric field $\boldsymbol{v} \times \boldsymbol{B}$ is given by

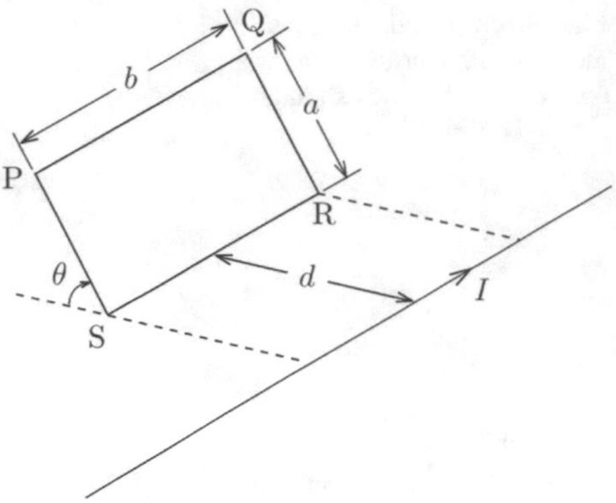

Fig. E10.5 Straight line carrying constant current and rectangular coil rotating around side RS

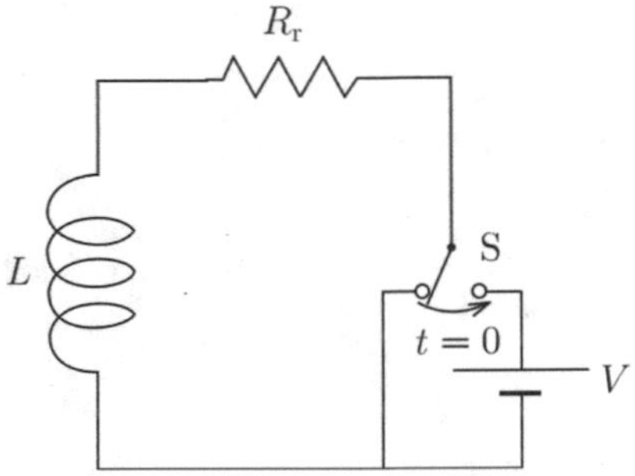

Fig. E10.6 Electric circuit composed of resistor and coil

$$\boldsymbol{I} \cdot (\boldsymbol{v} \times \boldsymbol{B}) = -(\boldsymbol{I} \times \boldsymbol{B}) \cdot \boldsymbol{v}.$$

This is equal to the negative of the work done by the Lorentz force in unit time, Eq. (10.16). Explain what it means. The fact that the Lorentz force does mechanical work seems to contradict the statement in Example 6.4. Discuss whether these are really contradictory.

10.10. Suppose that we apply an AC electric field of amplitude E_0 and angular frequency ω along the z-axis parallel to the surface of a semi-infinite conductor of electric conductivity σ_c that occupies $x \geq 0$. Determine the electric field and magnetic flux density in the conductor.

10.11. Suppose that we apply current I to an infinitely long cylindrical thin conductor of radius a. Calculate the magnetic energy in this condition from the work necessary to carry the current from the position at $R = R_\infty$ sufficiently far from the conductor. We assume that the return current flows uniformly at $R = R_\infty$.

Chapter 11
Displacement Current and Maxwell's Equations

Abstract This chapter covers the phenomena associated with the displacement current, which makes Ampere's law applicable to the case in which the current changes with time. The displacement current is proved experimentally in various cases. The set of Maxwell's equations is completed by introducing the displacement current. The breaking of symmetry of Maxwell's equations is discussed, based on the difference between a scalar source (electric charge) and vector source (current), which give the irrotational electric field and the solenoidal magnetic flux density, respectively. The electromagnetic fields are determined by the electric potential and the vector potential, and the set of these potentials is called the electromagnetic potential. We introduce the Poynting vector that describes the flow of electromagnetic energy.

11.1 Displacement Current

In a steady state, Ampere's law, Eq. (9.14), holds for a closed line, C, with different surfaces on it, S_1 and S_2, as shown in Fig. 11.1a, b. If the magnetic field $\boldsymbol{H}$ is integrated on C in opposite directions as drawn in Fig. 11.1a, b, the sum of the two integrations is naturally zero. At the same time, the sum of the surface integrals of the current density $\boldsymbol{i}$ on S_1 and S_2 is also zero. This sum becomes the surface integral on closed surface S_{12} composed of S_1 and S_2 (see Fig. 11.1c). Thus, we have

$$\int_{S_{12}} \boldsymbol{i} \cdot \mathrm{d}\boldsymbol{S} = 0. \tag{11.1}$$

This agrees with Eq. (5.8) in a steady state. However, it means that Ampere's law contradicts Eq. (5.8) in a non-steady state. In such a general case, the law of conservation of electric charge, Eq. (5.8), must be satisfied. Substituting Eq. (4.14) into this equation, we have

T. Matsushita, *Electricity and Magnetism*, Undergraduate Lecture Notes in Physics,
https://doi.org/10.1007/978-3-030-82150-0_11

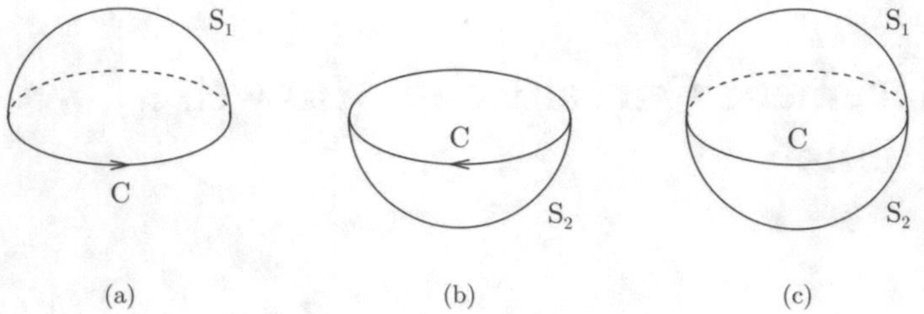

Fig. 11.1 **a** Surface S_1 and **b** surface S_2 surrounded by closed line C, and **c** closed surface composed of S_1 and S_2

$$\int_{S_{12}} \left(\boldsymbol{i} + \frac{\partial \boldsymbol{D}}{\partial t} \right) \cdot \mathrm{d}\boldsymbol{S} = 0. \tag{11.2}$$

This strongly suggests that, for generalizing to a non-steady state, we can assume

$$\oint_{C} \boldsymbol{H} \cdot \mathrm{d}\boldsymbol{s} = \int_{S} \left(\boldsymbol{i} + \frac{\partial \boldsymbol{D}}{\partial t} \right) \cdot \mathrm{d}\boldsymbol{S} \tag{11.3}$$

instead of Ampere's law. In the above, S is the surface surrounded by C. The second term on the right side, $\partial \boldsymbol{D}/\partial t$, is called **displacement current** and has the same unit as electric current density [A/m^2]. In a steady state, Eq. (11.3) reduces to the usual form of Ampere's law, Eq. (9.14), and there is no problem. Equation (11.3) is called the **generalized form of Ampere's law**. The corresponding **generalized differential form of Ampere's law** is

$$\nabla \times \boldsymbol{H} = \boldsymbol{i} + \frac{\partial \boldsymbol{D}}{\partial t}. \tag{11.4}$$

Here, we show the validity of the displacement current. Suppose that a capacitor is energized using an electric power source. When we apply current I to the capacitor, as shown in Fig. 11.2a, the electric charge Q in the electrode changes. We assume a closed line, C, around a wire through which the current flows and a surface, S_1, as in the figure. We apply Eq. (11.3) to C and S_1. The displacement current is zero on S_1, and the right side is equal to the current I applied to the capacitor. Next, we assume another surface S_2 that does not contain the wire, as shown in Fig. 11.2b. In this case, while the left side does not change, $\boldsymbol{i}$ is zero on S_2, and the right side is

$$\frac{\partial}{\partial t} \int_{S_2} \boldsymbol{D} \cdot \mathrm{d}\boldsymbol{S} = \frac{\mathrm{d}Q}{\mathrm{d}t}. \tag{11.5}$$

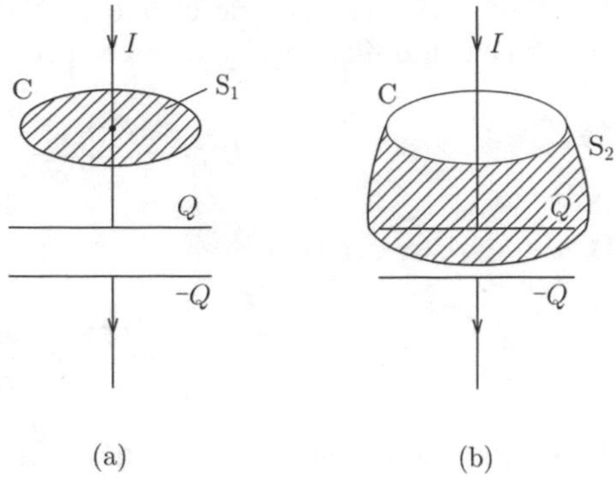

Fig. 11.2 Closed line C around a current-carrying wire and surfaces surrounded by C: **a** surface that includes the wire and **b** surface that does not include the wire

In the above, we changed the order of the time differential and surface integral, since S_2 does not change with time. Hence,

$$I = \frac{dQ}{dt} \tag{11.6}$$

and no contradiction results from Eq. (11.4).

When current changes with time as in the case of alternating current (AC), electric charges are stored in the electrodes of a capacitor, and hence, current flows through the capacitor. Thus, the time variation in the electric flux in the space between the electrodes generates a magnetic field as well as current.

Example 11.1. We apply AC $I(t) = I_0\sin\omega t$ to a capacitor with circular electroplates of radius a separated by d, as shown in Fig. 11.3. Determine the displacement current and magnetic field in the space between the electroplates.

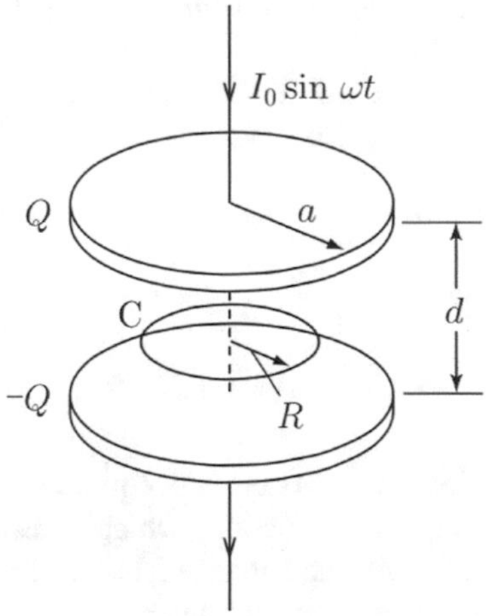

Fig. 11.3 AC flowing through circular parallel-plate capacitor

Solution 11.1. The electric charge on the electroplate is $Q(t) = -(I_0/\omega)\cos\omega t$, and the electric flux density is directed downward with magnitude

$$D(t) = \frac{Q(t)}{\pi a^2} = -\frac{I_0}{\pi a^2 \omega}\cos\omega t.$$

Hence, the displacement current is

$$\frac{\partial D(t)}{\partial t} = \frac{I_0}{\pi a^2}\sin\omega t.$$

This is similar to a virtual situation in which the current of the same density flows uniformly in the space between electroplates, suggesting the continuity of current. In fact, this situation is realized if the space is occupied by a material with electric resistivity sufficiently higher than that of the electroplates.

The magnetic field produced by the displacement current is concentric around the central axis of the capacitor. If the magnetic field on a circle of radius R from the axis is $H(R)$, the left side of Eq. (11.3) is $2\pi RH(R)$. The right side is $\pi R^2 \partial D(t)/\partial t = (R^2 I_0/a^2)\sin\omega t$. Thus, we have

$$H(t) = \frac{I_0 R}{2\pi a^2}\sin\omega t.$$

◇

11.2 Maxwell's Equations

All of the equations that describe electromagnetic phenomena have been introduced. They are summarized as follows:

$$\nabla \times \boldsymbol{E} = -\frac{\partial \boldsymbol{B}}{\partial t}, \tag{11.7}$$

$$\nabla \times \boldsymbol{H} = \boldsymbol{i} + \frac{\partial \boldsymbol{D}}{\partial t}, \tag{11.8}$$

$$\nabla \cdot \boldsymbol{D} = \rho, \tag{11.9}$$

$$\nabla \cdot \boldsymbol{B} = 0. \tag{11.10}$$

These are **Maxwell's equations**. Equation (11.7) describes the law of electromagnetic induction, which directly connects the first fundamental variables in electricity and magnetism. Equation (11.8) is the generalized differential form of Ampere's law that directly connects the second fundamental variables in electricity and magnetism. Equations (11.9) and (11.10) are Gauss' laws on electric flux and magnetic flux, respectively, and represent the conditions of divergence.

When we look back at Eqs. (11.7)–(11.10), a break of formal symmetry may be found among them. However, if there are no sources ($\rho = 0$, $\boldsymbol{i} = 0$), a beautiful symmetry appears. This is the essential feature of electromagnetism. The break of symmetry comes from the difference in the nature of fields produced by different sources, electric charge density ρ (scalar) and current density $\boldsymbol{i}$ (vector). The electric field is an irrotational field with divergence, and the magnetic flux density is a solenoidal field with rotation. It should be noted that such a difference comes the from more fundamental nature of mathematics, i.e., the difference between the source being a scalar or vector.

The field (i.e., a distortion vector in space) that a potential due to a scalar source can produce is only a gradient. This field naturally has no rotation and has divergence. In contrast, the field that a potential due to a vector source can produce is only a rotation, which naturally has no divergence. Poisson's equation and its solution clearly show that a scalar or vector source produces a scalar or vector potential, respectively. Therefore, including the effects of these sources in the above equations automatically determines the corresponding equations to be modified. The electric charge density ρ is included in Eq. (11.9), which represents divergence, and the current density $\boldsymbol{i}$ is included in Eq. (11.8), which represents rotation. As a result, the effects are not included in Eqs. (11.7) and (11.10) for $\boldsymbol{E}$ and $\boldsymbol{B}$.

These equations are solved for electromagnetic fields with material relationships:

$$\boldsymbol{D} = \epsilon \boldsymbol{E}, \tag{11.11}$$

$$\boldsymbol{B} = \mu \boldsymbol{H}, \tag{11.12}$$

$$\boldsymbol{i} = \sigma_c \boldsymbol{E}. \tag{11.13}$$

The variables to be obtained are the electric field $\boldsymbol{E}$, magnetic flux density $\boldsymbol{B}$, electric flux density $\boldsymbol{D}$, magnetic field $\boldsymbol{H}$, and current density $\boldsymbol{i}$. These five variables are obtained with five equations, (11.7), (11.8), and (11.11)–(11.13). Equations (11.9) and (11.10) provide supplementary conditions.

We transform the differential Eqs. (11.7)–(11.10) to integral equations:

$$\oint_C \boldsymbol{E} \cdot \mathrm{d}\boldsymbol{s} = -\frac{\mathrm{d}}{\mathrm{d}t} \int_S \boldsymbol{B} \cdot \mathrm{d}\boldsymbol{S}, \tag{11.14}$$

$$\oint_C \boldsymbol{H} \cdot \mathrm{d}\boldsymbol{s} = \int_S \left(\boldsymbol{i} + \frac{\partial \boldsymbol{D}}{\partial t} \right) \cdot \mathrm{d}\boldsymbol{S}, \tag{11.15}$$

$$\int_S \boldsymbol{D} \cdot \mathrm{d}\boldsymbol{S} = \int_V \rho \, \mathrm{d}V, \tag{11.16}$$

$$\int_S \boldsymbol{B} \cdot \mathrm{d}\boldsymbol{S} = 0. \tag{11.17}$$

In the above, C is the closed line that surrounds the surface S in Eqs. (11.14) and (11.5), S is the closed surface, and V is its internal region in Eqs. (11.16) and (11.17).

Here, we show an example for solving Maxwell's equations. In terms of only the electric field and magnetic flux density, Eq. (11.8) is rewritten as

$$\frac{1}{\mu}\nabla \times \boldsymbol{B} = \sigma_c \boldsymbol{E} + \epsilon \frac{\partial \boldsymbol{E}}{\partial t}, \tag{11.18}$$

and Eq. (11.9) gives

$$\epsilon \nabla \cdot \boldsymbol{E} = \rho. \tag{11.19}$$

Substituting Eq. (11.18) into a curl of Eq. (11.7) gives

$$\nabla \times (\nabla \times \boldsymbol{E}) = -\frac{\partial}{\partial t}(\nabla \times \boldsymbol{B}) = -\mu\epsilon \frac{\partial^2 \boldsymbol{E}}{\partial t^2} - \mu\sigma_c \frac{\partial \boldsymbol{E}}{\partial t}. \tag{11.20}$$

It is common to solve Eq. (11.20) under the condition of Eq. (11.19). When there is no electric charge ($\rho = 0$) in a material, according to Eq. (A1.46), the left side of Eq. (11.20) is equal to $-\Delta \boldsymbol{E}$, and we have

$$\Delta \boldsymbol{E} - \mu\epsilon \frac{\partial^2 \boldsymbol{E}}{\partial t^2} - \mu\sigma_c \frac{\partial \boldsymbol{E}}{\partial t} = 0. \tag{11.21}$$

The same equations are obtained for the other four variables including $\boldsymbol{B}$. This equation is the **telegraphic equation**.

The second and third terms on the left side of Eq. (11.21) correspond to the second (displacement current) and first terms (electric current) on the right side of Eq. (11.18), respectively. Hence, in usual cases where we can neglect the displacement current, the second term in the telegraphic equation disappears, and the equation leads to the diffusion equation, Eq. (10.50). Here, we discuss the condition in which this approximation holds. Assume AC electromagnetic fields of angular frequency ω. The magnitudes of the second and third terms are $\mu\epsilon\omega^2|E|$ and $\mu\sigma_c\omega|E|$, respectively. The ratio of the second and third terms is

$$\frac{\epsilon\omega}{\sigma_c}. \tag{11.22}$$

This ratio is also directly derived from the displacement current and electric current. In usual metals, ϵ and σ_c are of the order of $\epsilon_0 \simeq 1 \times 10^{-11}$ C^2/Nm^2 and 1×10^7 S/m, respectively. Hence, this ratio is as small as 6×10^{-8} even for a microwave of 10 GHz ($\omega = 2\pi \times 10^{10}$). In usual metals, we can therefore safely neglect the displacement current even at a very high frequency. In contrast, for insulating materials such as mica, typical material constants are $\epsilon \simeq 7\epsilon_0 \simeq 6 \times 10^{-11}$ C^2/Nm^2

and $\sigma_c \simeq 1 \times 10^{-14}$ S/m. Hence, the ratio takes a value as large as 2×10^6 even for a low frequency like a commercial one of 50 Hz. In this case, we can neglect the current. It is easily understood that a large difference in the electric conductivity dramatically affects the electromagnetic phenomena. In the latter case, Eq. (11.21) reduces to

$$\Delta \boldsymbol{E} - \mu\epsilon \frac{\partial^2 \boldsymbol{E}}{\partial t^2} = 0. \tag{11.23}$$

This differential equation of the second order is called the **wave equation**. This will be investigated in Chap. 12, in which we learn the electromagnetic wave.

Example 11.2. Show that the equation for the magnetic flux density $\boldsymbol{B}$ is also the telegraphic equation.

Solution 11.2. Substituting Eq. (11.7) into a curl of Eq. (11.8) gives

$$\nabla \times (\nabla \times \boldsymbol{B}) = \mu\sigma_c \nabla \times \boldsymbol{E} + \mu\epsilon \frac{\partial}{\partial t} \nabla \times \boldsymbol{E} = -\mu\sigma_c \frac{\partial \boldsymbol{B}}{\partial t} - \mu\epsilon \frac{\partial^2 \boldsymbol{B}}{\partial t^2}.$$

Using Eqs. (A1.46) and (11.10), the left side reduces to $-\Delta \boldsymbol{B}$ independently of the existence of electric charges, and we have

$$\Delta \boldsymbol{B} - \mu\epsilon \frac{\partial^2 \boldsymbol{B}}{\partial t^2} - \mu\sigma_c \frac{\partial \boldsymbol{B}}{\partial t} = 0.$$

◇

11.3 Boundary Conditions

Since the equation for the magnetic field $\boldsymbol{H}$ changes from Eqs. (9.12) to (11.4), we investigate the boundary condition for $\boldsymbol{H}$ here.

We denote the magnetic fields in materials 1 and 2 in the vicinity of an interface by $\boldsymbol{H}_1$ and $\boldsymbol{H}_2$, respectively. Consider a plane that is normal to the boundary and contains $\boldsymbol{H}_1$ and $\boldsymbol{H}_2$ (see Fig. 9.12a, b). Integrating the magnetic field on the small rectangle ΔC with two sides parallel to the boundary, we have

$$\oint_{\Delta C} \boldsymbol{H} \cdot \mathrm{d}\boldsymbol{s} = \int_{\Delta S} \nabla \times \boldsymbol{H} \cdot \mathrm{d}\boldsymbol{S} = \int_{\Delta S} \boldsymbol{i} \cdot \mathrm{d}\boldsymbol{S} + \frac{\partial}{\partial t} \int_{\Delta S} \boldsymbol{D} \cdot \mathrm{d}\boldsymbol{S}, \tag{11.24}$$

where we have used Stokes' theorem and Eq. (11.4), and ΔS is the surface surrounded by ΔC. If the height Δh is sufficiently small, only the surface current contributes to the first term on the right side, and we can neglect the displacement current of a finite density. Hence, the boundary condition to be satisfied is the same as Eq. (9.24); i.e., if the surface density of current flowing on the boundary is $\boldsymbol{\tau}$, Eq. (11.24) gives

$$\boldsymbol{n} \times (\boldsymbol{H}_1 - \boldsymbol{H}_2) = \boldsymbol{\tau}, \tag{11.25}$$

where $\boldsymbol{n}$ is the unit vector normal to the boundary and is directed from material 1 to material 2. The difference in the parallel component of the magnetic field is equal to the surface density of current flowing on the boundary.

11.4 Electromagnetic Potential

Here, we summarize the potentials that describe the electromagnetic fields. The electric field is given by Eq. (10.25) with the electric potential (scalar potential) ϕ and the vector potential $\boldsymbol{A}$:

$$\boldsymbol{E} = -\nabla\phi - \frac{\partial \boldsymbol{A}}{\partial t}. \tag{11.26}$$

On the other hand, the magnetic flux density $\boldsymbol{B}$ always satisfies Eq. (11.10) and is given by Eq. (6.29) with the vector potential $\boldsymbol{A}$:

$$\boldsymbol{B} = \nabla \times \boldsymbol{A}. \tag{11.27}$$

Thus, the electric field and magnetic flux density are given by ϕ and $\boldsymbol{A}$, and the set of these potentials are called the **electromagnetic potential**.

There is no change in the magnetic flux density, even if we add a gradient of an arbitrary scalar function ψ to the vector potential $\boldsymbol{A}$. However, it is necessary to make the change

$$\phi \to \phi - \frac{\partial \psi}{\partial t}, \quad \boldsymbol{A} \to \boldsymbol{A} + \nabla\psi, \tag{11.28}$$

so that the electric field does not change. This transformation is the **gauge transformation**. Since the electromagnetic fields do not change under this transformation, the electromagnetic potential is arbitrary. Hence, it is necessary to impose a condition to determine the electromagnetic potential uniquely.

When an electric charge of density ρ and a current of density $\boldsymbol{i}$ coexist in space, it is common to use the condition

$$\epsilon\mu \frac{\partial \phi}{\partial t} + \nabla \cdot \boldsymbol{A} = 0. \tag{11.29}$$

This condition is the **Lorentz gauge**. The equations associated with the electric charge and current are Eqs. (11.8) and (11.9), respectively. The left side of Eq. (11.9) is written with the electromagnetic potential as

$$-\epsilon\left(\Delta\phi + \frac{\partial}{\partial t}\nabla\cdot\boldsymbol{A}\right) = -\epsilon\left(\Delta\phi - \epsilon\mu\frac{\partial^2\phi}{\partial t^2}\right). \tag{11.30}$$

Hence, the equation that ϕ should satisfy is

$$\Delta\phi - \epsilon\mu\frac{\partial^2\phi}{\partial t^2} = -\frac{\rho}{\epsilon}. \tag{11.31}$$

Equation (11.8) leads similarly to

$$\Delta\boldsymbol{A} - \epsilon\mu\frac{\partial^2\boldsymbol{A}}{\partial t^2} = -\mu\boldsymbol{i}. \tag{11.32}$$

Under the Lorentz gauge, the scalar function ψ must satisfy

$$\Delta\psi - \epsilon\mu\frac{\partial^2\psi}{\partial t^2} = 0. \tag{11.33}$$

When there is neither electric charge nor current, all these equations reduce to the same form as Eq. (11.23), i.e., the wave equation. This type of equation expresses the electromagnetic wave as will be shown in Chap. 12.

Example 11.3. Derive Eq. (11.32).

Solution 11.3. The left side of Eq. (11.8) is

$$\frac{1}{\mu}\nabla\times(\nabla\times\boldsymbol{A}) = \frac{1}{\mu}[\nabla(\nabla\cdot\boldsymbol{A}) - \Delta\boldsymbol{A}] = -\epsilon\nabla\frac{\partial\phi}{\partial t} - \frac{1}{\mu}\Delta\boldsymbol{A},$$

where Eq. (11.29) is used. The right side is

$$\boldsymbol{i} - \epsilon\nabla\frac{\partial\phi}{\partial t} - \epsilon\frac{\partial^2\boldsymbol{A}}{\partial t^2}.$$

Thus, Eq. (11.32) is derived.

◇

11.5 The Poynting Vector

The total energy density of electromagnetic fields varying with time is given by

$$u = \frac{1}{2}\epsilon\boldsymbol{E}^2 + \frac{1}{2\mu}\boldsymbol{B}^2 + \int \boldsymbol{i}\cdot\boldsymbol{E}\,\mathrm{d}t. \tag{11.34}$$

The first, second, and third terms are the electric energy, magnetic energy, and mechanical energy of charged particles, respectively. Hence, the variation in the total energy in space V with time is

$$\begin{aligned}\frac{\partial U}{\partial t} &= \frac{\partial}{\partial t}\int_{\mathrm{V}} u\,\mathrm{d}V \\ &= \int_{\mathrm{V}}\left(\epsilon\boldsymbol{E}\cdot\frac{\partial \boldsymbol{E}}{\partial t}+\frac{\boldsymbol{B}}{\mu}\cdot\frac{\partial \boldsymbol{B}}{\partial t}+\boldsymbol{i}\cdot\boldsymbol{E}\right)\mathrm{d}V.\end{aligned}\tag{11.35}$$

The first and third terms are transformed using Eq. (11.8), and $\partial \boldsymbol{B}/\partial t$ in the second term is eliminated by substituting Eq. (11.7). The right side of the above equation then becomes

$$\int_{\mathrm{V}}[\boldsymbol{E}\cdot(\nabla\times\boldsymbol{H})-\boldsymbol{H}\cdot(\nabla\times\boldsymbol{E})]\mathrm{d}V = -\int_{\mathrm{V}}\nabla\cdot(\boldsymbol{E}\times\boldsymbol{H})\mathrm{d}V = -\int_{\mathrm{S}}(\boldsymbol{E}\times\boldsymbol{H})\cdot\mathrm{d}\boldsymbol{S},\tag{11.36}$$

where S is the surface of V. Here, we define

$$\boldsymbol{S}_{\mathrm{P}} = \boldsymbol{E}\times\boldsymbol{H}.\tag{11.37}$$

We then rewrite Eq. (11.35) as

$$\frac{\partial}{\partial t}\int_{\mathrm{V}} u\,\mathrm{d}V + \int_{\mathrm{S}}\boldsymbol{S}_{\mathrm{P}}\cdot\mathrm{d}\boldsymbol{S} = 0.\tag{11.38}$$

The vector $\boldsymbol{S}_{\mathrm{P}}$ is called the **Poynting vector**. This equation says that the variation in the energy in space V with time is equal to the Poynting vector that enters V through the surface S. Hence, we understand that the Poynting vector represents the flow of electromagnetic energy, and Eq. (11.38) represents the law of conservation of energy. When this equation is written in differential form, we have

$$\frac{\partial u}{\partial t} + \nabla\cdot\boldsymbol{S}_{\mathrm{P}} = 0.\tag{11.39}$$

This gives the **continuity equation of energy**.

However, there is an extraordinary case where it is difficult to understand that the Poynting vector gives a real energy flow. See Column (2) in this chapter.

Example 11.4. The electric field is given by Eq. (10.56) in the case of the skin effect discussed in Sect. 10.5. Prove that the total energy loss $\int \mathrm{d}t \int_{\mathrm{V}} \boldsymbol{E}\cdot\boldsymbol{i}\mathrm{d}V$ is equal to $-\int \mathrm{d}t \int_{\mathrm{S}} \boldsymbol{S}_{\mathrm{P}}\cdot\mathrm{d}\boldsymbol{S}$, where the integration with time is carried out over one period of the AC field.

Solution 11.4. The current density corresponding to Eq. (10.56) is

$$i_y(x,t) = \sigma_c E_y(x,t) = B_0 \left(\frac{\sigma_c \omega}{\mu}\right)^{1/2} e^{-x/\delta} \cos\left(\omega t - \frac{x}{\delta} + \frac{\pi}{4}\right)$$

and the power loss in a unit area of the y-z plane is

$$\int_0^\infty E_y(x,t) i_y(x,t) \mathrm{d}x = \frac{B_0^2 \omega}{\mu} \int_0^\infty \exp\left(-\frac{2x}{\delta}\right) \cos^2\left(\omega t - \frac{x}{\delta} + \frac{\pi}{4}\right) \mathrm{d}x$$

$$= \frac{B_0^2 \omega \delta}{4\mu} \left[1 + \frac{1}{\sqrt{2}} \cos\left(2\omega t + \frac{\pi}{4}\right)\right].$$

Integrating this over unit time, the total loss energy density is

$$W = \frac{B_0^2 \omega \delta}{4\mu} = \frac{B_0^2}{2\mu} \left(\frac{\omega}{2\mu\sigma_c}\right)^{1/2}.$$

On the other hand, the Poynting vector on the surface is $\boldsymbol{S}_\mathrm{P}(x=0) = \boldsymbol{i}_x \left(E_y B_z\right)_{x=0}/\mu$. Noting that $\mathrm{d}\boldsymbol{S} = -\boldsymbol{i}_x \mathrm{d}S$, the surface integral of the Poynting vector is

$$-\int_\mathrm{S} \boldsymbol{S}_\mathrm{P} \cdot \mathrm{d}\boldsymbol{S} = \left(\frac{E_y B_z}{\mu}\right)_{x=0} = \frac{B_0^2}{\mu} \left(\frac{\omega}{\mu\sigma_c}\right)^{1/2} \cos\omega t \cos\left(\omega t + \frac{\pi}{4}\right).$$

Integrating this over unit time, we have

$$\frac{B_0^2}{2\mu} \left(\frac{\omega}{2\mu\sigma_c}\right)^{1/2},$$

which agrees with the above result. The reason for the same result is that the conditions of electric field and magnetic flux density are the same before and after one period, and hence, the electric and magnetic energies are unchanged. All the energy that flows into the conductor in one period is consumed and converted to Joule heat.

◇

Example 11.5. We apply magnetic flux density B parallel to the thin cylinder of radius a and height h shown in Fig. 11.4. Determine the magnetic energy stored in the cylinder using the Poynting vector.

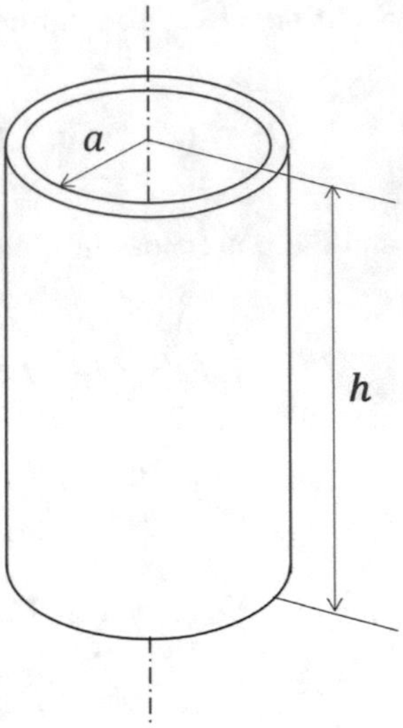

Fig. 11.4 Thin cylinder

Solution 11.5. We denote the magnetic flux density applied to the cylinder by b. The electromotive force induced on the cylinder surface is

$$V_{\mathrm{em}} = -\pi a^2 \frac{\mathrm{d}b}{\mathrm{d}t}.$$

The electric field is

$$E = \frac{V_{\mathrm{em}}}{2\pi a} = -\frac{a}{2} \cdot \frac{\mathrm{d}b}{\mathrm{d}t}$$

and is directed counterclockwise (in the direction of increasing azimuthal angle) from the top view. The magnetic field is $H = b/\mu_0$ and is directed upward. Thus, the Poynting vector directed inward into the cylinder is

$$S_{\mathrm{P}} = -EH = \frac{a}{2\mu_0} b \frac{\mathrm{d}b}{\mathrm{d}t}.$$

The magnetic energy penetrating the cylinder until b reaches B is determined to be

$$U_{\mathrm{m}} = \int 2\pi a h S_{\mathrm{p}} \mathrm{d}t = \frac{\pi a^2 h}{\mu_0} \int_0^B b \mathrm{d}b = \frac{\pi a^2 h}{2\mu_0} B^2.$$

This is equal to the product of the magnetic energy density $B^2/2\mu_0$ and the volume of the cylinder $\pi a^2 h$.

◇

Example 11.6. We apply current I to a wide parallel-plate transmission line using an electric power source of output voltage V, as shown in Fig. 11.5a. Determine the Poynting vector, and discuss the flow of energy. Neglect the electric resistance of the conductor. The width of a plate is w, and the plate separation is d. When we cannot neglect the electric resistance, how does the energy flow?

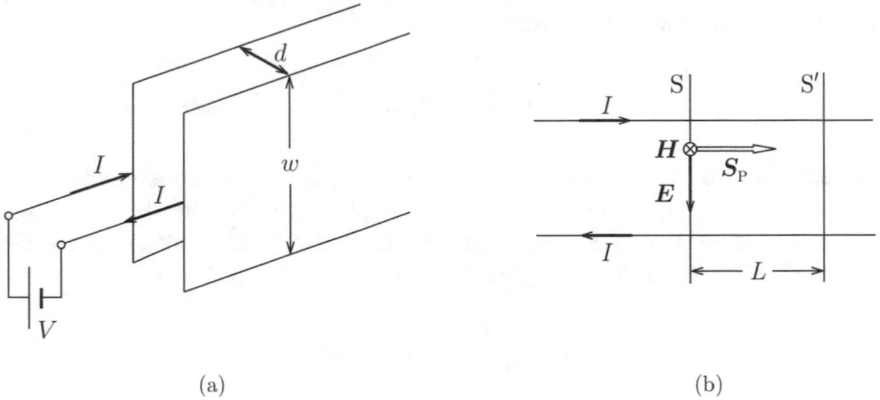

Fig. 11.5 **a** Parallel-plate transmission line and **b** the Poynting vector in the transmission line

Solution 11.6. First, we consider the case where the electric resistance can be neglected. We determine the Poynting vector on a surface, S, at some distance from the electric power source (see Fig. 11.5b). The electric field of strength $E = V/d$ is directed downward, and the magnetic field of strength $H = I/w$ is directed backward. Hence, the magnitude of the Poynting vector is $S_{\mathrm{P}} = VI/(wd)$, and the vector is directed from the electric power source to the terminal of the transmission line. Hence, the energy that flows from the electric power source to the transmission line in unit time is

$$S_{\mathrm{P}}wd = VI$$

and is equal to the electric power as well known. This value is constant and independent of the position of surface S.

Second, we consider the case where the electric resistance cannot be neglected. We use R'_{r} and V' to denote the electric resistance of the conducting plate in a unit length and the potential difference between the conductors on surface S, respectively. The potential difference on surface S′ at distance L from S is then $V' - 2LR'_{\mathrm{r}}I$. The electric power that enters through surface S is $V'I$, and the electric power that exits through S′ is $(V' - 2LR'_{\mathrm{r}}I)I = V'I - 2LR'_{\mathrm{r}}I^2$. Hence, the difference, $2LR'_{\mathrm{r}}I^2$, is the power consumed as the Joule heat in the region between S and S′. Here, we did not discuss the detailed energy flow into the conductor. See Exercise 11.10 for this discussion.

◇

Example 11.7. We assume that current I is applied to the material shown in Fig. 5.8 in Example 5.2 and that the voltage between the top and bottom is V. Prove that the dissipated power is given by $P = IV$ using the Poynting vector.

Solution 11.7 We define the ξ-axis on the side of the truncated cone that connects the points on the top and bottom of the same azimuthal angle, as shown in Fig. 11.6. The radius of the cone at distance ξ from the bottom is denoted by $R(\xi)$. The magnetic field at this position is

$$H(\xi) = \frac{I}{2\pi R(\xi)}.$$

The electric field denoted as $E(\xi)$ is directed along the ξ-axis and is normal to the magnetic field. Thus, the Poynting vector is $E(\xi)H(\xi)$ and is directed inward into the cone. The elementary surface vector is $|\mathrm{d}\boldsymbol{S}| = \mathrm{d}s\mathrm{d}\xi$ with $\mathrm{d}s$ denoting the elementary line vector on the perimeter. Thus, the dissipated power is

$$P = \iint E(\xi)H(\xi)\mathrm{d}s\mathrm{d}\xi.$$

Since $H(\xi)$ is constant on the perimeter, we have

$$\int H(\xi)\mathrm{d}s = H(\xi)\int \mathrm{d}s = H(\xi)2\pi R(\xi) = I.$$

Thus, the dissipated power is determined to be

$$P = I\int E(\xi)\mathrm{d}\xi = IV.$$

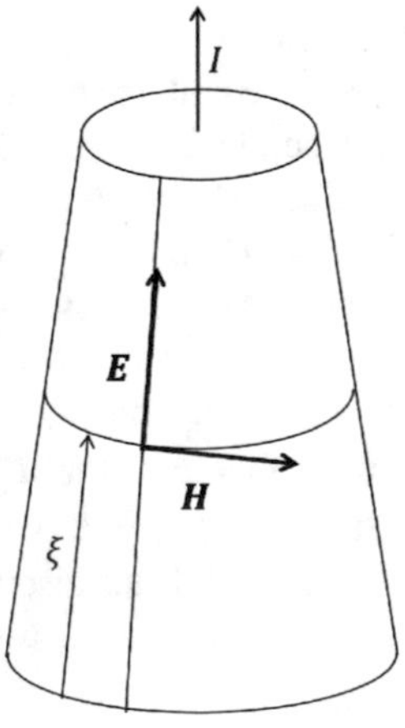

Fig. 11.6 Electric and magnetic fields on the surface of a truncated cone

◇

Column: (1) Polarization Current Density
Using the electric polarization $\boldsymbol{P}$, the displacement current is written as

$$\frac{\partial \boldsymbol{D}}{\partial t} = \epsilon_0 \frac{\partial \boldsymbol{E}}{\partial t} + \frac{\partial \boldsymbol{P}}{\partial t}.$$

The second term is the **polarization current density** due to the movement of the polarization charge of density ρ_{p}:

$$\frac{\partial \boldsymbol{P}}{\partial t} = \boldsymbol{i}_{\mathrm{p}}.$$

Using this equation and Eq. (4.7), the continuity equation for the polarization charge is obtained as

$$\nabla \cdot \boldsymbol{i}_{\mathrm{p}} + \frac{\partial \rho_{\mathrm{p}}}{\partial t} = 0$$

The first term of the displacement current is independent of the movement of charges and is not a current in the usual sense. The displacement current discussed in Fig. 11.2 and that in the vacuum region of the capacitor in Example 11.1 are examples of this component.

When a capacitor is occupied by a dielectric material, the polarization current given by the second term also flows. For a dielectric material with a larger dielectric constant, most of the displacement current is polarization current.

Column: (2) The Poynting Vector and Flow of Energy
We open the end terminal of the parallel-plate transmission line in Fig. 11.5b and apply voltage V and uniform magnetic flux density B_0 directed normally into the sheet. In this case, the electric field applied between the two plate conductors is $E = V/d$, and a Poynting vector of magnitude $S_{\mathrm{P}} = VB_0/\mu_0 d$ is directed from the electric power source to the terminal. Does this mean there is a steady energy flow from the electric power source even in this static condition?

It is difficult to consider that the energy flows continuously, since there is no electric power from the source. How should we understand Eq. (11.39), which is considered to represent the energy flow? If we substitute

$$\boldsymbol{S}_{\mathrm{P}} = \boldsymbol{S}_{\mathrm{P}0} + \nabla \times \boldsymbol{K}.$$

into Eq. (11.39), we have

$$\frac{\partial u}{\partial t} + \nabla \cdot \boldsymbol{S}_{\mathrm{P}0} = 0.$$

If $\boldsymbol{S}_{\mathrm{P}0}$ represents the real energy flow, the Poynting vector differs by a curl of an aribitrary vector function from the real energy flow [1]. Thus, the Poynting vector does not necessarily give the energy flow, and there is always arbitrariness in determining the energy flow. The above case is one example.

Literature

1. T. Matsushita, K. Funaki (2004) TEION KOGAKU (J. Cryo. Soc. Jpn.) **39**:2 [in Japanese]

Exercises

11.1. Derive the equations that the electric potential and vector potential satisfy when we use the Coulomb gauge, Eq. (6.30).

11.2. Suppose we apply an external AC electric field $E_0 \cos \omega t$ along the z-axis parallel to the surface of a semi-infinite dielectric material (dielectric constant: ϵ, magnetic permeability: μ) that occupies $x \geq 0$, as shown in Fig. E11.1. Determine the electric field and magnetic flux density in the dielectric material.

11.3. Discuss the energy flow in Exercise 11.2.

11.4. Suppose that an electric power source supplies electric charges to the electrodes of the circular parallel-plate capacitor in Fig. 11.3. Assume that the distance d between the electroplates is much smaller than a. Determine the Poynting vector that enters the capacitor when the electric charges increase to $\pm q(t)$. Then, calculate the energy stored in the capacitor, when the electric charges are $\pm Q$.

11.5. We apply direct current I to a long cylindrical conductor of radius a and electric conductivity σ_c. Determine the Poynting vector, and discuss the energy flow.

11.6. We apply current I to a long square prism of side length a and length l, as shown in Fig. E11.2, and the voltage between the top and the bottom is V. Prove that the dissipated power is given by $P = IV$ using the Poynting vector.

11.7. We apply current to a cylindrical one-turn coil made of a thin conducting plate shown in Fig. E11.3. Determine the Poynting vector as the current increases from zero to I, and discuss the flow of the energy. The height h is sufficiently greater, and the width of the gap δ is sufficiently smaller than the radius a. Neglect the electric resistance of the conducting plate.

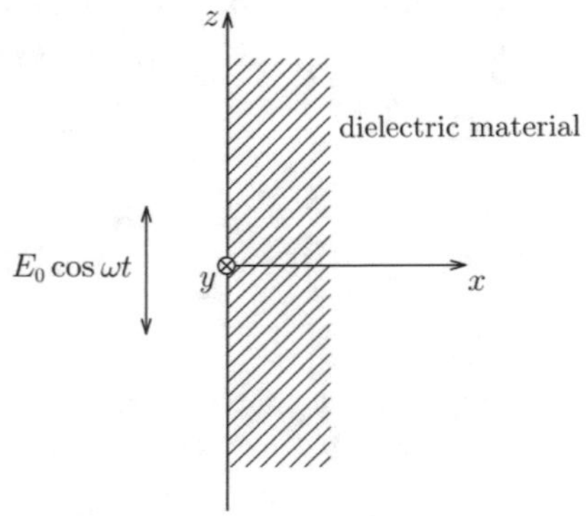

Fig. E11.1 AC electric field applied parallel to the surface of semi-infinite dielectric material

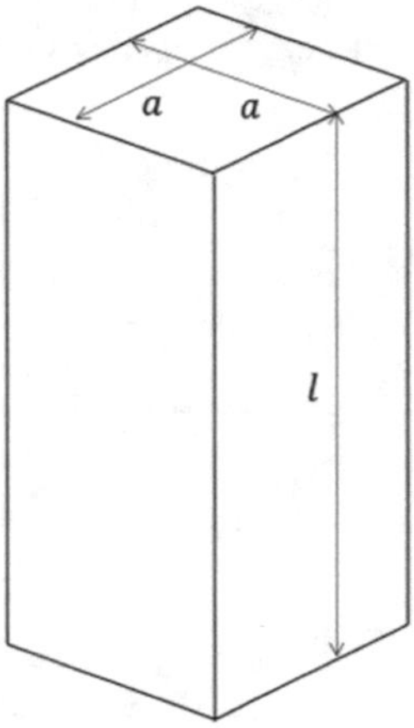

Fig. E11.2 Long square prism

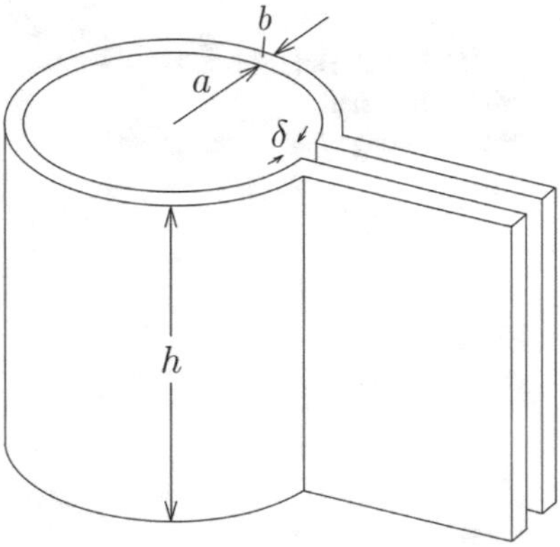

Fig. E11.3 Cylindrical one-turn coil

11.8. Discuss the case where we cannot neglect the electric resistance of the conductor in Exercise 11.7. The electric resistivity of the conductor is ρ_{r}, and the thickness b of the conductor is sufficiently less than the diameter a so that a slab approximation holds for the conductor.

11.9. Prove that the dissipated power is given by $P = IV$ using the Poynting vector, when the applied voltage and current are V and I, respectively, for a resistive material with an arbitrary shape such as shown in Fig. E11.4. (Hint: Use equipotential lines on the surface).

11.10. Discuss the energy flow into the conductor with electric resistance that is not discussed in Example 11.6.

11.11. Suppose we apply a uniform magnetic flux density B_0 along the z-axis parallel to a wide superconducting slab occupying $-d \leq x \leq d$ and virtually change the magnetic flux distribution as $B(x) = B_0 + b_0 x/d$ in the region $0 \leq x \leq d$ (see Fig. E11.5) by increasing the external magnetic flux density from B_0 to $B_0 + b_0$. In this case, a current flows along the y-axis, and it is known that the Lorentz force acts on the current. Note that

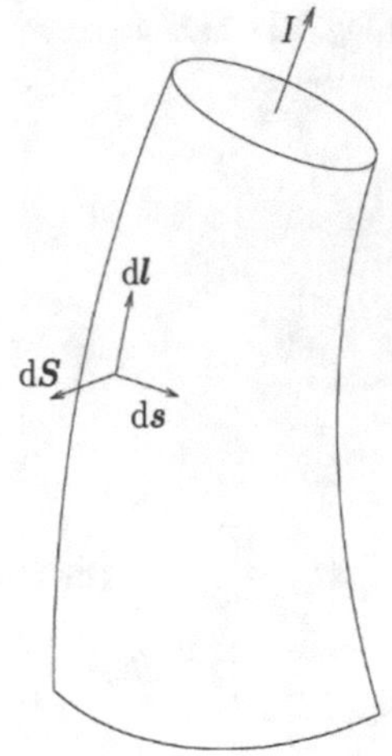

Fig. E11.4 Resistive material with arbitrary shape

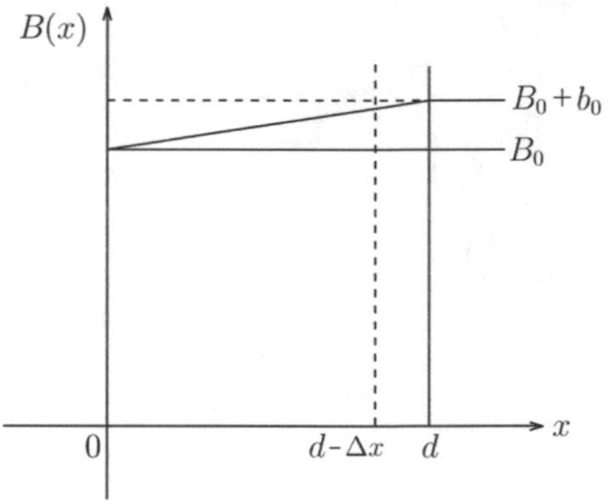

Fig. E11.5 Magnetic flux distribution in half a superconducting slab

we can presume that this force acts on flux lines whose distortion produces the current. Calculate the energy that flows into the region from $d - \Delta x$ to d through a unit area in the y-z plane as the external magnetic flux density increases. Derive an expression of the Lorentz force from the difference between the energy input and the enhancement of magnetic energy. (Hint: To derive the Lorentz force, it is necessary to determine the displacement of flux lines, u. For this purpose, use the continuity equation of magnetic flux, Eq. (10.20): $\mathrm{d}u/\mathrm{d}x \simeq -\Delta b/B_0$ with Δb denoting a variation in the magnetic flux density).

Chapter 12
Electromagnetic Wave

Abstract This chapter covers electromagnetic waves. The general solution of the wave equation is composed of waves propagating forward and backward with light speed. It is shown that the electric field and the magnetic flux density are perpendicular to each other and are perpendicular to the propagation direction. We discuss the refraction and reflection of a plane electromagnetic wave on an interface between two substances with different material constants using the boundary conditions. The Poynting vector of the electromagnetic wave has a magnitude equal to the electromagnetic energy density multiplied by the light speed and points in the propagation direction. The propagation of an electromagnetic wave in a wave guide is also discussed, and we learn about stable standing modes in the wave guide.

12.1 Planar Electromagnetic Wave

Electromagnetic fields in a dielectric material follow the wave equation, Eq. (11.23), since there is no electric charge in the material. This chapter covers the property of electromagnetic fields described by this equation. For simplicity, we focus only on the electric field and assume that it has only a y-component varying only along the x-axis. Thus, Eq. (11.23) reduces to

$$\frac{\partial^2 E_y}{\partial x^2} - \mu\epsilon \frac{\partial^2 E_y}{\partial t^2} = 0. \tag{12.1}$$

We assume that E_y varies with time as $\mathrm{e}^{\mathrm{i}\omega t}$ with ω denoting the angular frequency. Equation (12.1) then leads to

$$\frac{\partial^2 E_y}{\partial x^2} + \epsilon\mu\omega^2 E_y = 0. \tag{12.2}$$

This equation is easily solved as

T. Matsushita, *Electricity and Magnetism*, Undergraduate Lecture Notes in Physics,
https://doi.org/10.1007/978-3-030-82150-0_12

$$E_y = E_1 \mathrm{e}^{\mathrm{i}(\omega t + kx)} + E_2 \mathrm{e}^{\mathrm{i}(\omega t - kx)}, \tag{12.3}$$

where E_1 and E_2 are constants determined by initial and boundary conditions, and

$$k = (\epsilon\mu)^{1/2}\omega \equiv \frac{\omega}{c} \tag{12.4}$$

is the wave number. As will be shown later,

$$c = \left(\frac{1}{\epsilon\mu}\right)^{1/2} \tag{12.5}$$

is the speed of electromagnetic waves or the **light speed**, in the dielectric material. The wavelength is given by

$$\lambda = \frac{2\pi}{k}. \tag{12.6}$$

The practical electric field is given by the real part of the complex solution.

In the first term of Eq. (12.3),

$$\omega t + kx = \text{const.} \tag{12.7}$$

gives the position at which the phase of the wave is constant. Hence,

$$\frac{\mathrm{d}x}{\mathrm{d}t} = -\frac{\omega}{k} = -c \tag{12.8}$$

shows that the first term in Eq. (12.3) represents a wave that propagates with the velocity c along the negative x-axis. Namely this wave is an **electromagnetic wave**. Similarly, the second term in Eq. (12.3) gives the electromagnetic wave propagating along the positive x-axis. Such an electromagnetic wave, whose same phase is on a plane as in Eq. (12.3), is generally called a **plane wave**.

Since the left side of Eq. (11.7) is given by $\boldsymbol{i}_z \partial E_y / \partial x$, the magnetic flux density has only a z-component. Assuming the same time-dependent factor, the magnetic flux density is given by

$$\begin{aligned} B_z &= -B_1 \mathrm{e}^{\mathrm{i}(\omega t + kx)} + B_2 \mathrm{e}^{\mathrm{i}(\omega t - kx)} \\ &= -\frac{1}{c} E_1 \mathrm{e}^{\mathrm{i}(\omega t + kx)} + \frac{1}{c} E_2 \mathrm{e}^{\mathrm{i}(\omega t - kx)}. \end{aligned} \tag{12.9}$$

The first and second terms in this equation correspond to the first and second terms in Eq. (12.3) and represent electromagnetic waves propagating along the negative and positive x-axis, respectively. Thus, the electric field and magnetic flux density coexist in electromagnetic waves. That is, the time variation in a magnetic flux density induces an electric field, and the time variation in the produced electric field induces again a magnetic flux density; the process is repeated, and the variation propagates as a wave. Since current does not flow, there is no energy dissipation, and the electromagnetic wave does not decay with time.

We can also easily show that, if the electric field has only a z-component, the magnetic flux density has only a y-component, similarly to the above case. Hence, the planar electromagnetic wave is a **transverse wave** in which the electric field and magnetic flux density are perpendicular to each other and directed perpendicularly to the propagation direction. The ratio of these amplitudes is

$$\frac{E_1}{B_1} = \frac{E_2}{B_2} = c. \tag{12.10}$$

The magnetic field has been commonly used instead of the magnetic flux density to describe electromagnetic waves. In this case, the following equation is used:

$$H_z = -H_1 e^{i(\omega t + kx)} + H_2 e^{i(\omega t - kx)}. \tag{12.11}$$

The ratio of the amplitudes

$$\frac{E_1}{H_1} = \frac{E_2}{H_2} = \mu c = \left(\frac{\mu}{\epsilon}\right)^{1/2} \equiv Z, \tag{12.12}$$

is the **characteristic impedance** or **wave impedance**. This value is inherent to each medium, and its unit is $[\Omega]$.

The planar electromagnetic wave is generally expressed in the form

$$\exp[i(\omega t - \boldsymbol{k} \cdot \boldsymbol{r})]. \tag{12.13}$$

Using the unit vector along the propagation direction (the normalized wave number vector $\boldsymbol{i}_k = \boldsymbol{k}/|\boldsymbol{k}|$), the relationship between the electric field and magnetic field is expressed as

$$\boldsymbol{E} = Z(\boldsymbol{H} \times \boldsymbol{i}_k), \quad \boldsymbol{H} = Z^{-1}(\boldsymbol{i}_k \times \boldsymbol{E}). \tag{12.14}$$

In other words, the propagation direction of electromagnetic waves coincides with that of the Poynting vector representing the direction of the energy flow. For example, the first terms in Eqs. (12.3) and (12.11) correspond to each other, and since the electric field and magnetic field are directed along the positive y- and negative z-axes, respectively, and the Poynting vector is directed along the negative x-axis. Thus, the above relation holds. For the second terms in these equations, a similar relation holds.

In vacuum, the speed of the electromagnetic wave is

Table 12.1 Classification of electromagnetic waves

Name	Wavelength
Electromagnetic wave	Above 10^{-1} mm
Infrared ray	1 mm–0.76 μm
Visible ray	0.76–0.38 μm
Ultraviolet ray	0.38 μm–1 nm
X-ray	Several 10–10^{-3} nm
γ ray	Below 10^{-1} nm

$$c_0 = \frac{1}{(\epsilon_0 \mu_0)^{1/2}} = 2.997925 \times 10^8 \text{ m/s}, \tag{12.15}$$

and the characteristic impedance is

$$Z_0 = \left(\frac{\mu_0}{\epsilon_0}\right)^{1/2} = 376.730\ \Omega. \tag{12.16}$$

The electromagnetic wave is generally classified depending on its wavelength. Table 12.1 shows the classifications, although the classification ranges somewhat overlap.

Example 12.1. Prove that there are no components of the electric field or magnetic flux density parallel to the propagation direction of a planar electromagnetic wave.

Solution 12.1. Assume a planar electromagnetic wave propagating along the positive x-axis. The functional form of variations with respect to time and space for the electric field and magnetic flux density is $e^{i(\omega t - kx)}$. Hence, from Eq. (11.7), the x-component of the magnetic flux density is

$$-i\omega B_x = \frac{\partial E_z}{\partial y} - \frac{\partial E_y}{\partial z} = 0,$$

and from Eq. (11.8), the x-component of the electric field is

$$i\omega\epsilon E_x = \frac{1}{\mu}\left(\frac{\partial B_z}{\partial y} - \frac{\partial B_y}{\partial z}\right) = 0.$$

Thus, we prove that there are no components along the propagation direction for the electric field or magnetic flux density.

◇

Equations (12.3) and (12.9) deal with the case where each of the electric field and magnetic flux density remains in its own direction perpendicular to the other. Such a direction is called the direction of polarization. A polarization whose direction is fixed is referred to as **linear polarization**. In general, a superposition of components is possible, such as

$$E_y = E_1 \cos(\omega t - kx), \quad E_z = E_2 \cos(\omega t - kx + \delta). \tag{12.17}$$

The corresponding components of the magnetic flux density are

$$B_y = -\frac{E_2}{c}\cos(\omega t - kx + \delta) = -\frac{1}{c}E_z, \quad B_z = \frac{E_1}{c}\cos(\omega t - kx) = \frac{1}{c}E_y. \tag{12.18}$$

Usually, the direction of polarization is designated as that of the electric field, $\boldsymbol{E}$. Eliminating t in the above equations, we obtain the relationship between E_y and E_z as

$$\left(\frac{E_y}{E_1}\right)^2 - 2\cos\delta\,\frac{E_y E_z}{E_1 E_2} + \left(\frac{E_z}{E_2}\right)^2 = \sin^2\delta. \tag{12.19}$$

In the range $0 < \delta < \pi$, the direction of polarization turns to the right (clockwise) from the view of an observer directed along the propagation of the electromagnetic wave. This is called right-hand polarization (see Fig. 12.1). In the range $-\pi < \delta < 0$, the direction of polarization turns to the left (counterclockwise). When $\delta = 0$ or π, the electric field $\boldsymbol{E}$ is fixed in one direction and is linearly polarized. When $\delta = \pm\pi/2$ and $E_1 = E_2$, the trace of $\boldsymbol{E}$ is a circle, and this polarization is **circular polarization**. In other cases, the trace is an ellipse, and the polarization is **elliptical polarization**. Such phenomena that the directions of $\boldsymbol{E}$ and $\boldsymbol{B}$ in the electromagnetic

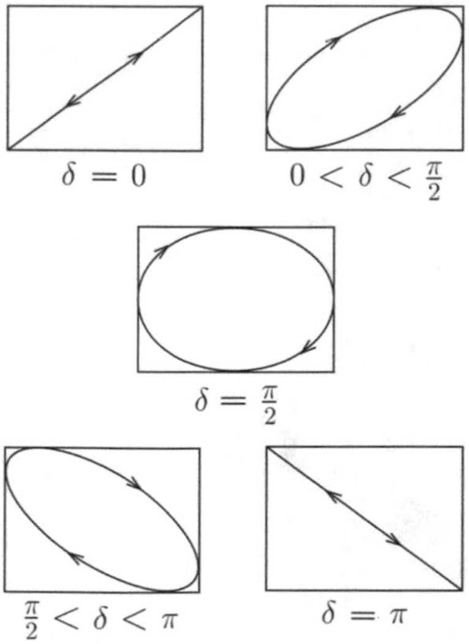

Fig. 12.1 Linear polarization and right-hand elliptical polarization: The abscissa and ordinate are, respectively, E_y and E_z

wave are not uniform but biased are referred to as the **polarization of a wave**, and such a wave is called a **polarized wave**.

12.2 Reflection and Refraction of the Planar Electromagnetic Wave

Electromagnetic wave is a family of waves that includes visible light. Reflection and refraction are well-known processes for light in optics. Here, we investigate reflection and refraction using electromagnetism.

Media 1 and 2 with dielectric constants ϵ_1 and ϵ_2 and magnetic permeabilities μ_1 and μ_2 face each other on the plane $z = 0$, as shown in Fig. 12.2a. Suppose that a planar electromagnetic wave propagates from medium 1 to the boundary. The plane formed by the propagation direction and the direction (z-axis) normal to the boundary is called a plane of incidence. We define the x-axis on the line on which the plane of incidence and the boundary meet and the y-axis on the boundary in such a way that it is normal to both the x- and z-axes.

In this case, the incident wave and reflected wave remain in medium 1, and the transmitted wave remains in medium 2. We use $\boldsymbol{k}$, $\boldsymbol{k}''$, and $\boldsymbol{k}'$ to denote the wave number vectors of the incident, reflected, and transmitted waves, respectively. Each wave propagates along the wave number vector. These vectors lie on the plane of incidence, the x-z plane. We denote the angles of the incident, reflected, and transmitted waves from the z-axis by θ, θ'', and θ', respectively (see Fig. 12.2b). The factor that represents the variation with time is commonly given by $\mathrm{e}^{\mathrm{i}\omega t}$. The incident, reflected, and transmitted waves are then expressed as

$$\exp[\mathrm{i}(\omega t - \boldsymbol{k} \cdot \boldsymbol{r})], \quad \exp[\mathrm{i}(\omega t - \boldsymbol{k}'' \cdot \boldsymbol{r})], \quad \exp[\mathrm{i}(\omega t - \boldsymbol{k}' \cdot \boldsymbol{r})],$$

with $\boldsymbol{r}$ representing the position vector.

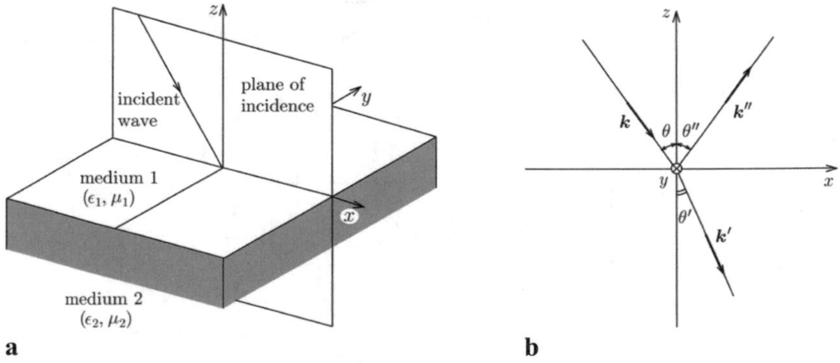

Fig. 12.2 Definition of **a** axes on the boundary and **b** angles of waves

The electric field $\boldsymbol{E}$ and magnetic flux density $\boldsymbol{B}$ are perpendicular to each other and lie on the plane normal to the propagation direction, as shown in the last section. For example, $\boldsymbol{E}$ and $\boldsymbol{B}$ for the incident wave are normal to $\boldsymbol{k}$. Here we consider the boundary conditions for $\boldsymbol{E}$ and $\boldsymbol{B}$: Eqs. (4.19), (10.27), (9.22), and (11.25). When there is neither electric charge nor current on the boundary, the parallel components of the electric field and magnetic field are continuous, and the normal components of the electric flux density and magnetic flux density are continuous across the boundary. That is,

$$\boldsymbol{n} \times (\boldsymbol{E}_1 - \boldsymbol{E}_2) = 0, \tag{12.20}$$

$$\boldsymbol{n} \cdot (\epsilon_1 \boldsymbol{E}_1 - \epsilon_2 \boldsymbol{E}_2) = 0, \tag{12.21}$$

$$\boldsymbol{n} \times \left(\frac{\boldsymbol{B}_1}{\mu_1} - \frac{\boldsymbol{B}_2}{\mu_2} \right) = 0, \tag{12.22}$$

$$\boldsymbol{n} \cdot (\boldsymbol{B}_1 - \boldsymbol{B}_2) = 0, \tag{12.23}$$

with $\boldsymbol{n}$ denoting the vector normal to the boundary. In the above, the subscripts 1 and 2 represent the variables in media 1 and 2, respectively.

Considering the orthogonality between $\boldsymbol{E}$ and $\boldsymbol{B}$, the incident wave is given by

$$\boldsymbol{E} = \boldsymbol{E}_0 \exp[\mathrm{i}(\omega t - \boldsymbol{k} \cdot \boldsymbol{r})], \tag{12.24a}$$

$$\boldsymbol{B} = \frac{\boldsymbol{k}}{k} \times \frac{\boldsymbol{E}_0}{c_1} \exp[\mathrm{i}(\omega t - \boldsymbol{k} \cdot \boldsymbol{r})], \tag{12.24b}$$

where $k = |\boldsymbol{k}|$ and c_1 is the light speed in medium 1. The magnetic flux density $\boldsymbol{B}$ is normal to both $\boldsymbol{E}$ and $\boldsymbol{k}$, and its magnitude is equal to the magnitude of $\boldsymbol{E}$ divided by the corresponding light speed. The reflected wave is similarly given by

$$\boldsymbol{E}'' = \boldsymbol{E}_0'' \exp[\mathrm{i}(\omega t - \boldsymbol{k}'' \cdot \boldsymbol{r})], \tag{12.25a}$$

$$\boldsymbol{B}'' = \frac{\boldsymbol{k}''}{k''} \times \frac{\boldsymbol{E}_0''}{c_1} \exp[\mathrm{i}(\omega t - \boldsymbol{k}'' \cdot \boldsymbol{r})], \tag{12.25b}$$

and the transmitted wave is given by

$$\boldsymbol{E}' = \boldsymbol{E}_0' \exp[\mathrm{i}(\omega t - \boldsymbol{k}' \cdot \boldsymbol{r})], \tag{12.26a}$$

$$\boldsymbol{B}' = \frac{\boldsymbol{k}'}{k'} \times \frac{\boldsymbol{E}_0'}{c_2} \exp[\mathrm{i}(\omega t - \boldsymbol{k}' \cdot \boldsymbol{r})]. \tag{12.26b}$$

In the above, $k'' = |\boldsymbol{k}''|$, $k' = |\boldsymbol{k}'|$, and c_2 is the light speed in medium 2. Thus, the electric field and magnetic flux density in medium 1 are

$$\boldsymbol{E}_1 = \boldsymbol{E} + \boldsymbol{E}'', \quad \boldsymbol{B}_1 = \boldsymbol{B} + \boldsymbol{B}'', \tag{12.27}$$

and those in medium 2 are

$$\boldsymbol{E}_2 = \boldsymbol{E}', \quad \boldsymbol{B}_2 = \boldsymbol{B}'. \tag{12.28}$$

Since Eqs. (12.20)–(12.23) should be satisfied at the boundary ($z = 0$) at any time, the phase must be the same for the three waves. This condition is given by

$$\boldsymbol{k} \cdot \boldsymbol{r}|_{z=0} = \boldsymbol{k}'' \cdot \boldsymbol{r}|_{z=0} = \boldsymbol{k}' \cdot \boldsymbol{r}|_{z=0}. \tag{12.29}$$

Equation (12.29) is expressed as

$$\boldsymbol{k} \cdot \boldsymbol{r}_0 = \boldsymbol{k}'' \cdot \boldsymbol{r}_0 = \boldsymbol{k}' \cdot \boldsymbol{r}_0 \tag{12.30}$$

in terms of an arbitrary position vector $\boldsymbol{r}_0$ on the boundary. If $\boldsymbol{r}_0$ is given by

$$\boldsymbol{r}_0 = x\boldsymbol{i}_x + y\boldsymbol{i}_y, \tag{12.31}$$

we have

$$k \sin\theta = k'' \sin\theta'' = k' \sin\theta', \tag{12.32}$$

since the wave number vectors are perpendicular to the y-axis. The speeds of the incident and reflected waves in the same medium are the same, and the wave numbers of these waves are also the same $k = k''$. Hence, we have

$$\theta = \theta''. \tag{12.33}$$

That is, the incident and reflection angles are the same, and this is called the **law of reflection**. We also have the relationship between the incident and transmission angles as

$$\frac{\sin\theta}{\sin\theta'} = \frac{k'}{k} = \frac{c_1}{c_2} = \left(\frac{\epsilon_2\mu_2}{\epsilon_1\mu_1}\right)^{1/2}. \tag{12.34}$$

This is called **Snell's law** for refraction.

Using the above results, Eqs. (12.20)–(12.23) are rewritten as

$$\boldsymbol{n} \times (\boldsymbol{E}_0 + \boldsymbol{E}_0'' - \boldsymbol{E}_0') = 0, \tag{12.35}$$

$$\boldsymbol{n} \cdot [\epsilon_1(\boldsymbol{E}_0 + \boldsymbol{E}_0'') - \epsilon_2\boldsymbol{E}_0'] = 0, \tag{12.36}$$

$$\boldsymbol{n} \times \left[\frac{1}{\mu_1}\left(\frac{\boldsymbol{k} \times \boldsymbol{E}_0}{kc_1} + \frac{\boldsymbol{k}'' \times \boldsymbol{E}_0''}{k''c_1}\right) - \frac{1}{\mu_2} \cdot \frac{\boldsymbol{k}' \times \boldsymbol{E}_0'}{k'c_2}\right] = 0, \tag{12.37}$$

$$\boldsymbol{n} \cdot \left(\frac{\boldsymbol{k} \times \boldsymbol{E}_0}{kc_1} + \frac{\boldsymbol{k}'' \times \boldsymbol{E}_0''}{k''c_1} - \frac{\boldsymbol{k}' \times \boldsymbol{E}_0'}{k'c_2} \right) = 0. \tag{12.38}$$

Although the electric field of the incident wave is directed in various directions, we focus for simplicity on the case where the electric field is directed parallel to the *y*-axis (i.e., normal to the plane of incidence) as shown in Fig. 12.3. In this case, Eq. (12.35) reduces to

$$E_0 + E_0'' - E_0' = 0. \tag{12.39}$$

Since the electric field is perpendicular to the normal vector, $\boldsymbol{n}$, Eq. (12.36) is already satisfied. Equation (12.37) becomes

$$\left(\frac{\epsilon_1}{\mu_1} \right)^{1/2} (E_0 - E_0'') \cos\theta - \left(\frac{\epsilon_2}{\mu_2} \right)^{1/2} E_0' \cos\theta' = 0. \tag{12.40}$$

Equation (12.38) is written as

$$\frac{E_0}{c_1} \sin\theta + \frac{E_0''}{c_1} \sin\theta - \frac{E_0'}{c_2} \sin\theta' = 0, \tag{12.41}$$

which is found to reduce to Eq. (12.39) using Eq. (12.34). Thus, from Eqs. (12.39) and (12.40), we obtain the amplitudes of the electric fields of the refracted and reflected waves as

$$E_0' = \frac{2(\epsilon_1/\mu_1)^{1/2} \cos\theta}{(\epsilon_1/\mu_1)^{1/2} \cos\theta + (\epsilon_2/\mu_2)^{1/2} \cos\theta'} E_0, \tag{12.42a}$$

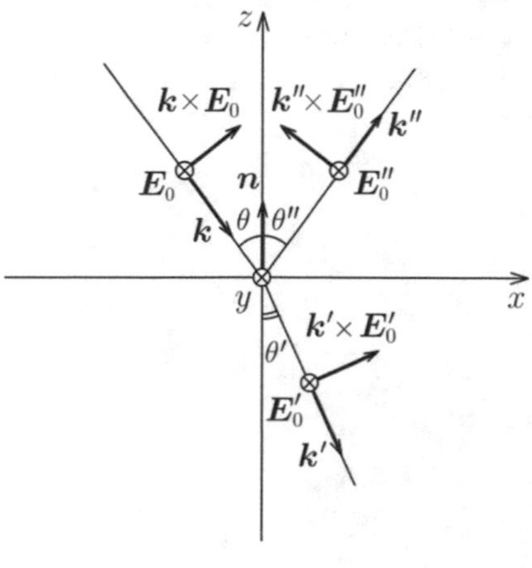

Fig. 12.3 Case where the electric field of the incident wave is normal to the plane of incidence, i.e., parallel to the *y*-axis

$$E_0'' = \frac{(\epsilon_1/\mu_1)^{1/2}\cos\theta - (\epsilon_2/\mu_2)^{1/2}\cos\theta'}{(\epsilon_1/\mu_1)^{1/2}\cos\theta + (\epsilon_2/\mu_2)^{1/2}\cos\theta'}E_0. \tag{12.42b}$$

The amplitudes of magnetic flux densities are

$$B_0' = \frac{E_0'}{c_2} = (\epsilon_2\mu_2)^{1/2}E_0', \tag{12.43a}$$

$$B_0'' = \frac{E_0''}{c_1} = (\epsilon_1\mu_1)^{1/2}E_0''. \tag{12.43b}$$

Example 12.2. Solve Eqs. (12.35)–(12.38) when the electric field of the incident wave is parallel to the plane of incidence, i.e., the magnetic flux density is parallel to the y-axis.

Solution 12.2. Under this condition, $(\boldsymbol{k} \times \boldsymbol{E}_0)$, $(\boldsymbol{k}' \times \boldsymbol{E}_0')$, and $(\boldsymbol{k}'' \times \boldsymbol{E}_0'')$ are directed along the y-axis, as shown in Fig. 12.4. Hence, Eqs. (12.35) and (12.36) become

$$(E_0 - E_0'')\cos\theta - E_0'\cos\theta' = 0,$$
$$\epsilon_1(E_0 + E_0'')\sin\theta - \epsilon_2 E_0'\sin\theta' = 0,$$

respectively. The latter equation is rewritten as

$$\left(\frac{\epsilon_1}{\mu_1}\right)^{1/2}(E_0 + E_0'') - \left(\frac{\epsilon_2}{\mu_2}\right)^{1/2}E_0' = 0.$$

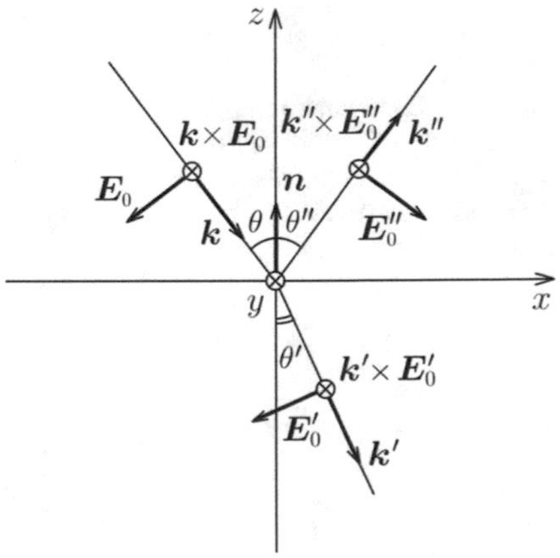

Fig. 12.4 Case where the electric field of the incident wave is parallel to the plane of incidence, i.e., normal to the y-axis

In Eq. (12.37), $\boldsymbol{n} \times (\boldsymbol{k} \times \boldsymbol{E})$ is parallel to the x-axis, and this equation agrees with the above result from Eq. (12.36). The condition given by Eq. (12.38) is satisfied. From the above two equations, we obtain the electric fields of the refracted and reflected waves as

$$E_0' = \frac{2(\epsilon_1/\mu_1)^{1/2}\cos\theta}{(\epsilon_2/\mu_2)^{1/2}\cos\theta + (\epsilon_1/\mu_1)^{1/2}\cos\theta'}E_0,$$

$$E_0'' = \frac{(\epsilon_2/\mu_2)^{1/2}\cos\theta - (\epsilon_1/\mu_1)^{1/2}\cos\theta'}{(\epsilon_2/\mu_2)^{1/2}\cos\theta + (\epsilon_1/\mu_1)^{1/2}\cos\theta'}E_0.$$

The corresponding magnetic flux densities are derived by substituting these results into Eqs. (12.43a) and (12.43b).

◇

12.3 Energy of the Electromagnetic Wave

Here we discuss the energy of the planar electromagnetic wave described in Sect. 12.1. For simplicity, we treat the second terms in Eqs. (12.3) and (12.9). In this case, the electric field and magnetic flux density are given by

$$\boldsymbol{E} = E_2\cos(\omega t - kx)\boldsymbol{i}_y, \tag{12.44}$$

$$\boldsymbol{B} = \frac{E_2}{c}\cos(\omega t - kx)\boldsymbol{i}_z. \tag{12.45}$$

Hence, the electric energy density and magnetic energy density are equal to each other and given by

$$\frac{1}{2}\epsilon\boldsymbol{E}^2 = \frac{1}{2\mu}\boldsymbol{B}^2 = \frac{1}{2}\epsilon E_2^2\cos^2(\omega t - kx). \tag{12.46}$$

Since there is no current, from Eq. (11.34) the total energy density is

$$u = \epsilon E_2^2\cos^2(\omega t - kx). \tag{12.47}$$

On the other hand, from Eq. (11.37) the Poynting vector is

$$\boldsymbol{S}_{\mathrm{P}} = \boldsymbol{E} \times \frac{\boldsymbol{B}}{\mu} = \left(\frac{\epsilon}{\mu}\right)^{1/2} E_2^2\cos^2(\omega t - kx)\boldsymbol{i}_x = cu\boldsymbol{i}_x. \tag{12.48}$$

Thus, we find that the Poynting vector has a magnitude equal to the total energy density multiplied by the light speed and is directed along the x-axis, i.e., the propagation direction of the electromagnetic wave. This holds for all planar electromagnetic waves including the elliptically polarized wave. Hence, the Poynting vector expresses the energy that flows through a unit area in unit time as defined in Sect. 11.5.

Example 12.3. Assume that the y- and z-components of the electric field of a polarized wave propagating along the positive x-axis are given by Eq. (12.17). Determine the energy density and the Poynting vector.

Solution 12.3. The electric energy density is

$$\begin{aligned}\frac{1}{2}\epsilon \boldsymbol{E}^2 &= \frac{1}{2}\epsilon\left(E_y^2 + E_z^2\right) \\ &= \frac{1}{2}\epsilon\left[E_1^2\cos^2(\omega t - kx) + E_2^2\cos^2(\omega t - kx + \delta)\right].\end{aligned}$$

The magnetic flux density is given by Eq. (12.18), and the magnetic energy density is

$$\begin{aligned}\frac{1}{2\mu}\boldsymbol{B}^2 &= \frac{1}{2\mu}\left(B_y^2 + B_z^2\right) \\ &= \frac{1}{2}\epsilon\left[E_1^2\cos^2(\omega t - kx) + E_2^2\cos^2(\omega t - kx + \delta)\right].\end{aligned}$$

Thus, the energy density of the electromagnetic wave is

$$u = \frac{1}{2}\epsilon \boldsymbol{E}^2 + \frac{1}{2\mu}\boldsymbol{B}^2 = \epsilon\left[E_1^2\cos^2(\omega t - kx) + E_2^2\cos^2(\omega - kx + \delta)\right].$$

The Poynting vector is

$$\begin{aligned}\boldsymbol{S}_{\mathrm{P}} &= \boldsymbol{E} \times \frac{\boldsymbol{B}}{\mu} = \frac{1}{\mu}\left(E_y\boldsymbol{i}_y + E_z\boldsymbol{i}_z\right) \times \left(B_y\boldsymbol{i}_y + B_z\boldsymbol{i}_z\right) = \frac{1}{\mu}\left(E_yB_z - E_zB_y\right)\boldsymbol{i}_x \\ &= \frac{1}{\mu c}\left[E_1^2\cos^2(\omega t - kx) + E_2^2\cos^2(\omega t - kx + \delta)\right]\boldsymbol{i}_x = cu\boldsymbol{i}_x.\end{aligned}$$

. ◇

12.4 Wave Guide

Hollow metal tubes called **wave guides** are used to transmit electromagnetic waves such as microwaves. The cross section of a wave guide is usually rectangular or circular. Here we treat a rectangular wave guide for simplicity. Assume that the wave guide is uniformly extended along the z-axis, and the internal vacuum region is $0 \leq x \leq a$ and $0 \leq y \leq b$, as shown in Fig. 12.5.

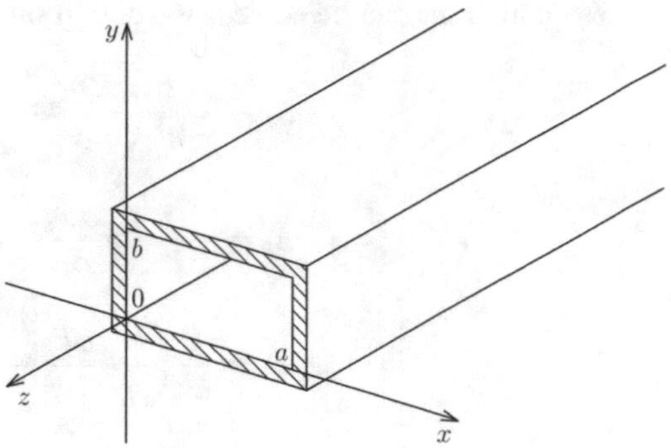

Fig. 12.5 Rectangular wave guide

We assume that the factors for time variation and spatial variation are given by $e^{i\omega t}$ and $e^{-i\gamma z}$, respectively. Equations (11.7) and (11.8) then reduce to

$$\frac{\partial E_z}{\partial y} + i\gamma E_y = -i\omega B_x, \tag{12.49a}$$

$$-i\gamma E_x - \frac{\partial E_z}{\partial x} = -i\omega B_y, \tag{12.49b}$$

$$\frac{\partial E_y}{\partial x} - \frac{\partial E_x}{\partial y} = -i\omega B_z, \tag{12.49c}$$

and

$$\frac{\partial B_z}{\partial y} + i\gamma B_y = i\frac{\omega}{c_0^2}E_x, \tag{12.50a}$$

$$-i\gamma B_x - \frac{\partial B_z}{\partial x} = i\frac{\omega}{c_0^2}E_y, \tag{12.50b}$$

$$\frac{\partial B_y}{\partial x} - \frac{\partial B_x}{\partial y} = i\frac{\omega}{c_0^2}E_z. \tag{12.50c}$$

Using these equations, the equations for the z-components, E_z and B_z, are obtained as

$$\frac{\partial^2 E_z}{\partial x^2} + \frac{\partial^2 E_z}{\partial y^2} + k^2 E_z = 0, \tag{12.51a}$$

$$\frac{\partial^2 B_z}{\partial x^2} + \frac{\partial^2 B_z}{\partial y^2} + k^2 B_z = 0. \tag{12.51b}$$

If these equations can be solved, we obtain other components from

$$E_x = -\frac{\mathrm{i}}{k^2}\left(\gamma\frac{\partial E_z}{\partial x} + \omega\frac{\partial B_z}{\partial y}\right), \tag{12.52a}$$

$$E_y = -\frac{\mathrm{i}}{k^2}\left(\gamma\frac{\partial E_z}{\partial y} - \omega\frac{\partial B_z}{\partial x}\right), \tag{12.52b}$$

$$B_x = \frac{\mathrm{i}}{k^2}\left(\frac{\omega}{c_0^2}\cdot\frac{\partial E_z}{\partial y} - \gamma\frac{\partial B_z}{\partial x}\right), \tag{12.52c}$$

$$B_y = -\frac{\mathrm{i}}{k^2}\left(\frac{\omega}{c_0^2}\cdot\frac{\partial E_z}{\partial x} + \gamma\frac{\partial B_z}{\partial y}\right), \tag{12.52d}$$

where

$$k^2 = \left(\frac{\omega}{c_0}\right)^2 - \gamma^2, \tag{12.53}$$

and $\nabla \cdot \boldsymbol{E} = 0$ is used. Equations (12.52a)–(12.52d) hold for $k \neq 0$.

In the case of $k = 0$, we have $\gamma = \pm\omega/c_0$, and we may consider that there is an electromagnetic wave propagating along the z-axis at light speed. For example, we assume an electromagnetic wave without z-components ($E_z = B_z = 0$) similar to a planar electromagnetic wave. This is called the **transverse electromagnetic (TEM) wave**. If we choose $\gamma = \omega/c_0$, Eqs. (12.49a) and (12.49b) lead to

$$E_x = c_0 B_y, \quad E_y = -c_0 B_x, \tag{12.54}$$

showing that the electric field $\boldsymbol{E}$ and magnetic flux density $\boldsymbol{B}$ are perpendicular to each other. However, Eq. (12.49c) gives

$$\frac{\partial E_y}{\partial x} - \frac{\partial E_x}{\partial y} = 0. \tag{12.55}$$

This shows that the electric field is a dynamic two-dimensional field with no rotation. The spatial structure of the irrotational field is the same as that of the electrostatic field, and hence, we conclude that such an electric field cannot exist in a space surrounded by a conductor like a rectangular wave guide. That is, TEM wave cannot exist in simple rectangular or circular wave guides. Such a field can exist only when the guide is composed of two or more conductors like those in Fig. 12.6, and a potential difference can appear between conductors with electric field lines extending from one conductor to another.

From the above discussion, we know that either the electric field $\boldsymbol{E}$ or magnetic flux density $\boldsymbol{B}$ has at least one component in the propagation direction. The electromagnetic wave with a zero longitudinal component of the electric field is called

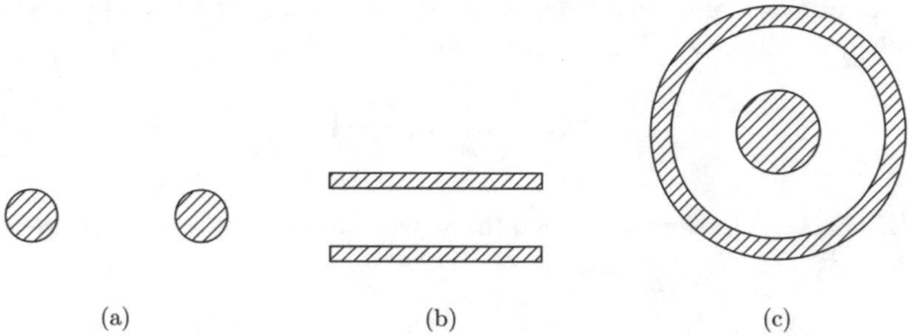

Fig. 12.6 Examples of the cross section of a wave guide in which a TEM wave exists: **a** parallel cylindrical conductor, **b** parallel-plate conductor, and **c** coaxial conductor

the **transverse electric (TE) wave** and that with a zero longitudinal component of the magnetic flux density is called the **transverse magnetic (TM) wave**. The general electromagnetic wave is given by a linear combination of these waves.

Here we consider a TM wave. In this case, $B_z = 0$. Since any electromagnetic wave of high frequency does not penetrate the conductor, the electric field is perpendicular to and the magnetic flux density is parallel to the conductor surface. That is,

$$\begin{aligned} E_y = E_z = B_x = 0; \quad x = 0, a, \\ E_x = E_z = B_y = 0; \quad y = 0, b. \end{aligned} \tag{12.56}$$

The general solution of Eq. (12.51a) is given by

$$E_z(x, y, z, t) = K \exp[\pm \mathrm{i}(k_x x + k_y y)] \exp[\mathrm{i}(\omega t - \gamma z)] \tag{12.57}$$

with

$$k_x^2 + k_y^2 = k^2. \tag{12.58}$$

The dependence on x can be written as

$$E_z = K_1 \mathrm{e}^{\mathrm{i}k_x x} + K_2 \mathrm{e}^{-\mathrm{i}k_x x}. \tag{12.59}$$

From Eq. (12.56), the following conditions should be satisfied:

$$K_1 + K_2 = 0, \quad \mathrm{e}^{\mathrm{i}2k_x a} = 1. \tag{12.60}$$

The latter condition gives

$$k_x = \frac{m\pi}{a}; \quad m = 1, 2, \ldots. \tag{12.61}$$

The case of $m = 0$ also satisfies this condition. However, we have $E_z = 0$, which is meaningless. Thus, Eq. (12.59) reduces to

$$E_z = 2K_1' \sin\left(\frac{m\pi x}{a}\right) \tag{12.62}$$

with $K_1' = \mathrm{i}K_1$. We similarly obtain the y-dependence with $k_y = n\pi/b$ (n = 1, 2, …), and Eq. (12.57) is rewritten as

$$E_z(x, y, z, t) = A \sin\left(\frac{m\pi x}{a}\right) \sin\left(\frac{n\pi y}{b}\right) \exp[\mathrm{i}(\omega t - \gamma z)], \tag{12.63}$$

where A is a constant and there is a relationship between m and n written as

$$\left(\frac{m\pi}{a}\right)^2 + \left(\frac{n\pi}{b}\right)^2 = \left(\frac{\omega}{c_0}\right)^2 - \gamma^2. \tag{12.64}$$

The mode of the electromagnetic wave is different depending on the set of integers, (m, n), and each mode of the TM wave is expressed as TM_{mn}. The mode of the TE wave is represented similarly as TE_{mn}.

So that the TM wave propagates through the wave guide along the z-axis without damping, γ must be a real number and we have

$$\left(\frac{\omega}{c_0}\right)^2 - \left(\frac{m\pi}{a}\right)^2 - \left(\frac{n\pi}{b}\right)^2 \geq 0. \tag{12.65}$$

That is, the angular frequency ω should be larger than the **cut-off frequency** given by

$$\omega_0 = c_0 \left[\left(\frac{m\pi}{a}\right)^2 + \left(\frac{n\pi}{b}\right)^2\right]^{1/2}. \tag{12.66}$$

Example 12.4. Determine other components of electromagnetic fields of the above TM wave.

Solution 12.4. Substituting E_z in Eq. (12.63) and $B_z = 0$ into Eqs. (12.52a)–(12.52d), we have

$$E_x = -\mathrm{i}A \frac{m\pi\gamma}{k^2 a} \cos\left(\frac{m\pi x}{a}\right) \sin\left(\frac{n\pi y}{b}\right),$$

$$E_y = -\mathrm{i}A \frac{n\pi\gamma}{k^2 b} \sin\left(\frac{m\pi x}{a}\right) \cos\left(\frac{n\pi y}{b}\right),$$

$$B_x = \mathrm{i}A \frac{n\pi\omega}{k^2 c_0^2 b} \sin\left(\frac{m\pi x}{a}\right) \cos\left(\frac{n\pi y}{b}\right),$$

$$B_y = -\mathrm{i}A\frac{m\pi\omega}{k^2c_0^2a}\cos\left(\frac{m\pi x}{a}\right)\sin\left(\frac{n\pi y}{b}\right),$$

where the factor $\exp[\mathrm{i}(\omega t - \gamma z)]$ is neglected. The real parts of these expressions give practical physical quantities. That is, the above factor is replaced by $\cos(\omega t - \gamma z)$ for E_z, and $\mathrm{i}\exp[\mathrm{i}(\omega t - \gamma z)]$ is replaced by $\sin(\omega t - \gamma z)$ for the above quantities. We easily find that $\boldsymbol{E} \cdot \boldsymbol{B} = 0$ is satisfied, indicating that the electric field and magnetic flux density are perpendicular to each other.

◇

Example 12.5. Determine the TEM wave that propagates along the length (z-axis) in a coaxial conductor, as shown in Fig. 12.6c, and discuss the relationship between the electric charge and the current. Assume $\mathrm{e}^{\mathrm{i}\omega t}$ and $\mathrm{e}^{-\mathrm{i}\gamma z}$ for the time and spatial variation, respectively, and that the vacuum space in the coaxial conductor is $a < R < b$.

Solution 12.5. From the properties of electric and magnetic fields, we can assume that the electric field has only the radial component, E_R, and that the magnetic flux density has only the azimuthal component, B_φ. The differentials with respect to t and z can be replaced by $\mathrm{i}\omega$ and $-\mathrm{i}\gamma$, respectively.. Thus, we have

$$\gamma E_R = \omega B_\varphi$$

and

$$\gamma B_\varphi = \omega\epsilon_0\mu_0 E_R.$$

These conditions lead to

$$\frac{\omega}{\gamma} = \frac{1}{(\epsilon_0\mu_0)^{1/2}} = c_0.$$

If the amplitude of the electric field at $R = a$ is denoted by E_0, we have

$$E_R = \frac{a}{R}E_0\exp[\mathrm{i}(\omega t - \gamma z)], \quad B_\varphi = \frac{a}{c_0R}E_0\exp[\mathrm{i}(\omega t - \gamma z)].$$

The densities of the electric charge and the current flowing along the z-axis that appear on the surface $R = a$ are, respectively, given by

$$\sigma = \epsilon_0 E_R(R = a) = \epsilon_0 E_0\exp[\mathrm{i}(\omega t - \gamma z)],$$

$$\tau = \frac{B_\varphi(R = a)}{\mu_0} = \left(\frac{\epsilon_0}{\mu_0}\right)^{1/2} E_0\exp[\mathrm{i}(\omega t - \gamma z)].$$

The densities of the electric charge and the current on the surface $R = b$ are

$$\sigma = -\epsilon_0 E_R(R = b) = -\frac{a}{b}\epsilon_0 E_0 \exp[i(\omega t - \gamma z)],$$

$$\tau = -\frac{B_\varphi(R = b)}{\mu_0} = -\frac{a}{b}\left(\frac{\epsilon_0}{\mu_0}\right)^{1/2} E_0 \exp[i(\omega t - \gamma z)].$$

It can be easily shown that the continuity equation of current given by Eq. (5.10),

$$\frac{\partial \tau}{\partial z} + \frac{\partial \sigma}{\partial t} = 0,$$

holds on the two surfaces.

◇

12.5 Spherical Wave

Here we consider the case in which an electromagnetic wave propagates radially. This wave is a **spherical wave**. We assume that the factor of time variation is given by $e^{i\omega t}$. Then, from Eq. (11.23), we write the equation for the electric field as

$$\Delta \boldsymbol{E} + \frac{\omega^2}{c^2}\boldsymbol{E} = 0. \tag{12.67}$$

We assume spherical symmetry of the electromagnetic quantities except in the vicinity of the source of the electromagnetic wave. We also assume that the electric field has only the zenithal component, E_θ, dependent only on the radius r:

$$E_\theta(r, t) = E_\theta(r)e^{i\omega t}. \tag{12.68}$$

Then, the above equation leads to

$$\frac{1}{r}\cdot\frac{\partial^2}{\partial r^2}(rE_\theta) + \left(\frac{\omega}{c}\right)^2 E_\theta = 0, \tag{12.69}$$

and we obtain a general solution as

$$E_\theta(r, t) = \frac{K_1}{r}\exp\left[i\omega\left(t + \frac{r}{c}\right)\right] + \frac{K_2}{r}\exp\left[i\omega\left(t - \frac{r}{c}\right)\right]. \tag{12.70}$$

The first term represents an electromagnetic wave propagating to the origin at speed c, but such a wave concentrating to one point is unrealistic. On the other hand, the second term represents an electromagnetic wave radiating at speed c from the origin, and this is the solution to be obtained. Taking the real part, we have

$$E_\theta(r,t) = \frac{K_2}{r}\cos\left[\omega\left(t - \frac{r}{c}\right)\right]. \tag{12.71}$$

Equation (11.7) shows that the magnetic flux density has only the azimuthal component, B_φ, and reduces to

$$\frac{1}{r}\cdot\frac{\partial}{\partial r}(rE_\theta) = -\frac{\partial B_\varphi}{\partial t}. \tag{12.72}$$

This is easily solved, and we have

$$B_\varphi(r,t) = \frac{K_2}{cr}\cos\left[\omega\left(t - \frac{r}{c}\right)\right]. \tag{12.73}$$

From the above results, we obtain the Poynting vector as

$$\boldsymbol{S}_\mathrm{P} = \frac{1}{\mu}E_\theta B_\varphi \boldsymbol{i}_r = \left(\frac{\epsilon}{\mu}\right)^{1/2}\frac{K_2^2}{r^2}\cos^2\left[\omega\left(t - \frac{r}{c}\right)\right]\boldsymbol{i}_r, \tag{12.74}$$

showing that it is directed radially. The energy density is

$$u = \frac{1}{2}\epsilon E_\theta^2 + \frac{1}{2\mu}B_\varphi^2 = \epsilon\frac{K_2^2}{r^2}\cos^2\left[\omega\left(t - \frac{r}{c}\right)\right] = \frac{1}{c}\boldsymbol{S}_\mathrm{P}\cdot\boldsymbol{i}_r. \tag{12.75}$$

This relationship is similar to that for a planar electromagnetic wave. The energy density and Poynting vector are proportional to r^{-2} because the wave front expands and yield constants when integrated on the spherical surface.

12.6 Retarded Potential

When electric charge (of density ρ) or current (of density $\boldsymbol{i}$) changes with time, an electromagnetic wave is emitted into the surrounding space. At a distance far from the source, the electromagnetic wave propagates as a spherical wave as treated in Sect. 12.5. The electromagnetic potential describing such time-dependent electromagnetic fields is given by Eqs. (11.31) and (11.32). In the static limit, these reduce to Eqs. (4.17) and (9.21), and the solutions are given by Eqs. (1.27) and (6.33) when ϵ_0 and μ_0 are replaced with ϵ and μ.

Using polar coordinates, Eq. (11.31) for the electric potential reduces to

$$\frac{1}{r}\cdot\frac{\partial^2}{\partial r^2}(r\phi) - \frac{1}{c^2}\cdot\frac{\partial^2\phi}{\partial t^2} = -\frac{\rho}{\epsilon}. \tag{12.76}$$

We denote by $\phi(r) = f(t)/r$ the solution for the corresponding quasi-static equation, i.e., the equation with the second term omitted. Referring to the solution of

Eq. (12.70) for Eq. (11.23) of the same form, we can prove that the general solution of Eq. (12.76) is given by

$$\phi(r,t) = \frac{1}{r} f\left(t - \frac{r}{c}\right). \tag{12.77}$$

Hence, we expect that the solution of Eq. (11.31) is given by

$$\phi(\boldsymbol{r},t) = \frac{1}{4\pi\epsilon} \int_{\mathrm{V}} \frac{\rho(\boldsymbol{r}', t - |\boldsymbol{r} - \boldsymbol{r}'|/c)}{|\boldsymbol{r} - \boldsymbol{r}'|} \mathrm{d}V' \tag{12.78}$$

(see Exercise 12.10). We also obtain the vector potential as

$$\boldsymbol{A}(\boldsymbol{r},t) = \frac{\mu}{4\pi} \int_{\mathrm{V}} \frac{\boldsymbol{i}(\boldsymbol{r}', t - |\boldsymbol{r} - \boldsymbol{r}'|/c)}{|\boldsymbol{r} - \boldsymbol{r}'|} \mathrm{d}V'. \tag{12.79}$$

The above results show that, since the speed of propagation of a variation in the fields due to the change in the source is finite (light speed c), the change is transmitted with a delay in time of R/c at distance $R = |\boldsymbol{r} - \boldsymbol{r}'|$. For this reason, the above potentials are called **retarded potentials**. When the variation with time is slow, we can neglect this delay, resulting in the static potentials given by Eqs. (1.27) and (6.33).

Column: To the Theory of Relativity

Newton's equation of motion is unchanged between coordinate system K with spatial coordinates (x, y, z) and time t and another coordinate system K* with the dimensions

$$x' = x - vt, \quad y' = y, \quad z' = z, \quad t' = t.$$

The coordinate systems K and K* move with constant velocity relatively to each other, and this transformation is called the Galilean transformation. That is, Newton's equation is unchanged under the Galilean transformation.

We assume that the equation for the electromagnetic potential [e.g., Eq. (11.31)] holds in coordinate system K:

$$\Delta\phi - \epsilon\mu \frac{\partial^2 \phi}{\partial t^2} = -\frac{\rho}{\epsilon}.$$

Under the Galilean transformation, however, this equation does not hold in the same form in coordinate system K*:

$$\Delta'\phi' - \epsilon\mu \frac{\partial^2 \phi'}{\partial t'^2} = -\frac{\rho}{\epsilon},$$

where Δ' represents the Laplacian with respect to x', y', and z'.

Since there is no rotation of space, the scalar potential ϕ' is equal to ϕ. From each relationship such as

$$\frac{\partial}{\partial x'} = \frac{\partial}{\partial x} \cdot \frac{\partial x}{\partial x'} + \frac{\partial}{\partial y} \cdot \frac{\partial y}{\partial x'} + \frac{\partial}{\partial z} \cdot \frac{\partial z}{\partial x'} + \frac{\partial}{\partial t} \cdot \frac{\partial t}{\partial x'} = \frac{\partial}{\partial x},$$

$$\frac{\partial}{\partial t'} = \frac{\partial}{\partial x} \cdot \frac{\partial x}{\partial t'} + \frac{\partial}{\partial y} \cdot \frac{\partial y}{\partial t'} + \frac{\partial}{\partial z} \cdot \frac{\partial z}{\partial t'} + \frac{\partial}{\partial t} \cdot \frac{\partial t}{\partial t'} = v\frac{\partial}{\partial x} + \frac{\partial}{\partial t},$$

we have

$$\Delta' = \Delta$$

and

$$\frac{\partial^2}{\partial t'^2} = v^2 \frac{\partial^2}{\partial x^2} + 2v\frac{\partial^2}{\partial x \partial t} + \frac{\partial^2}{\partial t^2} \neq \frac{\partial^2}{\partial t^2}.$$

The above results indicate that Maxwell's equations hold correctly only in one coordinate system. However, it was experimentally demonstrated that Maxwell's equations hold in all coordinate systems that move relatively to each other with constant velocities. This means that the Galilean transformation is not a correct transformation connecting two coordinate systems. The correct transformation is the Lorentz transformation, and, in this case, the time is no longer independent of the velocity of the coordinate system.

In 1905, Einstein proposed the special theory of relativity based on the principle of relativity that all coordinate systems are equivalent and the principle of the constancy of light speed that the light speed observed in any coordinate system is the same.

Exercises

12.1. Discuss the reflection and refraction of a planar electromagnetic wave at the boundary when medium 1 is a vacuum and medium 2 is a conductor. Assume that the electric field of the incident wave is normal to the plane of incidence.

12.2. Discuss the same problem as in Exercise 12.1 when the electric field of the incident wave is parallel to the plane of incidence.

12.3. Discuss the energy flow using the Poynting vector for the reflection and refraction of a planar electromagnetic wave treated in Sect. 12.2. Assume that the electric field in the incident wave is normal to the place of incidence.

12.4. Determine the TEM wave that propagates along the length (z-axis) in the parallel-plate conductor in Fig. E12.1 and discuss the relationship between the electric charge and the current induced on the conductor surface. Assume $e^{i\omega t}$ and $e^{-i\gamma z}$ for the time and spatial variation, respectively, and that the vacuum space in the parallel-plate conductor is $0 < x < a$ and $0 < y < b$.

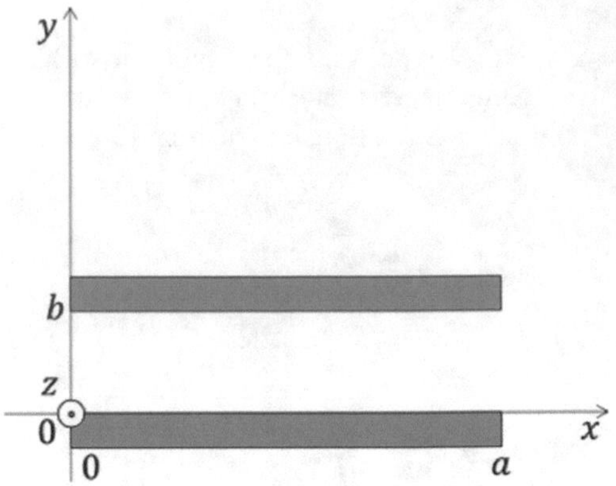

Fig. E12.1 Parallel-plate conductor

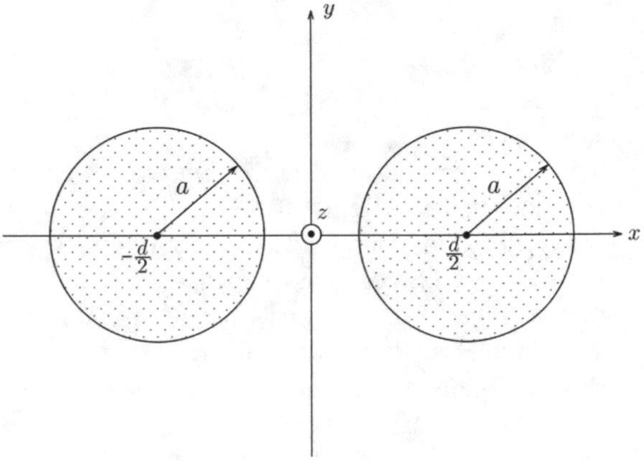

Fig. E12.2 Parallel cylindrical conductor

12.5. Discuss the energy flow for the TM wave in the wave guide treated in Sect. 12.4 and Example 12.4.
12.6. Determine the electromagnetic fields of the TE wave in the rectangular wave guide in Fig. 12.5.
12.7. Determine the electric charge density and current density on the inner surface $x = 0$ of the rectangular wave guide for the TM wave discussed in Example 12.4. Discuss the relation between them.
12.8. Determine the electromagnetic fields of a TEM wave propagating along the length (z-axis) for the case of two parallel cylindrical conductors of radius a and mean distance d in Fig. E12.2. Disregard the factor $\mathrm{e}^{\mathrm{i}(\omega t - \gamma z)}$ with $\gamma/\omega = 1/c_0$. (Hint: Since the arrangement of conductors is the same as that in Exercise 5.10, the electric field has the same form as in that case.)
12.9. Discuss the spherical wave described by Eq. (12.67) for the case in which the electric field has only the azimuthal component.
12.10. Prove that Eq. (12.78) satisfies Eq. (11.31). See Sect. A2.1 in the Appendix.

Correction to: Electricity and Magnetism

Correction to:
T. Matsushita, *Electricity and Magnetism*, Undergraduate Lecture Notes in Physics, Electricity and Magnetism, https://doi.org/10.1007/978-3-030-82150-0

The figure sizes in this book have been corrected and harmonized compared to the initially published version. These corrections led to a re-pagination of all chapters. The scientific content has not been changed. The publisher would like to apologize for this error to the author Teruo Matsushita and to the customers.

The updated version of the book can be found at
https://doi.org/10.1007/978-3-030-82150-0

T. Matsushita, *Electricity and Magnetism*, Undergraduate Lecture Notes in Physics,
https://doi.org/10.1007/978-3-030-82150-0_13

Appendix A

A1 Vector Analysis

A1.1 Scalars and Vectors

A quantity that is only specified by its magnitude is a **scalar**. Mass, energy, electric charge, and temperature are examples of scalars. On the other hand, there is a kind of quantity that needs to be specified not only by magnitude but also direction; such a quantity is a **vector**. Force, velocity, and moment in dynamics and the electric field, magnetic flux density, and current in electromagnetism are examples of vectors.

A vector is commonly denoted by a bold character such as $\boldsymbol{F}$ or a character with an arrow such as $\overrightarrow{F}$. A vector is specified in space by drawing a straight arrow, as shown in Fig. A1.1. The length of the line represents the magnitude of the vector, and the direction of the arrow represents the direction of the vector. The magnitude of vector $\boldsymbol{F}$ is written as $|\boldsymbol{F}|$.

In many cases, the effect of vector is unchanged even when it is displaced in parallel, i.e., moved without changing the direction in which it points, as shown in Fig. A1.2. Such a vector is called a free vector. On the other hand, there is a kind of vector that gives a different effect when displaced (like a force); such a vector is called a bound vector. For example, when a force is given on the center of gravity of an object, it causes a simple translation motion of the

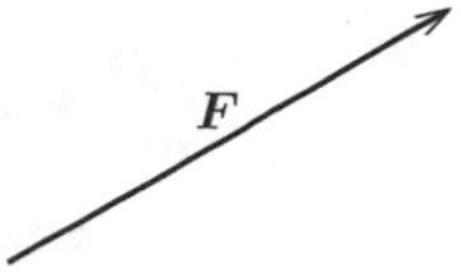

Fig. A1.1 Specifying a vector in space

T. Matsushita, *Electricity and Magnetism*, Undergraduate Lecture Notes in Physics,
https://doi.org/10.1007/978-3-030-82150-0

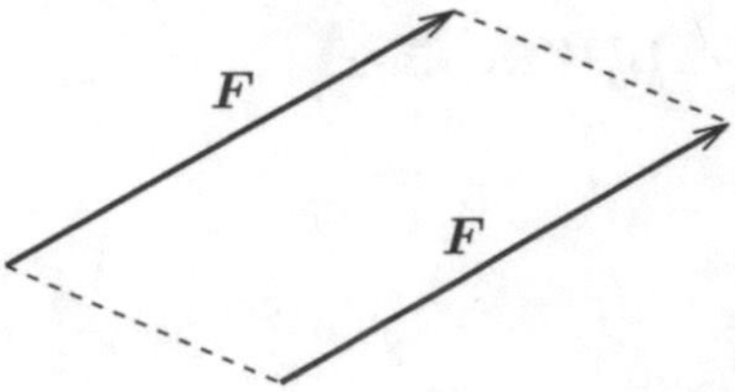

Fig. A1.2 Parallel displacement of vector

object. However, when a force is given on a point other than the center of gravity, it causes both straight motion of the center of gravity and a rotational motion around that center.

A1.2 Addition of Vectors

Suppose the sum of two free vectors $\boldsymbol{A}$ and $\boldsymbol{B}$, as shown in Fig. A1.3. The sum $\boldsymbol{A}+\boldsymbol{B}$ is obtained graphically by translating $\boldsymbol{B}$ so that its starting point reaches the end point of $\boldsymbol{A}$, as shown in Fig. A1.4a. In this case, the sum is obtained as a vector that connects the starting point of $\boldsymbol{A}$ and the end point of $\boldsymbol{B}$. The sum $\boldsymbol{B}+\boldsymbol{A}$ is similarly obtained by translating $\boldsymbol{A}$, and we can see it is equal to the sum $\boldsymbol{A}+\boldsymbol{B}$ (see Fig. A1.4b). Thus, the relation holds generally:

$$\boldsymbol{A}+\boldsymbol{B}=\boldsymbol{B}+\boldsymbol{A}. \tag{A1.1}$$

This is called the exchange law. For a sum of three vectors, the following relation holds:

$$(\boldsymbol{A}+\boldsymbol{B})+\boldsymbol{C}=\boldsymbol{A}+(\boldsymbol{B}+\boldsymbol{C}), \tag{A1.2}$$

where the operation inside parentheses has priority. This is called the combination law.

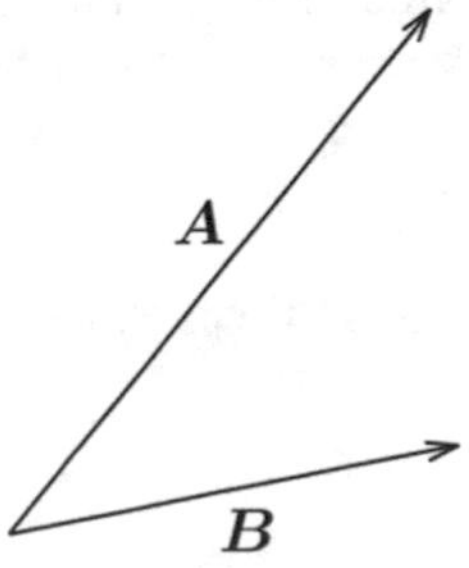

Fig. A1.3 Two vectors

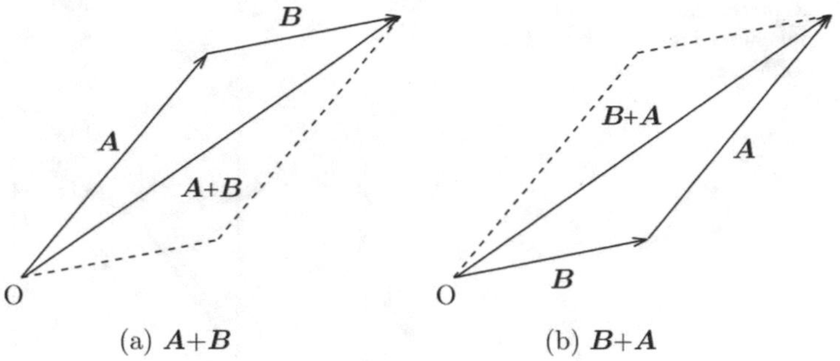

Fig. A1.4 Sum of two vectors: **a** $\boldsymbol{A}+\boldsymbol{B}$ and **b** $\boldsymbol{B}+\boldsymbol{A}$

A1.3 Products of Vectors and Scalars

The product of a vector $\boldsymbol{A}$ and scalar a gives a vector of magnitude equal to $a|\boldsymbol{A}|$ with the same direction as $\boldsymbol{A}$. For $a < 0$, the direction is reversed. The following relations hold for this kind of product:

$$m(n\boldsymbol{A}) = (mn)\boldsymbol{A}, \tag{A1.3}$$

$$(m+n)\boldsymbol{A} = m\boldsymbol{A} + n\boldsymbol{A}, \tag{A1.4}$$

$$m(\boldsymbol{A}+\boldsymbol{B}) = m\boldsymbol{A} + m\boldsymbol{B}. \tag{A1.5}$$

Equation (A1.5) is called the distribution law.

A1.4 Analytic Expression of a Vector

We use Cartesian coordinates (x, y, z) and denote unit vectors with a unit magnitude along the x-, y-, and z-axes by $\boldsymbol{i}_x$, $\boldsymbol{i}_y$, and $\boldsymbol{i}_z$, respectively. If vector $\boldsymbol{A}$ is expressed as

$$\boldsymbol{A} = A_x\boldsymbol{i}_x + A_y\boldsymbol{i}_y + A_z\boldsymbol{i}_z \tag{A1.6}$$

(see Fig. A1.5), A_x, A_y, and A_z are called the x-, y-, and z-components of $\boldsymbol{A}$, respectively. Using these components, $\boldsymbol{A}$ is also expressed as

$$(A_x, A_y, A_z). \tag{A1.7}$$

The magnitude of $\boldsymbol{A}$ is given by

$$|\boldsymbol{A}| = A = (A_x^2 + A_y^2 + A_z^2)^{1/2}. \tag{A1.8}$$

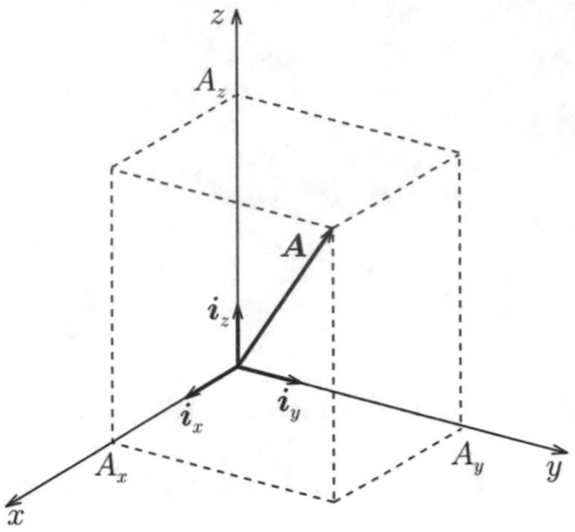

Fig. A1.5 Representation of vector components using Cartesian coordinates

When the components of vectors $\boldsymbol{A}$ and $\boldsymbol{B}$ are (A_x, A_y, A_z) and (B_x, B_y, B_z), respectively, the components of $\boldsymbol{A}+\boldsymbol{B}$ are

$$(A_x + B_x, A_y + B_y, A_z + B_z). \tag{A1.9}$$

The components of $a\boldsymbol{A}$ are

$$(aA_x, aA_y, aA_z). \tag{A1.10}$$

Using these methods, we can prove the laws of exchange, combination, and distribution.

A1.5 Products of Vectors

When the angle of vector $\boldsymbol{B}$ measured from vector $\boldsymbol{A}$ is θ $(-\pi < \theta \leq \pi)$, $AB\cos\theta$ is called a **scalar product** of $\boldsymbol{A}$ and $\boldsymbol{B}$ and is written as $\boldsymbol{A} \cdot \boldsymbol{B}$. Hence, we have

$$\boldsymbol{A} \cdot \boldsymbol{B} = \boldsymbol{B} \cdot \boldsymbol{A} = AB\cos\theta. \tag{A1.11}$$

If $\boldsymbol{A}$ and $\boldsymbol{B}$ are perpendicular to each other $(\theta = \pm\pi/2)$, $\boldsymbol{A} \cdot \boldsymbol{B} = 0$. When the components of $\boldsymbol{A}$ and $\boldsymbol{B}$ are (A_x, A_y, A_z) and (B_x, B_y, B_z), we have

$$\boldsymbol{A} \cdot \boldsymbol{B} = A_x B_x + A_y B_y + A_z B_z. \tag{A1.12}$$

The **vector product** of $\boldsymbol{A}$ and $\boldsymbol{B}$ is a vector of magnitude $AB\sin\theta$ that points along the direction of a screw when we rotate it from $\boldsymbol{A}$ to $\boldsymbol{B}$ and is written as $\boldsymbol{A} \times \boldsymbol{B}$. Thus,

$$\boldsymbol{A} \times \boldsymbol{B} = -\boldsymbol{B} \times \boldsymbol{A}. \tag{A1.13}$$

Vector $\boldsymbol{A} \times \boldsymbol{B}$ is normal to $\boldsymbol{A}$ and $\boldsymbol{B}$, and when $\boldsymbol{A}$ and $\boldsymbol{B}$ point in the same or opposite direction ($\theta = 0$ or π), $\boldsymbol{A} \times \boldsymbol{B} = 0$. Using the components, the vector product is expressed as

$$\boldsymbol{A} \times \boldsymbol{B} = (A_yB_z - A_zB_y)\boldsymbol{i}_x + (A_zB_x - A_xB_z)\boldsymbol{i}_y + (A_xB_y - A_yB_x)\boldsymbol{i}_z. \tag{A1.14}$$

We can also represent the vector product with a determinant

$$\boldsymbol{A} \times \boldsymbol{B} = \begin{vmatrix} \boldsymbol{i}_x & \boldsymbol{i}_y & \boldsymbol{i}_z \\ A_x & A_y & A_z \\ B_x & B_y & B_z \end{vmatrix}. \tag{A1.15}$$

Now, we treat a product of three vectors. Among the conceivable products of three vectors $\boldsymbol{A}$, $\boldsymbol{B}$, and $\boldsymbol{C}$, it is clear that $\boldsymbol{A} \cdot \boldsymbol{B} \cdot \boldsymbol{C}$ and $(\boldsymbol{A} \cdot \boldsymbol{B}) \times \boldsymbol{C}$ do not make sense, and the order of operation is not defined for $\boldsymbol{A} \times \boldsymbol{B} \times \boldsymbol{C}$. One meaningful product is

$$\boldsymbol{A}(\boldsymbol{B} \cdot \boldsymbol{C}) = a\boldsymbol{A}; \quad \boldsymbol{B} \cdot \boldsymbol{C} = a. \tag{A1.16}$$

The next one is

$$\boldsymbol{A} \cdot (\boldsymbol{B} \times \boldsymbol{C}) = (A_x\boldsymbol{i}_x + A_y\boldsymbol{i}_y + A_z\boldsymbol{i}_z) \cdot \begin{vmatrix} \boldsymbol{i}_x & \boldsymbol{i}_y & \boldsymbol{i}_z \\ B_x & B_y & B_z \\ C_x & C_y & C_z \end{vmatrix} = \begin{vmatrix} A_x & A_y & A_z \\ B_x & B_y & B_z \\ C_x & C_y & C_z \end{vmatrix}. \tag{A1.17}$$

This represents the volume of a parallelepiped composed of $\boldsymbol{A}$, $\boldsymbol{B}$, and $\boldsymbol{C}$ (see Fig. A1.6). We can easily prove the following equation:

$$\boldsymbol{A} \cdot (\boldsymbol{B} \times \boldsymbol{C}) = \boldsymbol{B} \cdot (\boldsymbol{C} \times \boldsymbol{A}) = \boldsymbol{C} \cdot (\boldsymbol{A} \times \boldsymbol{B}). \tag{A1.18}$$

This product is called a **scalar triple product**.

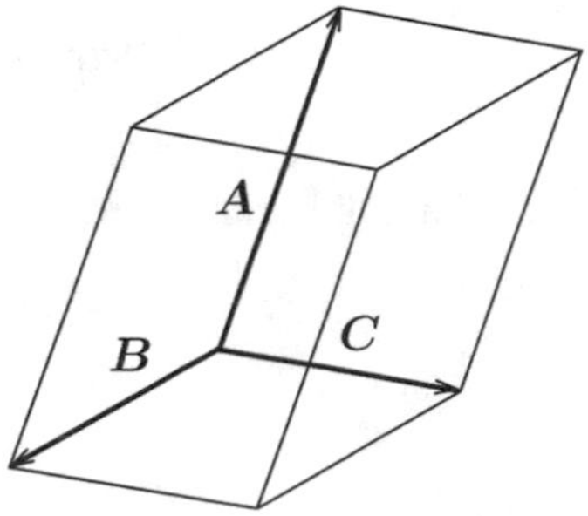

Fig. A1.6 Parallelepiped composed of $\boldsymbol{A}$, $\boldsymbol{B}$, and $\boldsymbol{C}$

The final meaningful product of three vectors is a **vector triple product**:

$$\boldsymbol{A} \times (\boldsymbol{B} \times \boldsymbol{C}) = (\boldsymbol{A} \cdot \boldsymbol{C})\boldsymbol{B} - (\boldsymbol{A} \cdot \boldsymbol{B})\boldsymbol{C}. \tag{A1.19}$$

A1.6 Differentiation of Vectors

When vector $\boldsymbol{A}$ changes spatially, we express it as a function of coordinates such as $\boldsymbol{A}(x, y, z)$. If position vector $\boldsymbol{r}$ corresponds to point (x, y, z), it is possible to write it as $\boldsymbol{A}(\boldsymbol{r})$. Suppose that only x changes by a small amount, Δx, with no change in y and z. The corresponding change in $\boldsymbol{A}$ is

$$\Delta \boldsymbol{A} = \boldsymbol{A}(x + \Delta x, y, z) - \boldsymbol{A}(x, y, z). \tag{A1.20}$$

If the limit of $\Delta \boldsymbol{A}/\Delta x$ exists in the limit $\Delta x \to 0$, this is called a **partial differential coefficient** and is written as

$$\frac{\partial \boldsymbol{A}}{\partial x}. \tag{A1.21}$$

If the components of $\boldsymbol{A}$ are (A_x, A_y, A_z), this coefficient is

$$\frac{\partial \boldsymbol{A}}{\partial x} = \frac{\partial A_x}{\partial x}\boldsymbol{i}_x + \frac{\partial A_y}{\partial x}\boldsymbol{i}_y + \frac{\partial A_z}{\partial x}\boldsymbol{i}_z. \tag{A1.22}$$

The partial differential coefficients with respect to y and z

$$\frac{\partial \boldsymbol{A}}{\partial y}, \quad \frac{\partial \boldsymbol{A}}{\partial z} \tag{A1.23}$$

are similarly defined.

When (x, y, z) changes to $(x + \Delta x, y + \Delta y, z + \Delta z)$ or $\boldsymbol{r}$ changes to $\boldsymbol{r} + \Delta \boldsymbol{r}$, the variation in $\boldsymbol{A}$ is

$$\begin{aligned} \Delta \boldsymbol{A} &= \boldsymbol{A}(x + \Delta x, y + \Delta y, z + \Delta z) - \boldsymbol{A}(x, y, z) \\ &\simeq \frac{\partial \boldsymbol{A}}{\partial x}\Delta x + \frac{\partial \boldsymbol{A}}{\partial y}\Delta y + \frac{\partial \boldsymbol{A}}{\partial z}\Delta z. \end{aligned} \tag{A1.24}$$

In the limit of small Δx, Δy, and Δz, these are written as $\mathrm{d}x$, $\mathrm{d}y$, and $\mathrm{d}z$, and then, $\Delta \boldsymbol{A}$ leads to

$$dA = \frac{\partial A}{\partial x}dx + \frac{\partial A}{\partial y}dy + \frac{\partial A}{\partial z}dz. \tag{A1.25}$$

This is called **total differentiation**. If $\boldsymbol{A}$ is a function not only of (x, y, z) but also of time t, the total differentiation of $\boldsymbol{A}$ also includes $(\partial \boldsymbol{A}/\partial t)dt$ on the right side of Eq. (A1.25).

If vectors $\boldsymbol{A}$ and $\boldsymbol{B}$ are functions of scalar φ, the following relations hold:

$$\frac{d}{d\varphi}(\boldsymbol{A}+\boldsymbol{B}) = \frac{d\boldsymbol{A}}{d\varphi} + \frac{d\boldsymbol{B}}{d\varphi}, \tag{A1.26}$$

$$\frac{d}{d\varphi}(m\boldsymbol{A}) = \frac{dm}{d\varphi}\boldsymbol{A} + m\frac{d\boldsymbol{A}}{d\varphi}, \tag{A1.27}$$

$$\frac{d}{d\varphi}(\boldsymbol{A}\cdot\boldsymbol{B}) = \frac{d\boldsymbol{A}}{d\varphi}\cdot\boldsymbol{B} + \boldsymbol{A}\cdot\frac{d\boldsymbol{B}}{d\varphi}, \tag{A1.28}$$

$$\frac{d}{d\varphi}(\boldsymbol{A}\times\boldsymbol{B}) = \frac{d\boldsymbol{A}}{d\varphi}\times\boldsymbol{B} + \boldsymbol{A}\times\frac{d\boldsymbol{B}}{d\varphi}. \tag{A1.29}$$

A1.7 Gradient of a Scalar

When $f(x, y, z)$ is a given scalar function, the following vector is called the **gradient** of f:

$$\frac{\partial f}{\partial x}\boldsymbol{i}_x + \frac{\partial f}{\partial y}\boldsymbol{i}_y + \frac{\partial f}{\partial z}\boldsymbol{i}_z. \tag{A1.30}$$

This is written as grad f, and the operation, grad, is also called the gradient. The function grad f is a vector that points in the direction of maximum variation with a magnitude equal to the maximum variation. If we use the operator defined by

$$\nabla = \boldsymbol{i}_x\frac{\partial}{\partial x} + \boldsymbol{i}_y\frac{\partial}{\partial y} + \boldsymbol{i}_z\frac{\partial}{\partial z}, \tag{A1.31}$$

Eq. (A1.30) is also written as

$$\mathrm{grad}f = \nabla f = \left(\boldsymbol{i}_x\frac{\partial}{\partial x} + \boldsymbol{i}_y\frac{\partial}{\partial y} + \boldsymbol{i}_z\frac{\partial}{\partial z}\right)f. \tag{A1.32}$$

The operator ∇ is called **nabla**.

If the unit vector along some direction is $\boldsymbol{s}$, the variation rate of a given function f in this direction is $\boldsymbol{s}\cdot\nabla f$. For example, the variation rate along the x-axis is

$$i_x \cdot \nabla f = \frac{\partial f}{\partial x}. \tag{A1.33}$$

Thus, the gradient is an operator that operates on a scalar to result in a vector. An example is the relation between temperature and heat flow. The temperature is a scalar, and the heat flows along the opposite direction of its gradient, i.e., from a position with a higher temperature to a position with a lower one. When the temperature and heat conductivity are T and K, the heat that flows across a unit area in unit time because of the temperature gradient is $-K\nabla T$.

A1.8 Divergence of a Vector

When $\boldsymbol{A}(x, y, z)$ is a given vector function, the following scalar is called the **divergence** of $\boldsymbol{A}$:

$$\frac{\partial A_x}{\partial x} + \frac{\partial A_y}{\partial y} + \frac{\partial A_z}{\partial z}. \tag{A1.34}$$

This is written as div$\boldsymbol{A}$, and the operation, div, is also called the divergence. Using the vector operator ∇, this is also written as $\nabla \cdot \boldsymbol{A}$. Namely,

$$\mathrm{div}\boldsymbol{A} = \nabla \cdot \boldsymbol{A} = \frac{\partial A_x}{\partial x} + \frac{\partial A_y}{\partial y} + \frac{\partial A_z}{\partial z}. \tag{A1.35}$$

The mathematical definition of divergence will be given later in Eq. (A1.67).

Thus, the divergence is an operator that operates on a vector to result in a scalar. For example, if the velocity of a fluid is $\boldsymbol{v}$, $\nabla \cdot \boldsymbol{v}$ is the volume of the fluid that comes out through a unit area in unit time.

A1.9 Curl of a Vector

When $\boldsymbol{A}(x, y, z)$ is a given vector function, the following vector is called the **curl** or **rotation** of $\boldsymbol{A}$:

$$\boldsymbol{i}_x\left(\frac{\partial A_z}{\partial y} - \frac{\partial A_y}{\partial z}\right) + \boldsymbol{i}_y\left(\frac{\partial A_x}{\partial z} - \frac{\partial A_z}{\partial x}\right) + \boldsymbol{i}_z\left(\frac{\partial A_y}{\partial x} - \frac{\partial A_x}{\partial y}\right). \tag{A1.36}$$

This is written as curl$\boldsymbol{A}$ or rot$\boldsymbol{A}$, and the operation, curl or rot, is also called curl or rotation. Using the vector operator ∇, this is also written as $\nabla \times \boldsymbol{A}$. We can also use a determinant to express Eq. (A1.36) as

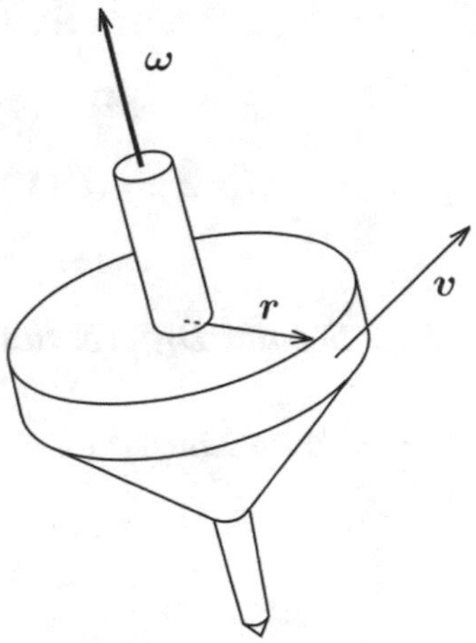

Fig. A1.7 Rotation of rigid body

$$\mathrm{curl}\boldsymbol{A} = \nabla \times \boldsymbol{A} = \begin{vmatrix} \boldsymbol{i}_x & \boldsymbol{i}_y & \boldsymbol{i}_z \\ \partial/\partial x & \partial/\partial y & \partial/\partial z \\ A_x & A_y & A_z \end{vmatrix}$$
$$= \boldsymbol{i}_x\left(\frac{\partial A_z}{\partial y} - \frac{\partial A_y}{\partial z}\right) + \boldsymbol{i}_y\left(\frac{\partial A_x}{\partial z} - \frac{\partial A_z}{\partial x}\right) + \boldsymbol{i}_z\left(\frac{\partial A_y}{\partial x} - \frac{\partial A_x}{\partial y}\right). \tag{A1.37}$$

The mathematical definition of curl will be given later in Eq. (A1.74).

Thus, the curl is an operator that operates on a vector to result in a vector. We consider a rotation of a rigid body. When the rigid body rotates with an angular velocity $\boldsymbol{\omega}$ around an axis through the center of gravity, as shown in Fig. A1.7, the velocity of a point located at $\boldsymbol{r}$ is $\boldsymbol{v} = \boldsymbol{\omega} \times \boldsymbol{r}$. Thus, we obtain $\boldsymbol{\omega} = (1/2)\nabla \times \boldsymbol{v}$.

A1.10 Differentiation of Products of Vectors

The following relations hold for various products:

$$\nabla(\phi\psi) = \phi\nabla\psi + \psi\nabla\phi, \tag{A1.38}$$

$$\nabla(\boldsymbol{A}\cdot\boldsymbol{B}) = (\boldsymbol{A}\cdot\nabla)\boldsymbol{B} + (\boldsymbol{B}\cdot\nabla)\boldsymbol{A} + \boldsymbol{A}\times(\nabla\times\boldsymbol{B}) + \boldsymbol{B}\times(\nabla\times\boldsymbol{A}), \tag{A1.39}$$

$$\nabla\cdot(\phi\boldsymbol{A}) = \phi\nabla\cdot\boldsymbol{A} + \nabla\phi\cdot\boldsymbol{A}, \tag{A1.40}$$

$$\nabla \cdot (\boldsymbol{A} \times \boldsymbol{B}) = \boldsymbol{B} \cdot (\nabla \times \boldsymbol{A}) - \boldsymbol{A} \cdot (\nabla \times \boldsymbol{B}), \tag{A1.41}$$

$$\nabla \times (\phi \boldsymbol{A}) = \phi \nabla \times \boldsymbol{A} - \boldsymbol{A} \times \nabla \phi, \tag{A1.42}$$

$$\nabla \times (\boldsymbol{A} \times \boldsymbol{B}) = (\boldsymbol{B} \cdot \nabla)\boldsymbol{A} - (\boldsymbol{A} \cdot \nabla)\boldsymbol{B} + \boldsymbol{A}\nabla \cdot \boldsymbol{B} - \boldsymbol{B}\nabla \cdot \boldsymbol{A}. \tag{A1.43}$$

A1.11 Second Differentiation

There are three formulae for second differentiation. One of them is for an arbitrary vector $\boldsymbol{A}$:

$$\mathrm{div}(\mathrm{curl}\boldsymbol{A}) = \nabla \cdot (\nabla \times \boldsymbol{A}) = 0. \tag{A1.44}$$

The second one is for an arbitrary scalar ϕ:

$$\mathrm{curl}(\mathrm{grad}\phi) = \nabla \times \nabla \phi = 0. \tag{A1.45}$$

The last one is

$$\mathrm{curl}(\mathrm{curl}\boldsymbol{A}) = \nabla \times (\nabla \times \boldsymbol{A}) = \nabla(\nabla \cdot \boldsymbol{A}) - \nabla^2 \boldsymbol{A}. \tag{A1.46}$$

In Cartesian coordinates, the second term is written as

$$\nabla^2 \boldsymbol{A} = \left(\frac{\partial^2}{\partial x^2} + \frac{\partial^2}{\partial y^2} + \frac{\partial^2}{\partial z^2} \right) \boldsymbol{A}. \tag{A1.47}$$

Formulae for cylindrical and polar coordinates are given in Sects. A1.17 and A1.18, respectively. It should be noted that the operator ∇^2 is different between scalars and vectors:

$$\nabla^2 \phi = (\nabla \cdot \nabla)\phi = \nabla \cdot (\nabla \phi), \tag{A1.48}$$

$$\nabla^2 \boldsymbol{A} = (\nabla \cdot \nabla)\boldsymbol{A} \neq \nabla(\nabla \cdot \boldsymbol{A}). \tag{A1.49}$$

A1.12 Curvilinear Integral of a Vector

We denote the tangential component of a vector $\boldsymbol{F}(x, y, z)$ on a smooth curve, C, by $F_t(x, y, z)$ and the elementary line vector on C by $\mathrm{d}\boldsymbol{s}$ (see Fig. A1.8). The following integral is called a **curvilinear integral**:

Fig. A1.8 Curvilinear integral of vector $\boldsymbol{F}$ on curve C

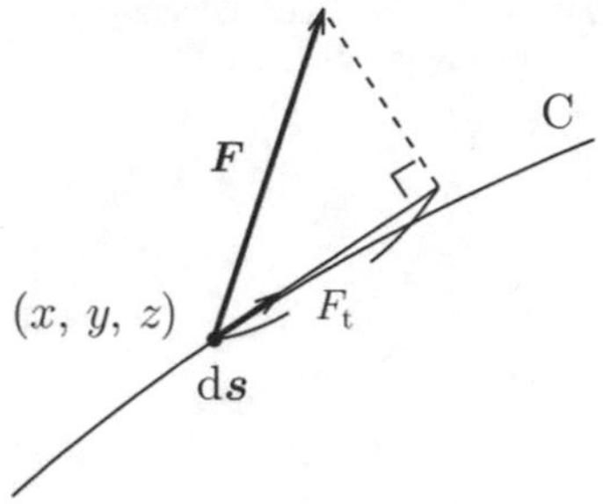

$$\int_{\mathrm{C}} \boldsymbol{F}(x,y,z)\cdot \mathrm{d}\boldsymbol{s} = \int_{\mathrm{C}} F_{\mathrm{t}}(x,y,z)\mathrm{d}s. \tag{A1.50}$$

If we divide this into components, it becomes

$$\int_{\mathrm{C}} \boldsymbol{F}\cdot \mathrm{d}\boldsymbol{s} = \int_{\mathrm{C}} F_x \mathrm{d}x + \int_{\mathrm{C}} F_y \mathrm{d}y + \int_{\mathrm{C}} F_z \mathrm{d}z. \tag{A1.51}$$

When $\boldsymbol{F}$ is a force on a matter particle, its curvilinear integral given by Eq. (A1.50) is the work to move the particle along C.

When $\boldsymbol{F} = \nabla\phi$, its curvilinear integral is

$$\int_{\mathrm{C}} \nabla\phi\cdot \mathrm{d}\boldsymbol{s} = \int_{\mathrm{C}} \frac{\partial\phi}{\partial x}\mathrm{d}x + \int_{\mathrm{C}} \frac{\partial\phi}{\partial y}\mathrm{d}y + \int_{\mathrm{C}} \frac{\partial\phi}{\partial z}\mathrm{d}z = \int_{\mathrm{C}} \mathrm{d}\phi, \tag{A1.52}$$

where $\mathrm{d}\phi$ is a total differential. Hence, the curvilinear integral of $\boldsymbol{F}$ from point P of position $\boldsymbol{r}_{\mathrm{P}}$ to point Q of position $\boldsymbol{r}_{\mathrm{Q}}$ along C leads to

$$\int_{\boldsymbol{r}_{\mathrm{P}}}^{\boldsymbol{r}_{\mathrm{Q}}} \nabla\phi\cdot \mathrm{d}\boldsymbol{s} = \int_{\boldsymbol{r}_{\mathrm{P}}}^{\boldsymbol{r}_{\mathrm{Q}}} \mathrm{d}\phi = \phi(\boldsymbol{r}_{\mathrm{Q}}) - \phi(\boldsymbol{r}_{\mathrm{P}}). \tag{A1.53}$$

Thus, we find that the inverse operation of the gradient is the curvilinear integral. It is obvious that the curvilinear integral of a gradient is determined only by the starting point P and terminating point Q and is independent of the integral path. If there are two integral paths C_1 and C_2 that connect P and Q, as shown in Fig. A1.9, the following relation holds:

$$\int_{\mathrm{C}_1} \nabla\phi\cdot \mathrm{d}\boldsymbol{s} = \int_{\mathrm{C}_2} \nabla\phi\cdot \mathrm{d}\boldsymbol{s} = -\int_{\mathrm{C}_2'} \nabla\phi\cdot \mathrm{d}\boldsymbol{s}, \tag{A1.54}$$

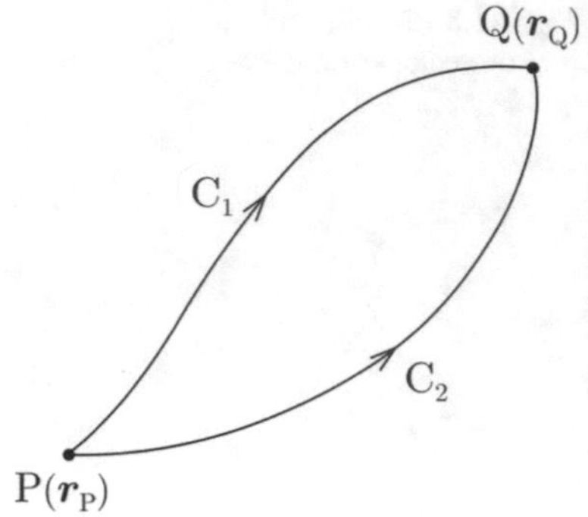

Fig. A1.9 Two integral paths connecting points P and Q

where C_2' is the integral path from Q to P along the opposite direction on path C_2. Hence, the circular integral on closed line C composed of C_1 and C_2' is

$$\oint_C \nabla\phi \cdot d\boldsymbol{s} = 0. \tag{A1.55}$$

This holds for an arbitrary closed line C.

A1.13 Surface Integral of a Vector

We denote the normal component of vector $\boldsymbol{F}(x, y, z)$ on a curved surface, S, by $F_n(x, y, z)$ and the elementary surface vector on S by $d\boldsymbol{S}$ (see Fig. A1.10). In this case, the following integral is called a **surface integral**:

$$\int_S \boldsymbol{F} \cdot d\boldsymbol{S} = \int_S F_n dS. \tag{A1.56}$$

When $\boldsymbol{F}$ is the velocity of a fluid and S is a cross section, the surface integral given by Eq. (A1.56) is the amount of fluid that flows through the cross section in unit time.

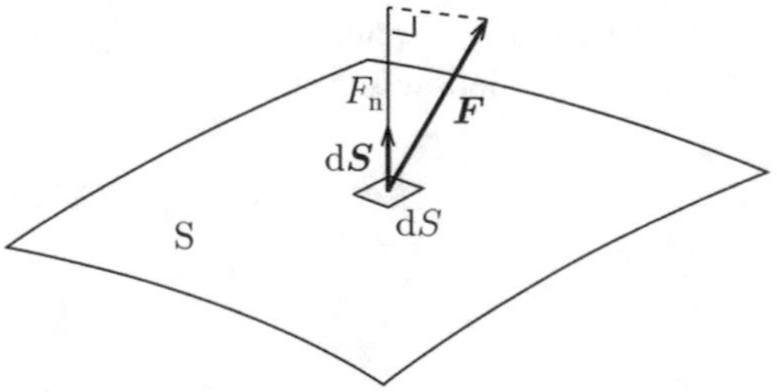

Fig. A1.10 Surface integral of vector $\boldsymbol{F}$ on curved surface S

A1.14 Gauss' Theorem

We suppose a small parallelepiped in the region x to $x+\Delta x$, y to $y+\Delta y$, and z to $z+\Delta z$ of volume $\Delta V = \Delta x \Delta y \Delta z$ and integrate vector $\boldsymbol{A}$ on the surface of this region:

$$\int_{\Delta S} \boldsymbol{A} \cdot \mathrm{d}\boldsymbol{S}, \tag{A1.57}$$

where $\mathrm{d}\boldsymbol{S}$ is directed outward from this region. This integral is divided into six surface integrals. First, we treat integrals on two surfaces parallel to the y-z plane at x and $x+\Delta x$ (see Fig. A1.11). On the surface at x, $\mathrm{d}\boldsymbol{S} = -\boldsymbol{i}_x \mathrm{d}S$, and the integral on this surface is

$$-\int A_x(x, y, z)\mathrm{d}S \simeq -A_x\left(x,\ y+\frac{\Delta y}{2},\ z+\frac{\Delta z}{2}\right)\Delta y \Delta z, \tag{A1.58}$$

where we have used the mean value of A_x on this surface. On the surface at $x+\Delta x$, $\mathrm{d}\boldsymbol{S} = \boldsymbol{i}_x \mathrm{d}S$, and the integral on this surface is

$$\int A_x(x+\Delta x,\ y,\ z)\mathrm{d}S \simeq A_x\left(x+\Delta x,\ y+\frac{\Delta y}{2},\ z+\frac{\Delta z}{2}\right)\Delta y \Delta z. \tag{A1.59}$$

We expand this as

$$A_x\left(x+\Delta x,\ y+\frac{\Delta y}{2},\ z+\frac{\Delta z}{2}\right) \simeq A_x\left(x,\ y+\frac{\Delta y}{2},\ z+\frac{\Delta z}{2}\right) + \frac{\partial A_x}{\partial x}\Delta x. \tag{A1.60}$$

Then, the sum of both surface integrals yields

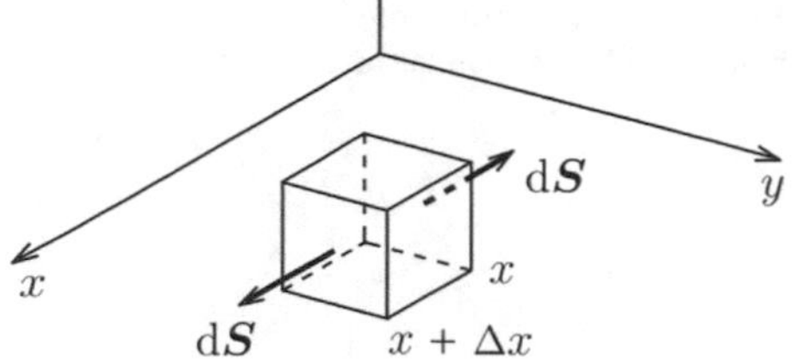

Fig. A1.11 Small parallelepiped with sides parallel to x-, y-, and z-axes

$$\left[A_x\left(x+\Delta x,\ y+\frac{\Delta y}{2},\ z+\frac{\Delta z}{2}\right) - A_x\left(x,\ y+\frac{\Delta y}{2},\ z+\frac{\Delta z}{2}\right)\right]\Delta y\Delta z = \frac{\partial A_x}{\partial x}\Delta x\Delta y\Delta z = \frac{\partial A_x}{\partial x}\Delta V. \tag{A1.61}$$

We similarly obtain the contributions from the sets of two surfaces parallel to the z-x and x-y planes as

$$\frac{\partial A_y}{\partial y}\Delta V,\quad \frac{\partial A_z}{\partial z}\Delta V. \tag{A1.62}$$

Thus, we have

$$\int_{\Delta S} \boldsymbol{A}\cdot \mathrm{d}\boldsymbol{S} = \left(\frac{\partial A_x}{\partial x}+\frac{\partial A_y}{\partial y}+\frac{\partial A_z}{\partial z}\right)\Delta V = \nabla\cdot\boldsymbol{A}\Delta V = \int_{\Delta V} \nabla\cdot\boldsymbol{A}\,\mathrm{d}V, \tag{A1.63}$$

where ΔV is the region surrounded by ΔS.

Here, we divide region V surrounded by closed surface S into a set of small regions $\{\Delta V_i\}$ and denote the surface of the i-th region ΔV_i as ΔS_i. The surface integral of $\boldsymbol{A}$ on S is

$$\int_S \boldsymbol{A}\cdot \mathrm{d}\boldsymbol{S} = \sum_i \int_{\Delta S_i} \boldsymbol{A}\cdot \mathrm{d}\boldsymbol{S}, \tag{A1.64}$$

since the surface integrals on a common surface between two adjacent regions cancel each other because $\mathrm{d}\boldsymbol{S}$ has an opposite direction in each integral (see Fig. A1.12). On the other hand, using Eq. (A1.63), this is equal to

$$\sum_i \int_{\Delta V_i} \nabla\cdot\boldsymbol{A}\,\mathrm{d}V = \int_V \nabla\cdot\boldsymbol{A}\,\mathrm{d}V. \tag{A1.65}$$

Thus, we have

$$\int_S \boldsymbol{A}\cdot \mathrm{d}\boldsymbol{S} = \int_V \nabla\cdot\boldsymbol{A}\mathrm{d}V. \tag{A1.66}$$

This is called **Gauss' theorem**.

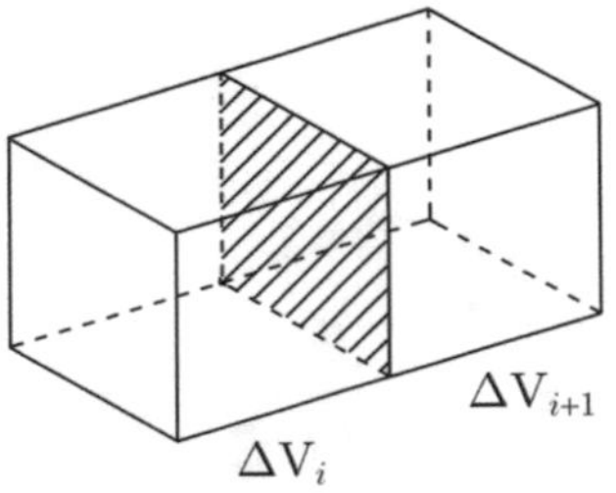

Fig. A1.12 Two adjacent regions with common surface

In the limit $\Delta V \to 0$ in Eq. (A1.63), the right side leads to $\Delta V \nabla \cdot \boldsymbol{A}$. Hence, we obtain the relationship

$$\lim_{\Delta V \to 0} \frac{1}{\Delta V} \int_{\Delta S} \boldsymbol{A} \cdot \mathrm{d}\boldsymbol{S} = \nabla \cdot \boldsymbol{A}. \tag{A1.67}$$

This gives the definition of divergence.

A1.15 Stokes' Theorem

We consider a small rectangle in the region y to $y + \Delta y$ and z to $z + \Delta z$ on the y-z plane. We integrate vector $\boldsymbol{A}$ along this small rectangle. The integral path is directed counterclockwise, as shown in Fig. A1.13, so that a screw points in the direction of the x-axis when we rotate a screwdriver along the direction of the integral. On the path from P to Q, $\mathrm{d}\boldsymbol{s} = -\boldsymbol{i}_y \mathrm{d}y$, and the contribution from this region to the integral is $A_y(x, y + (1/2)\Delta y, z)\Delta y$ using the mean value. On the path from R to S, $\mathrm{d}\boldsymbol{s} = -\boldsymbol{i}_y \mathrm{d}y$, and the contribution from this region is $-A_y(x, y + (1/2)\Delta y, z + \Delta z)\Delta y$. Their sum is

$$\begin{aligned} &-\left[A_y\left(x,\ y + \frac{\Delta y}{2},\ z + \Delta z\right) - A_y\left(x,\ y + \frac{\Delta y}{2},\ z\right)\right]\Delta y \\ &\simeq -\frac{\partial A_y}{\partial z}\Delta y \Delta z = -\frac{\partial A_y}{\partial z}\Delta S, \end{aligned} \tag{A1.68}$$

where $\Delta S = \Delta y \Delta z$ is the area of the small rectangle.

On the path from Q to R, $\mathrm{d}\boldsymbol{s} = \boldsymbol{i}_z \mathrm{d}z$ with a contribution of $A_z(x, y + \Delta y, z + (1/2)\Delta z)\Delta z$, and on the path from S to P, $\mathrm{d}\boldsymbol{s} = -\boldsymbol{i}_z \mathrm{d}z$ with a contribution of $-A_z(x, y, z + (1/2)\Delta z)\Delta z$. Their sum is

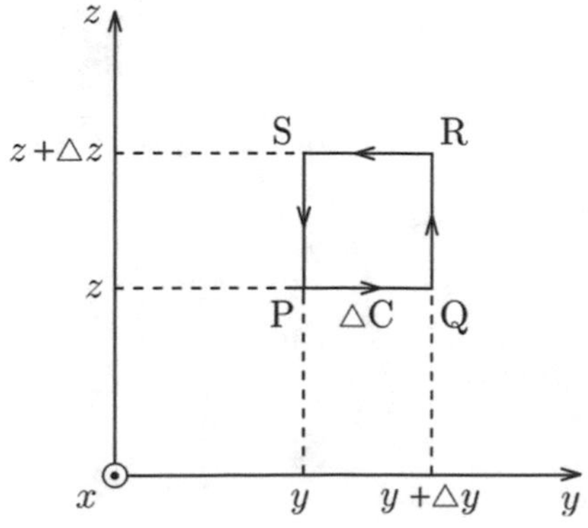

Fig. A1.13 Small rectangle with sides parallel to y- and z-axes

$$\left[A_z\left(x,\ y+\Delta y,\ z+\frac{\Delta z}{2}\right)-A_z\left(x,\ y,\ z+\frac{\Delta z}{2}\right)\right]\Delta z\simeq\frac{\partial A_z}{\partial y}\Delta S. \tag{A1.69}$$

Thus, the curvilinear integral on ΔC is

$$\oint\limits_{\Delta C} \boldsymbol{A}\cdot \mathrm{d}\boldsymbol{s}=\left(\frac{\partial A_z}{\partial y}-\frac{\partial A_y}{\partial z}\right)\Delta S=(\nabla\times\boldsymbol{A})_x\Delta S=\int\limits_{\Delta S}(\nabla\times\boldsymbol{A})_x\mathrm{d}S_x. \tag{A1.70}$$

We consider a curvilinear integral of $\boldsymbol{A}$ on an arbitrary closed line C. The surface surrounded by C is divided into a set of infinitesimal rectangular regions normal to each of the x-, y-, and z-axes (see Fig. A1.14). We can show that the curvilinear integral on C is equal to the sum of the curvilinear integrals of all rectangular regions. That is,

$$\oint\limits_{C}\boldsymbol{A}\cdot\mathrm{d}\boldsymbol{s}=\sum_i\oint\limits_{\Delta C_i}\boldsymbol{A}\cdot\mathrm{d}\boldsymbol{s}. \tag{A1.71}$$

This is because every curvilinear integral on the common side of two adjacent regions cancels out. Using Eq. (A1.70), the right side of Eq. (A1.71) leads to

$$\sum_i\int\limits_{\Delta S_i}(\nabla\times\boldsymbol{A})\cdot\mathrm{d}\boldsymbol{S}=\int\limits_{S}(\nabla\times\boldsymbol{A})\cdot\mathrm{d}\boldsymbol{S}, \tag{A1.72}$$

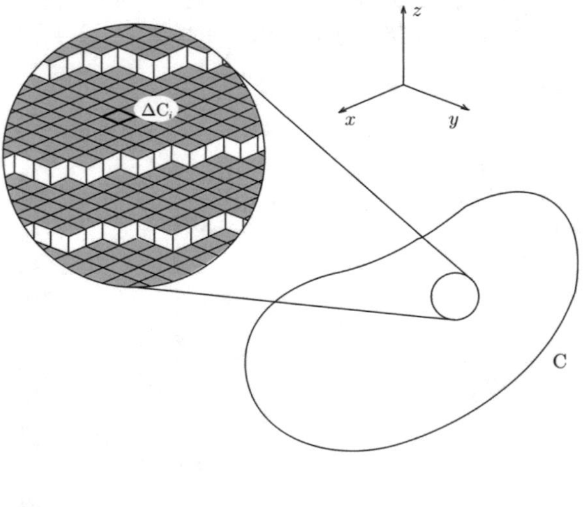

Fig. A1.14 Closed line C and small rectangular segments $\{\Delta C_i\}$

where ΔS_i is the surface surrounded by ΔC_i, and S is the curved surface surrounded by C. Thus, we have

$$\oint_{C} \boldsymbol{A} \cdot \mathrm{d}\boldsymbol{s} = \int_{S} (\nabla \times \boldsymbol{A}) \cdot \mathrm{d}\boldsymbol{S}. \tag{A1.73}$$

This is called **Stokes' theorem.**

In the limit $\Delta S \to 0$ in Eq. (A1.70), its right side leads to $(\nabla \times \boldsymbol{A})_x \Delta S$. Hence, if the unit vector normal to the small surface is $\boldsymbol{n}$, we have

$$\lim_{\Delta S \to 0} \frac{1}{\Delta S} \oint_{\Delta C} \boldsymbol{A} \cdot \mathrm{d}\boldsymbol{s} = (\nabla \times \boldsymbol{A}) \cdot \boldsymbol{n}. \tag{A1.74}$$

This gives the definition of curl.

A1.16 Green's Theorem

Substituting $\boldsymbol{A} = \nabla\psi$ into Eq. (A1.40) leads to

$$\nabla \cdot (\phi \nabla \psi) = \phi \nabla^2 \psi + (\nabla \phi) \cdot (\nabla \psi). \tag{A1.75}$$

Integrating this over region V and transforming the left side into a surface integral using Gauss' theorem, we have

$$\int_{S} \phi \nabla \psi \cdot \mathrm{d}\boldsymbol{S} = \int_{V} \phi \nabla^2 \psi \, \mathrm{d}V + \int_{V} (\nabla \phi) \cdot (\nabla \psi) \mathrm{d}V. \tag{A1.76}$$

Then, the subtraction between this and the quantity in which ϕ and ψ are exchanged yields

$$\int_{S} (\phi \nabla \psi - \psi \nabla \phi) \cdot \mathrm{d}\boldsymbol{S} = \int_{V} (\phi \nabla^2 \psi - \psi \nabla^2 \phi) \mathrm{d}V. \tag{A1.77}$$

Equations (A1.76) and (A1.77) are called **Green's theorem**.

A1.17 Cylindrical Coordinates

In Sect. A1.4, we used Cartesian coordinates. However, it is convenient to use **cylindrical coordinates** when we calculate electromagnetic properties for long cylindrical objects. When we use cylindrical coordinates, we first define the central axis (the z-axis). Then, we express the target position with the distance from this axis (R), the azimuthal angle (φ) on the plane normal to the z-axis, and the position on this axis (z): (R, φ, z). When we use the common z-axis with Cartesian coordinates, as shown in Fig. A1.15, the relationships between the two sets of coordinates are

$$R = (x^2 + y^2)^{1/2}, \quad \varphi = \tan^{-1}\frac{y}{x}, \quad z = z. \tag{A1.78}$$

If we use $\boldsymbol{i}_R$, $\boldsymbol{i}_\varphi$, and $\boldsymbol{i}_z$ to denote the unit vectors along the radial, azimuthal, and z-axial directions, respectively, these are perpendicular to each other and follow the right-hand rule in the order $\boldsymbol{i}_R \to \boldsymbol{i}_\varphi \to \boldsymbol{i}_z \to \boldsymbol{i}_R$.

The gradient, divergence, and curl in cylindrical coordinates are

$$\nabla f = \boldsymbol{i}_R \frac{\partial f}{\partial R} + \boldsymbol{i}_\varphi \frac{1}{R} \cdot \frac{\partial f}{\partial \varphi} + \boldsymbol{i}_z \frac{\partial f}{\partial z}, \tag{A1.79}$$

$$\nabla \cdot \boldsymbol{A} = \frac{1}{R} \cdot \frac{\partial (RA_R)}{\partial R} + \frac{1}{R} \cdot \frac{\partial A_\varphi}{\partial \varphi} + \frac{\partial A_z}{\partial z}, \tag{A1.80}$$

$$\nabla \times \boldsymbol{A} = \boldsymbol{i}_R \left(\frac{1}{R} \cdot \frac{\partial A_z}{\partial \varphi} - \frac{\partial A_\varphi}{\partial z} \right) + \boldsymbol{i}_\varphi \left(\frac{\partial A_R}{\partial z} - \frac{\partial A_z}{\partial R} \right) + \boldsymbol{i}_z \frac{1}{R} \left[\frac{\partial (RA_\varphi)}{\partial R} - \frac{\partial A_R}{\partial \varphi} \right]. \tag{A1.81}$$

The second differentiation of a scalar function is given by

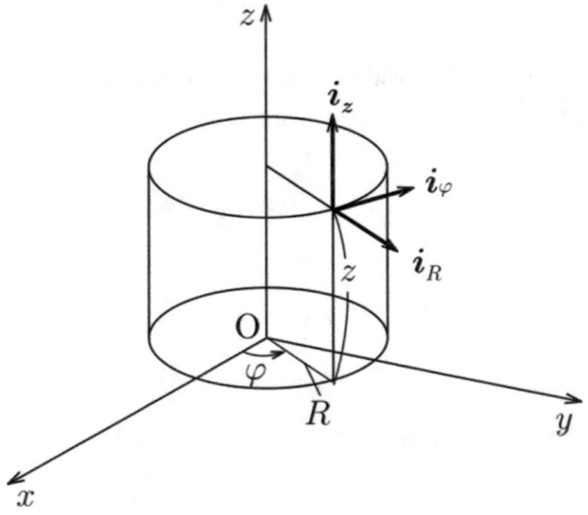

Fig. A1.15 Cylindrical coordinates

$$\nabla^2 f = \frac{1}{R} \cdot \frac{\partial}{\partial R}\left(R\frac{\partial f}{\partial R}\right) + \frac{1}{R^2} \cdot \frac{\partial^2 f}{\partial \varphi^2} + \frac{\partial^2 f}{\partial z^2}$$
$$= \frac{\partial^2 f}{\partial R^2} + \frac{1}{R} \cdot \frac{\partial f}{\partial R} + \frac{1}{R^2} \cdot \frac{\partial^2 f}{\partial \varphi^2} + \frac{\partial^2 f}{\partial z^2}. \quad \text{(A1.82)}$$

A1.18 Polar Coordinates

It is convenient to use **polar coordinates** when we calculate electromagnetic properties for spherical objects. When we use polar coordinates, we first define the center with an axis that determines the two poles. Then, we express the target position with the distance from the center (r), the zenithal angle (θ) measured from the north pole, and the azimuthal angle (φ) on the plane normal to the axis: (r, θ, φ). When we use the common center and z-axis with Cartesian coordinates, as shown in Fig. A1.16, the relationships between the two sets of coordinates are

$$r = (x^2 + y^2 + z^2)^{1/2}, \quad \theta = \tan^{-1}\frac{(x^2 + y^2)^{1/2}}{z}, \quad \varphi = \tan^{-1}\frac{y}{x}. \quad \text{(A1.83)}$$

If we use $\boldsymbol{i}_r$, $\boldsymbol{i}_\theta$, and $\boldsymbol{i}_\varphi$ to denote the unit vectors along the radial, zenithal, and azimuthal directions, respectively, these are perpendicular to each other and follow the right-hand rule in the order $\boldsymbol{i}_r \to \boldsymbol{i}_\theta \to \boldsymbol{i}_\varphi \to \boldsymbol{i}_r$.

The gradient, divergence, and curl in polar coordinates are

$$\nabla f = \boldsymbol{i}_r \frac{\partial f}{\partial r} + \boldsymbol{i}_\theta \frac{1}{r} \cdot \frac{\partial f}{\partial \theta} + \boldsymbol{i}_\varphi \frac{1}{r \sin\theta} \cdot \frac{\partial f}{\partial \varphi}, \quad \text{(A1.84)}$$

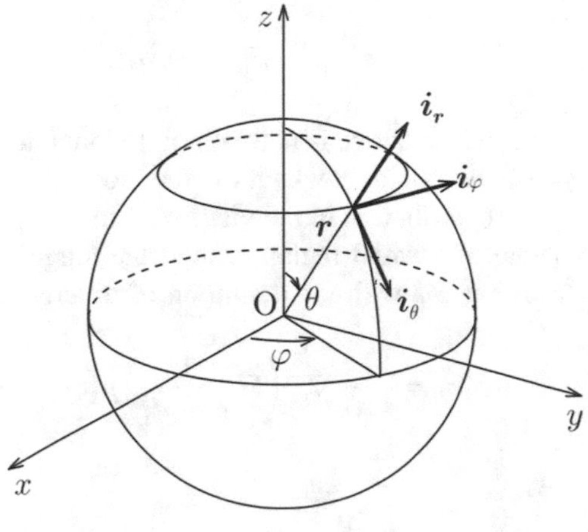

Fig. A1.16 Polar coordinates

$$\nabla\cdot\boldsymbol{A}=\frac{1}{r^2}\cdot\frac{\partial}{\partial r}(r^2A_r)+\frac{1}{r\sin\theta}\cdot\frac{\partial}{\partial\theta}(\sin\theta A_\theta)+\frac{1}{r\sin\theta}\cdot\frac{\partial A_\varphi}{\partial\varphi}, \tag{A1.85}$$

$$\begin{aligned}\nabla\times\boldsymbol{A}=\boldsymbol{i}_r\frac{1}{r\sin\theta}\left[\frac{\partial}{\partial\theta}(\sin\theta A_\varphi)-\frac{\partial A_\theta}{\partial\varphi}\right]+\boldsymbol{i}_\theta\frac{1}{r}\left[\frac{1}{\sin\theta}\cdot\frac{\partial A_r}{\partial\varphi}-\frac{\partial}{\partial r}(rA_\varphi)\right]\\+\boldsymbol{i}_\varphi\frac{1}{r}\left[\frac{\partial}{\partial r}(rA_\theta)-\frac{\partial A_r}{\partial\theta}\right].\end{aligned} \tag{A1.86}$$

The second differentiation of a scalar function is given by

$$\begin{aligned}\nabla^2 f&=\frac{1}{r^2}\cdot\frac{\partial}{\partial r}\left(r^2\frac{\partial f}{\partial r}\right)+\frac{1}{r^2\sin\theta}\cdot\frac{\partial}{\partial\theta}\left(\sin\theta\frac{\partial f}{\partial\theta}\right)+\frac{1}{r^2\sin^2\theta}\cdot\frac{\partial^2 f}{\partial\varphi^2}\\&=\frac{\partial^2 f}{\partial r^2}+\frac{2}{r}\cdot\frac{\partial f}{\partial r}+\frac{1}{r^2}\cdot\frac{\partial^2 f}{\partial\theta^2}+\frac{\cos\theta}{r^2\sin\theta}\cdot\frac{\partial f}{\partial\theta}+\frac{1}{r^2\sin^2\theta}\cdot\frac{\partial^2 f}{\partial\varphi^2}.\end{aligned} \tag{A1.87}$$

A2 Proofs

A2.1 Proof of Eq. (1.37)

For the electric potential given by Eq. (1.27), we have

$$\Delta\phi(\boldsymbol{r})=\frac{1}{4\pi\epsilon_0}\int_{\mathrm{V}}\left(\Delta\frac{1}{|\boldsymbol{r}-\boldsymbol{r}'|}\right)\rho(\boldsymbol{r}')\mathrm{d}V'.$$

Here, we put $\boldsymbol{R}=\boldsymbol{r}-\boldsymbol{r}'$ with $R=|\boldsymbol{R}|$ and define Δ_R as the Laplacian with respect to $\boldsymbol{R}$. Then, applying the formula might suggest

$$\Delta\frac{1}{|\boldsymbol{r}-\boldsymbol{r}'|}=\Delta_R\frac{1}{R}=\frac{1}{R^2}\cdot\frac{\partial}{\partial R}\left(R^2\frac{\partial}{\partial R}\cdot\frac{1}{R}\right)=0.$$

However, the above is not valid in the vicinity of $\boldsymbol{R}=0(\boldsymbol{r}=\boldsymbol{r}')$, since this position is a singular point at which the function $1/|\boldsymbol{r}-\boldsymbol{r}'|$ diverges. Hence, the integral has a finite contribution in the vicinity of this singular point. We denote the surface and volume of a small region around the singular point as ΔS and ΔV, respectively. Rewriting $\Delta\phi$ with the definition of divergence, Eq. (A1.67), we have

$$\Delta\phi(\boldsymbol{r})=\frac{1}{4\pi\epsilon_0}\nabla\cdot\left(\nabla\frac{1}{|\boldsymbol{r}-\boldsymbol{r}'|}\right)\rho(\boldsymbol{r})\,\Delta V=\lim_{\Delta V\to 0}\frac{\rho(\boldsymbol{r})}{4\pi\epsilon_0}\int_{\Delta\mathrm{S}}\nabla\frac{1}{|\boldsymbol{r}-\boldsymbol{r}'|}\cdot\mathrm{d}\boldsymbol{S}.$$

Here, we assume a spherical surface of radius a with the center at $\boldsymbol{r}'$ for small ΔS; then,

$$\int_{\Delta S} \nabla \frac{1}{|\boldsymbol{r}-\boldsymbol{r}'|} \cdot d\boldsymbol{S} = -\int_{\Delta S} \frac{\boldsymbol{r}-\boldsymbol{r}'}{|\boldsymbol{r}-\boldsymbol{r}'|^3} \cdot d\mathbf{S} = -\frac{4\pi a^2}{a^2} = -4\pi.$$

Thus, we obtain Eq. (1.37)

$$\Delta\phi(\boldsymbol{r}) = -\frac{\rho(\boldsymbol{r})}{\epsilon_0}.$$

We find that the Laplacian of $1/|\boldsymbol{r}-\boldsymbol{r}'|$ can be expressed in terms of the three-dimensional delta function as

$$\Delta \frac{1}{|\boldsymbol{r}-\boldsymbol{r}'|} = -4\pi\delta(\boldsymbol{r}-\boldsymbol{r}').$$

A2.2 Proof of Eq. (6.21)

Since the divergence is a differentiation with respect to $\boldsymbol{r}$ for the magnetic flux density in Eq. (6.7), we have

$$\nabla \cdot \boldsymbol{B}(\boldsymbol{r}) = \frac{\mu_0}{4\pi} \int_V \nabla \cdot \left[\frac{\boldsymbol{i}(\boldsymbol{r}') \times (\boldsymbol{r}-\boldsymbol{r}')}{|\boldsymbol{r}-\boldsymbol{r}'|^3}\right] dV'.$$

Using Eq. (A1.41) with

$$\boldsymbol{A} \to \boldsymbol{i}(\boldsymbol{r}'), \quad \boldsymbol{B} \to \frac{\boldsymbol{r}-\boldsymbol{r}'}{|\boldsymbol{r}-\boldsymbol{r}'|^3},$$

the integrand in the above equation leads to

$$\nabla \cdot \left[\frac{\boldsymbol{i}(\boldsymbol{r}') \times (\boldsymbol{r}-\boldsymbol{r}')}{|\boldsymbol{r}-\boldsymbol{r}'|^3}\right] = -\boldsymbol{i}(\boldsymbol{r}') \cdot \left(\nabla \times \frac{\boldsymbol{r}-\boldsymbol{r}'}{|\boldsymbol{r}-\boldsymbol{r}'|^3}\right),$$

where we have used $\nabla \times \boldsymbol{i}(\boldsymbol{r}') = 0$. If we use

$$\frac{\boldsymbol{r}-\boldsymbol{r}'}{|\boldsymbol{r}-\boldsymbol{r}'|^3} = -\nabla \frac{1}{|\boldsymbol{r}-\boldsymbol{r}'|}$$

and Eq. (A1.45), we obtain Eq. (6.21)

$$\nabla \times \boldsymbol{B} = 0.$$

A2.3 Proof of Eq. (6.27)

The curl of Eq. (6.7) is

$$\nabla \times \boldsymbol{B}(\boldsymbol{r}) = \frac{\mu_0}{4\pi} \int_{\mathrm{V}} \nabla \times \left[\frac{\boldsymbol{i}(\boldsymbol{r}') \times (\boldsymbol{r} - \boldsymbol{r}')}{|\boldsymbol{r} - \boldsymbol{r}'|^3} \right] \mathrm{d}V'.$$

Substituting

$$\boldsymbol{A} \to \boldsymbol{i}(\boldsymbol{r}'), \quad \boldsymbol{B} \to \frac{\boldsymbol{r} - \boldsymbol{r}'}{|\boldsymbol{r} - \boldsymbol{r}'|^3} = -\nabla \frac{1}{|\boldsymbol{r} - \boldsymbol{r}'|}$$

into Eq. (A1.43) yields

$$\nabla \times \left[\frac{\boldsymbol{i}(\boldsymbol{r}') \times (\boldsymbol{r} - \boldsymbol{r}')}{|\boldsymbol{r} - \boldsymbol{r}'|^3} \right] = -\boldsymbol{i}(\boldsymbol{r}') \nabla \cdot \left(\nabla \frac{1}{|\boldsymbol{r} - \boldsymbol{r}'|} \right) + [\boldsymbol{i}(\boldsymbol{r}') \cdot \nabla] \nabla \frac{1}{|\boldsymbol{r} - \boldsymbol{r}'|}.$$

Here, we denote the differential operator with respect to $\boldsymbol{r}'$ by ∇'. Then,

$$\nabla \frac{1}{|\boldsymbol{r} - \boldsymbol{r}'|} = -\nabla' \frac{1}{|\boldsymbol{r} - \boldsymbol{r}'|}.$$

That is, ∇ is equivalent to $-\nabla'$. Using this relationship in part of the above equation, we have

$$\nabla \times \left[\frac{\boldsymbol{i}(\boldsymbol{r}') \times (\boldsymbol{r} - \boldsymbol{r}')}{|\boldsymbol{r} - \boldsymbol{r}'|^3} \right] = -\boldsymbol{i}(\boldsymbol{r}') \nabla'^2 \frac{1}{|\boldsymbol{r} - \boldsymbol{r}'|} - [\boldsymbol{i}(\boldsymbol{r}') \cdot \nabla'] \nabla \frac{1}{|\boldsymbol{r} - \boldsymbol{r}'|}.$$

Here, we use the relationship shown in Sect. A2.1:

$$-\nabla'^2 \frac{1}{|\boldsymbol{r} - \boldsymbol{r}'|} = 4\pi\delta(\boldsymbol{r}' - \boldsymbol{r}).$$

Then, the integral of the first term on the right side leads to

$$-\int_{\mathrm{V}} \boldsymbol{i}(\boldsymbol{r}') \nabla'^2 \frac{1}{|\boldsymbol{r} - \boldsymbol{r}'|} \mathrm{d}V' = 4\pi \boldsymbol{i}(\boldsymbol{r}).$$

Changing the order of differentials in the second term, we have

$$-[\boldsymbol{i}(\boldsymbol{r}') \cdot \nabla'] \nabla \frac{1}{|\boldsymbol{r} - \boldsymbol{r}'|} = -\nabla [\boldsymbol{i}(\boldsymbol{r}') \cdot \nabla'] \frac{1}{|\boldsymbol{r} - \boldsymbol{r}'|} = -\nabla \left(\boldsymbol{i}(\boldsymbol{r}') \cdot \nabla' \frac{1}{|\boldsymbol{r} - \boldsymbol{r}'|} \right).$$

The scalar function operated on by ∇ is written as

$$\boldsymbol{i}(\boldsymbol{r}')\cdot\nabla'\frac{1}{|\boldsymbol{r}-\boldsymbol{r}'|}=\nabla'\cdot\frac{\boldsymbol{i}(\boldsymbol{r}')}{|\boldsymbol{r}-\boldsymbol{r}'|}-\frac{1}{|\boldsymbol{r}-\boldsymbol{r}'|}\nabla'\cdot\boldsymbol{i}(\boldsymbol{r}').$$

The condition of a steady current gives $\nabla'\cdot\boldsymbol{i}(\boldsymbol{r}')=0$. Hence, the integral of the second term becomes

$$-\nabla\int_{\mathrm{V}}\nabla'\cdot\frac{\boldsymbol{i}(\boldsymbol{r}')}{|\boldsymbol{r}-\boldsymbol{r}'|}\mathrm{d}V'=-\nabla\int_{\mathrm{S}}\frac{\boldsymbol{i}(\boldsymbol{r}')}{|\boldsymbol{r}-\boldsymbol{r}'|}\cdot\mathrm{d}\boldsymbol{S}'.$$

If we suppose an infinitely large sphere for S, $\boldsymbol{i}$ decreases to zero on S. Thus, the relationship $\nabla\times\boldsymbol{B}=\mu_0\boldsymbol{i}$, Eq. (6.27), is valid.

A2.4 Proof of Eq. (6.33)

The curl of Eq. (6.33) is

$$\nabla\times\boldsymbol{A}(\boldsymbol{r})=\frac{\mu_0}{4\pi}\int_{\mathrm{V}}\nabla\times\frac{\boldsymbol{i}(\boldsymbol{r}')}{|\boldsymbol{r}-\boldsymbol{r}'|}\mathrm{d}V'.$$

Noting that the curl is a derivative with respect to $\boldsymbol{r}$, Eq. (A1.42) leads to

$$\nabla\times\frac{\boldsymbol{i}(\boldsymbol{r}')}{|\boldsymbol{r}-\boldsymbol{r}'|}=-\boldsymbol{i}(\boldsymbol{r}')\times\nabla\frac{1}{|\boldsymbol{r}-\boldsymbol{r}'|}=\boldsymbol{i}(\boldsymbol{r}')\times\frac{\boldsymbol{r}-\boldsymbol{r}'}{|\boldsymbol{r}-\boldsymbol{r}'|^3}.$$

Thus, the above equation is written as

$$\nabla\times\boldsymbol{A}(\boldsymbol{r})=\frac{\mu_0}{4\pi}\int_{\mathrm{V}}\frac{\boldsymbol{i}(\boldsymbol{r}')\times(\boldsymbol{r}-\boldsymbol{r}')}{|\boldsymbol{r}-\boldsymbol{r}'|^3}\mathrm{d}V'.$$

Since this agrees with the right side of Eq. (6.7), Eq. (6.33) is valid.

A2.5 Proof of Eq. (6.46)

For Eq. (6.46) to hold, only the following equation has to be satisfied:

$$\nabla\times\left(\frac{\boldsymbol{m}\times\boldsymbol{r}}{r^3}\right)=-\nabla\left(\frac{\boldsymbol{m}\cdot\boldsymbol{r}}{r^3}\right).$$

Here, we put

$$a = \frac{r}{r^3} = -\nabla \frac{1}{r}.$$

Then, from Eq. (A1.43), we have

$$\nabla \times (m \times a) = (a \cdot \nabla)m - (m \cdot \nabla)a + m(\nabla \cdot a) - a(\nabla \cdot m).$$

The first and fourth differentiation terms of the constant vector m are zero. Substituting $a = (1/r^2)i_r$ leads to

$$\nabla \cdot a = \frac{1}{r^2} \cdot \frac{\partial}{\partial r}\left(r^2 \frac{1}{r^2}\right) = 0.$$

Thus, only the second term remains. On the other hand, Eq. (A1.39) gives

$$\nabla(m \cdot a) = (m \cdot \nabla)a + (a \cdot \nabla)m + m \times (\nabla \times a) + a \times (\nabla \times m).$$

The second and fourth differentiation terms of the constant vector m are zero. We can easily show that $\nabla \times a = 0$ in the third term from Eq. (A1.45). Thus, only the first term remains. As a result, the target equation is valid, and we prove Eq. (6.46).

A2.6 Proof of Eq. (9.5)

Using the relationship,

$$\frac{r - r'}{|r - r'|^3} = -\nabla \frac{1}{|r - r'|} = \nabla' \frac{1}{|r - r'|},$$

on the right side of Eq. (9.4), this equation is written as

$$A(r) = \frac{\mu_0}{4\pi} \int_V M(r') \times \nabla' \frac{1}{|r - r'|} \, dV'.$$

If we put $\phi = 1/|r - r'|$ in Eq. (A1.42), this leads to

$$M \times \nabla' \phi = \phi(\nabla' \times M) - \nabla' \times (\phi M).$$

Thus, Eq. (9.4) becomes

$$A(r) = \frac{\mu_0}{4\pi} \int_V \frac{\nabla' \times M(\mathbf{r}')}{|r - r'|} \, dV' - \frac{\mu_0}{4\pi} \int_V \nabla' \times \frac{M(r')}{|r - r'|} \, dV'.$$

Equation (A1.41) becomes

$$\nabla \cdot (\boldsymbol{a} \times \boldsymbol{b}) = \boldsymbol{b} \cdot (\nabla \times \boldsymbol{a}) - \boldsymbol{a} \cdot (\nabla \times \boldsymbol{b}) = \boldsymbol{b} \cdot (\nabla \times \boldsymbol{a})$$

for a constant vector $\boldsymbol{b}$. The volume integral on the left side leads to $\int_V \nabla \cdot (\boldsymbol{a} \times \boldsymbol{b}) \mathrm{d}V = \int_S (\boldsymbol{a} \times \boldsymbol{b}) \cdot \boldsymbol{n} \mathrm{d}S$, where $\boldsymbol{n}$ is the unit vector normal to the surface denoted by S. Since vector $\boldsymbol{b}$ is constant, the above surface integral reduces to $\int_S \boldsymbol{b} \cdot (\boldsymbol{n} \times \boldsymbol{a}) \mathrm{d}S = \boldsymbol{b} \cdot \int_S \boldsymbol{n} \times \boldsymbol{a}\, \mathrm{d}S$. On the other hand, the volume integral on the right side leads to $\int_V \boldsymbol{b} \cdot (\nabla \times \boldsymbol{a}) \mathrm{d}V = \boldsymbol{b} \cdot \int_V \nabla \times \boldsymbol{a}\, \mathrm{d}V$. Since $\boldsymbol{b}$ is an arbitrary vector, we obtain the general relationship

$$\int_V \nabla \times \boldsymbol{a}\, \mathrm{d}V = \int_S \boldsymbol{n} \times \boldsymbol{a}\, \mathrm{d}S.$$

Using this relationship, the second integral of the vector potential leads to

$$-\frac{\mu_0}{4\pi} \int_S \boldsymbol{n}' \times \frac{\boldsymbol{M}(\boldsymbol{r}')}{|\boldsymbol{r} - \boldsymbol{r}'|} \mathrm{d}S'.$$

If we assume an infinitely large sphere for S ($|\boldsymbol{r}'| = r' \to \infty$), $|\boldsymbol{M}|$ and $|\boldsymbol{r} - \boldsymbol{r}'|^{-1}$ are of the orders of $1/r'^2$ and $1/r'$, and the surface integral is of the order of r'^2 in magnitude. As a result, this surface integral is of the order of $1/r'$, and we can neglect it. Thus, the vector potential is

$$\boldsymbol{A}(\boldsymbol{r}) = \frac{\mu_0}{4\pi} \int_V \frac{\nabla' \times \boldsymbol{M}(\boldsymbol{r}')}{|\boldsymbol{r} - \boldsymbol{r}'|} \mathrm{d}V',$$

and we derive Eq. (9.5).

A3 Superconductivity

A3.1 Phenomenological Electromagnetism

Here, we introduce the phenomenological London theory that describes the electromagnetic phenomenon associated with the Meissner-Ochsenfeld effect in superconductors. We denote the mass, electric charge, and velocity of a superconducting electron by m^*, $-e^*$, and $\boldsymbol{v}_s$, respectively. The theory assumes the equation of motion of the superconducting electron:

$$m^* \frac{\mathrm{d}\boldsymbol{v}_\mathrm{s}}{\mathrm{d}t} = -e^* \boldsymbol{E}.$$

The right side is the Coulomb force. The viscous force in Eq. (5.21) does not act on superconducting electrons. The above equation requires a superconducting current to flow without decaying in a steady state where there is no electric field. It is known that the superconducting electron is a pair of two electrons, thus $e^* = 2e$. If we denote the density of superconducting electrons by n_s, from Eq. (5.24), the superconducting current density is

$$\boldsymbol{i} = -e^* n_\mathrm{s} \boldsymbol{v}_\mathrm{s}.$$

Eliminating $\boldsymbol{v}_\mathrm{s}$ using this equation yields

$$\boldsymbol{E} = \frac{m^*}{n_\mathrm{s} e^{*2}} \cdot \frac{\mathrm{d}\boldsymbol{i}}{\mathrm{d}t}. \qquad \text{(A3.1)}$$

Taking a curl of this equation and using Eqs. (10.39) and (6.27) for $\nabla \times \boldsymbol{E}$ and $\boldsymbol{i}$, we have

$$\frac{\partial}{\partial t}\left(\boldsymbol{B} + \frac{m^*}{\mu_0 n_\mathrm{s} e^{*2}} \nabla \times \nabla \times \boldsymbol{B} \right) = 0.$$

Hence, we find that the quantity in the parentheses is a constant value. When it is zero, the Meissner-Ochsenfeld effect can be explained. Namely, it leads to

$$\boldsymbol{B} + \lambda^2 \nabla \times \nabla \times \boldsymbol{B} = 0, \qquad \text{(A3.2)}$$

where

$$\lambda = \left(\frac{m^*}{\mu_0 n_\mathrm{s} e^{*2}} \right)^{1/2}$$

is a quantity with the dimension of length called the **penetration depth** of the magnetic field. Equations (A3.1) and (A3.2) are called the **London equations**. Since $\nabla \cdot \boldsymbol{B} = 0$, using Eq. (A1.46) reduces Eq. (A3.2) to

$$\nabla^2 \boldsymbol{B} - \frac{1}{\lambda^2} \boldsymbol{B} = 0. \qquad \text{(A3.3)}$$

Here, we show that Eq. (A3.3) describes the Meissner-Ochsenfeld effect. Suppose a semi-infinitesuperconductor that occupies the region $x \geq 0$ with the surface at $x = 0$. We apply an external magnetic field of magnetic flux density B_0 parallel to the z-axis. In this case, it is reasonable to assume that the internal magnetic flux density has only the z-component and is uniform in the y-z plane. Thus, Eq. (A3.3) reduces to

$$\frac{d^2B}{dx^2} - \frac{1}{\lambda^2}B = 0.$$

From the boundary conditions that $B(0) = B_0$ and B should be finite at infinity ($x \to \infty$), we have

$$B(x) = B_0 \exp\left(-\frac{x}{\lambda}\right).$$

This shows that the magnetic flux penetrates from the surface to a depth of about λ (see Fig. A3.1). For this reason, λ is called the penetration depth. Usually λ takes on a value of the order of tens of nm and can be neglected in comparison with the specimen size, and hence, the Meissner-Ochsenfeld effect can be explained. From Eq. (6.27), a current of density

$$i(x) = \frac{B_0}{\mu_0 \lambda} \exp\left(-\frac{x}{\lambda}\right)$$

flows along the y-axis. That is, the diamagnetism in the superconductor is caused by the current flowing on the surface, and this current is called the **Meissner current**. If we regard this as a real surface current, the current that flows within a unit width along the z-axis is given by

$$\tau = \int_0^\infty i(x)dx = \frac{B_0}{\mu_0},$$

which satisfies Eq. (7.8).

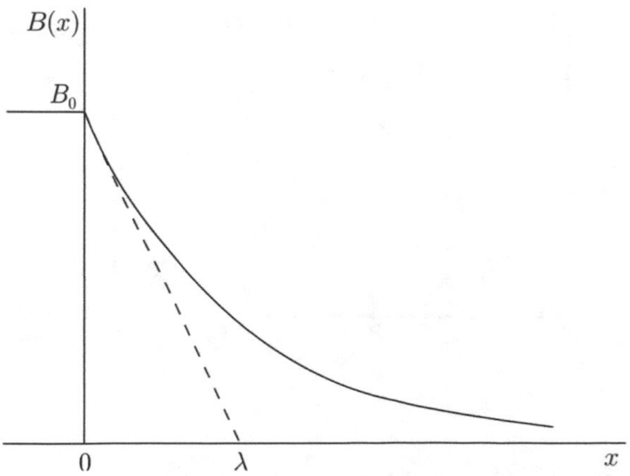

Fig. A3.1 Distribution of magnetic flux density in the vicinity of the superconductor surface

As was just stated, the London equation is assumed to explain the Meissner-Ochsenfeld effect. It is easy to show that Eqs. (A3.1) and (A3.2) are derived from the following equation:

$$\boldsymbol{i} = -\frac{n_s e^{*2}}{m^*}\boldsymbol{A}. \tag{A3.4}$$

Namely, we obtain Eq. (A3.1) by differentiating this equation with respect to time using the fact that there is no electrostatic field$(-\nabla\phi = 0)$ in the superconductor. The curl of Eq. (A3.4) directly derives Eq. (A3.2). Equation (A3.4) is derived from the rigorous Ginzburg-Landau theory, and hence, we are justified in assuming the London equations.

We can also derive the London equation, (A3.3), by minimizing a suitable energy. This is given by

$$\frac{1}{2\mu_0}\boldsymbol{B}^2 + \frac{\lambda^2}{2\mu_0}(\nabla \times \boldsymbol{B})^2. \tag{A3.5}$$

The first term is the magnetic energy, and the second term is the kinetic energy of superconducting electrons.

A3.2 Mixed State

There are two kinds of superconductors, i.e., type 1 and type 2 superconductors. When there is no geometrical effect as in the case of a long slab superconductor in a parallel magnetic flux density, the magnetizations of these superconductors are like those shown in Fig. A3.2a, b. Most superconducting elements are classified into type 1 in (a). In this case, when the external magnetic flux density B_0 is small, the superconductor is in the Meissner state, a perfect diamagnetic state. When B_0

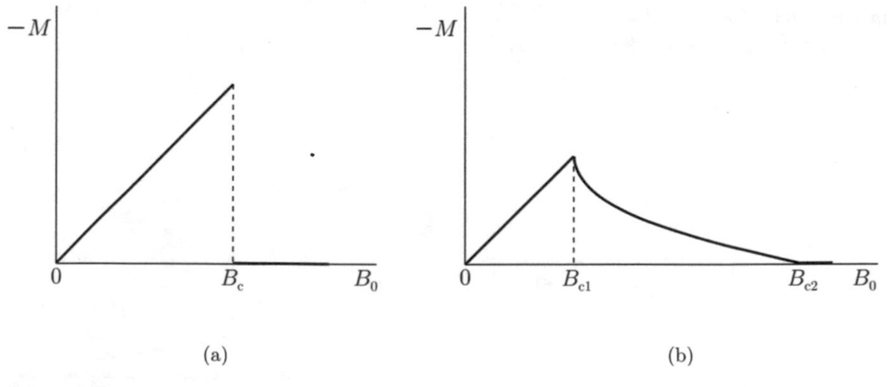

Fig. A3.2 Magnetization of **a** type 1 and **b** type 2 superconductors

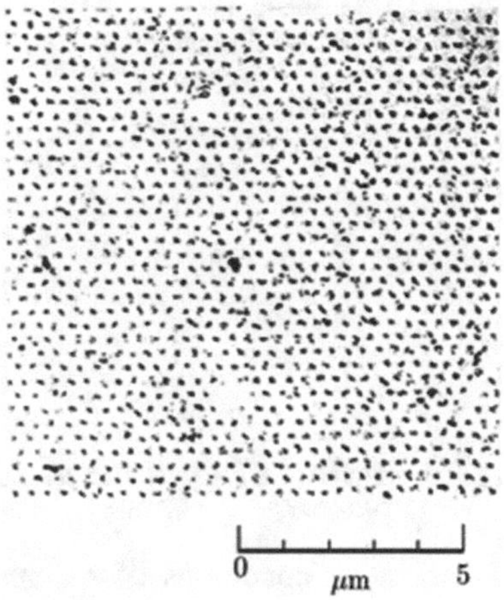

Fig. A3.3 Micrograph of quantized magnetic fluxes in superconducting Pb-Tl (courtesy of Dr. B. Obst at the Research Center in Karlsruhe). *Black dots* are ferromagnetic particles attached to the central part of each quantized magnetic flux

exceeds the critical value B_c, the superconductivity disappears with a jump in magnetization to zero, and the superconductor enters in the normal state with an electric resistivity. We call B_c the critical magnetic flux density, or $B_c/\mu_0 = H_c$ is called the critical magnetic field.

Alloy or compound superconductors, including practical superconductors, are classified into type 2 shown in Fig. A3.2b. When B_0 is below B_{c1}, the superconductor is in the Meissner state, but when B_0 exceeds B_{c1}, the superconductor enters an imperfectly diamagnetic state called the **mixed state** with a penetration of magnetic flux. When B_0 exceeds B_{c2}, the superconductor enters the normal state with zero magnetization. The quantities B_{c1} and B_{c2} are called the lower and upper critical magnetic flux density.

In the mixed state, the magnetic flux is quantized as shown in Fig. A3.3, and each has a magnetic quantum of

$$\phi_0 = \frac{h_P}{2e} = 2.0678 \times 10^{-15}\ \text{Wb},$$

where h_P is Planck's constant. The central part of each quantized magnetic flux is in the normal state, and the magnetic flux is concentrated in the region about λ from the center. Hence, the circular current flows stably around the center, and the quantized magnetic flux is also called a vortex.

The structure of each quantized magnetic flux is much smaller than the size of a superconductor specimen, and the internal magnetic flux density can be regarded as uniform, as schematically shown in Fig. A3.4. If this magnetic flux density is B, the magnetization of the superconductor is given by Eq. (7.38).

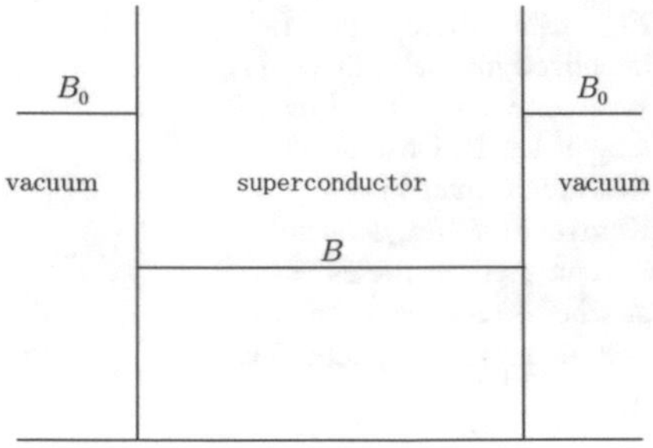

Fig. A3.4 Magnetic flux distribution in type 2 superconductor in the mixed state

A3.3 Motion of Quantized Magnetic Flux

In operating conditions of superconducting equipment, the superconductors are generally in the mixed state with penetration of quantized magnetic fluxes. Thus, the Lorentz force

$$\boldsymbol{F}' = \boldsymbol{i} \times \boldsymbol{B}$$

is exerted on quantized magnetic fluxes in a unit volume under the transport current. In this case, the current is not localized only near the surface but flows uniformly inside the superconductor. When quantized magnetic fluxes are driven to move with velocity $\boldsymbol{V}$ by the Lorentz force, the electric field given by Eq. (10.21) is induced:

$$\boldsymbol{E} = \boldsymbol{B} \times \boldsymbol{V}. \tag{A3.6}$$

This is Josephson's relation. Since the condition is steady without a change in the magnetic flux density with time, this induced electric field satisfies Eq. (1.28).

The relationship between the current density and electric field in the superconductor in the mixed state is similar to Ohm's law for a usual metal, as shown by the solid line in Fig. A3.5. That is, when the current density increases, the Lorentz force

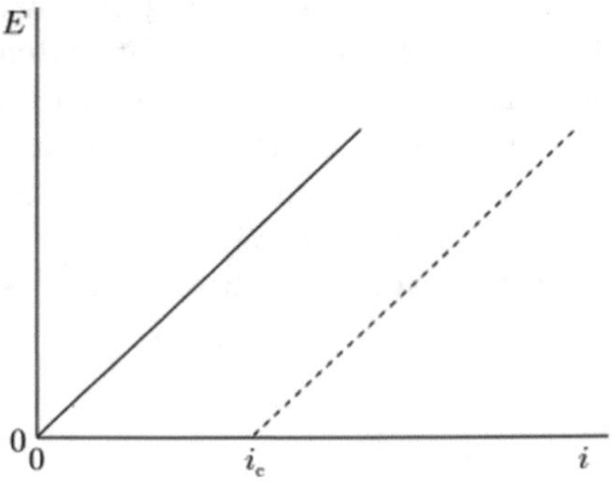

Fig. A3.5 Relationship between current density and induced electric field in superconductor. The *solid* and *dashed lines* show the characteristics for the superconductor without and with pinning centers

increases, and the velocity of quantized magnetic flux increases, resulting in an increase in the electric field. In such a resistive state, heat is generated because of the energy dissipation. This occurs because the central part of each quantized magnetic flux is in the normal state. The induced electric field drives normal electrons, resulting in energy dissipation similar to that in a normal metal. Thus, when the magnetic flux density increases, the number density of quantized magnetic flux also increases, and the electric resistance increases. When the magnetic flux density reaches B_{c2}, all the area in the superconductor reaches the normal state, and the electric resistance reaches the normal value.

To transport a current without appearance of electric resistance, the motion of quantized magnetic fluxes needs to stop ($\boldsymbol{V} = 0$) even under the Lorentz force. This action is called flux pinning, and defects such as normal precipitates or grain boundaries are known to act effectively. These defects are called pinning centers. Practical superconductors contain such pinning centers dispersed with a high concentration. The condition of the force equilibrium on quantized magnetic fluxes is given by

$$\boldsymbol{i} \times \boldsymbol{B} + \boldsymbol{F}_{\mathrm{p}} = 0, \tag{A3.7}$$

where $\boldsymbol{F}_{\mathrm{p}}$ is the pinning force density. The corresponding relationship between the current density and electric field under the influence of flux pinning is shown by the dashed line in Fig. A3.5. The current density i_{c} at which the electric field starts to appear is called the critical current density. In this condition, Eq. (A3.7) gives

$$i_{\mathrm{c}} = \frac{F_{\mathrm{p}}}{B}. \tag{A3.8}$$

To transport a current of high density without appearance of electric resistance, the strength of the pinning force needs to be enhanced.

A3.4 Electromagnetism and Superconductivity

Here, we carefully look at the fundamental factors that construct electromagnetism. The independent principles are as follows:

(a) The Coulomb force (with Coulomb's law),
(b) The Lorentz force (with the Biot-Savart law),
(c) The law of electromagnetic induction,
(d) The displacement current.

(a) gives Eq. (11.9), (b) gives Eq. (11.10) and a part of Eq. (11.8), (c) gives Eq. (11.7), and (d) gives a part of Eq. (11.8). Thus, the above four principles are arranged into Maxwell's equations. From the comprehensive Maxwell theory based on these equations, the Coulomb force and Lorentz force are derived in terms of the

Maxwell stress tensor. In fact, the Lorentz force is derived from a theoretical investigation of the energy in Exercise 11.11.

However, it should be noted that these principles are not enough to describe electromagnetic phenomena completely. That is, we need empirical Ohm's law for a system in which current flows. This law is not derived theoretically. Hence, electromagnetic theory is not complete in this sense.

Here, we discuss electromagnetic phenomena in superconductors. These phenomena are independent of Ohm's law, and the mechanism that determines the current is obtained by minimizing the free energy. Hence, we can say that electromagnetic theory is complete for superconductors including the case where current flows. In addition, if we include the cases of pressurization or films, more than half of the elements become superconducting at low temperatures, and most metallic compounds and some organic compounds are superconductors. That is, superconductors are fairly common substances. This textbook shows that superconductor has its own place in the ***E*-*B*** analogy. In principle, it was even possible to predict the existence of superconductors in the 19th Century.

Second, we discuss electromagnetic phenomena in a superconductor with pinning centers in the mixed state. In many cases, we can neglect the kinetic energy in Eq. (A3.5), and the suitable energy density to be minimized is

$$\frac{1}{2\mu_0}\boldsymbol{B}^2 + U_\mathrm{p}, \tag{A3.9}$$

where U_p is the pinning energy. Minimizing this energy with respect to the displacement of quantized magnetic fluxes leads to Eq. (A3.7) for an isolated superconductor.[(1)] That is, the variation in the magnetic energy due to the deformation of magnetic structure brings about the Lorentz force, and the variation in the pinning energy gives the pinning force density. We can extend this relationship to a non-isolated case and then to a general irreversible case. Thus, we obtain the force balance equation that describes practical electromagnetic phenomena in superconductors. The Lorentz force given by the first term in Eq. (A3.7) is transformed to

$$\boldsymbol{i} \times \boldsymbol{B} = \frac{1}{\mu_0}\left[(\boldsymbol{B}\cdot\nabla)\boldsymbol{B} - \frac{1}{2}\nabla\boldsymbol{B}^2\right]. \tag{A3.10}$$

Each term on the right side is expressed as an elastic restoring force against the distortion of quantized magnetic fluxes. The first term gives the line tension for the bent magnetic fluxes in Fig. A3.6a, and the second term gives the magnetic pressure to make the magnetic flux density uniform in Fig. A3.6b. The Lorentz force derived in Exercise 11.11 is the magnetic pressure.

It is known that various peculiar electromagnetic phenomena called longitudinal magnetic field effects are observed when we apply a current to a long superconducting wire or slab in a parallel magnetic field.[(2)] In this condition, the current and magnetic flux density are parallel to each other, and the Lorentz force on the quantized magnetic flux is zero ($\boldsymbol{i} \times \boldsymbol{B} = 0$). This state is called the force-free state.

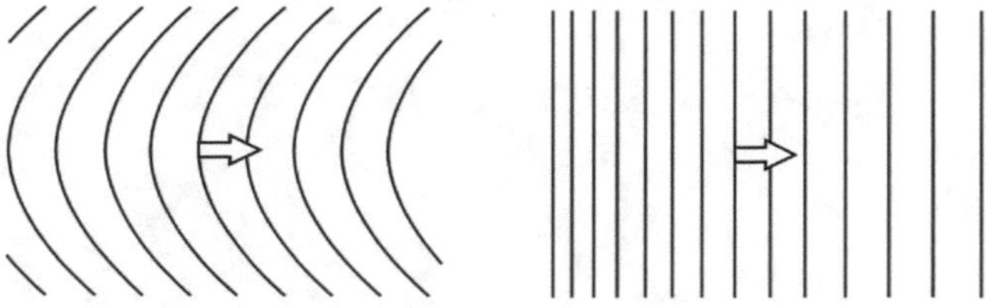

Fig. A3.6 Distortions of quantized magnetic fluxes: **a** bending of magnetic fluxes and **b** gradient of magnetic flux density. The Lorentz forces shown by the *arrows* to reduce the distortion are the line tension and magnetic pressure, respectively

Fig. A3.7 Distortion of magnetic flux lines in the force-free state. Current flows parallel to the magnetic flux lines. Restoring torque is predicted to work on the flux lines as shown by the *arrows*

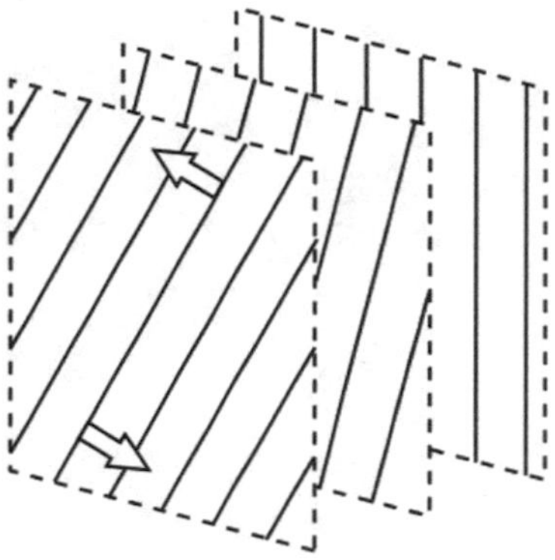

However, the magnetic structure contains a twisted distortion produced by the current, as shown in Fig. A3.7. We can expect that some restoring torque works to reduce the distortion, as shown by the arrows in the figure. In fact, we can derive the **force-free torque** using a similar method in Exercise 11.11.[(3)] We can explain that the rotational motion of quantized magnetic flux driven by the restoring torque causes the peculiar electromagnetic phenomena of the longitudinal magnetic field effects[(2)].

Such a torque in a static condition is not known in electromagnetism. We can easily show that $\nabla \times \boldsymbol{J} \neq 0$ when a current flows as in Fig. A3.7. Hence, this situation cannot be realized in normal conductors (see Sect. 7.4). To say this in more detail, the helicity given by $\boldsymbol{J} \cdot \boldsymbol{B}$ or $\boldsymbol{A} \cdot \boldsymbol{B}$ is not zero in this condition.

In this textbook, we showed that a superconductor can be considered a general material in electromagnetism. Here, we showed that a superconductor is a more purely physical material described by a complete theory. We can even expect that superconductors will open the door to electromagnetic phenomena that people have never yet experienced.

Answers to Exercises

Chapter 1

1.1. We presume the electric charge in a small region between x and x + dx from point A, $\mathrm{d}Q = (Q/L)\mathrm{d}x$, to be a point charge. Then, the Coulomb force on point charge q by this charge is $\mathrm{d}F = qQ\mathrm{d}x/[4\pi\epsilon_0 L(L+d-x)^2]$, and all the forces from each position point in the same direction. Thus, the total force is

$$F = \int_0^L \frac{qQ\mathrm{d}x}{4\pi\epsilon_0 L(L+d-x)^2} = \frac{qQ}{4\pi\epsilon_0 d(L+d)}.$$

1.2. The electric field strength due to the electric charge λdy in the region y to y + dy from the lower edge of the bar is $\mathrm{d}E = \lambda\mathrm{d}y/[4\pi\epsilon_0(y^2+b^2)]$. We define angle θ as shown in Fig. B1.1. The x- and y-components of the electric field are $\mathrm{d}E\cos\theta$ and $-\ \mathrm{d}E\sin\theta$, respectively, and here we put $y = b\tan\theta$ with $\theta_a = \tan^{-1}(a/b)$. Thus, we obtain

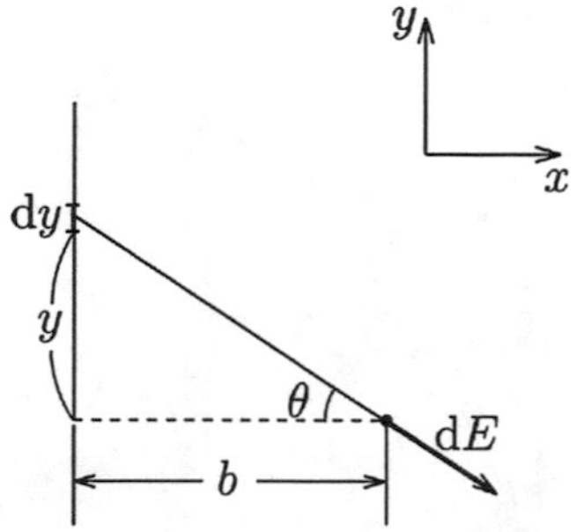

Fig. B1.1 Definition of angle θ

T. Matsushita, *Electricity and Magnetism*, Undergraduate Lecture Notes in Physics,
https://doi.org/10.1007/978-3-030-82150-0

$$E_x = \frac{\lambda}{4\pi\epsilon_0 b}\int_0^{\theta_a} \cos\theta \mathrm{d}\theta = \frac{\lambda a}{4\pi\epsilon_0 b(a^2+b^2)^{1/2}},$$

$$E_y = -\frac{\lambda}{4\pi\epsilon_0 b}\int_0^{\theta_a} \sin\theta \mathrm{d}\theta = -\frac{\lambda}{4\pi\epsilon_0 b}\left[1-\frac{b}{(a^2+b^2)^{1/2}}\right].$$

1.3. The distance from one side to point P is $r = [(a^2/4)+z^2]^{1/2}$, and the electric field strength due to the electric charge on one side is $E' = \lambda a/\{4\pi\epsilon_0 r [(a^2/4)+z^2]^{1/2}\}$. From symmetry, only the vertical component of the electric field remains (see Fig. B1.2), and we obtain the electric field by summing the contributions from the four sides as

$$E = 4E' \sin\beta = \frac{\lambda a z}{\pi\epsilon_0[(a^2/4)+z^2][(a^2/2)+z^2]^{1/2}}.$$

1.4. We define the coordinates as shown in Fig. B1.3. Although we cannot directly apply Gauss' law, it can be used to estimate the electric field produced by the line charge of density $\sigma \mathrm{d}x$ in a thin region between x and x + dx. The electric field strength at point A due to this line charge is $\mathrm{d}E'_{\mathrm{A}} = \sigma \mathrm{d}x/[2\pi\epsilon_0(x^2+b^2)^{1/2}]$. From symmetry, only the z-component, $\mathrm{d}E_{\mathrm{A}} = \mathrm{d}E'_{\mathrm{A}} \cos\theta$, remains. Integration yields the electric field at A:

$$E_{\mathrm{A}} = \int_{-a}^{a} \frac{\sigma\cos\theta \mathrm{d}x}{2\pi\epsilon_0(x^2+b^2)^{1/2}} = \frac{\sigma}{2\pi\epsilon_0}\int_{-\theta_a}^{\theta_a} \mathrm{d}\theta = \frac{\sigma\theta_a}{\pi\epsilon_0},$$

where we have transformed as $x = b\tan\theta$ with $\theta_a = \tan^{-1}(a/b)$.

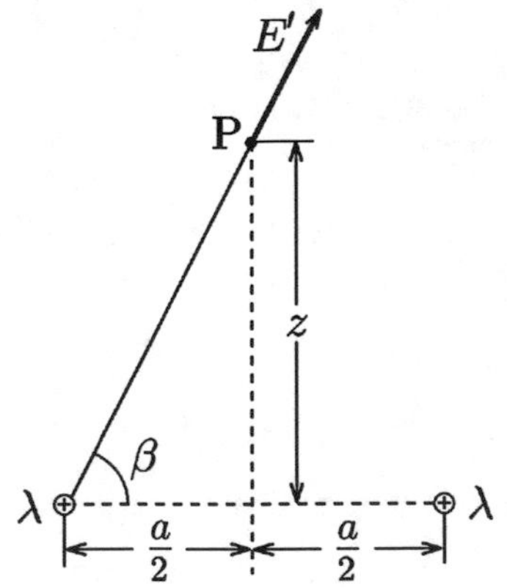

Fig. B1.2 Electric field produced by electric charge on one side

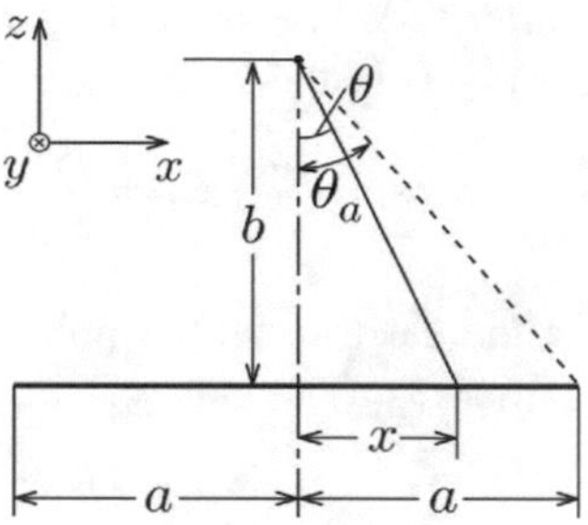

Fig. B1.3 Electric field produced by line charge in a part of slab

The electric field at point B produced by the line charge in a thin region between x and x + dx has a strength of $\mathrm{d}E_{\mathrm{B}} = \sigma \mathrm{d}x/[2\pi\epsilon_0(d-x)]$ and is directed along the x-axis. Thus, a simple summation yields

$$E_{\mathrm{B}} = \frac{\sigma}{2\pi\epsilon_0}\int_{-a}^{a}\frac{\mathrm{d}x}{d-x} = \frac{\sigma}{2\pi\epsilon_0}\log\frac{d+a}{d-a}.$$

1.5. Although we cannot directly obtain the electric field using Gauss' law, we can solve this problem using a superposition of two solvable cases. The given condition can be realized by superposing the situation where the electric charge is uniformly distributed with density ρ in the whole region of the sphere as shown in Fig. B1.4a, and that where the electric charge is uniformly distributed with density $-\rho$ in the vacant region as shown in Fig. B1.4b. We can calculate the electric field in each case using Gauss' law. First we determine the electric field at the center A of the vacancy. In (a), the total electric charge is $Q' = (4\pi/3)d^3\rho$, and the electric field is

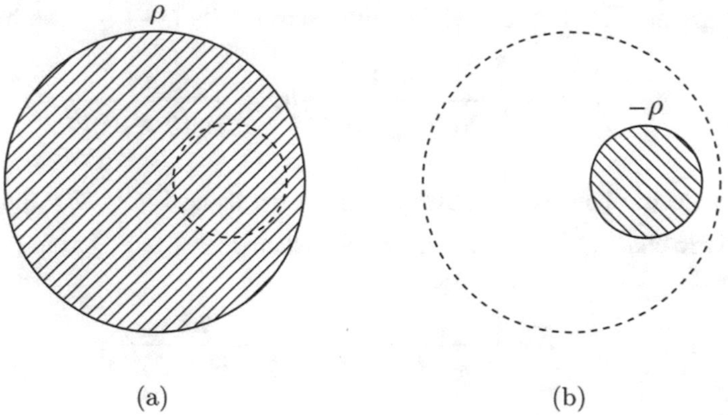

Fig. B1.4 Superposition of **a** uniformly distributed electric charge with density ρ over the whole sphere and **b** uniformly distributed electric charge with density $-\rho$ in the vacancy

$E_1 = Q'/(4\pi\epsilon_0 d^2) = \rho d/(3\epsilon_0)$. We similarly determine the electric field in (b) to be $E_2 = \rho r/(3\epsilon_0) \rightarrow 0$ in the limit $r \rightarrow 0$. Thus, we have

$$E_A = E_1 + E_2 = \frac{\rho d}{3\epsilon_0}.$$

Second, we determine the electric field at point B. The contributions from (a) and (b) are $E_3 = \rho a^3/(3\epsilon_0 r^2)$ and $E_4 = -\rho b^3/[3\epsilon_0 (r-d)^2]$, respectively. Thus, we have

$$E_B = E_3 + E_4 = \frac{\rho}{3\epsilon_0}\left[\frac{a^3}{r^2} - \frac{b^3}{(r-d)^2}\right].$$

1.6. From Eq. (1.25), we obtain the electric potential due to the electric charge $\lambda \mathrm{d}y$ in a small region between y and $y + \mathrm{d}y$ from origin O as $\mathrm{d}\phi = \lambda \mathrm{d}y/[4\pi\epsilon_0 (y^2+b^2)^{1/2}]$. Integrating this with respect to y from $-a$ to a yields

$$\phi = \int_{-a}^{a} \frac{\lambda \mathrm{d}y}{4\pi\epsilon_0 (y^2+b^2)^{1/2}} = \frac{\lambda}{2\pi\epsilon_0}\int_0^{\theta_a} \frac{\mathrm{d}\theta}{\cos\theta} = \frac{\lambda}{2\pi\epsilon_0}\log\frac{(a^2+b^2)^{1/2}+a}{b},$$

where we have transformed as $y = b\tan\theta$ with $\theta_a = \tan^{-1}(a/b)$.

1.7. We easily obtain the electric potential from Eq. (1.27) in this case. Since all the electric charge is located at distance a from the center, the electric potential is

$$\phi = \frac{1}{4\pi\epsilon_0 a}\int_V \rho(\boldsymbol{r}')\mathrm{d}V' = \frac{Q}{4\pi\epsilon_0 a} = \frac{\sigma a}{\epsilon_0},$$

where $Q = 4\pi a^2\sigma$ is the total electric charge. We can also calculate the electric potential from the electric field [see Eq. (2.9b)].

1.8. Integrating the electric field strength from infinity ($z \rightarrow \infty$), we have

$$\phi = -\int_{\infty}^{z} \frac{az'\lambda}{2\epsilon_0 (z'^2+a^2)^{3/2}}\mathrm{d}z' = \frac{a\lambda}{2\epsilon_0 (z^2+a^2)^{1/2}}.$$

The electric potential can also be calculated using Eq. (1.27). Since all the electric charge $Q = 2\pi a\lambda$ at the same distance $(z^2+a^2)^{1/2}$ from point P, we have

$$\phi = \frac{Q}{4\pi\epsilon_0 (z^2+a^2)^{1/2}} = \frac{a\lambda}{2\epsilon_0 (z^2+a^2)^{1/2}}.$$

1.9. First, we determine the electric field strength produced at P by the charge distributed in a thin circle R to $R+\mathrm{d}R$. Since $\sigma \mathrm{d}R$ corresponds to the line density λ in Example 1.3, the electric field strength is

$$\mathrm{d}E = \frac{z\sigma R \mathrm{d}R}{2\epsilon_0 \left(z^2+R^2\right)^{3/2}}.$$

Hence, the total electric field strength is determined to be

$$E = \int_0^a \frac{z\sigma R}{2\epsilon_0 \left(z^2+R^2\right)^{3/2}} \mathrm{d}R = \frac{\sigma}{2\epsilon_0}\left[1 - \frac{z}{\left(z^2+a^2\right)^{1/2}}\right].$$

We can determine the electric potential by integrating the electric field strength from infinity ($z \to \infty$). The direct integration of each term from infinity leads to divergence, however. So, we integrate the electric field strength from the point at distance z_0 from the observation point:

$$\begin{aligned}\phi_0 &= -\frac{\sigma}{2\epsilon_0}\int_{z_0}^{z}\left[1 - \frac{z}{\left(z^2+a^2\right)^{1/2}}\right]\mathrm{d}z \\ &= \frac{\sigma}{2\epsilon_0}\left[\left(z^2+a^2\right)^{1/2} - z - \left(z_0^2+a^2\right)^{1/2} + z_0\right].\end{aligned}$$

When z_0 is very large, $\left({z_0}^2+a^2\right)^{1/2} \simeq z_0 + a^2/2z_0$. Hence, the electric potential is obtained to be

$$\phi = \frac{\sigma}{2\epsilon_0}\left[\left(z^2+a^2\right)^{1/2} - z - \lim_{z_0\to\infty}\frac{a^2}{2z_0}\right] = \frac{\sigma}{2\epsilon_0}\left[\left(z^2+a^2\right)^{1/2} - z\right].$$

The electric potential can also be determined using Eq. (1.27). The contribution from the charge in a thin circle R to $R+\mathrm{d}R$ is

$$\mathrm{d}\phi = \frac{R\sigma \mathrm{d}R}{2\epsilon_0 \left(z^2+R^2\right)^{1/2}}.$$

Integrating this for $0 \leq R \leq a$, we have

$$\phi = \int_0^a \frac{R\sigma \mathrm{d}R}{2\epsilon_0 \left(z^2+R^2\right)^{1/2}} = \frac{\sigma}{2\epsilon_0}\left[\left(z^2+a^2\right)^{1/2} - z\right].$$

1.10. We assume a closed parallelepiped S, one plane of which stays on the central plane, as shown in Fig. B1.5. The electric field strength has only a x-component (E_x), and its value must be zero at $x = 0$. Hence, the electric field is parallel to the surface on the four surfaces parallel to the x-axis, and the

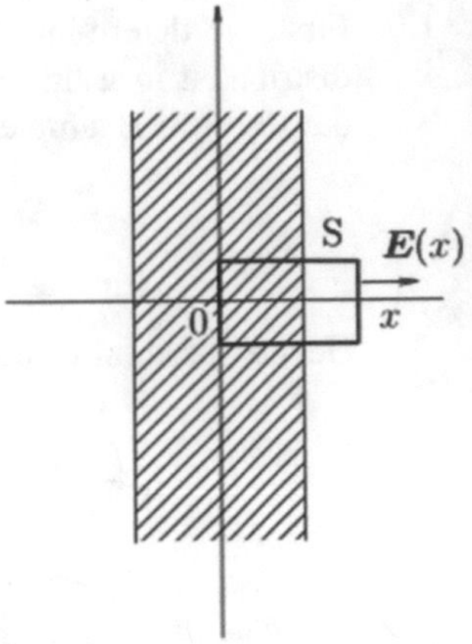

Fig. B1.5 Closed parallelepiped S for $x > a$ on which Gauss' law is applied

surface integral of the electric field strength on these surfaces is zero. The surface integral only has a nonzero value on the remaining surface. The position and the electric field on this surface are denoted by x and $E_x(x)$, respectively. Then, the surface integral of the electric field strength in Eq. (1.19) is $AE_x(x)$, where A is the area of the surface parallel to the y-z plane. Now, we estimate the total electric charge inside S. A simple calculation leads to $Ax\rho_0$ for $0 \leq x \leq a$ and $Aa\rho_0$ for $x > a$. From symmetry with respect to $x = 0$, we have

$$
\begin{aligned}
E_x &= -\frac{a\rho_0}{\epsilon_0}; && x < -a, \\
&= \frac{x\rho_0}{\epsilon_0}; && -a \leq x \leq a, \\
&= \frac{a\rho_0}{\epsilon_0}; && x > a,
\end{aligned}
$$

where we have used the symmetry condition with respect to $x = 0$.

When this result is substituted into Eq. (1.21), we have

$$
\begin{aligned}
\epsilon_0 \nabla \cdot \boldsymbol{E} &= \rho_0; && -a \leq x \leq a, \\
&= 0; && x < -a, x > a.
\end{aligned}
$$

Thus, the electric charge is distributed with density ρ_0 inside the slab, as assumed in the beginning.

1.11. We denote the distance from the sphere center by r. Then, the electric potential outside the sphere ($r > a$) is $\phi(r) = Q/(4\pi\epsilon_0 r)$. Thus, from Eq. (1.33), the work is determined to be

$$
W = q[\phi(r_{\mathrm{B}}) - \phi(r_{\mathrm{A}})] = \frac{qQ}{4\pi\epsilon_0}\left(\frac{1}{r_{\mathrm{B}}} - \frac{1}{r_{\mathrm{A}}}\right).
$$

1.12. The equipotential surface is expressed as

$$\frac{\cos\varphi}{R} = c,$$

where c is a constant. Using Cartesian coordinates, $x = R\cos\varphi$ and $y = R\sin\varphi$, so this equation leads to

$$\frac{x}{x^2 + y^2} = c.$$

Thus, we have

$$\left(x - \frac{1}{2c}\right)^2 + y^2 = \frac{1}{4c^2}.$$

The equipotential surface is a cylindrical surface that contains the central axis $(x = 0, y = 0)$.

Chapter 2

2.1. Electric charge Q_1 is distributed uniformly on the surface of the inner sphere $(r = a)$, and the electric charge $-Q_1$ induced by the electrostatic induction is distributed uniformly on the inner surface of the outer sphere $(r = b)$. Thus, the electric charge $Q_1 + Q_2$ appears on the outer surface of the outer sphere $(r = c)$, following the principle of conservation of charge. The electric field is directed radially, and its strength is

$$\begin{aligned} E_r &= 0; & 0 \le r < a, \\ &= \frac{Q_1}{4\pi\epsilon_0 r^2}; & a < r < b, \\ &= 0; & b < r < c, \\ &= \frac{Q_1 + Q_2}{4\pi\epsilon_0 r^2}; & r > c. \end{aligned}$$

The electric potential is determined to be

$$\begin{aligned} \phi &= \frac{Q_1 + Q_2}{4\pi\epsilon_0 r}; & r > c, \\ &= \frac{Q_1 + Q_2}{4\pi\epsilon_0 c}; & b < r < c, \\ &= \frac{Q_1}{4\pi\epsilon_0}\left(\frac{1}{r} - \frac{1}{b} + \frac{1}{c}\right) + \frac{Q_2}{4\pi\epsilon_0 c}; & a < r < b, \\ &= \frac{Q_1}{4\pi\epsilon_0}\left(\frac{1}{a} - \frac{1}{b} + \frac{1}{c}\right) + \frac{Q_2}{4\pi\epsilon_0 c}; & 0 \le r < a. \end{aligned}$$

2.2. We denote by Q_0 the electric charge induced on the surface of the inner conductor ($r = a$). Using Gauss' law, the electric charge on the inner surface of the outer sphere ($r = b$) is determined to be $-Q_0$. Hence, the electric charge on the outer surface ($r = c$) is $Q + Q_0$. If we define the electric potential to be zero at infinity, the electric potential of the outer sphere is

$$\phi = \frac{Q + Q_0}{4\pi\epsilon_0 c}.$$

On the other hand, the electric field is $E = Q_0/(4\pi\epsilon_0 r^2)$ in the region $a < r < b$, and the electric potential there is $\phi = Q_0/(4\pi\epsilon_0 r) + C$ with C denoting a constant. From the condition that $\phi = 0$ at $r = a$ because of grounding, we have $C = -Q_0/(4\pi\epsilon_0 a)$. Thus, the electric potential of the outer sphere is

$$\phi = -\frac{Q_0}{4\pi\epsilon_0}\left(\frac{1}{a} - \frac{1}{b}\right).$$

The requirement that this is equal to the electric potential determined from infinity yields

$$Q_0 = -\left(\frac{1}{a} - \frac{1}{b} + \frac{1}{c}\right)^{-1}\frac{Q}{c}.$$

2.3. We denote by Q_1 and Q_2 the electric charges on the surfaces at $x = -a$ and $x = b$, respectively. Then, the electric charges on the surfaces at $x = -b$ and $x = a$ are $Q - Q_1$ and $-Q_2$, respectively. So, the electric field in each region is shown in Table B2.1. That is, the electric field strengths in the left and right conductors are $(Q - 2Q_1)/2\epsilon_0$ and $(Q - 2Q_2)/2\epsilon_0$, respectively. Since these should be zero, we have $Q_1 = Q/2$ and $Q_2 = Q/2$. Then, the electric charge on each surface is

Table B2.1 Electric field strength in each region caused by electric charges

Position of charge	Electric field strength in each region caused by the charge on the surfaces				
	$x < -b$	$-b < x < -a$	$-a < x < a$	$a < x < b$	$x > b$
$x = -b$	$(Q_1 - Q)/2\epsilon_0$	$(Q - Q_1)/2\epsilon_0$	$(Q - Q_1)/2\epsilon_0$	$(Q - Q_1)/2\epsilon_0$	$(Q - Q_1)/2\epsilon_0$
$x = -a$	$-Q_1/2\epsilon_0$	$-Q_1/2\epsilon_0$	$Q_1/2\epsilon_0$	$Q_1/2\epsilon_0$	$Q_1/2\epsilon_0$
$x = a$	$Q_2/2\epsilon_0$	$Q_2/2\epsilon_0$	$Q_2/2\epsilon_0$	$-Q_2/2\epsilon_0$	$-Q_2/2\epsilon_0$
$x = b$	$-Q_2/2\epsilon_0$	$-Q_2/2\epsilon_0$	$-Q_2/2\epsilon_0$	$-Q_2/2\epsilon_0$	$Q_2/2\epsilon_0$
Sum	$-Q/2\epsilon_0$	$(Q - 2Q_1)/2\epsilon_0$	$Q/2\epsilon_0$	$(Q - 2Q_2)/2\epsilon_0$	$Q/2\epsilon_0$

$$\begin{array}{ll} Q/2; & x=-b, \\ Q/2; & x=-a, \\ -Q/2; & x=a, \\ Q/2; & x=b. \end{array}$$

The electric field strength along the x-axis is

$$\begin{aligned} E &= -Q/2\epsilon_0; \quad x<-b, \\ &= 0; \quad -b<x<-a, \\ &= Q/2\epsilon_0; \quad -a<x<a, \\ &= 0; \quad a<x<b, \\ &= Q/2\epsilon_0; \quad x>b. \end{aligned}$$

2.4. The reason why the electric field produced by the electric charge distributed on the conductor surface is doubled is that there are other electric field contributions from electric charges in other areas. For the same reason, the electric field inside the conductor cancels to zero. Examples are found in the case where an electric charge of different kind is distributed on the surface of the opposite electrode of a capacitor, as shown in Fig. B2.1a, or in the case where an electric charge of the same kind stays on the opposite surface of the conductor, as shown in Fig. B2.1b. The situation in Example 1.5 corresponds to the thin limit of the conductor in Fig. B2.1b.

2.5. We define two-dimensional polar coordinates (R, φ) on the conductor surface with the origin on the point at which the vertical line from point charge q meets the surface. We consider a thin ring of radius R to $R + \mathrm{d}R$ and presume the electric charge in a small part of the azimuthal angle φ to $\varphi + \mathrm{d}\varphi$, $\mathrm{d}Q = -qaR\mathrm{d}R\mathrm{d}\varphi/[2\pi(R^2 + a^2)^{3/2}]$, as a point electric charge. The Coulomb force on q caused by this point charge is $\mathrm{d}F = q\mathrm{d}Q/$

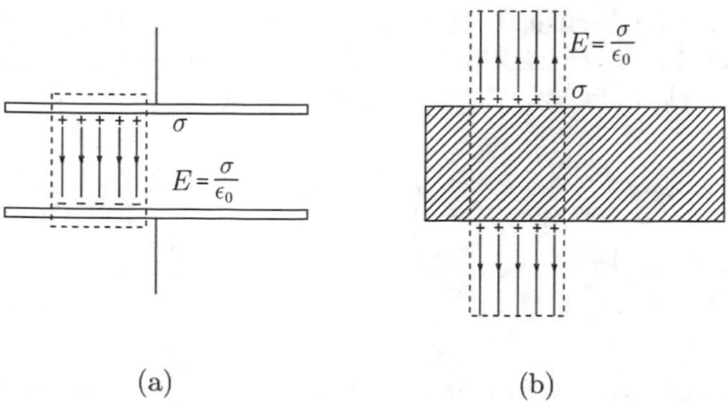

Fig. B2.1 Examples of doubled electric field strength: **a** parallel-plate capacitor and **b** distribution of electric charge of the same kind on the opposite surface of the conductor

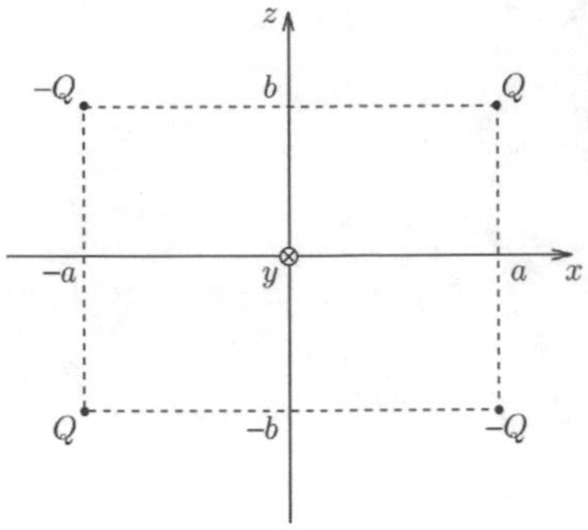

Fig. B2.2 True electric charge Q and three image charges

$[4\pi\epsilon_0(R^2+a^2)]$, and only its vertical component, $\mathrm{d}F_z = [a/(R^2+a^2)^{1/2}]\mathrm{d}F$, remains from symmetry. Integrating this over the surface, we have

$$F_z = -\frac{q^2a^2}{8\pi^2\epsilon_0}\int_0^\infty \frac{2\pi R\mathrm{d}R}{(R^2+a^2)^3} = -\frac{q^2}{16\pi\epsilon_0 a^2}.$$

This agrees with the image force, Eq. (2.15).

2.6. We denote two conductor surfaces that are perpendicular to each other by the x-y and y-z planes, as shown in Fig. B2.2. Assume the given electric charge is located on the plane $y = 0$. We virtually remove the conductor and place three electric charges, $-Q$, $-Q$ and Q, at $(a, 0, -b)$, $(-a, 0, b)$ and $(-a, 0, -b)$, respectively. Then, the electric potential in the vacuum region $(x > 0, z > 0)$ is

$$\phi(x,y,z) = \frac{Q}{4\pi\epsilon_0}\left\{\frac{1}{[(x-a)^2+y^2+(z-b)^2]^{1/2}} - \frac{1}{[(x-a)^2+y^2+(z+b)^2]^{1/2}} - \frac{1}{[(x+a)^2+y^2+(z-b)^2]^{1/2}} + \frac{1}{[(x+a)^2+y^2+(z+b)^2]^{1/2}}\right\}.$$

This satisfies $\phi = 0$ on the surfaces $x = 0$ and $z = 0$, and hence, this gives the correct electric potential. We determine the electric charge density on the x-y and y-z planes to be

$$\sigma(x,y,0) = -\epsilon_0\left(\frac{\partial\phi}{\partial z}\right)_{z=0} = -\frac{Qb}{2\pi}\left\{\frac{1}{[(x-a)^2+y^2+b^2]^{3/2}} - \frac{1}{[(x+a)^2+y^2+b^2]^{3/2}}\right\},$$

$$\sigma(0,y,z) = -\epsilon_0\left(\frac{\partial\phi}{\partial x}\right)_{x=0} = -\frac{Qa}{2\pi}\left\{\frac{1}{[y^2+(z-b)^2+a^2]^{3/2}} - \frac{1}{[y^2+(z+b)^2+a^2]^{3/2}}\right\}.$$

2.7. We virtually remove the conductor and put an image electric charge of linear density $-\lambda$ on the line at $z = -a$, which is symmetrical to the given linear charge with respect to the conductor surface, $z = 0$. Then, the electric potential in the vacuum region ($z > 0$) is

$$\phi = \frac{\lambda}{4\pi\epsilon_0}\log\frac{x^2+(z+a)^2}{x^2+(z-a)^2}.$$

This satisfies the condition, $\phi(z = 0) = 0$. Thus, the electric field is

$$E_x = -\frac{\partial\phi}{\partial x} = \frac{2a\lambda xz}{\pi\epsilon_0\left[x^2+(z-a)^2\right]\left[x^2+(z+a)^2\right]},$$

$$E_y = -\frac{\partial\phi}{\partial y} = 0,$$

$$E_z = -\frac{\partial\phi}{\partial z} = \frac{a\lambda(-x^2+z^2-a^2)}{\pi\epsilon_0\left[x^2+(z-a)^2\right]\left[x^2+(z+a)^2\right]}.$$

It is evident that the electric field is perpendicular to the surface, $E_x = E_y = 0$, on the conductor surface ($z = 0$). The density of the electric charge on the conductor surface is

$$\sigma = \epsilon_0 E_z(z=0) = -\frac{a\lambda}{\pi(x^2+a^2)}.$$

The total electric charge in a unit length along the y-axis on the conductor surface is

$$-\int_{-\infty}^{\infty}\frac{a\lambda \mathrm{d}x}{\pi(x^2+a^2)} = -\frac{\lambda}{\pi}\int_{-\pi/2}^{\pi/2} d\theta = -\lambda,$$

which is equal to the image charge.

From the obtained electric potential, the equipotential surface is given by

$$\frac{x^2+(z+a)^2}{x^2+(z-a)^2} = K,$$

with K denoting a constant. This is transformed to

$$x^2 + \left(z - \frac{K+1}{K-1}a\right)^2 = \frac{4a^2K}{(K-1)^2}.$$

Thus, the equipotential surface is a cylindrical surface parallel to the y-axis.

2.8. A simple calculation gives

$$E_r = -\frac{\partial \phi}{\partial r} = \frac{q}{4\pi\epsilon_0}\left\{\frac{r - d\cos\theta}{(r^2 + d^2 - 2rd\cos\theta)^{3/2}} - \frac{a}{d}\cdot\frac{r - (a^2/d)\cos\theta}{[r^2 + (a^2/d)^2 - (2a^2r/d)\cos\theta]^{3/2}}\right\},$$

$$E_\theta = -\frac{1}{r}\cdot\frac{\partial \phi}{\partial \theta} = \frac{q\sin\theta}{4\pi\epsilon_0}\left\{\frac{d}{(r^2 + d^2 - 2rd\cos\theta)^{3/2}} - \frac{a^3}{d^2}\cdot\frac{1}{[r^2 + (a^2/d)^2 - (2a^2r/d)\cos\theta]^{3/2}}\right\},$$

$$E_\varphi = -\frac{1}{r\sin\theta}\cdot\frac{\partial \phi}{\partial \varphi} = 0.$$

2.9. We virtually remove the conductor and place a line electric charge of density λ' at the line located at distance h from the center O. As shown in Examplc 1.9, the electric potential at point P on the surface of cylindrical conductor is

$$\phi = \frac{\lambda}{2\pi\epsilon_0}\log\frac{R_0}{(a^2 + d^2 - 2ad\cos\varphi)^{1/2}} + \frac{\lambda'}{2\pi\epsilon_0}\log\frac{R_0'}{(a^2 + h^2 - 2ah\cos\varphi)^{1/2}},$$

where φ is the angle defined in Fig. B2.3, and R_0 and R'_0 are distances from O to reference points of the electric potential. So that the electric potential does not depend on φ, the conditions $\lambda' = -\lambda$ and $h = a^2/d$ must be fulfilled. In addition, $R_0' = (a/d)R_0$ is requited so that the electric potential of the conductor is zero. Thus, the electric potential outside the conductor is

$$\phi(R, \varphi) = \frac{\lambda}{2\pi\epsilon_0}\log\frac{d[R^2 + (a^2/d)^2 - 2(a^2R/d)\cos\varphi]^{1/2}}{a(R^2 + d^2 - 2Rd\cos\varphi)^{1/2}}.$$

The electric charge density on the conductor surface is

$$\sigma = -\epsilon_0\left(\frac{\partial\phi(R, \varphi)}{\partial R}\right)_{R=a} = -\frac{\lambda(d^2 - a^2)}{2\pi a(a^2 + d^2 - 2ad\cos\varphi)}.$$

This gives the electric charge in a unit length;

$$\int_0^{2\pi} \sigma a\, \mathrm{d}\varphi = -\frac{\lambda(d^2 - a^2)}{\pi}\int_0^{\pi}\frac{\mathrm{d}\varphi}{a^2 + d^2 - 2ad\cos\varphi} = -\lambda,$$

where we have used Eq. (7.26).

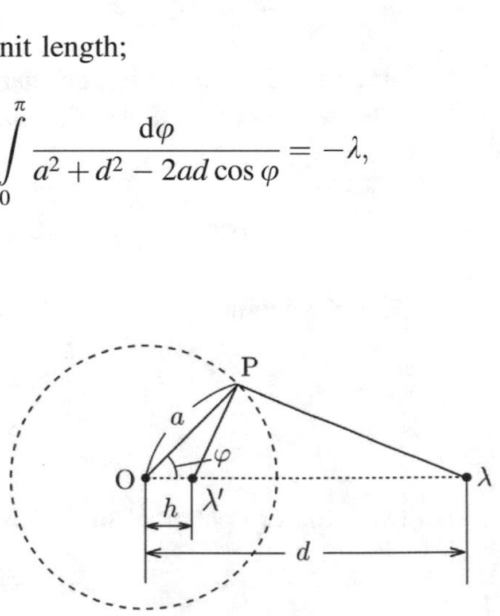

Fig. B2.3 Definition of angle φ

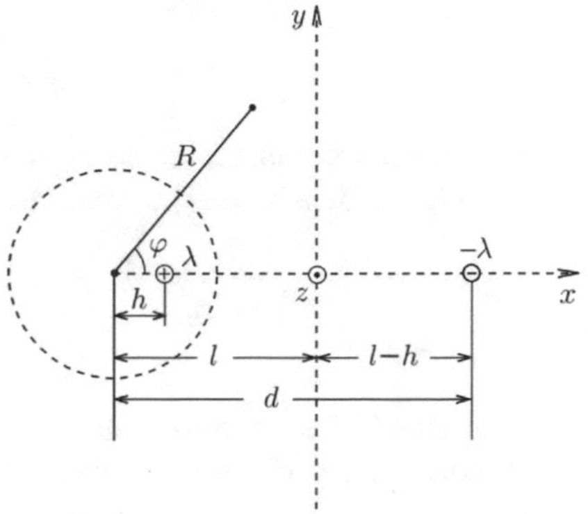

Fig. B2.4 Image line charges placed in two conductors

2.10. We assume that the electric field produced by the electric charge on the cylindrical conductor surface is the same as that produced by the line charge of density λ placed at distance h from the center of the cylinder after virtually removing the cylinder (see Fig. B2.4). If we place an image line charge of density $-\lambda$ in the infinite conductor at distance $l-h$ from its surface after virtually removing the infinite conductor, the infinite conductor surface is equipotential. Hence, if the distance $2l-h$ between the image charge $-\lambda$ and the cylinder center corresponds to d in Exercise 2.9, the cylindrical conductor surface is also equipotential, and all the required conditions are satisfied. The result in Exercise 2.9 gives $h = a^2/d$. From the above conditions, we have

$$d = l + \sqrt{l^2 - a^2}, \quad h = l - \sqrt{l^2 - a^2}.$$

Substituting these into the result in Exercise 2.9 yields the electric potential outside the conductors;

$$\phi(R, \varphi) = -\frac{\lambda}{4\pi\epsilon_0} \log \frac{R^2 + (l - \sqrt{l^2 - a^2})^2 - 2R(l - \sqrt{l^2 - a^2}) \cos\varphi}{R^2 + (l + \sqrt{l^2 - a^2})^2 - 2R(l + \sqrt{l^2 - a^2}) \cos\varphi}.$$

We find that the electric potential on the surface, $\phi(a, \varphi) = -[\lambda/(2\pi\epsilon_0)] \log [(l - \sqrt{l^2 - a^2})/a]$, is constant. The electric charge density on the cylindrical surface is

$$\sigma = -\epsilon_0 \left(\frac{\partial \phi}{\partial x}\right)_{R=a} = \frac{\lambda}{2\pi a} \cdot \frac{\sqrt{l^2 - a^2}}{l - a\cos\varphi}.$$

Next we define Cartesian coordinates with the y-z plane ($x = 0$) on the infinite conductor surface and the central axis of the cylindrical conductor at $y = 0$. From the relationships $R\cos\varphi = x + l$ and $R\sin\varphi = y$, the electric potential is also expressed as

$$\phi(x,y) = -\frac{\lambda}{4\pi\epsilon_0}\log\frac{(x+\sqrt{l^2-a^2})^2+y^2}{(x-\sqrt{l^2-a^2})^2+y^2}.$$

Thus, we can easily confirm that $\phi(x=0)=0$ is satisfied. The electric charge density on the infinite conductor surface is

$$\sigma = \epsilon_0\left(\frac{\partial\phi}{\partial x}\right)_{x=0} = -\frac{\lambda\sqrt{l^2-a^2}}{\pi(y^2+l^2-a^2)}.$$

It should be noted that the sign is opposite, since the normal vector on the conductor surface is directed along the negative x-axis.

2.11. The radial and zenithal components of the applied electric field outside the spherical conductor are $E_0\cos\theta$ and $-E_0\sin\theta$, respectively. The radial and zenithal components due to the electric dipole moment p at a point at distance d from the origin are $p\cos\theta/(2\pi\epsilon_0 r^3)$ and $p\sin\theta/(4\pi\epsilon_0 r^3)$, respectively. The condition that the zenithal component of the electric field just outside the surface is zero is written as

$$-E_0\sin\theta + \frac{p\sin\theta}{4\pi\epsilon_0 r^3} = 0,$$

which gives

$$p = 4\pi\epsilon_0 a^3 E_0.$$

The normal component of the electric field just outside the surface is equal to the surface electric charge density σ divided by ϵ_0. Thus, we have

$$\sigma = \epsilon_0\left(E_0\cos\theta + \frac{p\cos\theta}{2\pi\epsilon_0 a^3}\right) = 3\epsilon_0 E_0\cos\theta.$$

These results agree with Eqs. (2.26) and (2.29).

Chapter 3

3.1. First we determine the coefficients. Assuming $Q_1 = 1$ and $Q_2 = 0$, we have

$$\phi_1 = p_{11} = \frac{1}{4\pi\epsilon_0 a}, \quad \phi_2 = p_{21} = \frac{1}{4\pi\epsilon_0 d} = p_{12}.$$

When $Q_1 = 0$ and $Q_2 = q$, the electric potential of the spherical conductor is

$$\phi_1 = p_{11}Q_1 + p_{12}Q_2 = \frac{q}{4\pi\epsilon_0 d}.$$

We easily find this agrees with the result, $\phi(a,\ \theta)$, in Example 2.5.

3.2. We denote the cylindrical conductor and a thin linear conductor placed at the position of the line charge as conductors 1 and 2, respectively. We give a unit electric charge to conductor 1 of a unit length ($\lambda_1 = 1$) and no electric charge to conductor 2 ($\lambda_2 = 0$). Then, the electric potentials of conductors 1 and 2 are

$$\phi_1 = p'_{11} = \frac{1}{2\pi\epsilon_0}\log\frac{R_0}{a}, \quad \phi_2 = p'_{21} = \frac{1}{2\pi\epsilon_0}\log\frac{R_0}{d}.$$

Thus, we obtain the coefficients of electrostatic potential. In a general case where $\lambda_1 = \Lambda$ and $\lambda_2 = \lambda$, the electric potential of conductor 1 is

$$\phi_1 = p'_{11}\Lambda + p'_{12}\lambda.$$

When conductor 1 is grounded, $\phi_1 = 0$. This with $p'_{12} = p'_{21}$ yields

$$\Lambda = -\frac{p'_{21}}{p'_{11}}\lambda = -\frac{\log R_0 - \log a}{\log R_0 - \log d}\lambda.$$

If the reference point is infinity ($R_0 \to \infty$), this reduces to

$$\Lambda = -\lambda.$$

3.3. We denote the inner and outer conductors as conductors 1 and 2, respectively. In a general case where conductor 1 is not grounded, the assumptions $Q_1 = 1$ and $Q_2 = 0$ give

$$\phi_1 = p_{11} = \frac{1}{4\pi\epsilon_0}\left(\frac{1}{a} - \frac{1}{b} + \frac{1}{c}\right), \quad \phi_2 = p_{21} = p_{12} = \frac{1}{4\pi\epsilon_0 c}.$$

For $Q_1 = q$ and $Q_2 = Q$, the electric potential of conductor 1 is

$$\phi_1 = p_{11}q + p_{12}Q = \frac{q}{4\pi\epsilon_0}\left(\frac{1}{a} - \frac{1}{b} + \frac{1}{c}\right) + \frac{Q}{4\pi\epsilon_0 c}.$$

Hence, when conductor 1 is grounded ($\phi_1 = 0$), we have

$$q = -\left(\frac{1}{a} - \frac{1}{b} + \frac{1}{c}\right)^{-1}\frac{Q}{c}.$$

This agrees with the result obtained in Exercise 2.2.

3.4. The electric charge distributed in a unit length of the concentric conductor is λ on the surface of the inner conductor ($R = a$) and $-\lambda$ on the inner surface of the outer conductor ($R = b$). As a result, the electric field is $E = \lambda/(2\pi\epsilon_0 R)$ in the region $a < R < b$ and is zero in other regions. Hence, the electrostatic energy density in this region is $u_e = \epsilon_0 E^2/2 = \lambda^2/(8\pi^2\epsilon_0 R^2)$, and the electrostatic energy in the conductor of a unit length is

$$U'_e = \int_a^b \frac{\lambda^2}{8\pi^2\epsilon_0 R^2}\cdot 2\pi R\,\mathrm{d}R = \frac{\lambda^2}{4\pi\epsilon_0}\log\frac{b}{a}.$$

The electric potential of the outer conductor is zero and that of the inner conductor is

$$\phi = \frac{\lambda}{2\pi\epsilon_0}\log\frac{b}{a}.$$

Hence, we obtain the same electrostatic energy from Eq. (3.36) with $U'_e = \lambda\phi/2$.

Using this result and $U'_e = \lambda^2/(2C')$ corresponding to Eq. (3.38), the capacitance in a unit length is

$$C' = \frac{2\pi\epsilon_0}{\log(b/a)}.$$

3.5. (1) The electric field is $E(r) = Q/(4\pi\epsilon_0 r^2)$ in the region $a < r < b$ and is zero in other regions. The electrostatic energy density has a nonzero value, $u_e = Q^2/(32\pi^2\epsilon_0 r^4)$, only in the region $a < r < b$. We calculate the electrostatic energy as

$$U_e = \int_a^b u_e 4\pi r^2 \mathrm{d}r = \frac{(b-a)Q^2}{8\pi\epsilon_0 ab}.$$

(2) Using the electric field in (1), the electric potential of the outer conductor is zero and that of the inner conductor is

$$\phi = -\int_b^a \frac{Q}{4\pi\epsilon_0 r^2}\mathrm{d}r = \frac{(b-a)Q}{4\pi\epsilon_0 ab}.$$

The electrostatic energy is

$$U_e = \frac{1}{2}Q\phi = \frac{(b-a)Q^2}{8\pi\epsilon_0 ab}.$$

(3) We denote the inner and outer conductors as conductors 1 and 2, respectively. The coefficients of electric potential are

$$p_{11} = \frac{1}{4\pi\epsilon_0}\left(\frac{1}{a} - \frac{1}{b} + \frac{1}{c}\right), \quad p_{12} = p_{21} = p_{22} = \frac{1}{4\pi\epsilon_0 c}.$$

The electric charges are $Q_1 = Q$ and $Q_2 = -Q$. Thus, the electrostatic energy is

$$U_e = \frac{1}{2}p_{11}Q^2 - p_{12}Q^2 + \frac{1}{2}p_{22}Q^2 = \frac{(b-a)Q^2}{8\pi\epsilon_0 ab}.$$

3.6. Suppose that electric charges $\pm\lambda$ are given to each conductor in a unit length. We define the x-axis normal to these conductors in such a way that it passes through the centers of these conductors. We denote the positions of the centers of the conductors with negative and positive electric charges by $x = 0$ and $x = d$, respectively. Since the diameter of these conductors is much smaller than the interval, d, we can approximate the electric charges as being uniformly distributed on each surface. Hence, the electric field at position x is

$$E = -\frac{\lambda}{2\pi\epsilon_0}\left(\frac{1}{x} + \frac{1}{d-x}\right)$$

under the definition of positive electric field directed along the positive x-axis. The electric potential difference between the two conductors is

$$V = -\int_a^{d-a} E\mathrm{d}x = \frac{\lambda}{\pi\epsilon_0}\log\left(\frac{d}{a} - 1\right).$$

The capacitance in a unit length is

$$C' = \frac{\lambda}{V} = \frac{\pi\epsilon_0}{\log[(d/a) - 1]}.$$

3.7. (a) The coefficients of electric potential are

$$p_{11} = \frac{1}{4\pi\epsilon_0}\left(\frac{1}{r_0} - \frac{1}{r_1} + \frac{1}{r_2} - \frac{1}{r_3} + \frac{1}{r_4}\right),$$

$$p_{21} = p_{12} = p_{22} = \frac{1}{4\pi\epsilon_0}\left(\frac{1}{r_2} - \frac{1}{r_3} + \frac{1}{r_4}\right),$$

$$p_{31} = p_{32} = p_{13} = p_{23} = p_{33} = \frac{1}{4\pi\epsilon_0 r_4}.$$

(b) From Eq. (3.36), we calculate the electrostatic energy as

$$\begin{aligned} U_e &= \frac{1}{2}\left(p_{11}Q_1^2 + p_{22}Q_2^2 + p_{33}Q_3^2\right) + p_{12}Q_1Q_2 + p_{23}Q_2Q_3 + p_{31}Q_3Q_1 \\ &= \frac{(Q_1+Q_2+Q_3)^2}{8\pi\epsilon_0 r_4} + \frac{(Q_1+Q_2)^2}{8\pi\epsilon_0}\left(\frac{1}{r_2} - \frac{1}{r_3}\right) + \frac{Q_1^2}{8\pi\epsilon_0}\left(\frac{1}{r_0} - \frac{1}{r_1}\right). \end{aligned}$$

3.8. The electric potential of the cylindrical conductor ($R = a$) is given by

$$\phi = \frac{\lambda}{2\pi\epsilon_0}\log\frac{a}{l - \sqrt{l^2 - a^2}}.$$

Note that the electric potential is set to be zero when the cylindrical conductor touches the infinite conductor surface with zero electric potential ($l = a$). Hence, the electrostatic energy in a unit length is

$$U_\mathrm{e}' = \frac{1}{2}\phi\lambda = \frac{\lambda^2}{4\pi\epsilon_0}\log\frac{a}{l - \sqrt{l^2 - a^2}}.$$

Thus, the force on the cylindrical conductor of a unit length is

$$F' = -\frac{\partial U_\mathrm{e}'}{\partial l} = -\frac{\lambda^2}{4\pi\epsilon_0\sqrt{l^2 - a^2}}.$$

This is a negative force for increasing l. Hence, this is an attractive force. This force can also be directly derived for two linear image electric charges λ and $-\lambda$ separated by distance $2\sqrt{l^2 - a^2}$.

3.9. The electric field strength that the electric charge on one electrode gives to the other is $E = Q/2\epsilon_0 S$ and is independent of the distance between the electrodes x. The electrostatic force that one electrode exerts on the other is $F = QE = Q^2/2\epsilon_0 S$. Thus, the mechanical work needed to change the distance from d to l is

$$W = \int_d^l F\mathrm{d}x = \frac{Q^2}{2\epsilon_0 S}(l - d).$$

The electrostatic energy density in the space between the two electrodes is $u_\mathrm{e} = \epsilon_0 E^2/2$. Since the volume of this space changes by $S(l - d)$, the change in the electrostatic energy is

$$\Delta U_\mathrm{e} = u_\mathrm{e} S(l - d) = \frac{Q^2}{2\epsilon_0 S}(l - d),$$

3.10. The electric field in the space where the conductor is not inserted is $E = V/d$ and the densities of electric charges on the electrode surfaces in this region are $\pm\sigma_1 = \pm\epsilon_0 E = \pm\epsilon_0 V/d$. On the other hand, the electric field is concentrated only in the vacuum in other region, and its strength is $E = V/(d - t)$. Thus, the densities of electric charges on the electrode surfaces in this region are $\pm\sigma_2 = \pm\epsilon_0 V/(d - t)$. Hence, when the depth of insertion changes from x to $x + \Delta x$, the change in the electric charge in the electrode is $\Delta Q = \epsilon_0 btV\Delta x/[d(d-t)]$. The electrostatic energy of the capacitor is

$$U_e = \frac{1}{2}[(a-x)\sigma_1 + x\sigma_2]bV = \frac{\epsilon_0 bV^2}{2}\left(\frac{a-x}{d} + \frac{x}{d-t}\right).$$

The variation in the electrostatic energy when x increases by Δx is $\Delta U_e = \epsilon_0 btV^2\Delta x/[2d(d-t)]$.

If we denote the force on the conductor by F, the work done by the conductor is $F\Delta x$. The input energy from the electric power source to the system is $V\Delta Q$. Hence, from the relationship $\Delta U_e = -F\Delta x + V\Delta Q$, we obtain the force as

$$F = \lim_{\Delta x\to 0}\frac{V\Delta Q - \Delta U_e}{\Delta x} = \frac{\epsilon_0 btV^2}{2d(d-t)}.$$

Thus, the force is positive for increasing x and is attractive.

3.11. In the solution of Exercise 3.10, the electric charge,

$$Q = (a-x)b\sigma_1 + xb\sigma_2 = \epsilon_0 bV\left(\frac{a-x}{d} + \frac{x}{d-t}\right),$$

is kept constant, and there is no energy flow from the electric power source. Thus, the electrostatic energy is given by

$$U_e = \frac{1}{2}QV = \frac{Q^2}{2\epsilon_0 b}\left(\frac{a-x}{d} + \frac{x}{d-t}\right)^{-1}.$$

We obtain the force on the conductor as

$$F = -\frac{\partial U_e}{\partial x} = \frac{dt(d-t)Q^2}{2\epsilon_0 b[a(d-t)+tx]^2}.$$

Confirm that this force is identical with that in Exercise 3.10.

Chapter 4

4.1. Assume that electric charges Q and $-Q$ appear on the inner and outer electrodes, respectively, when we apply potential difference V between the two electrodes. The electric flux density is directed radially between the two electrodes, and its values are $D_1 = D_2 = Q/(4\pi r^2)$ in each region of different dielectric materials. Hence, the electric fields in each region are $E_1 = Q/(4\pi\epsilon_1 r^2)$ and $E_2 = Q/(4\pi\epsilon_2 r^2)$. The electric potential difference between the two electrodes is

$$V = \int_a^b E_1 \mathrm{d}r + \int_b^c E_2 \mathrm{d}r = \frac{Q}{4\pi\epsilon_1}\left(\frac{1}{a} - \frac{1}{b}\right) + \frac{Q}{4\pi\epsilon_2}\left(\frac{1}{b} - \frac{1}{c}\right).$$

We obtain the capacitance as

$$C = \frac{Q}{V} = \frac{4\pi\epsilon_1\epsilon_2 abc}{\epsilon_1 a(c-b) + \epsilon_2 c(b-a)}.$$

4.2. Assume that electric charges of density σ_1 and σ_2 appear on the inner electrode surface regions ($r = a$) faced to dielectric materials of ϵ_1 and ϵ_2, respectively, when we apply potential difference V between the two electrodes. The electric flux density is directed radially between the two electrodes, and its values in dielectric materials 1 and 2 are $D_1 = a^2\sigma_1/r^2$ and $D_2 = a^2\sigma_2/r^2$. The electric fields in respective regions are $E_1 = a^2\sigma_1/(\epsilon_1 r^2)$ and $E_2 = a^2\sigma_2/(\epsilon_2 r^2)$. Since the integration of these electric fields between the two electrodes is V, we have

$$\frac{a^2\sigma_1}{\epsilon_1}\left(\frac{1}{a} - \frac{1}{b}\right) = \frac{a^2\sigma_2}{\epsilon_2}\left(\frac{1}{a} - \frac{1}{b}\right) = V.$$

Thus, we determine the surface charge densities to be

$$\sigma_1 = \frac{b\epsilon_1 V}{a(b-a)}, \quad \sigma_2 = \frac{b\epsilon_2 V}{a(b-a)}.$$

This yields the total electric charge on the internal electrode,

$$Q = 2\pi a^2(\sigma_1 + \sigma_2) = \frac{2\pi ab(\epsilon_1 + \epsilon_2)V}{b-a}.$$

We obtain the capacitance as

$$C = \frac{Q}{V} = \frac{2\pi ab(\epsilon_1 + \epsilon_2)}{b-a}.$$

4.3. We denote the plane determined by the normal vector $\boldsymbol{n}$ on the interface and the electric field $\boldsymbol{E}_1$ in dielectric material 1 as S. Assume that the electric field $\boldsymbol{E}_2$ in dielectric material 2 does not lie on this plane. We consider a plane, S′, normal to both the interface and S and define a small rectangle on S′ that includes the interface. The two sides of the rectangle are parallel to the interface. When we integrate the electric field along this rectangle, the integral in dielectric material 2 is not zero, while that in dielectric material 1 is zero. This is contradictory, since Eq. (1.30) is not satisfied under this assumption. Thus, we prove that the electric field $\boldsymbol{E}_2$ also lies on plane S.

4.4. Since the parallel component of the electric field is continuous across the wide interface, the electric field inside the slit is also E_0, and the electric flux density is $D = \epsilon_0 E_0$.

4.5. Since the normal component of the electric flux density is continuous across the wide interface, the electric flux density inside the slit is also $D = \epsilon E_0$, and the electric field is $E = D/\epsilon_0 = (\epsilon/\epsilon_0)E_0$.

4.6. The electric field strength inside the dielectric material is $E = E_0$ from the continuity of its parallel component given by Eq. (4.22). Thus, the electric flux density is $D = \epsilon E = \epsilon E_0$. The electric polarization is determined to be

$$P = D - \epsilon_0 E = (\epsilon - \epsilon_0)E_0.$$

Since the electric field is parallel to the surface, there is no polarization charge on the surface ($\sigma_\mathrm{p} = 0$). If we look at this phenomenon on a much wider scale, the polarization charge appears in the upper and lower regions. The present situation corresponds to the part around $\theta = -\pi/2$ in Fig. 4.19 in Example 4.5, and $E_\theta = 3\epsilon_0 E_0/(\epsilon + 2\epsilon_0)$ in this case corresponds to the present applied electric field strength.

4.7. The electric field E is given by the sum of E_0 and the electric field produced by the polarization charge of surface density, $\sigma_\mathrm{p}(\theta) = [3\epsilon_0(\epsilon - \epsilon_0)/(\epsilon + 2\epsilon_0)]E_0 \cos\theta$, with θ denoting the zenithal angle. Since the electric charge of surface density $\sigma = 3\epsilon_0 E_0 \cos\theta$ in Eq. (2.29) produces the uniform electric field $-E_0$ inside the sphere, the above polarization charge produces the uniform electric field $-(\epsilon - \epsilon_0)E_0/(\epsilon + 2\epsilon_0)$. Thus, we have

$$E = E_0 - \frac{(\epsilon - \epsilon_0)E_0}{\epsilon + 2\epsilon_0} = \frac{3\epsilon_0}{\epsilon + 2\epsilon_0} E_0.$$

This agrees with the result in Example 4.5.

4.8. We define cylindrical coordinates with the z-axis at the central axis of the dielectric cylinder and the azimuthal angle measured from the direction of the applied electric field. We assume that the electric field outside the dielectric cylinder ($R > a$) produced by the polarized charge is given by the linear electric dipole of moment $\hat{p}$ in a unit length placed at the central axis after virtually removing the dielectric cylinder. The direction of the dipole moment is the same as that of the applied electric field. We assume that the electric field inside the dielectric cylinder ($R < a$) has a uniform strength E and is directed parallel to the applied electric field ($\varphi = 0$). The continuity conditions for the parallel (azimuthal) component of the electric field and the normal (radial) component of the electric flux density give

$$\hat{p} = \frac{\epsilon - \epsilon_0}{\epsilon + \epsilon_0} \cdot 2\pi\epsilon_0 a^2 E_0, \quad E = \frac{2\epsilon_0}{\epsilon + \epsilon_0} E_0.$$

The electric field is

$$E_R = \frac{D_R}{\epsilon_0} = \left(1 + \frac{\epsilon - \epsilon_0}{\epsilon + \epsilon_0} \cdot \frac{a^2}{R^2}\right) E_0 \cos\varphi,$$
$$E_\varphi = \frac{D_\varphi}{\epsilon_0} = -\left(1 - \frac{\epsilon - \epsilon_0}{\epsilon + \epsilon_0} \cdot \frac{a^2}{R^2}\right) E_0 \sin\varphi,$$

outside the dielectric cylinder ($R > a$) and

$$E_R = \frac{D_R}{\epsilon} = \frac{2\epsilon_0}{\epsilon + \epsilon_0} E_0 \cos\varphi, \qquad E_\varphi = \frac{D_\varphi}{\epsilon} = -\frac{2\epsilon_0}{\epsilon + \epsilon_0} E_0 \sin\varphi.$$

inside the dielectric cylinder ($0 \leq R < a$). The electric polarization inside the dielectric cylinder is

$$P = (\epsilon - \epsilon_0)E = \frac{2\epsilon_0(\epsilon - \epsilon_0)}{\epsilon + \epsilon_0} E_0.$$

Here, we apply Eq. (4.9) to a small shell that includes the surface of the dielectric cylinder, as shown in Fig. 4.18. Since there is no true electric charge on the surface, the surface polarization charge density is given by the difference in the normal component of the electric field on the surface multiplied by ϵ_0;

$$\sigma_\mathrm{p}(\varphi) = \frac{2\epsilon_0(\epsilon - \epsilon_0)}{\epsilon + \epsilon_0} E_0 \cos\varphi = P \cos\varphi.$$

4.9. It is assumed that the electric field is applied along the x-axis. Hence, its electric potential is given by

$$\phi_\mathrm{f} = -E_0 x = -E_0 r \cos\theta.$$

We put an electric dipole moment p directed along the applied electric field on the origin after the dielectric material is virtually removed. Its electric potential is given by

$$\phi_\mathrm{p} = \frac{p\cos\theta}{4\pi\epsilon_0 r^2}.$$

Thus, the electric potential outside the dielectric sphere is

$$\phi_1 = \phi_\mathrm{f} + \phi_\mathrm{p} = \left(-E_0 r + \frac{p}{4\pi\epsilon_0 r^2}\right)\cos\theta.$$

The electric field inside the dielectric sphere is uniform along the x-axis, and its value is denoted by E. Thus, the electric potential inside is

$$\phi_2 = -Er\cos\theta.$$

Equations (4.20) and (4.24) lead to

$$\epsilon_0\left(E_0 + \frac{p}{2\pi\epsilon_0 a^3}\right) = \epsilon E, \quad E_0 - \frac{p}{4\pi\epsilon_0 a^3} = E.$$

Thus, we have

$$p = \frac{\epsilon - \epsilon_0}{\epsilon + \epsilon_0} 4\pi\epsilon_0 a^3 E_0, \quad E = \frac{3\epsilon_0}{\epsilon + \epsilon_0} E_0.$$

4.10. We define the x-y plane ($z = 0$) on the dielectric material surface and the position of the line current as $x = 0$. To determine the electric potential in the vacuum region ($z > 0$), we assume that all the space is vacuum and the electric potential is produced by both the line charge of linear density λ and a virtual line charge of linear density λ' located at the symmetric position with respect to the dielectric material surface;

$$\phi_{\mathrm{v}}(x,z) = \frac{1}{2\pi\epsilon_0}\left\{\lambda \log\frac{R_0}{[x^2 + (z-a)^2]^{1/2}} + \lambda' \log\frac{R_0}{[x^2 + (z+a)^2]^{1/2}}\right\}.$$

In the above, R_0 is the distance of the reference point from the line at $x = 0$ on the surface. To determine the electric potential inside the dielectric material ($z < 0$), we assume that all the space is occupied by the dielectric material and the electric potential is given by a line charge of linear density λ'' placed at the original position. Hence, the electric potential at (x, z) inside the dielectric material is

$$\phi_{\mathrm{d}}(x,z) = \frac{1}{2\pi\epsilon}\lambda'' \log\frac{R_0}{[x^2 + (z-a)^2]^{1/2}}.$$

The continuity condition of the parallel component of the electric field, Eq. (4.24), gives $\phi_{\mathrm{v}}(z=0) = \phi_{\mathrm{d}}(z=0)$. This yields

$$\frac{\lambda + \lambda'}{\epsilon_0} = \frac{\lambda''}{\epsilon}.$$

Since there is no true electric charge on the surface, the normal component of the electric flux density is continuous. Then, $\epsilon_0(\partial\phi_{\mathrm{v}}/\partial z)_{z=0} = \epsilon(\partial\phi_{\mathrm{d}}/\partial z)_{z=0}$ given by Eq. (4.20) yields

$$\lambda - \lambda' = \lambda''.$$

From these conditions, we obtain the linear electric charge densities as

$$\lambda' = -\frac{\epsilon - \epsilon_0}{\epsilon + \epsilon_0}\lambda, \quad \lambda'' = \frac{2\epsilon}{\epsilon + \epsilon_0}\lambda.$$

The electric potential is

$$\begin{aligned}\phi &= \frac{\lambda}{2\pi\epsilon_0}\left\{\log\frac{R_0}{[x^2+(z-a)^2]^{1/2}} - \frac{\epsilon-\epsilon_0}{\epsilon+\epsilon_0}\log\frac{R_0}{[x^2+(z+a)^2]^{1/2}}\right\}; \quad z>0,\\ &= \frac{\lambda}{\pi(\epsilon+\epsilon_0)}\log\frac{R_0}{[x^2+(z-a)^2]^{1/2}}; \quad z<0.\end{aligned}$$

4.11. Assume that electric charges of density σ_1 and σ_2 appear on the inner electrode surface ($r = a$) facing dielectric materials with dielectric constants ϵ_1 and ϵ_2, respectively, when we apply electric charges $\pm Q'$ to the two electrodes. Thus, we have $Q' = \pi a(\sigma_1 + \sigma_2)$. The electric field strength inside dielectric materials 1 and 2 is

$$E_1 = \frac{\sigma_1 a}{\epsilon_1 R}, \quad E_2 = \frac{\sigma_2 a}{\epsilon_2 R}.$$

Since these must be equal to each other, we have

$$\frac{\sigma_1}{\epsilon_1} = \frac{\sigma_2}{\epsilon_2}.$$

From the above conditions, we determine the electric field strength to be

$$E(R) = \frac{Q'}{\pi(\epsilon_1+\epsilon_2)R}.$$

The electric potential difference between the two electrodes is

$$V = \int_a^b E(R)\mathrm{d}R = \frac{Q'}{\pi(\epsilon_1+\epsilon_2)}\log\frac{b}{a}.$$

Thus, the electrostatic energy in a unit length is determined to be

$$U_\mathrm{e}' = \frac{1}{2}Q'V = \frac{Q'^2}{2\pi(\epsilon_1+\epsilon_2)}\log\frac{b}{a}.$$

Chapter 5

5.1. We apply voltage V between the two edges. The electric field along the circle of radius R from the center is $E(R) = 2V/(\pi R)$ (see Fig. B5.1). Hence, the current density at this point is $i(R) = 2V/(\pi\rho_\mathrm{r}R)$. Here, we define the angle θ

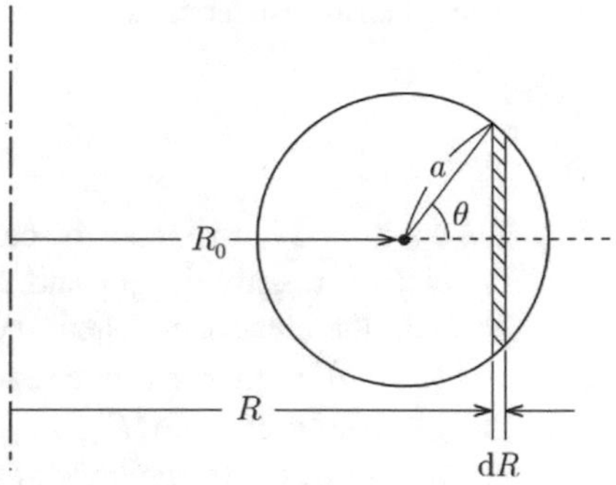

Fig. B5.1 Part in the region R to R + dR from the center

as in the figure. Then, $R = R_0 + a\cos\theta$. The current flowing in the region between R and R + dR is $i(R)2a\sin\theta \mathrm{d}R = 4Va^2\sin^2\theta \mathrm{d}\theta/[\pi\rho_\mathrm{r}(R_0 + a\cos\theta)]$. Hence, the total current is

$$I = \int_0^\pi \frac{4Va^2\sin^2\theta \mathrm{d}\theta}{\pi\rho_\mathrm{r}(R_0 + a\cos\theta)}.$$

We transform the integrand as

$$\frac{\sin^2\theta}{R_0 + a\cos\theta} = \frac{R_0}{a^2} - \frac{1}{a}\cos\theta - \left(\frac{R_0^2}{a^2} - 1\right)\frac{1}{R_0 + a\cos\theta}.$$

For integration of the third term, we use Eq. (7.26) with $\theta = \pi - \varphi$. A simple calculation gives

$$I = \frac{4V}{\rho_\mathrm{r}}\left[R_0 - (R_0^2 - a^2)^{1/2}\right].$$

Then, we obtain the electric resistance as

$$R_\mathrm{r} = \frac{\rho_\mathrm{r}}{4\left[R_0 - (R_0^2 - a^2)^{1/2}\right]}.$$

5.2. The cross-sectional area at height x from the bottom is $S(x) = a[b - (b - c)x/h]$, and the current density there is $i(x) = I/\{a[b - (b - c)x/h]\}$, when we apply current I. Since the electric field is $E(x) = \rho_\mathrm{r} i(x)$, the voltage between the two edges is

$$V = \int_0^h \frac{\rho_\mathrm{r} I}{a[b - (b - c)x/h]}\mathrm{d}x = \frac{h\rho_\mathrm{r} I}{a(b - c)}\log\frac{b}{c}.$$

The electric resistance is

$$R_{\mathrm{r}} = \frac{h\rho_{\mathrm{r}}}{a(b-c)}\log\frac{b}{c}.$$

5.3. We use I_1 and I_2 to denote the currents flowing in the respective regions with the electric resistivities $\rho_{\mathrm{r}1}$ and $\rho_{\mathrm{r}2}$, respectively, when we apply voltage V between the electrodes. Then, the current densities at positions at a distance $r(a \le r \le b)$ in the respective regions are $i_1 = I_1/2\pi r^2$ and $i_2 = I_2/2\pi r^2$, and the electric fields are $E_1 = \rho_{\mathrm{r}1} I_1/2\pi r^2$ and $E_2 = \rho_{\mathrm{r}2} I_2/2\pi r^2$. From the condition that the integrations of the electric fields from $r = a$ to $r = b$ are V, we have

$$I_1 = \frac{2\pi abV}{(b-a)\rho_{\mathrm{r}1}}, \quad I_2 = \frac{2\pi abV}{(b-a)\rho_{\mathrm{r}2}}.$$

Since $I_1 + I_2 = I$ is the total current, we have

$$R_{\mathrm{r}} = \frac{V}{I} = \frac{(b-a)\rho_{\mathrm{r}1}\rho_{\mathrm{r}2}}{2\pi ab(\rho_{\mathrm{r}1} + \rho_{\mathrm{r}2})}.$$

5.4. When we apply current I, the current density at distance R from the central axis is

$$i(R) = \frac{I}{2\pi lR}.$$

The electric field is $E(R) = \rho_{\mathrm{r}1} i(R)$ for $a < R < b$ and $E(R) = \rho_{\mathrm{r}2} i(R)$ for $b < R < c$. Hence, the voltage between the two electrodes is

$$V = \int_a^b \frac{\rho_{\mathrm{r}1} I}{2\pi lR}\mathrm{d}R + \int_b^c \frac{\rho_{\mathrm{r}2} I}{2\pi lR}\mathrm{d}R = \frac{I}{2\pi l}\left(\rho_{\mathrm{r}1}\log\frac{b}{a} + \rho_{\mathrm{r}2}\log\frac{c}{b}\right).$$

The resistance is

$$R_{\mathrm{r}} = \frac{V}{I} = \frac{1}{2\pi l}\left(\rho_{\mathrm{r}1}\log\frac{b}{a} + \rho_{\mathrm{r}2}\log\frac{c}{b}\right).$$

5.5. We use I_1 and I_2 to denote the currents flowing in the respective regions with the electric resistivities $\rho_{\mathrm{r}1}$ and $\rho_{\mathrm{r}2}$ when we apply voltage V between the electrodes. Then, the current densities at positions at distance $R(a \le R \le b)$ in the respective regions are $i_1(R) = I_1/(\pi Rl)$ and $i_2(R) = I_2/(\pi Rl)$, and the electric fields are $E_1(R) = \rho_{\mathrm{r}1} I_1/(\pi Rl)$ and $E_2(R) = \rho_{\mathrm{r}2} I_2/(\pi Rl)$. From the conditions that the integrations of the electric fields from $R = a$ to $R = b$ are V, we have

$$I_1 = \frac{\pi l V}{\rho_{r1} \log(b/a)}, \quad I_2 = \frac{\pi l V}{\rho_{r2} \log(b/a)}.$$

Since the total current is $I = I_1 + I_2$, we obtain the electric resistance as

$$R_r = \frac{\rho_{r1}\rho_{r2} \log(b/a)}{\pi l(\rho_{r1} + \rho_{r2})}.$$

5.6. When we apply voltage V between the two edges, the electric field at a point of radius R is $E(R) = 2V/(\pi R)$. Hence, the electric power density is $p(R) = 4V^2/(\pi^2 \rho_r R^2)$. The electric power in the region R to $R + \mathrm{d}R$ is

$$\mathrm{d}P = \frac{\pi w R \mathrm{d}R p(R)}{2} = \frac{2wV^2}{\pi \rho_r} \frac{\mathrm{d}R}{R}.$$

Thus, the total dissipated electric power is

$$P = \frac{2wV^2}{\pi \rho_r} \int_{R_0 - d/2}^{R_0 + d/2} \frac{\mathrm{d}R}{R} = \frac{2wV^2}{\pi \rho_r} \log \frac{R_0 + d/2}{R_0 - d/2}.$$

This is equal to IV.

5.7. When we apply voltage V between the electrodes of the capacitor with a dielectric material of dielectric constant ϵ in the space, the electric field is $E = V/d$, and the electric flux density is $D = \epsilon E = \epsilon V/d$. The surface charge density on the electrode is $\sigma = D = \epsilon V/d$, and the total electric charge is $Q = \sigma S = \epsilon S V/d$. Thus, the capacitance of the capacitor is $C = Q/V = \epsilon S/d$.

When we apply voltage V between the electrodes of the resistor with a substance of electric conductivity σ_c in the space, the electric field is $E = V/d$, and the current density is $i = \sigma_c E = \sigma_c V/d$. The total current is $I = iS = \sigma_c S V/d$. Thus, the electric resistance of the resistor is $R_r = V/I = d/(\sigma_c S)$.

From the above results, we obtain the same result as Eq. (5.38):

$$CR_r = \frac{\epsilon}{\sigma_c}.$$

5.8. We can use the answer to Exercise 4.8, if we convert the electric flux density $\boldsymbol{D}$ to the current density $\boldsymbol{i}$ with conversion of the dielectric constants ϵ_0 and ϵ to the electric conductivities σ_{c0} and σ_c. The uniform electric field E_0 corresponds to $\sigma_{c0} i_0$. We define cylindrical coordinates with the z-axis at the central axis of the cylinder and azimuthal angle φ measured from the direction of the applied uniform current. The current density outside the cylinder ($R > a$) is

$$i_R = \left(1 + \frac{\sigma_c - \sigma_{c0}}{\sigma_c + \sigma_{c0}} \cdot \frac{a^2}{R^2}\right) i_0 \cos\varphi,$$
$$i_\varphi = -\left(1 - \frac{\sigma_c - \sigma_{c0}}{\sigma_c + \sigma_{c0}} \cdot \frac{a^2}{R^2}\right) i_0 \sin\varphi,$$

and that inside the cylinder ($R < a$) is

$$i_R = \frac{2\sigma_c}{\sigma_c + \sigma_{c0}} i_0 \cos\varphi, \quad i_\varphi = -\frac{2\sigma_c}{\sigma_c + \sigma_{c0}} i_0 \sin\varphi.$$

5.9. The electric potentials in tapes A and B are denoted by $V_A(x)$ and $V_B(x)$, and the currents flowing in tapes A and B are denoted by I_A and I_B. Thus, we have $I = I_A + I_B$. The electric potentials at $x + dx$ in each tape are

$$V_A(x + dx) = V_A(x) - R_A' I_A dx,$$
$$V_B(x + dx) = V_B(x) - R_B' I_B dx.$$

The current that passes from A to B at $x + dx$ is $-dI_A$, which is given by $dI_A = -g'(V_A - V_B)dx$. Eliminating I_B, we have

$$\frac{d^2 I_A}{dx^2} = g'\left[(R_A' + R_B')I_A - R_B' I\right].$$

The solution is given by

$$I_A = K_1 e^{x/\alpha} + K_2 e^{-x/\alpha} + C$$

with K_1 and K_2 denoting constants and

$$\alpha = \left[g'\left(R_A' + R_B'\right)\right]^{-1/2}, \qquad C = \frac{R_B' I}{R_A' + R_B'}.$$

Using the boundary conditions, $I_A(0) = I_A(2L) = I$, we have

$$I_A = \frac{R_B' I}{R_A' + R_B'}\left\{1 + \frac{R_A' \cosh[(x - L)/\alpha]}{R_B' \cosh(L/\alpha)}\right\}.$$

If the length $2L$ is much longer than α, the current is approximately given by $I_A \simeq R_B' I/(R_A' + R_B')$ in the region of $\alpha \ll x \ll 2L - \alpha$, which is the same as the result for a simple parallel circuit.

Fig. B5.2 Virtual line charges

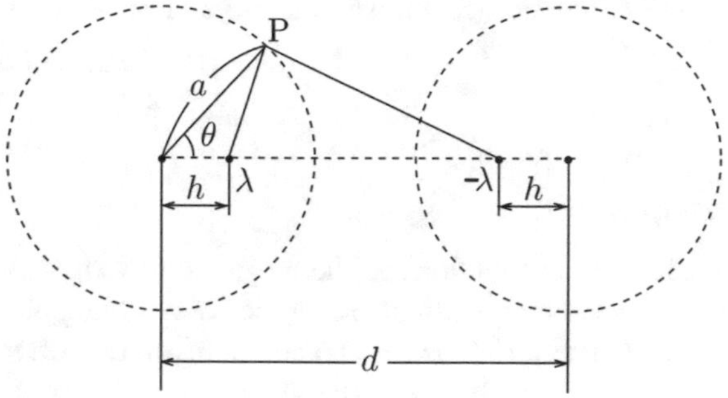

5.10. Suppose we replace a substance of electric conductivity σ_c with a dielectric material of dielectric constant ϵ and place virtual line charges $\pm\lambda$ at the positions shown in Fig. B5.2. Then, the electric potential at point P on the surface of the left conductor is

$$\phi(a,\theta) = \frac{\lambda}{2\pi\epsilon}\left\{\log\frac{R_0}{(a^2+h^2-2ah\cos\theta)^{1/2}} - \log\frac{R_0'}{[a^2+(d-h)^2-2a(d-h)\cos\theta]^{1/2}}\right\},$$

where R_0 and R_0' are constants. So that this electric potential is constant and independent of angle θ, the following condition should be satisfied;

$$\frac{ah}{a^2+h^2} = \frac{a(d-h)}{a^2+(d-h)^2},$$

which reduces to $h(d-h) = a^2$. This condition for h is simply solved as

$$h = \frac{d-\sqrt{d^2-4a^2}}{2}.$$

The symmetry condition of $\phi = 0$ on the central plane between the two conductors gives $R_0 = R_0'$. Then, the electric potential of the left conductor is

$$\phi_+ = \phi(a,\theta) = \frac{\lambda}{4\pi\epsilon}\log\frac{d+\sqrt{d^2-4a^2}}{d-\sqrt{d^2-4a^2}} = \frac{\lambda}{2\pi\epsilon}\log\frac{d+\sqrt{d^2-4a^2}}{2a}.$$

The electric potential of the right conductor is $\phi_- = -\phi(a,\theta)$. Hence, the capacitance in a unit length is

$$C' = \frac{\lambda}{2\phi(a,\theta)} = \frac{\pi\epsilon}{\log[(d+\sqrt{d^2+4a^2})/2a]}.$$

Using Eq. (5.38), we obtain the electric resistance in a unit length as

$$R_{\mathrm{r}}' = \frac{1}{\pi\sigma_{\mathrm{c}}}\log\frac{d+\sqrt{d^2+4a^2}}{2a}.$$

Chapter 6

6.1. All contributions to the magnetic flux density at the center O produced by elementary currents at respective regions point normal backward. The angle θ in Eq. (6.5) is zero on any point on the left straight section and is π on any point on the right straight section. Thus, there is no contribution to the resultant magnetic flux density from these sections. The angle θ is $\pi/2$ and $r = a$ on the semicircle. The contribution from the elementary current in this section is $\mathrm{d}B = \mu_0 I\mathrm{d}s/(4\pi a^2)$. Integrating this over the semicircle yields

$$B = \frac{\mu_0 I}{4\pi a^2}\cdot \pi a = \frac{\mu_0 I}{4a}.$$

6.2. The distance between one side and point P is $l = [(a^2/4)+b^2]^{1/2}$. Using the same method as in Example 6.2, we calculate the magnetic flux density produced by the current on one side as

$$B' = \frac{\mu_0 I}{4\pi l}\int_{\theta_1}^{\pi-\theta_1}\sin\theta\,\mathrm{d}\theta = \frac{\mu_0 I}{2\pi l}\cos\theta_1 = \frac{\mu_0 I a}{4\pi l[(a^2/4)+l^2]^{1/2}},$$

where θ_1 is the angle of point P measured from the edge of the side. Figure B6.1 schematically shows the magnetic flux density produced by the current on one side. Only the vertical component remains from symmetry, and we obtain as

$$B = 4B'\cos\alpha = \frac{\mu_0 I a^2}{2\pi l^2[(a^2/4)+l^2]^{1/2}} = \frac{\mu_0 I a^2}{2\pi[(a^2/4)+b^2][(a^2/2)+b^2]^{1/2}}.$$

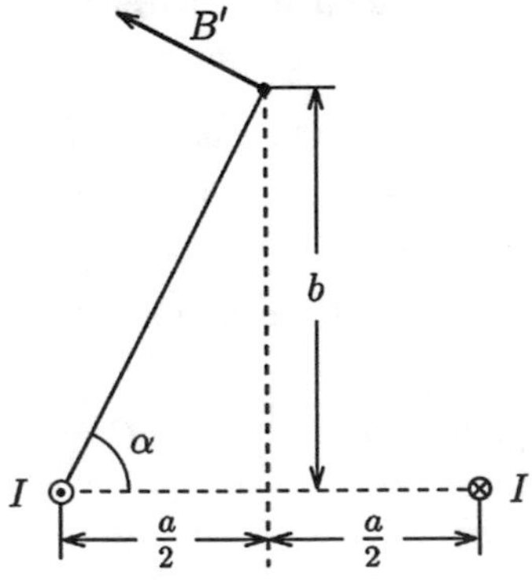

Fig. B6.1 Magnetic flux density produced at point P by current on one side

6.3. The current density is expressed as $i = nqv$ in terms of the velocity v of an electric charge. The force of $F = -qvB$ acts on the charge along the y-axis, resulting in the condition that the charges are accumulated on the side of the negative y-axis. Such an accumulation causes the electric field E along the y-axis, and the electric force $F' = qE$ works on the charge. In the steady state, we attain the balanced condition given by $F + F' = 0$. Hence, we obtain the electric field as

$$E = vB = \frac{iB}{nq}.$$

This is called the **Hall electric field**. The sign of the Hall electric field is determined by the sign of electric charge. That is, the sign of the Hall electric field clarifies whether the current-carrying charges are electrons ($q = -e$) or holes ($q = e$). This phenomenon, i.e., the induction of the electric field in the direction normal to the current and magnetic flux density is called the **Hall effect**.

6.4. Closed circuit C is projected on a plane normal to the current, as shown in Fig. B6.2. We denote the projected closed trajectory and elementary line vector d$\boldsymbol{s}$ as C′ and d$\boldsymbol{s}'$, respectively. Since the magnetic flux density $\boldsymbol{B}$ stays in a plane normal to the current, we have $\boldsymbol{B} \cdot \mathrm{d}\boldsymbol{s} = \boldsymbol{B} \cdot \mathrm{d}\boldsymbol{s}'$. That is, the following relationship holds

$$\int_{\mathrm{C}} \boldsymbol{B} \cdot \mathrm{d}\boldsymbol{s} = \int_{\mathrm{C}'} \boldsymbol{B} \cdot \mathrm{d}\boldsymbol{s}'.$$

Hence, Eq. (6.22) holds for an arbitrary closed line. We can similarly prove Eq. (6.23) when the current does not penetrate the closed line.

6.5. The magnetic flux density is directed along the z-axis, and its value is

$$\begin{aligned} B_z(x) &= 0; & x < -b, \\ &= \mu_0 i(x+b); & -b < x < -a, \\ &= \mu_0 i(b-a) : & -a < x < a, \\ &= \mu_0 i(b-x); & a < x < b, \\ &= 0; & x > b. \end{aligned}$$

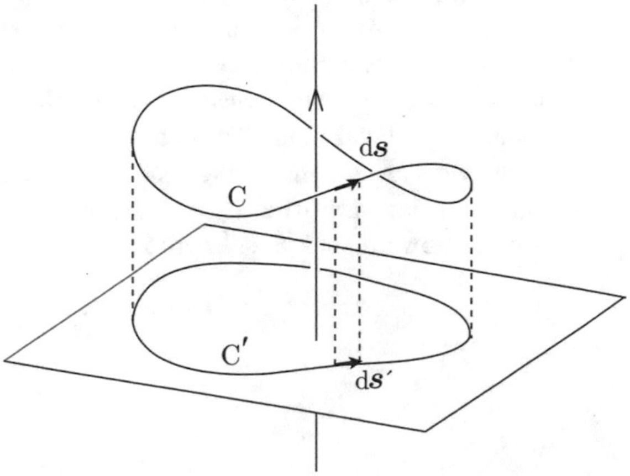

Fig. B6.2 Closed line C and its projection C′ on a plane perpendicular to straight current

The vector potential has only the y-component, A_y, and the from the relationship $B_z = \partial A_y/\partial x$, we have

$$\begin{aligned} A_y(x) &= -\frac{\mu_0 i}{2}(b^2 - a^2); & x < -b, \\ &= \frac{\mu_0 i}{2}(x^2 + 2bx + a^2); & -b < x < -a, \\ &= \mu_0 i(b - a)x : & -a < x < a, \\ &= \frac{\mu_0 i}{2}(-x^2 + 2bx - a^2); & a < x < b, \\ &= \frac{\mu_0 i}{2}(b^2 - a^2); & x > b. \end{aligned}$$

6.6. We define the x- and y-axes along the slab width and current, respectively, with $x = 0$ at the center of the slab. We presume the current $\mathrm{d}I = I\mathrm{d}x/w$ flowing in a thin region x to $x + \mathrm{d}x$ as a line current. The magnetic flux density at point P produced by this line current is directed along the negative z-axis, and its value is $\mathrm{d}B_z = -\mu_0 \mathrm{d}I/[2\pi(d - x)] = -\mu_0 I\mathrm{d}x/[2\pi w(d - x)]$. The magnetic flux density at point P is

$$B_z(d) = -\frac{\mu_0 I}{2\pi w}\int_{-w/2}^{w/2} \frac{\mathrm{d}x}{d - x} = -\frac{\mu_0 I}{2\pi w}\log\frac{d + w/2}{d - w/2}.$$

The vector potential at $x > w/2$ is directed along the y-axis, and from $B_z(x) = \partial A_y/\partial x$, we have

$$\begin{aligned} A_y(x) &= \int B_z(x)\mathrm{d}x \\ &= -\frac{\mu_0 I}{2\pi w}\left[\left(x + \frac{w}{2}\right)\log\left(x + \frac{w}{2}\right) - \left(x - \frac{w}{2}\right)\log\left(x - \frac{w}{2}\right) + C\right], \end{aligned}$$

where C is a constant determined by the position of reference point, and $A_y(d)$ is the value on point P.

6.7. The current density is $i = I/[\pi(a^2 - b^2)]$. We can solve this problem by superposing case (a) in which the current flows uniformly with density i in the whole cross section (see Fig. B6.3a) and case (b) in which the current flows uniformly with density i along the opposite direction inside the vacancy (see Fig. B6.3b). The contribution to the magnetic flux density at the vacancy center A from case (a) is $B_1 = \mu_0 \pi d^2 i/(2\pi d) = \mu_0 I d/[2\pi(a^2 - b^2)]$ and that from case (b) is $B_2 = 0$. Hence, the magnetic flux density at A is directed upward, and its strength is

$$B_{\mathrm{A}} = B_1 = \frac{\mu_0 I d}{2\pi(a^2 - b^2)}.$$

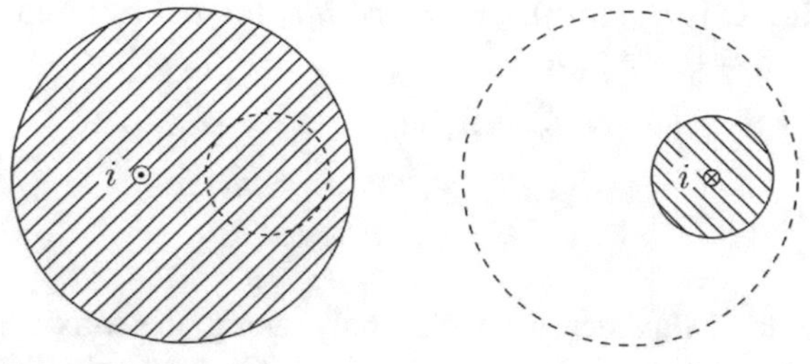

Fig. B6.3 Superposition of **a** current flowing uniformly with density i and **b** current flowing uniformly with density i along the opposite direction inside the vacancy

The contribution to the magnetic flux density at point B from (a) is $B_3 = \mu_0 a^2 i/(2R) = \mu_0 I a^2/[2\pi(a^2 - b^2)R]$ and that from (b) is $B_4 = -\mu_0 b^2 i/[2(R-d)] = -\mu_0 I b^2/[2\pi(a^2 - b^2)(R-d)]$. These point in the same direction. Thus, the magnetic flux density at B is directed upward, and its strength is

$$B_B = B_3 + B_4 = \frac{\mu_0 I}{2\pi(a^2 - b^2)}\left(\frac{a^2}{R} - \frac{b^2}{R-d}\right).$$

6.8. We assume a rectangle C with one side on the center on the x-z plane, as shown in Fig. B6.4. The magnetic flux density has only a z-component, and its value must be 0 at $x = 0$. The magnetic flux density is integrated along rectangle C, as shown by the arrows, so as to be consistent with the current flow along the positive y-axis. Since $\boldsymbol{B}$ is perpendicular to $d\boldsymbol{s}$ on the top and bottom sides of C, the integral is zero there. The left side of Eq. (6.25) is $-B_z(x)l$, where l is the length of the side of C along the z-axis. The current

Fig. B6.4 Rectangle C for $x > a$ on which Ampere's law is applied

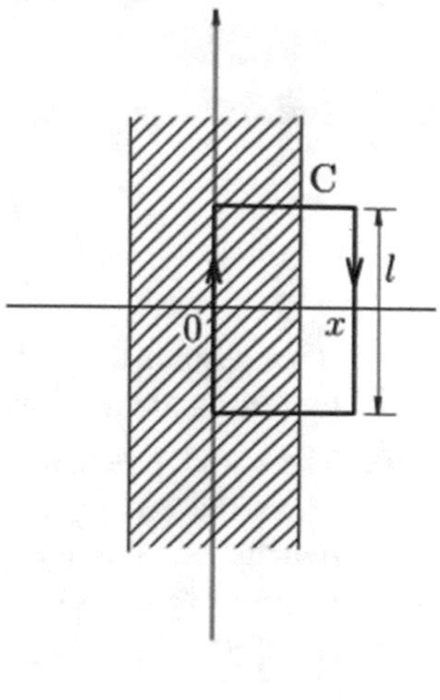

penetrating C is lxi_0 for $0 \le x \le a$ and lai_0 for $x > a$. From symmetry with respect to $x = 0$, we have

$$\begin{aligned} B_z &= \mu_0 i_0 a; \quad && x < -a, \\ &= -\mu_0 i_0 x; && -a \le x \le a, \\ &= -\mu_0 i_0 a; && x > a. \end{aligned}$$

The magnetic flux density varies only along the x-axis ($\partial/\partial y = 0$ and $\partial/\partial z = 0$), and the left side of Eq. (6.27) has only the y-component, $-\partial B_z/\partial x$. Thus, we have

$$\begin{aligned} i_y &= i_0; \quad && -a \le x \le a, \\ &= 0; && x < -a, x > a. \end{aligned}$$

This shows that the current flows with density i_0 in the slab, as assumed in the beginning.

6.9. The divergence of Eq. (6.33) is

$$\nabla \cdot \boldsymbol{A}(\boldsymbol{r}) = \frac{\mu_0}{4\pi} \int_V \nabla \cdot \frac{\boldsymbol{i}(\boldsymbol{r}')}{|\boldsymbol{r} - \boldsymbol{r}'|} \mathrm{d}V'.$$

Since ∇ is the differential operator with respect to $\boldsymbol{r}$, from Eq. (A1.40), we have

$$\nabla \cdot \frac{\boldsymbol{i}(\boldsymbol{r}')}{|\boldsymbol{r} - \boldsymbol{r}'|} = \boldsymbol{i}(\boldsymbol{r}') \cdot \nabla \frac{1}{|\boldsymbol{r} - \boldsymbol{r}'|}.$$

If we use the differential operator ∇' with respect to $\boldsymbol{r}'$, $\nabla|\boldsymbol{r} - \boldsymbol{r}'|^{-1} = -\nabla'|\boldsymbol{r} - \boldsymbol{r}'|^{-1}$. Thus, using Eq. (A1.40) again, the integrand is written as

$$-\boldsymbol{i}(\boldsymbol{r}') \cdot \nabla' \frac{1}{|\boldsymbol{r} - \boldsymbol{r}'|} = \frac{1}{|\boldsymbol{r} - \boldsymbol{r}'|} \nabla' \cdot \boldsymbol{i}(\boldsymbol{r}') - \nabla' \cdot \frac{\boldsymbol{i}(\boldsymbol{r}')}{|\boldsymbol{r} - \boldsymbol{r}'|}.$$

Since $\nabla' \cdot \boldsymbol{i}(\boldsymbol{r}') = 0$, applying Gauss' law yields

$$-\int_V \nabla' \cdot \frac{\boldsymbol{i}(\boldsymbol{r}')}{|\boldsymbol{r} - \boldsymbol{r}'|} \mathrm{d}V' = -\int_S \frac{\boldsymbol{i}(\boldsymbol{r}')}{|\boldsymbol{r} - \boldsymbol{r}'|} \cdot \mathrm{d}\boldsymbol{S}'.$$

If we assume the surface S of region V at infinity, the surface integral reduces to zero, and we prove Eq. (6.30).

6.10. The vector potential has only the azimuthal component, A_φ, since the current flows only along this direction. This is related to the axial magnetic flux density, B_z, through $(1/R)(\partial R A_\varphi/\partial R) = B_z$. Outside the coil ($R > a$), substituting $B_z = 0$ yields $A_\varphi = C_1/R$ with C_1 being a constant. Inside the coil ($0 \le R < a$), substituting $B_z = \mu_0 n I$ yields

$$A_\varphi = \frac{1}{2}\mu_0 nIR + \frac{C_2}{R}$$

with C_2 being a constant. Since the value of A_φ must be finite at $R = 0$, we find that $C_2 = 0$. The continuity at $R = a$ gives $C_1 = \mu_0 nIa^2/2$. Thus, the vector potential is

$$A_\varphi(R) = \frac{\mu_0 nIR}{2}; \quad 0 \leq R < a,$$
$$= \frac{\mu_0 nIa^2}{2R}; \quad R > a.$$

This agrees with the result in Example 6.8.

6.11. Here, we replace the magnetic moment of $m = Id^2$ by the small closed current by that of an equivalent pair of magnetic charges. We use the coordinates in Fig. 6.22 and place magnetic charges $\pm\, q_{\mathrm{m}}$ at points (0, 0, $\pm d/2$). Then, we obtain the magnetic potential due to positive and negative magnetic charges similarly to the calculation in Sect. 1.6 as $\phi_{\mathrm{m}\pm}(r) = \pm[\mu_0 q_{\mathrm{m}}/(4\pi r^2)][r \pm (d/2)\cos\theta]$. The magnetic potential due to the magnetic charge pair is

$$\phi_{\mathrm{m}}(r) = \phi_{\mathrm{m}+}(r) + \phi_{\mathrm{m}-}(r) = \frac{\mu_0 q_{\mathrm{m}} d}{4\pi r^2}\cos\theta = \frac{\mu_0 m}{4\pi r^2}\cos\theta.$$

The magnetic flux density, Eq. (6.45), is derived using Eq. (6.50).

6.12. The equivector potential surface is given by Eq. (6.54). Using the relationships $x = R\cos\varphi$ and $y = R\sin\varphi$, this is rewritten as

$$\frac{y}{x^2 + y^2} = c, \tag{1}$$

where c is a constant. Thus, we have

$$x^2 + \left(y - \frac{1}{2c}\right)^2 = \frac{1}{4c^2}.$$

This shows a cylindrical surface along the z-axis passing through $x = 0$ and $y = 0$.

The magnetic flux density is given by Eq. (6.57), and we have

$$B_x = B_R\cos\varphi - B_\varphi\sin\varphi = \frac{\mu_0 \hat{m}}{2\pi R^2}\cos 2\varphi,$$
$$B_y = B_R\sin\varphi + B_\varphi\cos\varphi = \frac{\mu_0 \hat{m}}{2\pi R^2}\sin 2\varphi.$$

This leads to

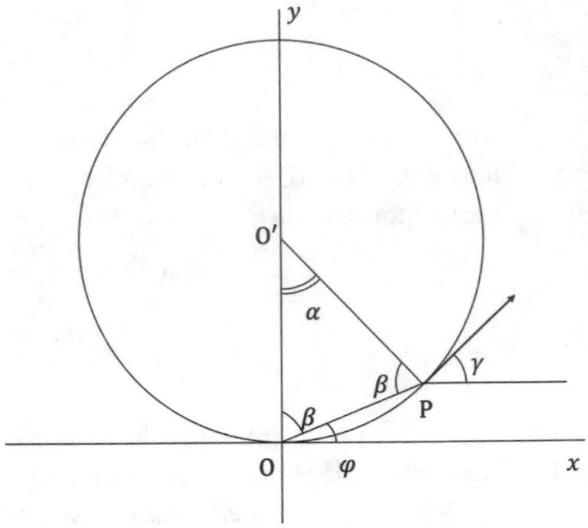

Fig. B6.5 Circle in the *x*-*y* plane that is of the same structure as the equivector potential surface and a tangential line vector at point P on the circle

$$\frac{B_y}{B_x} = \tan 2\varphi. \tag{2}$$

This gives the slope of the magnetic flux density in the *x*-*y* plane. Here, we prove that the slope of the equivector potential surface is equal to this. From Eq. (1), we have

$$\frac{\mathrm{d}y}{\mathrm{d}x} = -\frac{x}{y - (1/2c)}. \tag{3}$$

The right-hand side of Eq. (2) leads to

$$\tan 2\varphi = \frac{2xy}{x^2 - y^2}.$$

By eliminating x^2 on the right-hand side using Eq. (1), we can show that the above equation reduces to the right-hand side of Eq. (3). Thus, the magnetic flux density is parallel to the equivector potential surface.

This can also be solved geometrically. We assume a circle that is of the same structure as the equivector potential surface in the *x*-*y* plane, as shown in Fig. B6.5. The azimuthal angle at point P on the circle is denoted by φ. Other angles are defined as shown in this figure. Since $\beta + \varphi$ is equal to $\pi/2$, it is easy to show that $\alpha = \gamma = 2\varphi$. Hence, the angle of the tangential line vector on P, γ, is 2φ. This is equal to the angle of the magnetic flux density shown above, and it is proved that the magnetic flux density is parallel to the cylindrical surface.

Chapter 7

7.1. The current I_1 flows uniformly on the surface of the inner superconductor ($R = a$), and the induced current $-I_1$ flows uniformly on the inner surface of the outer superconductor ($R = b$). The current $I_1 + I_2$ flows on the outer surface of the outer superconductor ($R = c$), following the conservation law of current. The resultant magnetic flux density has the azimuthal component, and its value is

$$\begin{aligned} B_\varphi &= 0; && 0 \le R < a, \\ &= \frac{\mu_0 I_1}{2\pi R}; && a < R < b, \\ &= 0; && b < R < c, \\ &= \frac{\mu_0 (I_1 + I_2)}{2\pi R}; && R > c. \end{aligned}$$

The vector potential has the z-component, and its value is

$$\begin{aligned} A_z &= \frac{\mu_0 (I_1 + I_2)}{2\pi} \log \frac{R_0}{R}; && R > c, \\ &= \frac{\mu_0 (I_1 + I_2)}{2\pi} \log \frac{R_0}{c}; && b < R < c, \\ &= \frac{\mu_0 I_1}{2\pi} \log \frac{bR_0}{cR} + \frac{\mu_0 I_2}{2\pi} \log \frac{R_0}{c}; && a < R < b, \\ &= \frac{\mu_0 I_1}{2\pi} \log \frac{bR_0}{ac} + \frac{\mu_0 I_2}{2\pi} \log \frac{R_0}{c}; && 0 \le R < a, \end{aligned}$$

where $R = R_0 (> c)$ is the position of the reference point.

7.2. We denote the currents on the surfaces at $x = -a$ and $x = b$ by I_1 and I_2, respectively. Then, the currents on the surfaces at $x = -b$ and $x = a$ are $I - I_1$ and $-I_2$, respectively. So, the magnetic flux density in each region is shown in Table B7.1. That is, the magnetic flux densities in the left and right superconductors are $\mu_0(I - 2I_1)/2$ and $\mu_0(I - 2I_2)/2$, respectively. Since these should be zero, we have $I_1 = I/2$ and $I_2 = I/2$. Then, the current on each surface is

$$\begin{aligned} &I/2; && x = -b, \\ &I/2; && x = -a, \\ &-I/2; && x = a, \\ &I/2; && x = b. \end{aligned}$$

Table B7.1 Magnetic flux density in each region caused by currents

Position of current	Magnetic flux density in each region caused by the currents on the surfaces				
	$x < -b$	$-b < x < -a$	$-a < x < a$	$a < x < b$	$x > b$
$x = -b$	$\mu_0(I_1 - I)/2$	$\mu_0(I - I_1)/2$	$\mu_0(I - I_1)/2$	$\mu_0(I - I_1)/2$	$\mu_0(I - I_1)/2$
$x = -a$	$-\mu_0 I_1/2$	$-\mu_0 I_1/2$	$\mu_0 I_1/2$	$\mu_0 I_1/2$	$\mu_0 I_1/2$
$x = a$	$\mu_0 I_2/2$	$\mu_0 I_2/2$	$\mu_0 I_2/2$	$-\mu_0 I_2/2$	$-\mu_0 I_2/2$
$x = b$	$-\mu_0 I_2/2$	$-\mu_0 I_2/2$	$-\mu_0 I_2/2$	$-\mu_0 I_2/2$	$\mu_0 I_2/2$
Sum	$-\mu_0 I/2$	$\mu_0(I - 2I_1)/2$	$\mu_0 I/2$	$\mu_0(I - 2I_2)/2$	$\mu_0 I/2$

The magnetic flux density is

$$\begin{aligned} B &= -\frac{\mu_0 I}{2}; && x < -b, \\ &= 0; && -b < x < -a, \\ &= \frac{\mu_0 I}{2}; && -a < x < a, \\ &= 0; && a < x < b, \\ &= \frac{\mu_0 I}{2}; && x > b. \end{aligned}$$

7.3. The reason why the magnetic flux density produced by the current flowing on the superconductor surface is doubled is that there are other magnetic flux density contributions from currents flowing in other areas. For the same reason, the magnetic flux density inside the superconductor cancels to zero. Examples are found in the case where a current flows along the opposite direction on the surface of the opposite plate of a superconducting transmission line, as shown in Fig. B7.1a, or in the case where a current flows along the same direction on the opposite surface of the superconductor, as shown in Fig. B7.1b. The situation in Example 6.6 corresponds to the thin limit of the superconductor in Fig. B7.1b.

7.4. We define the x-axis on the superconductor surface along the direction normal to the current and the position of the current to be $x = 0$. The density of the current induced on the surface is given by Eq. (7.16). The force on I in a unit length caused by the current $\mathrm{d}I = \tau(x)\mathrm{d}x$ flowing in a thin region x to $x + \mathrm{d}x$ is $\mathrm{d}F' = \mu_0 I^2 a\mathrm{d}x/[2\pi^2(x^2 + a^2)^{3/2}]$. From symmetry, only the component normal to the surface remains: $\mathrm{d}F'_z = [a/(x^2 + a^2)^{1/2}]\mathrm{d}F'$. Thus, the total force in a unit length is

$$F'_z = \frac{\mu_0 I^2 a^2}{\pi^2} \int_0^\infty \frac{\mathrm{d}x}{(x^2 + a^2)^2} = \frac{\mu_0 I^2}{\pi^2 a} \int_0^{\pi/2} \cos^2\theta \mathrm{d}\theta = \frac{\mu_0 I^2}{4\pi a}.$$

This agrees with the image force, Eq. (7.18).

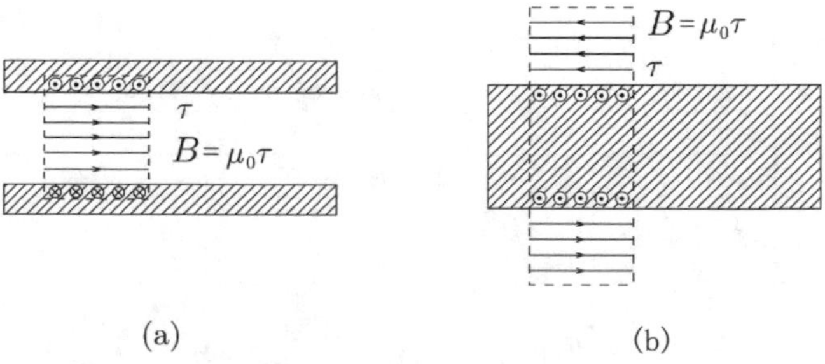

Fig. B7.1 Examples of doubled magnetic flux density: **a** superconducting transmission line with opposite currents and **b** superconductor with same currents in the both sides

7.5. We denote two superconductor surfaces that are perpendicular to each other by the x-y and y-z planes, as shown in Fig. B7.2. Assume the given current I is located at (a, b) on the x-z plane. We virtually remove the superconductor and place three image currents, $-I$, $-I$, and I, at $(a, -b)$, $(-a, b)$, and $(-a, -b)$, respectively. Then, the vector potential in the vacuum region ($x > 0$, $z > 0$) is

$$A_y(x, z) = \frac{\mu_0 I}{4\pi} \log \frac{[(x-a)^2 + (z+b)^2][(x+a)^2 + (z-b)^2]}{[(x-a)^2 + (z-b)^2][(x+a)^2 + (z+b)^2]},$$

neglecting a constant term associated with a choice of the reference point. This satisfies $A_y = 0$ on the surfaces $x = 0$ and $z = 0$, and hence, this gives the correct vector potential. We determine the current density on the x-y and y-z planes to be

$$\tau(x, y, 0) = -\frac{1}{\mu_0}\left(\frac{\partial A_y}{\partial z}\right)_{z=0} = -\frac{4Iabx}{\pi[(x-a)^2 + b^2][(x+a)^2 + b^2]},$$

$$\tau(0, y, z) = \frac{1}{\mu_0}\left(\frac{\partial A_y}{\partial x}\right)_{x=0} = -\frac{4Iabz}{\pi[(z-b)^2 + a^2][(z+b)^2 + a^2]}.$$

7.6. A simple calculation gives

$$B_R = \frac{1}{R}\cdot\frac{\partial A_z}{\partial \varphi} = \frac{\mu_0 I \sin\varphi}{2\pi}\left[\frac{a^2/d}{R^2 + (a^2/d)^2 - 2(a^2R/d)\cos\varphi} - \frac{d}{R^2 + d^2 - 2dR\cos\varphi}\right],$$

$$B_\varphi = -\frac{\partial A_z}{\partial R} = -\frac{\mu_0 I}{2\pi}\left[\frac{R - (a^2/d)\cos\varphi}{R^2 + (a^2/d)^2 - 2(a^2R/d)\cos\varphi} - \frac{R - d\cos\varphi}{R^2 + d^2 - 2dR\cos\varphi}\right],$$

$$B_z = 0.$$

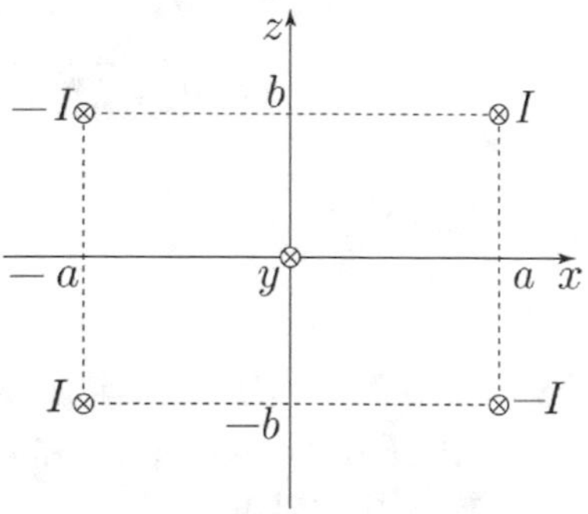

Fig. B7.2 Current I and three image currents

7.7. From Eq. (7.13), the equivector potential surface is given by

$$\frac{x^2+(z+a)^2}{x^2+(z-a)^2}=K,$$

with K denoting a constant. This is transformed to

$$x^2+\left(z-\frac{K+1}{K-1}\,a\right)^2=\frac{4a^2K}{(K-1)^2}.$$

This expresses a cylindrical surface parallel to the y-axis. This has the same structure with the equipotential surface in Exercise 2.7. Thus, we can see that the vector potential lines on the surface.

7.8. The direction of the magnetic flux density vector is parallel to the x-z plane and

$$\frac{B_z}{B_x}=\frac{2xz}{x^2-z^2+a^2}.$$

Since the vector potential of Eq. (7.13) is written as

$$A_y=\frac{\mu_0 I}{4\pi}\log\left[1+\frac{4az}{x^2+(z-a)^2}\right],$$

the equivector potential surface is given by

$$\frac{z}{x^2+(z-a)^2}=\text{const.}$$

This value does not change when the position changes as $(x,z)\rightarrow(x+\mathrm{d}x,z+\mathrm{d}z)$ on the equivector potential surface. Thus, we have

$$\frac{z+\mathrm{d}z}{x^2+(z-a)^2+2x\mathrm{d}x+2(z-a)\mathrm{d}z}=\frac{z}{x^2+(z-a)^2}.$$

This is reduced to

$$\frac{\mathrm{d}z}{\mathrm{d}x}=\frac{2xz}{x^2-z^2+a^2},$$

which is the same as the direction of the magnetic flux density vector. Thus, it is proved that the magnetic flux density is parallel to the equivector potential surface.

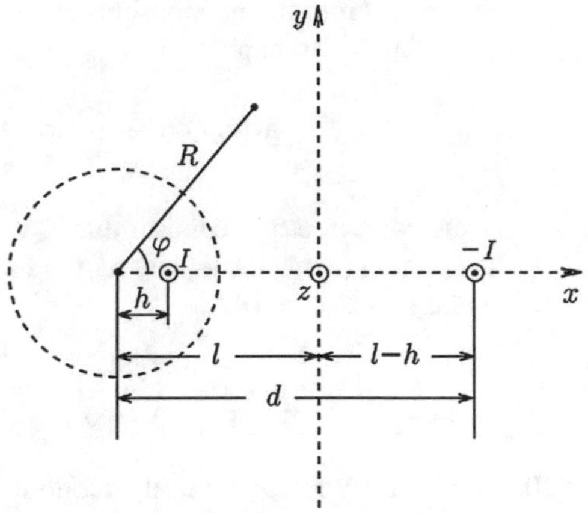

Fig. B7.3 Image currents placed in two superconductors

7.9. We assume that the magnetic flux density in the vacuum region is the same as that when we place an image current I in the superconducting cylinder at distance h from the center of the cylinder and an image current $-I$ in the infinite superconductor at distance $l - h$ from its surface after virtually removing the two superconductors (see Fig. B7.3), similarly to the answer to Exercise 2.10. In this case, the boundary condition on the infinite superconductor surface is satisfied. If the distance $2l - h$ between the image current $-I$ and the cylinder center corresponds to d in Fig. 7.14b, the boundary condition on the cylinder surface is also satisfied. From the above relationship and Eq. (7.22), we obtain

$$d = l + \sqrt{l^2 - a^2}, \quad h = l - \sqrt{l^2 - a^2}.$$

Substituting these into Eq. (7.23) yields thc vector potential outside the superconductors;

$$A_z(R, \varphi) = -\frac{\mu_0 I}{4\pi} \log \frac{R^2 + (l - \sqrt{l^2 - a^2})^2 - 2R(l - \sqrt{l^2 - a^2}) \cos\varphi}{R^2 + (l + \sqrt{l^2 - a^2})^2 - 2R(l + \sqrt{l^2 - a^2}) \cos\varphi}.$$

We find that the vector potential on the surface, $A_z(a, \varphi) = -[\mu_0 I/(2\pi)] \log[(l - \sqrt{l^2 - a^2})/a]$, is constant. The current density on the cylinder surface is

$$\tau = -\frac{1}{\mu_0}\left(\frac{\partial A_z}{\partial R}\right)_{R=a} = \frac{I}{2\pi a} \cdot \frac{\sqrt{l^2 - a^2}}{l - a\cos\varphi}.$$

Next we define Cartesian coordinates with the y-z plane ($x = 0$) on the infinite superconductor surface and the central axis of the cylindrical superconductor at y = 0. From the relationships $R\cos\varphi = x + l$ and $R\sin\varphi = y$, the vector potential is also expressed as

$$A_z(x,y) = -\frac{\mu_0 I}{4\pi}\log\frac{(x+\sqrt{l^2-a^2})^2+y^2}{(x-\sqrt{l^2-a^2})^2+y^2}.$$

Thus, we can easily confirm that $A_z(x=0)=0$ is satisfied. The density of the current (along the z-axis) on the infinite superconductor surface, which is equal to $-B_y(x=0)/\mu_0$, is

$$\tau = \frac{1}{\mu_0}\left(\frac{\partial A_z}{\partial x}\right)_{x=0} = -\frac{I\sqrt{l^2-a^2}}{\pi(y^2+l^2-a^2)}.$$

7.10. We virtually remove the superconductor and place an image current I' parallel to the current I on a plane including the central axis and the current I (see Fig. B7.4). We denote the distance between the central axis and the image current by d. The vector potential at Point P on the inner surface of the superconductor is

$$A_z(a,\varphi) = \frac{\mu_0 I'}{2\pi}\log\frac{R_0'}{(a^2+d^2-2ad\cos\varphi)^{1/2}} + \frac{\mu_0 I}{2\pi}\log\frac{R_0}{(a^2+h^2-2ah\cos\varphi)^{1/2}}.$$

The conditions that satisfy $A_z(a,\varphi)$ = const. give

$$I' = -I, \quad d = \frac{a^2}{h}.$$

Since the total current is zero $(I+I'=0)$, we can choose the infinity as the reference point of the vector potential, and we have $R_0 = R_0'$. The vector potential in the hollow is

$$A_z(R,\varphi) = \frac{\mu_0 I}{2\pi}\log\frac{[R^2+(a^2/h)^2-2(a^2R/h)\cos\varphi]^{1/2}}{(R^2+h^2-2Rh\cos\varphi)^{1/2}}.$$

Fig. B7.4 Image current I'

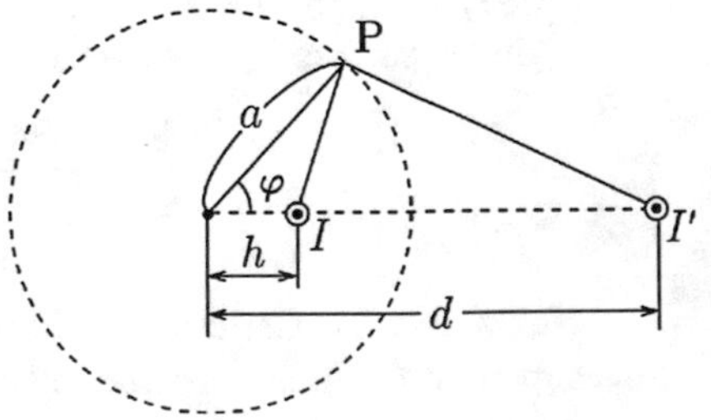

The current density on the inner surface is

$$\tau(\varphi) = \frac{B_{\varphi}(R = a)}{\mu_0} = -\frac{1}{\mu_0}\left(\frac{\partial A_z}{\partial R}\right)_{R=a} = -\frac{I(a^2 - h^2)}{2\pi a(a^2 + h^2 - 2ah\cos\varphi)}.$$

7.11. The radial and zenithal components of the applied magnetic flux density outside the spherical superconductor are $B_0\cos\theta$ and $-B_0\sin\theta$, respectively. The radial and zenithal components due to the magnetic moment m at a point at distance r from the origin are $\mu_0 m\cos\theta/(2\pi r^3)$ and $\mu_0 m\sin\theta/(4\pi r^3)$, respectively. The condition that the radial component of the magnetic flux density just outside the surface is zero is written as

$$B_0\cos\theta + \frac{\mu_0 m\cos\theta}{2\pi a^3} = 0,$$

which gives

$$m = -\frac{2\pi a^3}{\mu_0}B_0.$$

The zenithal component of the magnetic flux density just outside the surface is equal to the surface current density τ multiplied by μ_0. Thus, we have

$$\tau = \frac{1}{\mu_0}\left(-B_0\sin\theta + \frac{\mu_0 m\sin\theta}{4\pi a^3}\right) = -\frac{3}{2\mu_0}B_0\sin\theta.$$

These results agree with Eqs. (7.30) and (7.33).

7.12. The value of the magnetic moment in a unit length on the central axis is denoted by $\hat{m}$. Then, the magnetic scalar potential is given by Eq. (6.56):

$$\phi_{\mathrm{c}} = \frac{\mu_0\hat{m}}{2\pi R}\cos\varphi.$$

The magnetic scalar potential of the applied uniform magnetic flux density B_0 is given by

$$\phi_{\mathrm{f}} = -B_0 R\cos\varphi.$$

Thus, the magnetic scalar potential in the space outside the superconductor is

$$\phi_{\mathrm{m}} = \phi_{\mathrm{c}} + \phi_{\mathrm{f}} = -\left(B_0 R - \frac{\mu_0\hat{m}}{2\pi R}\right)\cos\varphi.$$

Since the magnetic flux density is normal to the superconductor surface, the following condition must be satisfied:

$$-\left(\frac{\partial \phi_{\mathrm{m}}}{\partial R}\right)_{R=a} = -\left(B_0 + \frac{\mu_0 \hat{m}}{2\pi a^2}\right)\cos\varphi = 0.$$

Thus, we have $\hat{m} = -2\pi a^2 B_0/\mu_0$, and the magnetic scalar potential is

$$\phi_{\mathrm{m}} = -B_0\left(R + \frac{a^2}{R}\right)\cos\varphi.$$

The magnetic flux density outside the superconductor is

$$B_R = -\frac{\partial \phi_{\mathrm{m}}}{\partial R} = B_0\left(1 - \frac{a^2}{R^2}\right)\cos\varphi,$$

$$B_\varphi = -\frac{1}{R}\cdot\frac{\partial \phi_{\mathrm{m}}}{\partial \varphi} = -B_0\left(1 + \frac{a^2}{R^2}\right)\sin\varphi,$$

$$B_z = -\frac{\partial \phi_{\mathrm{m}}}{\partial z} = 0.$$

Thus, the same result is obtained.

Chapter 8

8.1. When we apply current I to the left line, the magnetic flux that penetrates the coil is

$$\Phi_{\mathrm{l}} = \int_c^{b+c} \frac{\mu_0 I}{2\pi x} w(x)\mathrm{d}x,$$

where $w(x) = (a/b)(b+c-x)$ is the width of the triangle at distance x from the line. A simple calculation gives

$$\Phi_{\mathrm{l}} = \frac{\mu_0 I a}{2\pi}\left(\frac{b+c}{b}\log\frac{b+c}{c} - 1\right).$$

The magnetic flux produced by the current on the right line is similarly given by

$$\Phi_{\mathrm{r}} = \frac{\mu_0 I a}{2\pi}\left(1 - \frac{d-b-c}{b}\log\frac{d-c}{d-b-c}\right).$$

Thus, we obtain the mutual inductance as

$$M = \frac{\Phi_{\mathrm{l}} + \Phi_{\mathrm{r}}}{I} = \frac{\mu_0 a}{2\pi b}\left[(b+c)\log\frac{b+c}{c} - (d-b-c)\log\frac{d-c}{d-b-c}\right].$$

8.2. The magnetic flux stays only in the region between the two superconductors, and the density is $B = \mu_0 I/(2\pi R)$. Hence, the magnetic flux in a unit length is

$$\Phi' = \int_a^b \frac{\mu_0 I}{2\pi R}\,\mathrm{d}R = \frac{\mu_0 I}{2\pi}\log\frac{b}{a}.$$

We obtain the self-inductance in a unit length as

$$L' = \frac{\Phi'}{I} = \frac{\mu_0}{2\pi}\log\frac{b}{a}.$$

This agrees with the result calculated from the magnetic energy.

8.3. In the case of conductor, the current flows uniformly inside the conductor, and the magnetic flux densities in the regions $0 \le R < a$ and $b < R < c$ are, respectively, given by

$$\begin{aligned} B(R) &= \frac{\mu_0 IR}{2\pi a^2}; & 0 \le R < a, \\ &= \frac{\mu_0 I}{2\pi(c^2 - b^2)}\left(\frac{c^2}{R} - R\right); & b < R < c. \end{aligned}$$

Hence, in comparison with the case of superconductor, the magnetic energy increases by

$$\begin{aligned} \Delta U'_{\mathrm{m}} &= \frac{1}{2\mu_0}\int_0^a \left(\frac{\mu_0 IR}{2\pi a^2}\right)^2 \cdot 2\pi R\mathrm{d}R \\ &+ \frac{1}{2\mu_0}\int_b^c \left[\frac{\mu_0 I}{2\pi(c^2 - b^2)}\right]^2 \left(\frac{c^2}{R} - R\right)^2 \cdot 2\pi R\mathrm{d}R \\ &= \frac{\mu_0 c^2 I^2}{8\pi(c^2 - b^2)}\left(\frac{2c^2}{c^2 - b^2}\log\frac{c}{b} - 1\right). \end{aligned}$$

Adding this contribution to the result in Example 8.7, we obtain the self-inductance in a unit length as

$$L' = \frac{\mu_0}{2\pi}\log\frac{b}{a} + \frac{\mu_0 c^2}{4\pi(c^2 - b^2)}\left(\frac{2c^2}{c^2 - b^2}\log\frac{c}{b} - 1\right).$$

8.4. When current I flows in the parallel-wire transmission line as shown in Fig. B8.1, the magnetic flux that penetrates upward the coil by the right current is

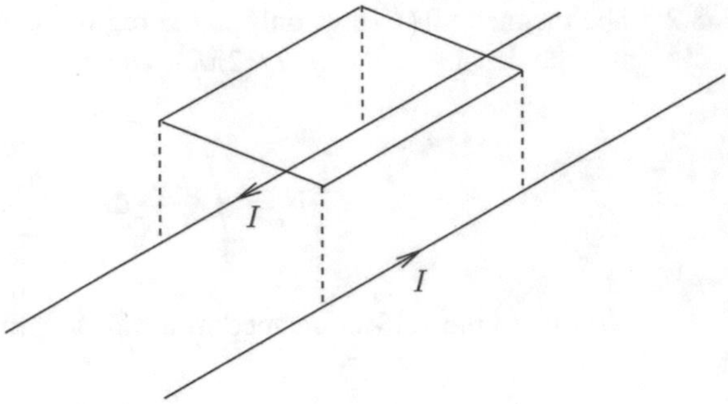

Fig. B8.1 Current in parallel-wire transmission line

$$\phi = w \int_{b}^{(a^2+b^2)^{1/2}} \frac{\mu_0 I}{2\pi r} \mathrm{d}r = \frac{\mu_0 I w}{2\pi} \log \frac{(a^2+b^2)^{1/2}}{b}.$$

The magnetic flux produced by the left current is the same, and the total magnetic flux is $\Phi = 2\phi$. The mutual inductance is

$$M = \frac{\Phi}{I} = \frac{\mu_0 w}{\pi} \log \frac{(a^2+b^2)^{1/2}}{b}.$$

8.5. When we apply current I to the outer coil, the magnetic flux density produced in the inner coil is $B = \mu_0 n_\mathrm{b} I$. Hence, the magnetic flux that penetrates one turn of the inner coil is $\phi = \pi a^2 B = \pi \mu_0 n_\mathrm{b} a^2 I$. The magnetic flux penetrating a unit length of this coil is

$$\Phi' = n_\mathrm{a} \phi = \pi \mu_0 n_\mathrm{a} n_\mathrm{b} a^2 I.$$

The mutual inductance in a unit length is

$$M' = \frac{\Phi'}{I} = \pi \mu_0 n_\mathrm{a} n_\mathrm{b} a^2.$$

8.6. First, we treat the conducting transmission line. The magnetic flux density is

$$\begin{aligned} B_y &= 0; & x < -b, x > b, \\ &= \frac{\mu_0 I (x+b)}{(b-a)l}; & -b < x < -a, \\ &= \frac{\mu_0 I}{l}; & -a < x < a, \\ &= \frac{\mu_0 I (b-x)}{(b-a)l}; & a < x < b. \end{aligned}$$

Hence, the magnetic energy in a unit length is

$$U'_{\mathrm{m}} = \frac{1}{2}\mu_0 I^2 \left[\frac{1}{(b-a)^2 l} \int_{-b}^{-a} (x+b)^2 \mathrm{d}x + \frac{2a}{l} + \frac{1}{(b-a)^2 l} \int_{a}^{b} (b-x)^2 \mathrm{d}x \right]$$
$$= \frac{(2a+b)\mu_0 I^2}{3l}.$$

Thus, the self-inductance in a unit length is

$$L' = \frac{2(2a+b)\mu_0}{3l}.$$

For the superconducting transmission line, the magnetic flux density is zero in the regions where $-b<x<-a$ and $a<x<b$. The magnetic energy in a unit length is

$$U'_{\mathrm{m}} = \frac{\mu_0 I^2 a}{l}$$

and the self-inductance in a unit length is

$$L' = \frac{2\mu_0 a}{l}.$$

The difference between the two cases comes from the magnetic energy inside the conducting regions. If we express b as $b = a(1+\delta)$, the effective distance between the two conducting regions is $2a(1+\delta/3)$ for the conducting case, while that for the superconducting case is $2a$. Note that this is not the mean distance $2a(1+\delta/2)$.

8.7. Since the magnetic flux penetrating a winding of the coil at the zenithal angle θ is $B_0\pi(a\sin\theta)^2$, from Eq. (6.35), the vector potential on this winding is

$$A_\varphi(\theta) = \frac{B_0\pi(a\sin\theta)^2}{2\pi a\sin\theta} = \frac{\mu_0 NI}{6}\sin\theta.$$

The current that flows in the area θ to $\theta+\mathrm{d}\theta$ of the coil is $\tau a \mathrm{d}\theta$ with the surface current density τ given by Eq. (8.22), and the length of one turn is $2\pi a\sin\theta$. Thus, the magnetic energy is

$$U_{\mathrm{m}} = \frac{1}{2}\int_0^{2\pi} A_\varphi(\theta)\tau a \cdot 2\pi a\sin\theta \mathrm{d}\theta = \frac{\pi}{12}\mu_0 N^2 I^2 \int_0^{2\pi} \sin^3\theta \mathrm{d}\theta = \frac{\pi}{9}\mu_0 a N^2 I^2.$$

8.8. We apply Ampere's law to circle C of radius R from the central axis (see Fig. B8.2). The magnetic flux density at this position is $B = \mu_0 NI/(2\pi R)$. If

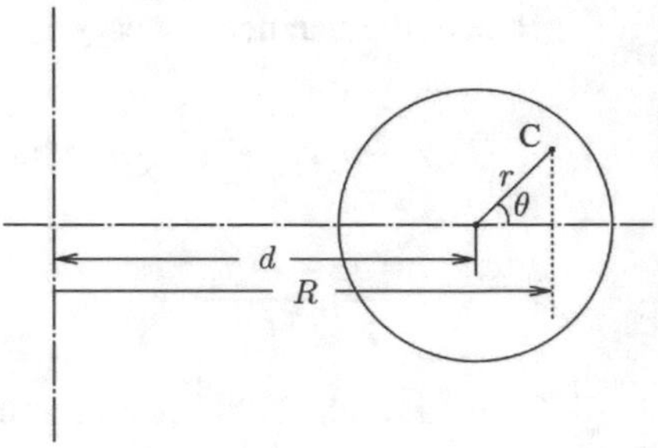

Fig. B8.2 Cross section of toroidal coil

we define the two-dimensional polar coordinates as in the figure, we have $R = d + r\cos\theta$. The magnetic energy is

$$U_{\mathrm{m}} = 2\int_0^{\pi} \mathrm{d}\theta \int_0^a \frac{\mu_0 N^2 I^2}{8\pi^2 (d + r\cos\theta)^2} \cdot 2\pi(d + r\cos\theta) r\mathrm{d}r$$
$$= \frac{\mu_0 N^2 I^2}{2\pi} \int_0^{\pi} \mathrm{d}\theta \int_0^a \frac{r\mathrm{d}r}{d + r\cos\theta}.$$

Using Eq. (7.26) for the integral with respect to θ, the magnetic energy leads to

$$U_{\mathrm{m}} = \frac{\mu_0 N^2 I^2}{2} \int_0^a \frac{r\mathrm{d}r}{(d^2 - r^2)^{1/2}} = \frac{\mu_0 N^2 I^2}{2} [d - (d^2 - a^2)^{1/2}].$$

The self-inductance is

$$L = \frac{2U_{\mathrm{m}}}{I^2} = \mu_0 N^2 [d - (d^2 - a^2)^{1/2}].$$

8.9. (a) The inductance coefficients are

$$L_{11} = \frac{\mu_0}{2\pi} \log \frac{R_1 R_3 R_\infty}{R_0 R_2 R_4},$$
$$L_{21} = L_{12} = L_{22} = \frac{\mu_0}{2\pi} \log \frac{R_3 R_\infty}{R_2 R_4},$$
$$L_{31} = L_{32} = L_{13} = L_{23} = L_{33} = \frac{\mu_0}{2\pi} \log \frac{R_\infty}{R_4}.$$

(b) From Eq. (8.35), we calculate the magnetic energy as

$$
\begin{aligned}
U_{\mathrm{m}} &= \frac{1}{2}L_{11}I_1^2 + \frac{1}{2}L_{22}I_2^2 + \frac{1}{2}L_{33}I_3^2 + L_{12}I_1I_2 + L_{23}I_2I_3 + L_{31}I_3I_1 \\
&= \frac{\mu_0}{4\pi}\left[I_1^2 \log\frac{R_1R_3R_\infty}{R_0R_2R_4} + (I_2^2 + 2I_1I_2)\log\frac{R_3R_\infty}{R_2R_4} + (I_3^2 + 2I_2I_3 + 2I_3I_1)\log\frac{R_\infty}{R_4}\right] \\
&= \frac{\mu_0}{4\pi}\left[(I_1 + I_2 + I_3)^2 \log\frac{R_\infty}{R_4} + (I_1 + I_2)^2 \log\frac{R_3}{R_2} + I_1^2 \log\frac{R_1}{R_0}\right].
\end{aligned}
$$

This result can also be obtained from Eq. (8.40).

8.10. The magnetic flux density in the vacuum region where the superconducting rod is not inserted is $B_1 = \mu_0 I'$. Thus, the magnetic flux that penetrates the superconducting hollow cylinder is $\Phi = \pi b^2 B_1 = \pi\mu_0 b^2 I'$. The magnetic flux is the same in the space of the region where the superconducting rod is inserted, and the magnetic flux density there is $B_2 = b^2 B_1/(b^2 - a^2) = \mu_0 b^2 I'/(b^2 - a^2)$. The current density flowing on the inner surface of the superconducting hollow cylinder is

$$
I_2' = \frac{B_2}{\mu_0} = \frac{b^2 I'}{b^2 - a^2}.
$$

On the surface of the inserted superconducting rod, the current of the same surface density flows along the opposite direction. Thus, the total magnetic energy is

$$
U_{\mathrm{m}} = \frac{1}{2}\Phi I'(l - x) + \frac{1}{2}\Phi I_2' x = \frac{\pi\mu_0 b^2 I'^2}{2}\left(l - x + \frac{b^2 x}{b^2 - a^2}\right),
$$

where l is the length of the superconducting hollow cylinder. We obtain the same result from Eq. (8.40). The force on the cylindrical rod is

$$
F = -\frac{\partial U_{\mathrm{m}}}{\partial x} = -\frac{\pi\mu_0 a^2 b^2 I'^2}{2(b^2 - a^2)},
$$

indicating a repulsive force, since it is negative for increasing x.

8.11. When the distance between the two coils, x, changes to $x + \Delta x$, we assume that I_1 and I_2 change to $I_1 + \Delta I_1$ and $I_2 + \Delta I_2$, respectively. If we neglect small terms of the second order, the conditions that the magnetic fluxes do not change in each coil are given by

$$
\begin{aligned}
\Delta\Phi_1 &= L_{11}\Delta I_1 + \Delta L_{21} I_2 + L_{21}\Delta I_2 = 0, \\
\Delta\Phi_2 &= \Delta L_{21} I_1 + L_{21}\Delta I_1 + L_{22}\Delta I_2 = 0,
\end{aligned}
$$

where L_{11}, L_{22}, and L_{21} are inductance coefficients, and ΔL_{21} is the change in the mutual inductance coefficient. The corresponding change in the magnetic energy is

$$\Delta U_{\mathrm{m}} = L_{11}I_1\Delta I_1 + L_{21}(I_1\Delta I_2 + I_2\Delta I_1) + \Delta L_{21}I_1I_2 + L_{22}I_2\Delta I_2.$$

Using the above two conditions, this reduces to $\Delta U_{\mathrm{m}} = -\Delta L_{21}I_1I_2$. The mutual inductance coefficient L_{21} is given by

$$L_{21} = -\frac{\mu_0 l}{2\pi}\log\frac{(x+a)(x+b)}{x(x+a+b)}$$

and the change in L_{21} due to the change in x is $\Delta L_{21} = (\partial L_{21}/\partial x)\Delta x$. Thus, we calculate the magnetic force as

$$\begin{aligned} F &= -\frac{\partial U_{\mathrm{m}}}{\partial x} = \frac{\partial L_{21}}{\partial x}I_1I_2 \\ &= \frac{\mu_0 l}{2\pi}\left(\frac{1}{x} - \frac{1}{x+a} - \frac{1}{x+b} + \frac{1}{x+a+b}\right)I_1I_2. \end{aligned}$$

We can easily confirm that this agrees with the Lorentz force between the two circuits.

Chapter 9

9.1. The magnetic flux density and magnetic field are parallel to the slab. We denote these values in magnetic materials 1 and 2 by B_1, H_1, B_2, and H_2, respectively. Ampere's law derives $H_1 = H_2 = I/w$, and these satisfy the continuity of the parallel component of the magnetic field on the boundary. These yield $B_1 = \mu_1 I/w$ and $B_2 = \mu_2 I/w$. The magnetic flux in a unit length is $\Phi' = d(B_1 + B_2)/2$, and the self-inductance in a unit length is

$$L' = \frac{(\mu_1 + \mu_2)d}{2w}.$$

9.2. We denote the distance from the central axis by R. When we apply current I to the transmission line, the magnetic field is $H(R) = I/(2\pi R)$ in the region $a<R<c$ and zero in other regions. Hence, the magnetic flux densities in magnetic materials 1 and 2 are $B_1 = \mu_1 I/(2\pi R)$ and $B_2 = \mu_2 I/(2\pi R)$, respectively. The magnetic flux in a unit length is

$$\Phi' = \int_a^b \frac{\mu_1 I}{2\pi R}\mathrm{d}R + \int_b^c \frac{\mu_2 I}{2\pi R}\mathrm{d}R = \frac{I}{2\pi}\left(\mu_1\log\frac{b}{a} + \mu_2\log\frac{c}{b}\right).$$

The self-inductance is

$$L' = \frac{1}{2\pi}\left(\mu_1\log\frac{b}{a} + \mu_2\log\frac{c}{b}\right).$$

The magnetic energy densities in magnetic materials 1 and 2 are $\mu_1(I/2\pi R)^2/2$ and $\mu_2(I/2\pi R)^2/2$, respectively. The magnetic energy in a unit length is

$$U'_{\rm m} = \int_a^b \frac{\mu_1}{2}\left(\frac{I}{2\pi R}\right)^2 2\pi R{\rm d}R + \int_b^c \frac{\mu_2}{2}\left(\frac{I}{2\pi R}\right)^2 2\pi R{\rm d}R$$
$$= \frac{I^2}{4\pi}\left(\mu_1 \log\frac{b}{a} + \mu_2 \log\frac{c}{b}\right).$$

9.3. We denote the plane determined by the normal vector $\boldsymbol{n}$ on the interface and the magnetic field $\boldsymbol{H}_1$ in magnetic material 1 as S. Assume that the magnetic field $\boldsymbol{H}_2$ in magnetic material 2 does not lie on this plane. We consider a plane, S$'$, normal to both the interface and S and define a small rectangle on S$'$ that includes the interface. The two sides of the rectangle are parallel to the interface. When we integrate the magnetic field along this rectangle, the integral in magnetic material 2 is not zero, while that in magnetic material 1 is zero. The circular integral of the magnetic field should be zero, since the planar current $\boldsymbol{\tau}$ flows on plane S$'$. Hence, the above assumption is contradictory, and we prove that the magnetic field $\boldsymbol{H}_2$ also lies on plane S.

9.4. Since the parallel component of the magnetic field is continuous across the interface, the magnetic field inside the slit is also B_0/μ, and the magnetic flux density is $B = (\mu_0/\mu)B_0$.

9.5. Since the normal component of the magnetic flux density is continuous across the interface, the magnetic flux density inside the slit is also $B = B_0$, and the magnetic field is $H = B/\mu_0 = B_0/\mu_0$.

9.6. The magnetic flux density inside the magnetic material is $B = B_0$ from the continuity of its normal component given by Eq. (9.22). Thus, the magnetic field is $H = B/\mu = B_0/\mu$. The magnetization is determined to be

$$M = \frac{B}{\mu_0} - H = \frac{B_0}{\mu_0} - \frac{B_0}{\mu} = \frac{(\mu - \mu_0)}{\mu_0 \mu} B_0.$$

Since the magnetic flux density is normal to the surface, there is no magnetizing current on the surface ($\tau_{\rm m} = 0$). If we look at this phenomenon on a much wider scale, the magnetizing current flows in the upper and lower regions. The present situation corresponds to a part around $\theta = \pi$ in Fig. 9.17 in Example 9.5, and $-B_r = 3\mu B_0/(\mu + 2\mu_0)$ in this case corresponds to the present applied magnetic flux density.

9.7. The magnetic flux density B is given by the sum of B_0 and the component produced by the magnetizing current of surface density, $\tau_{\rm m}(\theta) = 3(\mu - \mu_0)B_0 \sin\theta/[\mu_0(\mu + 2\mu_0)]$, where θ is the zenithal angle. Since the current of surface density $\tau = -(3B_0/2\mu_0)\sin\theta$ in Eq. (7.33) produces the uniform magnetic flux density $-B_0$ inside the sphere, the magnetizing current produces a uniform magnetic flux density $2(\mu - \mu_0)B_0/(\mu + 2\mu_0)$. Thus, we have

$$B = B_0 + \frac{2(\mu - \mu_0)B_0}{\mu + 2\mu_0} = \frac{3\mu}{\mu + 2\mu_0} B_0.$$

This agrees with Eq. (9.37) in Example 9.5.

9.8. We define cylindrical coordinates with the z-axis at the central axis of the cylindrical magnetic material and the azimuthal angle φ measured from the direction of the applied magnetic flux density. We assume that the magnetic flux density outside the magnetic material ($R > a$) due to its magnetization is given by the linear magnetic dipole of moment $\hat{m}$ in a unit length placed at the central axis after virtually removing the magnetic material. The magnetic flux density inside the magnetic material ($R < a$) B is assumed to be constant. The directions of the linear magnetic dipole and inner magnetic flux density are parallel to that of the applied magnetic flux density. The continuities of the normal (radial) component of the magnetic flux density and the parallel (azimuthal) component of the magnetic field at the surface ($R = a$) give

$$\hat{m} = \frac{\mu - \mu_0}{\mu + \mu_0} \cdot \frac{2\pi a^2 B_0}{\mu_0}, \qquad B = \frac{2\mu}{\mu + \mu_0} B_0.$$

Using these results, the magnetic flux density outside the magnetic material ($R > a$) is

$$B_R = \mu_0 H_R = \left(1 + \frac{\mu - \mu_0}{\mu + \mu_0} \cdot \frac{a^2}{R^2}\right) B_0 \cos\varphi,$$

$$B_\varphi = \mu_0 H_\varphi = -\left(1 - \frac{\mu - \mu_0}{\mu + \mu_0} \cdot \frac{a^2}{R^2}\right) B_0 \sin\varphi,$$

and that inside the magnetic material ($R < a$) is

$$B_R = \mu H_R = \frac{2\mu}{\mu + \mu_0} B_0 \cos\varphi,$$

$$B_\varphi = \mu H_\varphi = -\frac{2\mu}{\mu + \mu_0} B_0 \sin\varphi.$$

The magnetization of the magnetic material is

$$M = \left(\frac{1}{\mu_0} - \frac{1}{\mu}\right) B = \frac{2(\mu - \mu_0)}{\mu_0(\mu + \mu_0)} B_0.$$

Here, we apply the integral form of Eq. (9.10) to a small rectangle on a plane normal to the central axis that includes the surface of the magnetic material, as shown in Fig. 9.16. Since there is no true current on the surface, the surface magnetizing current density is given by the difference in the parallel component of the magnetic flux density on the surface divided by μ_0;

$$\tau_{\mathrm{m}}(\varphi) = \frac{2(\mu - \mu_0)}{\mu_0(\mu + \mu_0)} B_0 \sin\varphi = M \sin\varphi.$$

9.9. The magnetic scalar potential outside the spherical magnetic material is

$$\phi_{\mathrm{m}} = -\left(B_0 r - \frac{\mu_0 m}{4\pi r^2}\right)\cos\theta,$$

where the first and second components are the magnetic scalar potentials of the applied uniform magnetic flux density and the magnetic moment put on the center of the spherical magnetic material, respectively. The magnetic scalar potential inside the magnetic material with a uniform magnetic flux density is

$$\phi_{\mathrm{m}} = -Br\cos\theta.$$

The continuity condition of the normal component of the magnetic flux density gives

$$B_0 + \frac{\mu_0 m}{2\pi a^3} = B,$$

and that of the tangential component of the magnetic field gives

$$\frac{1}{\mu_0}\left(B_0 - \frac{\mu_0 m}{4\pi a^3}\right) = \frac{1}{\mu} B.$$

Thus, we have the same results:

$$m = \frac{\mu - \mu_0}{\mu + 2\mu_0} \cdot \frac{4\pi a^3 B_0}{\mu_0}, \quad B = \frac{3\mu}{\mu + 2\mu_0} B_0.$$

The vector potential on the surface is

$$A_{1\varphi} = A_{2\varphi} = \frac{3\mu a B_0}{2(\mu + 2\mu_0)} \sin\theta.$$

Thus, the vector potential is continuous on the boundary. On the other hand, the magnetic scalar potential on the surface is

$$\phi_{\mathrm{m}1} = -\frac{3\mu_0}{\mu + 2\mu_0} Ba\cos\theta, \quad \phi_{\mathrm{m}2} = -\frac{3\mu}{\mu + 2\mu_0} Ba\cos\theta.$$

It is found that the magnetic scalar potential is not continuous on the surface of the magnetic material. Such a difference is owing to the fact that the magnetic scalar potential cannot be used in the region where current flows. The magnetizing current flows on the surface, and this results in the difference on the boundary.

9.10. We use B to denote the uniform magnetic flux density inside the spherical superconductor. This is directed parallel to the applied magnetic flux density. The boundary conditions are

$$\left(B_0 + \frac{\mu_0 m}{2\pi a^3}\right)\cos\theta = B\cos\theta, \quad \frac{1}{\mu_0}\left(-B_0 + \frac{\mu_0 m}{4\pi a^3}\right)\sin\theta = -\frac{B}{\mu_0}\sin\theta + \tau.$$

From the former equation, we have $m = 2\pi a^3(B - B_0)/\mu_0$. The magnetic flux density on the superconductor surface is maximum on the equator ($\theta = \pm\pi/2$), and its absolute value is $B_0 - \mu_0 m/(4\pi a^3)$. The critical condition is that this value is equal to the critical magnetic flux density B_c. Thus, we have $m = 4\pi a^3(B_0 - B_c)/\mu_0$ or

$$-M = -\frac{m}{(4/3)\pi a^3} = \frac{3}{\mu_0}(B_c - B_0).$$

This characteristic shows the descending line in Fig. 7.23 in Column (2) in Chap. 7.

Using this result, we obtain B and τ as

$$B = 3B_0 - 2B_c, \quad \tau = -\frac{3}{\mu_0}(B_c - B_0)\sin\theta.$$

We can see that the values of B and τ agree with those in the Meissner state given by Eqs. (7.34) and (7.33) at $B_0 = (2/3)B_c$. The quantities M and τ decrease to zero at $B_0 = B_c$, showing the change to the normal state.

9.11. We suppose that currents of surface densities τ_0 and τ flow in the regions of the superconductor facing to the vacuum and the magnetic material, respectively. The magnetic field in the gap region is parallel to the superconductors, and its strength is $H_0 = \tau_0$ and $H = \tau$ in the vacuum and magnetic material, and the corresponding magnetic flux density is $B_0 = \mu_0\tau_0$ and $B = \mu\tau$. The boundary condition yields $\mu_0\tau_0 = \mu\tau$. Since the total current is $\tau_0(a - x) + \tau x = I$, we obtain the surface current densities as

$$\tau_0 = \frac{\mu I}{\mu a - (\mu - \mu_0)x},$$

$$\tau = \frac{\mu_0 I}{\mu a - (\mu - \mu_0)x}.$$

The magnetic flux density is

$$B_0 = B = \frac{\mu\mu_0 I}{\mu a - (\mu - \mu_0)x}.$$

Thus, we calculate the magnetic energy as

$$U_\mathrm{m} = bd\left[\frac{B_0^2}{2\mu_0}(a-x) + \frac{B^2}{2\mu}x\right] = \frac{\mu\mu_0 bdI^2}{2[\mu a - (\mu - \mu_0)x]}.$$

The force on the magnetic material is

$$F = -\frac{\partial U_\mathrm{m}}{\partial x} = -\frac{\mu\mu_0(\mu - \mu_0)bdI^2}{2[\mu a - (\mu - \mu_0)x]^2}.$$

Since F is negative ($\mu > \mu_0$), it is directed opposite to increasing x, i.e., repulsive. This agrees with the answer in Example 4.8, if we substitute $Q \rightarrow bI, S \rightarrow ab, t \rightarrow d, \epsilon_0 \rightarrow \mu_0^{-1}$ and $\epsilon \rightarrow \mu^{-1}$.

Chapter 10

10.1. The magnetic flux penetrating the coil is that staying in the region d to $(a^2+d^2)^{1/2}$ from the straight line:

$$\Phi = \frac{\mu_0 bI}{2\pi}\int_d^{(a^2+d^2)^{1/2}} \frac{\mathrm{d}R}{R} = \frac{\mu_0 bI}{2\pi}\log\frac{(a^2+d^2)^{1/2}}{d}.$$

The induced electromotive force is

$$\begin{aligned} V_\mathrm{em} &= -\frac{\mathrm{d}\Phi}{\mathrm{d}t} = -\frac{\mu_0 b}{2\pi}\log\frac{(a^2+d^2)^{1/2}}{d}\cdot\frac{\mathrm{d}I(t)}{\mathrm{d}t} \\ &= -\frac{\mu_0 I_\mathrm{m} b\omega}{2\pi}\log\frac{(a^2+d^2)^{1/2}}{d}\cos\omega t. \end{aligned}$$

10.2. First, we use the magnetic flux law to determine the induced electromotive force. The magnetic flux penetrating the closed circuit is $\Phi = -a(b+vt)$, when the magnetic flux produced by a current flowing along PQRS is defined as positive. The induced electromotive force is

$$V_\mathrm{em} = -\frac{\mathrm{d}\Phi}{\mathrm{d}t} = avB.$$

Second, we use the motional law. The electromotive force is induced only on side PQ, and $\boldsymbol{v} \times \boldsymbol{B}$ has magnitude vB and is directed from P to Q. Hence, the induced electromotive force is avB, and the result agrees with that from the magnetic flux law.

10.3. We define the origin at R and the x- and y-axes on sides RQ and RS, respectively. Under the given condition, the continuity equation leads to

$$\nabla \times (\boldsymbol{B} \times \boldsymbol{V}) = -\alpha \boldsymbol{i}_z.$$

The left side reduces to

$$B\left(\frac{\partial V_x}{\partial x} + \frac{\partial V_y}{\partial y}\right)\boldsymbol{i}_z.$$

The symmetry condition allows us to assume $(\partial V_x/\partial x) = (\partial V_y/\partial y)$. We can also assume the zero point of $\boldsymbol{V}$ at any point. Under the condition that $\boldsymbol{V} = 0$ at (0, 0), we have $V_x = -\alpha x/(2B)$ and $V_y = -\alpha y/(2B)$. On line PQ $(x = b + vt)$, $V_x = -\alpha(b+vt)/(2B)$ and $v_x = v$ give $V_x' = -\alpha(b+vt)/(2B) - v$, and the integral of the induced electric field from P to Q is

$$-\int_0^a (\boldsymbol{B} \times \boldsymbol{V}')_y \mathrm{d}y = \frac{\alpha a(b+vt)}{2} + Bva$$

On line SP$(y = a)$, $V_y = -\alpha a/(2B)$ and $v_y = 0$ give $V_y' = -\alpha a/(2B)$, and the integral of the induced electric field from S to P is

$$\int_0^{b+vt} (\boldsymbol{B} \times \boldsymbol{V}')_x \mathrm{d}x = \frac{\alpha a(b+vt)}{2}.$$

There are no contributions from sides QR and RS. Thus, the induced electromotive force is

$$V_{\mathrm{em}} = \alpha a(b+vt) + Bva.$$

10.4. Since the electric field is induced along the direction parallel to the applied current, there is no contribution to the electromotive force from sides QR and SP. The magnetic flux density on side PQ is $B = \mu_0 I/\{2\pi[R_0^2 + (a+d)^2]^{1/2}\}$, and the induced electric field $\boldsymbol{v} \times \boldsymbol{B}$ has a magnitude

$$E_{\mathrm{PQ}} = \frac{\mu_0 I v(a+d)}{2\pi[R_0^2 + (a+d)^2]},$$

and is directed from P to Q. The induced electric field on side RS has a magnitude

$$E_{\mathrm{RS}} = \frac{\mu_0 I v d}{2\pi(R_0^2 + d^2)},$$

and is directed from S to R. Thus, we obtain the induced electromotive force as

$$V_{\mathrm{em}} = b(E_{\mathrm{PQ}} - E_{\mathrm{RS}}) = \frac{\mu_0 I v a b[R_0^2 - (d_0 + vt)(a + d_0 + vt)]}{2\pi[R_0^2 + (d_0 + vt)^2][R_0^2 + (a + d_0 + vt)^2]}.$$

10.5. The magnetic flux law is used to determine the electromotive force first. The magnetic flux that the straight current produces in the circuit in the direction of ABCD is negative. This magnetic flux is

$$\Phi = -\frac{\mu_0 I b}{2\pi} \int_d^{d+a} \frac{\mathrm{d}x}{x} = -\frac{\mu_0 I b}{2\pi} \log \frac{d + a}{d},$$

where $d = d_0 + gt^2/2$ is the distance between the straight current and side AD. Hence, the electromotive force is

$$V_{\mathrm{em}} = -\frac{\mathrm{d}\Phi}{\mathrm{d}t} = -\frac{\partial \Phi}{\partial d} \cdot \frac{\partial d}{\partial t} = -\frac{\mu_0 I a b g t}{2\pi(d_0 + gt^2/2)(a + d_0 + gt^2/2)}.$$

Next, the electromotive force is determined using the motional law. On side AB, the induced electric field is normal to the direction of integration, and there is no contribution from this side. On side BC, the magnetic flux density is $\mu_0 I/2\pi(a + d)$, and the induced electric field is $\mu_0 I v/2\pi(a + d)$ directed from B to C. Hence, the contribution from this side is

$$V_{\mathrm{BC}} = \frac{\mu_0 I b v}{2\pi(a + d)} = \frac{\mu_0 I b g t}{2\pi(a + d_0 + gt^2/2)}.$$

On side CD, the induced electric field is normal to the direction of integration, and there is no contribution from this side. On side DA, the magnetic flux density is $\mu_0 I/2\pi d$, and the induced electric field is $\mu_0 I v/2\pi d$ directed from A to D. Hence, the contribution from this side is

$$V_{\mathrm{DA}} = -\frac{\mu_0 I b v}{2\pi d} = -\frac{\mu_0 I b g t}{2\pi(d_0 + gt^2/2)}.$$

As a result, the total electromotive force is

$$V_{\mathrm{em}} = V_{\mathrm{BC}} + V_{\mathrm{DA}} = -\frac{\mu_0 I a b g t}{2\pi(d_0 + gt^2/2)(a + d_0 + gt^2/2)}.$$

10.6. Using the distance $r = (d^2 + a^2 + 2ad\cos\theta)^{1/2}$ between side PQ and the straight line, the magnetic flux that penetrates the coil is $\Phi = -(\mu_0 Ib/2\pi)\log(r/d)$. Hence, the induced electromotive force is

$$V_{\text{em}} = -\frac{\mathrm{d}\Phi}{\mathrm{d}t} = -\frac{\mu_0 Iabd\omega\sin\omega t}{2\pi(d^2 + a^2 + 2ad\cos\omega t)}.$$

10.7. Since side RS does not move, this does not contribute to the induced electromotive force. Since $\boldsymbol{v} \times \boldsymbol{B}$ is parallel to sides QR and SP, there are no contributions from these sides. Using the distance $r = (d^2 + a^2 + 2ad\cos\theta)^{1/2}$ between side PQ and the straight line, the magnetic flux density on this side is $B = \mu_0 I/(2\pi r)$ and $v = a\omega$. We denote the angle between $\boldsymbol{v}$ and $\boldsymbol{B}$ and the angle from the line to side PQ by α and β (see Fig. B10.1), respectively. From relationships $a\sin\alpha = d\sin\beta$ and $a\sin\theta = r\sin\beta$, we have $\sin\alpha = (d/r)\sin\theta$. Hence, the magnitude of $\boldsymbol{v} \times \boldsymbol{B}$ is $vB\sin\alpha = \mu_0 adI\omega\sin\theta/(2\pi r^2)$, and this is directed from Q to P, i.e., opposite to the integration. Thus, the induced electromotive force is

$$V_{\text{em}} = -vBb\sin\alpha = -\frac{\mu_0 Iabd\omega\sin\omega t}{2\pi(d^2 + a^2 + 2ad\cos\omega t)}.$$

10.8. We denote the current flowing in the circuit by I. Since the electromotive force induced in the coil is $-L(\partial I/\partial t)$, the potential difference applied to the resistor is $V - L(\partial I/\partial t)$, which is equal to $R_r I$. Thus, we have

$$V - L\frac{\partial I}{\partial t} = R_r I.$$

The initial condition is $I(0) = 0$. The solution is

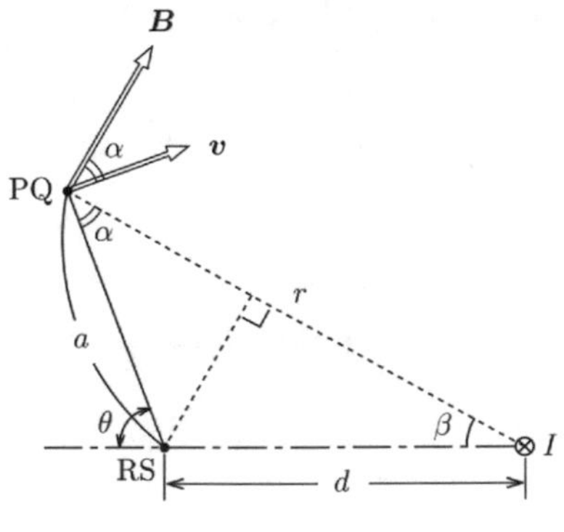

Fig. B10.1 Angles α and β

$$I(t) = \frac{V}{R_r}\left[1 - \exp\left(-\frac{t}{\tau}\right)\right]$$

with $\tau = L/R_r$.

10.9. So that the current $\boldsymbol{I}$ flows in the conductor of electric resistance R'_r in a unit length, the electric field strength that the electric power source supplies is

$$\boldsymbol{E}' = R'_r\boldsymbol{I} - \boldsymbol{v} \times \boldsymbol{B}.$$

The electric power inside the conductor is the sum of the component from the electric power source, $\boldsymbol{E}' \cdot \boldsymbol{I}$, and the component from the induced electric field, $\boldsymbol{I} \cdot (\boldsymbol{v} \times \boldsymbol{B})$. This leads to $R'_r\boldsymbol{I}^2$, i.e., the electric power consumed in the conductor. The remaining electric power from the source contributes to the mechanical work on the outside as shown in Eq. (10.17).

Thus, the work done by the induced electric field is virtual and cannot really be measured. The induced electric field prevents all the electric energy from the electric power source from being consumed as Joule heat in the conductor but converts part of it to the mechanical work on the outside.

The Lorentz force does no mechanical work on moving electric charges (electrons). However, the charges driven by the Lorentz force do mechanical work on ions in a material that force the charges to flow along the conductor through the Coulomb interaction. This is why the conductor moves. As a result, it looks as though the Lorentz force does mechanical work. In this case, the current decreases because of the law of conservation of energy, if the electric power from the source is not sufficient. The lost energy of the charges is the kinetic energy that is given by the electric power source in the initial state but not given by the Lorentz force.

10.10. We can assume that the derivatives with respect to y and z are zero from spatial symmetry and replace the time derivative by $\mathrm{i}\omega$. We can also assume that the inner electric field has only a z-component, E_z. Equation (10.39) leads to $\mathrm{d}E_z/\mathrm{d}x = \mathrm{i}\omega B_y$, showing that the magnetic flux density has only a y-component. Thus, Eq. (10.43) leads to $\mathrm{d}B_y/\mathrm{d}x = \mu\sigma_c E_z$. The above two equations yield

$$\frac{\mathrm{d}^2E_z}{\mathrm{d}x^2} - \mathrm{i}\omega\mu\sigma_c E_z = 0.$$

We can easily solve this equation under the boundary condition $E_z(x = 0) = E_0$. Taking the real part, we have

$$E_z(x,t) = E_0\mathrm{e}^{-x/\delta}\exp\left[\mathrm{i}\left(\omega t - \frac{x}{\delta}\right)\right] \to E_0\mathrm{e}^{-x/\delta}\cos\left(\omega t - \frac{x}{\delta}\right).$$

Substituting the complex solution into the first equation yields

$$B_y(x,t) = E_0\left(\frac{\mu\sigma_c}{\omega}\right)^{1/2} e^{-x/\delta} \exp\left[i\left(\omega t - \frac{x}{\delta} + \frac{3\pi}{4}\right)\right]$$
$$\rightarrow E_0\left(\frac{\mu\sigma_c}{\omega}\right)^{1/2} e^{-x/\delta} \cos\left(\omega t - \frac{x}{\delta} + \frac{3\pi}{4}\right).$$

10.11. We suppose that current I' flows uniformly on the thin conductor. When we carry a small current, $\Delta I'$, from the position $R = R_\infty$ to the conductor, an attractive force, $\mu_0 I' \Delta I'/(2\pi R)$, works on the small current of a unit length. Since the return current is uniformly distributed at R_∞, the force from the return current cancels. Hence, the work in a unit length necessary to carry the small current to the conductor is negative:

$$\Delta W_1 = \frac{\mu_0 I' \Delta I'}{2\pi} \int_{R_\infty}^{a} \frac{dR}{R} = -\frac{\mu_0 I' \Delta I'}{2\pi} \log \frac{R_\infty}{a}.$$

The electromotive force is induced to reduce the current in both the conductor circuit and the circuit composed of the small current. Hence, the electric power source in each circuit must supply an energy to maintain the current. For example, the magnetic flux penetrating a unit length of the circuit of the small current located at R is $\Phi' = (\mu_0 I'/2\pi)\log(R_\infty/R)$. The electromotive force induced in a unit length is $V_{em} = -d\Phi'/dt = [\mu_0 I'/(2\pi)]dR/dt$. Hence, the electric power necessary for the source to drive the current $\Delta I'$ continuously is $-V_{em}\Delta I'$, and the additional energy necessary to carry it from R_∞ to a is

$$\Delta W_2 = -\int V_{em} \Delta I' dt = -\frac{\mu_0 I' \Delta I'}{2\pi} \int_{R_\infty}^{a} \frac{dR}{R} = \frac{\mu_0 I' \Delta I'}{2\pi} \log \frac{R_\infty}{a}.$$

The same energy is also needed for the circuit composed of the conductor. Thus, the total energy needed to carry $\Delta I'$ is

$$\Delta W = \Delta W_1 + 2\Delta W_2 = \frac{\mu_0 I' \Delta I'}{2\pi} \log \frac{R_\infty}{a}.$$

The energy needed to carry the current I to the conductor is

$$W = \frac{\mu_0}{2\pi} \log \frac{R_\infty}{a} \int_0^I I' dI' = \frac{\mu_0 I^2}{4\pi} \log \frac{R_\infty}{a}.$$

Using the magnetic flux density $B(R) = \mu_0 I/(2\pi R)$ and Eq. (8.32), we can easily show that this is equal to the magnetic energy in the space of a unit length:

$$\int_a^{R_\infty} \frac{B^2(R)}{2\pi} 2\pi R \mathrm{d}R = \frac{\mu_0 I^2}{4\pi} \log \frac{R_\infty}{a}.$$

Thus, we can also derive the magnetic energy from the force between currents, if we correctly take into account the electromagnetic induction.

Chapter 11

11.1. The left side of Eq. (11.9) is $-\epsilon\Delta\phi$, and we obtain Poisson's equation for the electric potential,

$$\Delta\phi = -\frac{\rho}{\epsilon}.$$

The left side of Eq. (11.8) leads to

$$\frac{1}{\mu}\,\nabla \times (\nabla \times \boldsymbol{A}) = \frac{1}{\mu}[\nabla(\nabla \cdot \boldsymbol{A}) - \Delta\boldsymbol{A}] = -\frac{1}{\mu}\,\Delta\boldsymbol{A}.$$

The right side is the same as that shown in Example 11.3, and the equation for the vector potential is given by

$$\Delta\boldsymbol{A} - \epsilon\mu\,\frac{\partial^2 \boldsymbol{A}}{\partial t^2} - \epsilon\mu\,\nabla\,\frac{\partial\phi}{\partial t} = -\mu\boldsymbol{i}.$$

11.2. We use complex numbers and $\mathrm{e}^{\mathrm{i}\omega t}$ for the variation with time. We can assume that the internal electric field has only a z-component, E_z. Equation (10.39) leads to $\partial E_z/\partial x = \mathrm{i}\omega B_y$, showing that the magnetic flux density has only a y-component. Equation (11.4) leads to $\partial B_y/\partial x = \mathrm{i}\omega\mu\epsilon E_z$. Eliminating B_y yields

$$\frac{\partial^2 E_z}{\partial x^2} + \omega^2\mu\epsilon E_z = 0.$$

The general solution including the time dependence is given by

$$E_z(x,t) = K_1 \exp[\mathrm{i}(\omega t + kx)] + K_2 \exp[\mathrm{i}(\omega t - kx)],$$

where $k = \omega(\mu\epsilon)^{1/2}$. The first and second terms show electromagnetic waves propagating along the negative and positive directions of the x-axis, respectively. From causality, there is no wave propagating from infinity to the negative x-axis, and it is reasonable to assume $K_1 = 0$. Taking the real part, the boundary condition $E_z(x = 0, t) = E_0 \cos \omega t$ gives $K_2 = E_0$. Thus, we have

$$E_z(x, t) = E_0 \cos(\omega t - kx), \quad B_y(x, t) = -(\mu\epsilon)^{1/2} E_0 \cos(\omega t - kx).$$

11.3. From the answer to Exercise 11.2, the Poynting vector at depth x from the surface is

$$\boldsymbol{S}_{\mathrm{P}} = -\boldsymbol{i}_x \frac{E_z B_y}{\mu} = \frac{1}{(\mu\epsilon)^{1/2}} \cdot \epsilon E_0^2 \cos^2(\omega t - kx) \boldsymbol{i}_x.$$

It shows that the energy of density $\epsilon E_0^2 \cos^2(\omega t - kx)$ propagates with velocity $1/(\mu\epsilon)^{1/2}$ along the direction of the propagating electromagnetic wave. This does not decay with increasing x. This is because there is no energy dissipation due to electric resistivity.

11.4. When the electric charges on the electroplates are $\pm q(t)$, the electric field in the space between the electroplates is $E(t) = q(t)/\pi\epsilon_0 a^2$, and the displacement current there is $\partial D(t)/\partial t = (1/\pi a^2)[\partial q(t)/\partial t]$. The magnetic field on the surface of the space ($R = a$) is $H(t) = (1/2\pi a)[\partial q(t)/\partial t]$. Hence, the Poynting vector on the surface of the space is

$$S_{\mathrm{P}} = E(t)H(t) = \frac{q(t)}{2\pi^2 \epsilon_0 a^3} \cdot \frac{\partial q(t)}{\partial t}$$

and is directed inward the space. Integrating this with time gives

$$W = 2\pi a d \int S_{\mathrm{P}} \mathrm{d}t = \frac{d}{\pi\epsilon_0 a^2} \int_0^Q q \mathrm{d}q = \frac{dQ^2}{2\pi\epsilon_0 a^2} = \frac{Q^2}{2C},$$

where $C = \pi\epsilon_0 a^2/d$ is the capacitance of the capacitor. Thus, this energy is the electric energy stored in the capacitor.

11.5. Since the current density is $i = I/(\pi a^2)$, the electric field is $E = i/\sigma_{\mathrm{c}} = I/(\pi a^2 \sigma_{\mathrm{c}})$. The magnetic flux density on the surface is $B = \mu_0 I/(2\pi a)$. Thus, the Poynting vector on the surface has a magnitude

$$S_{\mathrm{P}} = \frac{EB}{\mu_0} = \frac{I^2}{2\pi^2 a^3 \sigma_{\mathrm{c}}}$$

and is directed normally inward the surface of the cylindrical conductor. The electric power penetrating into the conductor through a unit area is

$$P' = 2\pi a S_{\mathrm{P}} = \frac{I^2}{\pi a^2 \sigma_{\mathrm{c}}} = I^2 R'_{\mathrm{r}}$$

and is consumed in the conductor. In the above, $R'_{\mathrm{r}} = 1/(\pi a^2 \sigma_{\mathrm{c}})$ is the electric resistance in a unit length of the cylindrical conductor.

11.6. We define the z-axis along the length. The electric field is $\boldsymbol{E} = E\boldsymbol{i}_z$, and $E = V/l$ is uniform. The dissipated power is

$$P = -\int (\boldsymbol{E} \times \boldsymbol{H}) \cdot \mathrm{d}\boldsymbol{S} = -E \int \mathrm{d}z \oint (\boldsymbol{i}_z \times \boldsymbol{H}) \cdot (\mathrm{d}\boldsymbol{s} \times \boldsymbol{i}_z)$$
$$= -V \oint (\boldsymbol{i}_z \times \boldsymbol{H}) \cdot (\mathrm{d}\boldsymbol{s} \times \boldsymbol{i}_z),$$

where $\mathrm{d}\boldsymbol{s}$ is the elementary line vector on the perimeter (see Fig. B11.1), and $\mathrm{d}\boldsymbol{S} = \mathrm{d}\boldsymbol{s} \times \boldsymbol{i}_z \mathrm{d}z$ is the elementary surface vector directed outward the square prism. Using Eqs. (A1.18) and (A1.19) in the Appendix, we have

$$(\boldsymbol{i}_z \times \boldsymbol{H}) \cdot (\mathrm{d}\boldsymbol{s} \times \boldsymbol{i}_z) = [(\mathrm{d}\boldsymbol{s} \times \boldsymbol{i}_z) \times \boldsymbol{i}_z] \cdot \boldsymbol{H}$$
$$= [-(\boldsymbol{i}_z \cdot \boldsymbol{i}_z)\ \mathrm{d}\boldsymbol{s} + (\mathrm{d}\boldsymbol{s} \cdot \boldsymbol{i}_z)\boldsymbol{i}_z] \cdot \boldsymbol{H} = -\boldsymbol{H} \cdot \mathrm{d}\boldsymbol{s}.$$

Thus, the dissipated power leads to

$$P = V \oint \boldsymbol{H} \cdot \mathrm{d}\boldsymbol{s} = VI.$$

11.7. The magnetic flux density produced in the coil when the current I' flows is $B' = \mu_0 I'/h$, and the electric field induced in the conducting plate is $E'_{\mathrm{i}} = -(\mu_0 a/2h)\mathrm{d}I'/\mathrm{d}t$. The electric field provided by the electric power source to keep the current constant is $E'_{\mathrm{s}} = -E'_{\mathrm{i}}$. Thus, the electric field inside the conductor is $E' = E'_{\mathrm{i}} - E'_{\mathrm{s}} = 0$. Hence, the Poynting vector on the

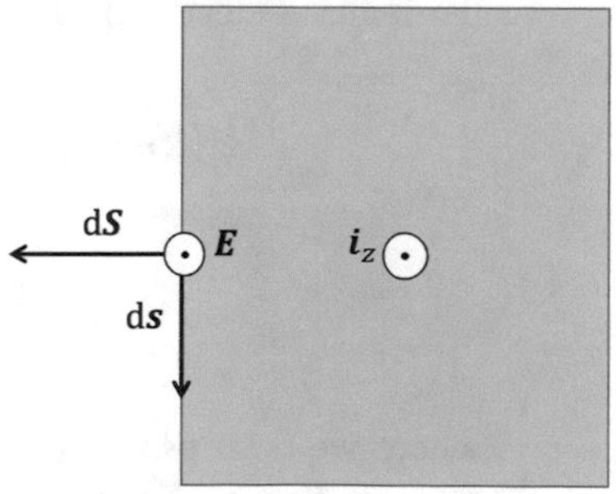

Fig. B11.1 Elementary vectors and unit vector

conductor surface is zero, and there is no energy flow into the conductor. On the other hand, since the voltage between the gap at the terminal is $V = 2\pi a E'_s$, the electric field there is

$$E' = \frac{V}{\delta} = \frac{\mu_0 \pi a^2}{h\delta} \cdot \frac{dI'}{dt}.$$

Hence, the Poynting vector at the terminal is directed inside the coil and the magnitude of the vector is

$$S_P = \frac{B'E'}{\mu_0} = \frac{\mu_0 \pi a^2}{h^2\delta} I' \frac{dI'}{dt}.$$

The energy supplied to the coil until the current reaches I is

$$U_m = h\delta \int_0^I \frac{\mu_0 \pi a^2}{h^2 \delta} I' dI' = \frac{\mu_0 \pi a^2}{2h} I^2 = \frac{B^2}{2\mu_0} \pi a^2 h,$$

where $B = \mu_0 I / h$ is the magnetic flux density in the final state, and $\pi a^2 h$ is the volume of the space in which the magnetic flux is stored. Hence, we can see that all the energy fed by the energy source is stored in the coil as the magnetic energy.

11.8. We denote the radius from the center by R. When the current applied to the coil is I', the magnetic flux density in the coil ($R < a$) is $B' = \mu_0 I'/h$. Because the conductor is sufficiently thin, we can assume that the current flows uniformly. Thus, the magnetic flux density in the conductor ($a \le R \le a + b$) is $B'(R) = \mu_0(a + b - R)I'/(bh)$. The induced electric field has an azimuthal component, and from the relationship

$$(\nabla \times \boldsymbol{E})_z = \frac{E'_i}{R} + \frac{dE'_i}{dR} \simeq \frac{dE'_i}{dR},$$

the induced electric field is given by

$$\begin{aligned} E'_i(R) &= -\frac{\mu_0}{bh} \cdot \frac{dI'}{dt} \int_a^R (a + b - R) dR + E'_i(a) \\ &= -\frac{\mu_0 [ab + 2b(R - a) - (R - a)^2]}{2bh} \cdot \frac{dI'}{dt}, \end{aligned}$$

where we have used $E'_i(a) = -(\mu_0 a / 2h) dI'/dt$. Averaging this in the sufficiently thin conductor gives $\langle E'_i \rangle = -[\mu_0(3a + 2b)/(6h)](dI'/dt)$. Hence, so that the current I' flows in the conductor, the sum of the electrostatic field, $E'_{\rm s}$, and $\langle E'_{\rm i}\rangle$ should be equal to $\rho_{\rm r}I'/(bh)$, and we have

$$E'_{\rm s} = \frac{\rho_{\rm r}I'}{bh} + \frac{\mu_0(3a+2b)}{6h}\cdot\frac{{\rm d}I'}{{\rm d}t}.$$

The electric field between the gap of the coil is $E'_0 = (2\pi a/\delta)E'_{\rm s}$, and the energy that enters the coil while the current increases linearly from 0 to I within period T is

$$U = h\delta\int\frac{E'_0B'}{\mu_0}{\rm d}t = \frac{\mu_0\pi a^2}{2h}I^2 + \frac{\mu_0\pi ab}{3h}I^2 + \frac{2\pi a}{3bh}\rho_{\rm r}I^2T.$$

The first term is the magnetic energy stored in the space of the coil (see Exercise 11.7). As will be shown later, the second and third terms are the magnetic energy stored in the conductor and the dissipated energy. We can show that these energies penetrate from the inner surface of the coil using the Poynting vector.

The magnetic energy in the conductor is

$$2\pi ah\int_a^{a+b}\frac{1}{2\mu_0}B^2(R){\rm d}R = \frac{\pi\mu_0 ab}{3h}I^2.$$

Assuming $I' = (t/T)I$, the dissipated energy is

$$\frac{2\pi a\rho_{\rm r}}{bh}\int_0^T\left(\frac{tI}{T}\right)^2{\rm d}t = \frac{2\pi a}{3bh}\rho_{\rm r}I^2T.$$

11.9. The electric field and magnetic field on the surface of the resistor are denoted by $\boldsymbol{E}$ and $\boldsymbol{H}$, respectively. The power consumed in the resistor is

$$P = -\int_{\rm S}(\boldsymbol{E}\times\boldsymbol{H})\cdot{\rm d}\boldsymbol{S},$$

where the elementary surface vector ${\rm d}\boldsymbol{S}$ is directed outwards. We can rewrite

$$(\boldsymbol{E}\times\boldsymbol{H})\cdot{\rm d}\boldsymbol{S} = ({\rm d}\boldsymbol{S}\times\boldsymbol{E})\cdot\boldsymbol{H}$$

The elementary surface vector is given by ${\rm d}\boldsymbol{S} = {\rm d}\boldsymbol{s}\times{\rm d}\boldsymbol{l}$, where ${\rm d}\boldsymbol{l}$ is the elementary vector along the direction of the current, and ${\rm d}\boldsymbol{s}$ is the elementary vector along the perimeter, which is given as a crossing line between the equipotential surface and the resistor surface. Hence, ${\rm d}\boldsymbol{s}$ is perpendicular to d$\boldsymbol{l}$. The quantity in the parentheses on the right-hand side is rewritten as

$$(\mathrm{d}\boldsymbol{s} \times \mathrm{d}\boldsymbol{l}) \times \boldsymbol{E} = (\boldsymbol{E} \cdot \mathrm{d}\boldsymbol{s})\mathrm{d}\boldsymbol{l} - (\boldsymbol{E} \cdot \mathrm{d}\boldsymbol{l})\mathrm{d}\boldsymbol{s} = -(E\mathrm{d}l)\mathrm{d}\boldsymbol{s},$$

where we have used the fact that $\boldsymbol{E}$ is parallel and normal to d$\boldsymbol{l}$ and d$\boldsymbol{s}$, respectively. Thus, the consumed power is given by

$$P = \oint \int E\boldsymbol{H} \cdot \mathrm{d}l\mathrm{d}\boldsymbol{s}.$$

Here, the surface is divided into small sections along the length with equipotential lines. We consider the m-th section, and the position vector along the perimeter and the magnetic field in this section are denoted by $\boldsymbol{s}_m$ and $\boldsymbol{H}_m$, respectively. The integration of E along the length gives the same potential difference ΔV_m, and the consumed power in this section is

$$\Delta P_m = \Delta V_m \oint \boldsymbol{H}_m \cdot \mathrm{d}\boldsymbol{s}_m = \Delta V_m I.$$

Hence, the total consumed power is given by

$$P = \sum_m \Delta V_m I = VI.$$

11.10. Exactly speaking, the conductors are not equipotential, and hence, the electric field is not perpendicular to the conductor surfaces (see Fig. B11.2). Thus, the Poynting vector is not parallel to the surface, and the dissipated energy enters the conductor.

11.11. The electric field induced along the y-axis while the magnetic flux density increases is

$$E_y(x) = -\int_0^x \frac{\partial B_z(x)}{\partial t}\mathrm{d}x = -\frac{x^2}{2d}\frac{\partial b_0}{\partial t},$$

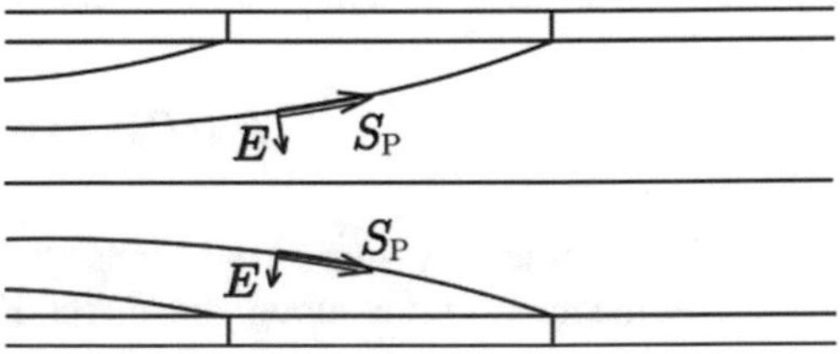

Fig. B11.2 Equipotential surface (*solid line*), electric field, and the Poynting vector

where we have used the symmetry condition, $E_y(0) = 0$. Thus, the Poynting vector at x directed along the positive x-axis is

$$S_{\mathrm{P}}(x) = -\frac{x^2}{2\mu_0 d}\left(B_0 + \frac{b_0 x}{d}\right)\frac{\partial b_0}{\partial t}.$$

The energy that penetrates into the region in unit time is

$$\Delta S_{\mathrm{P}} = -S_{\mathrm{P}}(d) + S_{\mathrm{P}}(d - \Delta x) = \frac{1}{\mu_0}\frac{\partial b_0}{\partial t}\left(B + \frac{3}{2}b_0\right)\Delta x.$$

Hence, the input energy while the magnetic flux density increases is

$$\Delta U = \int \Delta S_{\mathrm{P}} \mathrm{d}t = \frac{b_0}{\mu_0}\left(B_0 + \frac{3}{4}b_0\right)\Delta x.$$

On the other hand, the increase in the magnetic energy is

$$\Delta U_{\mathrm{m}} = \frac{1}{2\mu_0}\int_{d-\Delta x}^{d}\left[\left(B_0 + \frac{b_0}{d}x\right)^2 - B_0^2\right]\mathrm{d}x \simeq \frac{b_0}{\mu_0}\left(B_0 + \frac{b_0}{2}\right)\Delta x.$$

Hence, the work done by the expected restoring force to reduce the magnetic distortion is

$$\Delta W = \Delta U - \Delta U_{\mathrm{m}} = \frac{1}{4\mu_0}b_0^2\Delta x.$$

Here, we determine the displacement of the flux lines, u. Integrating the continuity equation of magnetic flux with time gives $\nabla \times (\boldsymbol{B} \times \boldsymbol{u}) = -\boldsymbol{b}$. This leads to $\mathrm{d}u/\mathrm{d}x = -b_0 x/(B_0 d)$. Under the symmetry condition $u(0) = 0$, we obtain the displacement to be

$$u(d) \simeq -\frac{b_0}{B_0 d}\int_0^d x\mathrm{d}x = -\frac{b_0 d}{2B_0}.$$

The work is written as $\Delta W = (B_0^2/\mu_0 d^2)u^2\Delta x$ in terms of the displacement. Hence, the force on this region is

$$f = \frac{\partial \Delta W}{\partial u} = -\frac{B_0 b_0}{\mu_0 d}\Delta x = -JB_0\Delta x,$$

where $J = b_0/\mu_0 d$ is the current density. Thus, we prove that the elastic restoring force is the Lorentz force. This force is directed along the negative x-axis to make the magnetic flux density uniform.

Thus, the Lorentz force is derived from the condition that the work on flux lines is equal to the difference between the input energy and stored energy. This should be equal to the dissipated energy. In this case, we assume that the Lorentz force is counterbalanced with a virtual force to stably maintain such a state with a higher energy. In reality, this virtual force is the pinning force (see Sect. A3.3 in the Appendix), and the work done by the Lorentz force is dissipated as the pinning loss.

Chapter 12

12.1. High frequency components of electromagnetic fields are completely shielded inside the conductor, and the electric charge and current are consequently induced on the conductor surface. Hence, the fulfilled boundary conditions are only Eqs. (12.20) and (12.23). Corresponding Eqs. (12.35) and (12.38) are

$$\boldsymbol{n} \times (\boldsymbol{E}_0 + \boldsymbol{E}_0'') = 0, \quad \boldsymbol{n} \cdot \left(\frac{1}{k} \boldsymbol{k} \times \boldsymbol{E}_0 + \frac{1}{k''} \boldsymbol{k}'' \times \boldsymbol{E}_0'' \right) = 0.$$

Since the electric field in the incident wave is normal to the plane of incidence (parallel to the y-axis in Fig. 12.3), the first equation leads to

$$E_0 + E_0'' = 0.$$

The second equation gives also the same result. In this case, taking the real part, the electric field in the vacuum region is

$$\begin{aligned} E_y &= E_0 \cos(\omega t - \boldsymbol{k} \cdot \boldsymbol{r}) - E_0 \cos(\omega t - \boldsymbol{k}'' \cdot \boldsymbol{r}) \\ &= -2E_0 \sin(kz \cos\theta) \sin(\omega t - kx \sin\theta). \end{aligned}$$

In the above, $k = k''$ and we have used Eq. (12.33) and the following relations:

$$\boldsymbol{k} \cdot \boldsymbol{r} = kx \sin\theta - kz \cos\theta, \quad \boldsymbol{k}'' \cdot \boldsymbol{r} = kx \sin\theta + kz \cos\theta.$$

In this configuration, the electric charge does not appear on the surface, since the electric field is parallel to the surface. The magnetic flux density is

$$
\begin{aligned}
B_x &= \frac{E_0}{c_0}\cos(\omega t - \boldsymbol{k}\cdot\boldsymbol{r})\cos\theta - \frac{E_0''}{c_0}\cos(\omega t - \boldsymbol{k}''\cdot\boldsymbol{r})\cos\theta \\
&= \frac{2E_0}{c_0}\cos\theta\cos(kz\cos\theta)\cos(\omega t - kx\sin\theta), \\
B_z &= \frac{E_0}{c_0}\cos(\omega t - \boldsymbol{k}\cdot\boldsymbol{r})\sin\theta + \frac{E_0''}{c_0}\cos(\omega t - \boldsymbol{k}''\cdot\boldsymbol{r})\sin\theta \\
&= -\frac{2E_0}{c_0}\sin\theta\sin(kz\cos\theta)\sin(\omega t - kx\sin\theta).
\end{aligned}
$$

The surface current density is given by

$$
\tau_y(x) = \frac{B_x(z=0)}{\mu_0} = 2\left(\frac{\epsilon_0}{\mu_0}\right)^{1/2} E_0\cos\theta\cos(\omega t - kx\sin\theta).
$$

12.2. The same two equations as in Exercise 12.1 appear. Using the definition in Fig. 12.4, the first equation reduces to $(E_0 - E_0'')\cos\theta = 0$, and we obtain $E_0'' = E_0$. The second equation is fulfilled. Hence, it is sufficient if the above equation is satisfied. The magnetic flux density has only a y-component:

$$
\begin{aligned}
B_y &= \frac{E_0}{c_0}\left[\cos(\omega t - \boldsymbol{k}\cdot\boldsymbol{r}) + \cos(\omega t - \boldsymbol{k}''\cdot\boldsymbol{r})\right] \\
&= \frac{2E_0}{c_0}\cos(kz\cos\theta)\cos(\omega t - kx\sin\theta).
\end{aligned}
$$

Since the parallel component of the magnetic flux density is not zero on the surface, the surface current density is

$$
\tau_x(x) = -\frac{B_y(z=0)}{\mu_0} = 2\left(\frac{\epsilon_0}{\mu_0}\right)^{1/2} E_0\cos(\omega t - kx\sin\theta).
$$

(Note the directions of the current and magnetic flux density.) The electric field is

$$
\begin{aligned}
E_x &= -E_0\cos(\omega t - \boldsymbol{k}\cdot\boldsymbol{r})\cos\theta + E_0''\cos(\omega t - \boldsymbol{k}''\cdot\boldsymbol{r})\cos\theta \\
&= 2E_0\cos\theta\sin(kz\cos\theta)\sin(\omega t - kx\sin\theta), \\
E_z &= -E_0\cos(\omega t - \boldsymbol{k}\cdot\boldsymbol{r})\sin\theta - E_0''\cos(\omega t - \boldsymbol{k}''\cdot\boldsymbol{r})\sin\theta \\
&= -2E_0\sin\theta\cos(kz\cos\theta)\cos(\omega t - kx\sin\theta).
\end{aligned}
$$

Since the normal component of the electric field is not zero on the surface, an electric charge appears on the surface, and its density is

$$\sigma(x) = \epsilon_0 E_z(z=0) = -2E_0\epsilon_0 \sin\theta \cos(\omega t - kx\sin\theta).$$

In this case, we can see that the following relationship holds between the surface current and surface charge:

$$\nabla \cdot \boldsymbol{\tau} + \frac{\partial \sigma}{\partial t} = 0,$$

which corresponds to Eq. (5.10) for a three-dimensional case. It should be noted that $\nabla \cdot \boldsymbol{\tau} = 0$ in Exercise 12.1.

12.3. From Eqs. (12.24) and (12.25), we obtain the electric powers flowing from medium 1 to medium 2 through a unit area as the incident and reflected waves as

$$-\frac{1}{\mu_1}[\boldsymbol{E}(z=0) \times \boldsymbol{B}(z=0)]_z = \frac{E_0^2}{c_1\mu_1}\cos^2(\omega t - \boldsymbol{k}\cdot\boldsymbol{r}_0)\cos\theta,$$
$$-\frac{1}{\mu_1}[\boldsymbol{E}''(z=0) \times \boldsymbol{B}''(z=0)]_z = -\frac{E_0''^{\,2}}{c_1\mu_1}\cos^2(\omega t - \boldsymbol{k}''\cdot\boldsymbol{r}_0)\cos\theta'',$$

respectively. From Eq. (12.26), the electric power penetrating into medium 2 as the transmitted wave is

$$-\frac{1}{\mu_1}[\boldsymbol{E}'(z=0) \times \boldsymbol{B}'(z=0)]_z = \frac{E_0'^{\,2}}{c_2\mu_2}\cos^2(\omega t - \boldsymbol{k}'\cdot\boldsymbol{r}_0)\cos\theta'.$$

Because of Eq. (12.30), the factors dependent on time and space such as $\cos^2(\omega t - \boldsymbol{k}\cdot\boldsymbol{r}_0)$ are the same. Neglecting these factors, the rate of energy flow from medium 1 is

$$\frac{1}{\mu_1 c_1}(E_0^2 - E_0''^{\,2})\cos\theta = \frac{4\alpha\cos^2\theta\cos\theta' E_0^2}{\mu_1 c_1(\cos\theta + \alpha\cos\theta')^2},$$

where $\alpha = (\epsilon_2\mu_1/\epsilon_1\mu_2)^{1/2}$, and we have used Eqs. (12.33) and (12.42b). On the other hand, Eq. (12.42a) yields the rate of energy penetration into medium 2:

$$\frac{1}{\mu_2 c_2}E_0'^{\,2}\cos\theta' = \frac{4\cos^2\theta\cos\theta' E_0^2}{\mu_2 c_2(\cos\theta + \alpha\cos\theta')^2}.$$

We can easily show that this is equal to the rate of energy flow from medium 1.

12.4. From the properties of electric and magnetic fields, we can assume that the electric field has only a y-component, E_y, and that the magnetic flux density has only a x-component, B_x. The differentials with respect to t and z can be replaced by iω and $-$iγ, respectively. Thus, we have

$$\gamma E_y = \omega B_x$$

and

$$\gamma B_x = \omega \epsilon_0 \mu_0 E_y.$$

These conditions lead to

$$\frac{\omega}{\gamma} = \frac{1}{(\epsilon_0 \mu_0)^{1/2}} = c_0.$$

If the amplitude of the electric field is denoted by E_0, we have

$$E_y = E_0 \exp[\mathrm{i}(\omega t - \gamma z)], \quad B_x = \frac{E_0}{c_0} \exp[\mathrm{i}(\omega t - \gamma z)].$$

The densities of the electric charge and the current flowing along the z-axis that appear on the surface at $y = 0$ are

$$\sigma = \epsilon_0 E_y(y = 0) = \epsilon_0 E_0 \exp[\mathrm{i}(\omega t - \gamma z)],$$
$$\tau = \frac{B_x(y = 0)}{\mu_0} = \left(\frac{\epsilon_0}{\mu_0}\right)^{1/2} E_0 \exp[\mathrm{i}(\omega t - \gamma z)].$$

It can be easily shown that the continuity equation of current given by Eq. (5.10)

$$\frac{\partial \tau}{\partial z} + \frac{\partial \sigma}{\partial t} = 0$$

holds. The densities of the electric charge and the current on the surface at $y = b$ are equal to the above quantities with the opposite sign.

12.5. The x- and y-components of the Poynting vector are $-E_z B_y / \mu_0$ and $E_z B_x / \mu_0$, respectively. From the condition of Eq. (12.56), these are zero on the surfaces of the wave guide, (x = 0, a) and (y = 0, b). Hence, there is no energy flow through these surfaces. Taking the real parts of the electric field and magnetic flux density, the z-component of the Poynting vector is

$$S_{\mathrm{P}z} = A^2 \frac{\pi^2 \epsilon_0 \gamma \omega}{k^4} \left[\frac{m^2}{a^2} \cos^2\left(\frac{m\pi x}{a}\right) \sin^2\left(\frac{n\pi y}{b}\right) + \frac{n^2}{b^2} \sin^2\left(\frac{m\pi x}{a}\right) \cos^2\left(\frac{n\pi y}{b}\right) \right] \sin^2(\omega t - \gamma z).$$

Integrating this in the x-y plane, the electric power through a unit area along the z-axis is

$$P = A^2 \frac{\pi^2 \epsilon_0 \gamma \omega (n^2 a^2 + m^2 b^2)}{4k^4 ab} \sin^2(\omega t - \gamma z)$$
$$= A^2 \frac{\epsilon_0 \gamma \omega ab}{4k^2} \sin^2(\omega t - \gamma z).$$

12.6. The boundary conditions on E_x, E_y, B_x, and B_y are given by Eq. (12.56). The boundary conditions on B_z are $\partial B_z / \partial x = 0$ at $x = 0$ and a from Eqs. (12.52b) and (12.52c), and $\partial B_z / \partial y = 0$ at $y = 0$ and b from Eqs. (12.52a) and (12.52d). Using these conditions, the general solution of Eq. (12.51b) is given by

$$B_z(x, y, z, t) = A' \cos\left(\frac{m\pi x}{a}\right) \cos\left(\frac{n\pi y}{b}\right).$$

Substituting this with $E_z = 0$ into Eqs. (12.52a)–(12.52d) yields

$$E_x = \mathrm{i}A' \frac{n\pi\omega}{k^2 b} \cos\left(\frac{m\pi x}{a}\right) \sin\left(\frac{n\pi y}{b}\right),$$
$$E_y = -\mathrm{i}A' \frac{m\pi\omega}{k^2 a} \sin\left(\frac{m\pi x}{a}\right) \cos\left(\frac{n\pi y}{b}\right),$$
$$B_x = \mathrm{i}A' \frac{m\pi\gamma}{k^2 a} \sin\left(\frac{m\pi x}{a}\right) \cos\left(\frac{n\pi y}{b}\right),$$
$$B_y = \mathrm{i}A' \frac{n\pi\gamma}{k^2 b} \cos\left(\frac{m\pi x}{a}\right) \sin\left(\frac{n\pi y}{b}\right).$$

For simplicity, the factor $\exp[\mathrm{i}(\omega t - \gamma z)]$ is omitted. The Poynting vector along the z-axis is

$$S_{Pz} = A'^2 \frac{\pi^2 \gamma \omega}{\mu_0 k^4} \left[\frac{n^2}{b^2} \cos^2\left(\frac{m\pi x}{a}\right) \sin^2\left(\frac{n\pi y}{b}\right) + \frac{m^2}{a^2} \sin^2\left(\frac{m\pi x}{a}\right) \cos^2\left(\frac{n\pi y}{b}\right)\right] \sin^2(\omega t - \gamma z).$$

Integrating this in the x-y plane yields the electric power through a unit area along the z-axis:

$$P = A'^2 \frac{\pi^2 \gamma \omega (n^2 a^2 + m^2 b^2)}{4\mu_0 k^4 ab} \sin^2(\omega t - \gamma z)$$
$$= A'^2 \frac{\gamma \omega ab}{4\mu_0 k^2} \sin^2(\omega t - \gamma z).$$

12.7. Using the real parts of the electric field and magnetic flux density, the surface densities of electric charge and current on the plane $x = 0$ are, respectively, given by

$$\sigma(x=0) = \epsilon_0 E_x(x=0) = \epsilon_0 A \frac{m\pi\gamma}{k^2 a} \sin\left(\frac{n\pi y}{b}\right) \sin(\omega t - \gamma z),$$

$$\tau_z(x=0) = \frac{1}{\mu_0} B_y(x=0) = \epsilon_0 A \frac{m\pi\omega}{k^2 a} \sin\left(\frac{n\pi y}{b}\right) \sin(\omega t - \gamma z),$$

where we have used $c_0 = 1/(\epsilon_0\mu_0)^{1/2}$. From the above results, we have

$$\frac{\partial}{\partial z}\tau_z(x=0) + \frac{\partial}{\partial t}\sigma(x=0) = 0.$$

Thus, the continuity equation of current holds. Similar relationships are obtained for other surfaces.

12.8. The electric field in the plane normal to the conductors is similar to that in the case where the line charges $\pm\lambda$ are given at the image axes ($\pm l$, 0) of the left and right conductors, respectively. The electric potential,

$$\phi(x,y) = \frac{\lambda}{4\pi\epsilon_0} \log \frac{(x-l)^2 + y^2}{(x+l)^2 + y^2},$$

gives

$$E_x = \frac{\lambda}{2\pi\epsilon_0}\left[\frac{x+l}{(x+l)^2 + y^2} - \frac{x-l}{(x-l)^2 + y^2}\right],$$

$$E_y = \frac{\lambda}{2\pi\epsilon_0}\left[\frac{y}{(x+l)^2 + y^2} - \frac{y}{(x-l)^2 + y^2}\right],$$

where $l = (d/2) - h = [(d/2)^2 - a^2]^{1/2}$ (see Fig. B5.2). For the TEM wave, λ is an arbitrary parameter associated with the electric field strength. The magnetic flux density is

$$B_x = \frac{E_y}{c_0}, \quad B_y = -\frac{E_x}{c_0}.$$

Although a detailed calculation is not shown, the total electric charges that appear on the surface of each conductor of a unit length are equal to $\pm\lambda$, and the continuity equation of current holds with the surface charges.

12.9. The electric field is given in the form of

$$E_\varphi(r,t) = E_\varphi(r)\mathrm{e}^{\mathrm{i}\omega t}.$$

Thus, Eq. (12.67) is reduced to

$$\frac{1}{r}\cdot\frac{\partial^2}{\partial r^2}(rE_\varphi)+\left(\frac{\omega}{c}\right)^2E_\varphi=0.$$

The real part of the realistic solution is

$$E_\varphi(r,t)=\frac{K}{r}\cos\left[\omega\left(t-\frac{r}{c}\right)\right].$$

The magnetic flux density has only a zenithal component, B_θ, and Eq. (11.7) is reduced to

$$\frac{1}{r}\cdot\frac{\partial}{\partial r}(rE_\varphi)=\frac{\partial B_\theta}{\partial t}.$$

The real part of the magnetic flux density is determined to be

$$B_\theta(r,t)=-\frac{K}{r}\cos\left[\omega\left(t-\frac{r}{c}\right)\right].$$

The Poynting vector is

$$\boldsymbol{S}_\mathrm{P}=-\frac{1}{\mu}E_\varphi B_\theta\boldsymbol{i}_r=\left(\frac{\epsilon}{\mu}\right)^{1/2}\frac{K^2}{r^2}\cos^2\left[\omega\left(t-\frac{r}{c}\right)\right]\boldsymbol{i}_r,$$

and the energy density is

$$u=\frac{1}{2}\epsilon E_\varphi^2+\frac{1}{2\mu}B_\theta^2=\epsilon\frac{K^2}{r^2}\cos^2\left[\omega\left(t-\frac{r}{c}\right)\right]=\frac{1}{c}\boldsymbol{S}_\mathrm{P}\cdot\boldsymbol{i}_r.$$

12.10. Using $\phi(\mathbf{r},t)$ in Eq. (12.78), the first term on the left side of Eq. (11.31) is

$$\Delta\phi(\boldsymbol{r},t)=\frac{1}{4\pi\epsilon}\int_\mathrm{V}\Delta\left[\frac{\rho(\boldsymbol{r}',t-R/c)}{R}\right]\mathrm{d}V',$$

where $R=|\boldsymbol{r}-\boldsymbol{r}'|$. Since Δ is the derivative with respect to $\boldsymbol{r}$, we have

$$\Delta\frac{\rho}{R}=\rho\Delta\frac{1}{R}+2\nabla\rho\cdot\nabla\frac{1}{R}+\frac{\Delta\rho}{R}.$$

The volume integral of the first term on the right side including the abnormal point gives $-\rho/\epsilon$ in Eq. (11.31), as shown in Sect. A2.1. The second term is written as $-(2/R^2)(\partial\rho/\partial R)$, and $\Delta\rho$ in the third term is

$$\Delta\rho = \frac{1}{R}\cdot\frac{\partial^2}{\partial R^2}(R\rho) = \frac{\partial^2\rho}{\partial R^2} + \frac{2}{R}\cdot\frac{\partial\rho}{\partial R}.$$

Hence, only $(\partial^2\rho/\partial R^2)/R$ remains from the second and third terms. It is obvious that the following equation holds:

$$\frac{\partial^2\phi}{\partial t^2} = \frac{1}{4\pi\epsilon}\int_{\mathrm{V}}\frac{1}{R}\cdot\frac{\partial^2\rho}{\partial t^2}\mathrm{d}V'.$$

Hence, we find that Eq. (11.31) is proved, if the following equation holds

$$\frac{\partial^2\rho}{\partial R^2} = \frac{1}{c^2}\cdot\frac{\partial^2\rho}{\partial t^2}.$$

The new definition $\xi = t - R/c$ gives $\partial\rho/\partial R = -(1/c)\partial\rho/\partial\xi$ and $\partial\rho/\partial t = \partial\rho/\partial\xi$. Thus, the following relations are resulted:

$$\frac{\partial^2\rho}{\partial R^2} = \frac{1}{c^2}\cdot\frac{\partial^2\rho}{\partial\xi^2}, \quad \frac{\partial^2\rho}{\partial t^2} = \frac{\partial^2\rho}{\partial\xi^2}.$$

Hence, the above relation holds, and we derive Eq. (11.31).

Literature

1. T. Matsushita, Jpn. J. Appl. Phys. **51**, 010109 (2012a)
2. T. Matsushita, Jpn. J. Appl. Phys. **51,** 010111 (2012b)
3. T. Matsushita, J. Phys. Soc. Jpn. **54**, 1054 (1985)

T. Matsushita, *Electricity and Magnetism*, Undergraduate Lecture Notes in Physics,
https://doi.org/10.1007/978-3-030-82150-0

Index

T. Matsushita, *Electricity and Magnetism*, Undergraduate Lecture Notes in Physics,
https://doi.org/10.1007/978-3-030-82150-0

Printed in the United States
by Baker & Taylor Publisher Services